Geology: An Earth System's Approach

Geology: An Earth System's Approach

Edited by Alexander Woods

STATES
ACADEMIC PRESS
www.statesacademicpress.com

Published by States Academic Press,
109 South 5th Street,
Brooklyn, NY 11249, USA
www.statesacademicpress.com

Geology: An Earth System's Approach
Edited by Alexander Woods

International Standard Book Number: 978-1-63989-237-2 (Hardback)

Cataloging-in-Publication Data

Geology : an earth system's approach / edited by Alexander Woods.
p. cm.
Includes bibliographical references and index.
ISBN 978-1-63989-237-2
1. Geology. 2. Earth sciences. I. Woods. Alexander.

QE26.3 .G46 2022
550--dc23

TABLE OF CONTENTS

PREFACE

Geology is the study of Earth processes, Earth materials, and Earth history, and how they change over time. It is an Earth science and is treated as one major aspect of integrated Earth system science and planetary science as it overlaps with hydrology and atmospheric sciences. It also studies the solid components of a terrestrial planet or natural satellites such as the Mars or the Moon. Geology describes the structure of the Earth, its surface and crust, and the processes that have shaped that structure. It also helps in determining the relative and absolute age and history of the rocks, and of the Earth. It provides the primary evidence for plate tectonics, the evolutionary history of life, and the Earth's past climates. This book provides comprehensive insights into the field of geology. It presents researches and studies performed by experts across the globe. This book will serve as a reference to a broad spectrum of readers.

The information contained in this book is the result of intensive hard work done by researchers in this field. All due efforts have been made to make this book serve as a complete guiding source for students and researchers. The topics in this book have been comprehensively explained to help readers understand the growing trends in the field.

I would like to thank the entire group of writers who made sincere efforts in this book and my family who supported me in my efforts of working on this book. I take this opportunity to thank all those who have been a guiding force throughout my life.

Editor

1

Shaking conditions required for flame structure formation in a water-immersed granular medium

Nao Yasuda and Ikuro Sumita*

Abstract

Flame structures found in sedimentary rocks may have formed from liquefaction and gravitational instability when the sediments were still unconsolidated and were subject to shaking caused by earthquakes. However, the details of the process that leads to the formation of the flame structure, and the conditions required for the instability to initiate and grow remain unclear. Here, we conduct a series of small-scale laboratory experiments by vertically shaking a case containing a water-immersed layered granular medium. The upper granular layer consists of finer particles and forms a permeability barrier against the interstitial water which percolates upwards. We shake the case sinusoidally at different combinations of acceleration and frequency. We find that there is a critical acceleration above which the instability develops at the two-layer interface. This is because the upward percolating water temporarily accumulates beneath the permeability barrier. For larger acceleration, the instability grows faster and the plumes grow to form a flame structure, which however do not completely penetrate through the upper layer. We classify the experimental results according to the final amplitude of the instability and construct a regime diagram in the parameter space of acceleration and frequency. We find that above a critical acceleration, the instability grows and its amplitude increases. Moreover, we find that the critical acceleration is frequency dependent and is smallest at approximately 100 Hz. The frequency dependence of the critical acceleration can be interpreted from the combined conditions of energy and jerk (i.e., the time derivative of acceleration) of shaking, exceeding their respective critical values. These results suggest that flame structures observed in sedimentary rocks may be used to constrain the shaking conditions of past earthquakes.

Keywords: Liquefaction; Earthquake; Flame structure; Permeability barrier; Laboratory experiment

Background

Liquefaction and related phenomena, such as the outflow of ground water, sand boils, mud volcanism, and ground subsidence, are commonly observed after earthquakes (Manga and Wang 2007). These phenomena reflect the consequences of loosening of particle contacts, pore pressure increase, upward percolation of pore water, and subsequent compaction of particles. When a layer consisting of fine particles (e.g., clay, mud) exists above a coarse particle (e.g., sand) layer, it forms a permeability barrier. This can help increase pore pressure and result in liquefaction because the expelled water temporarily accumulates at the barrier (Allen 1985). These phenomena can occur whenever liquid is present; indeed, topographic features which suggest mud volcanism have even been discovered on Mars (Oehler and Allen 2010; Pondrelli et al. 2011). Evidence of past earthquake-induced liquefaction may also be preserved in sedimentary rocks, and the resulting structures are commonly known as seismites (e.g., Montenat et al. 2007). One possible example is the flame structure (see Figure 1), which may have formed from liquefaction and subsequent gravitational instability when the sediments were still at the ocean floor.

There is considerable evidence which indicate that there is a critical condition for liquefaction and subsequent instability to occur. Compilation of liquefaction-related phenomena shows that there is a minimum seismic magnitude, which increases with epicentral distance, required for these phenomena to occur (Manga and Brodsky 2006;

*Correspondence: isumita@staff.kanazawa-u.ac.jp
Earth and Environmental Sciences Course, Division of Natural System, Graduate School of Natural Science and Technology, Kanazawa University, Kakuma, Kanazawa 920-1192, Japan

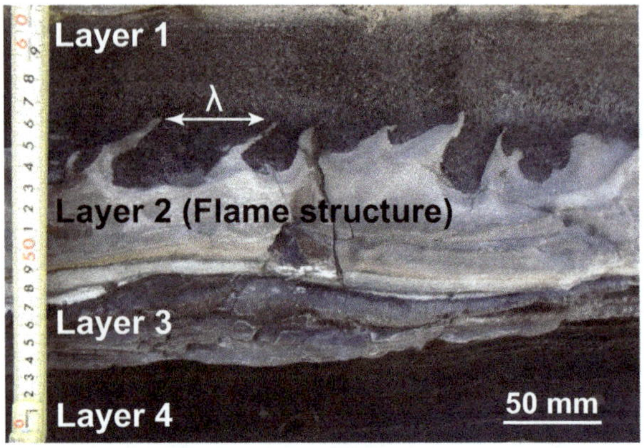

Figure 1 Flame structure in a sedimentary rock. Flame structure in a tuff layer (layer 2) observed in an outcrop on Jogashima Island, Kanagawa, Japan. The thickness of the layer 2 is approximately 60 mm, and the wavelength of the flame structure is $\lambda \sim 40$ mm. The measured densities of the particles comprising these layers are as follows: layer 1 = 2,340; layer 2 = 2,140; layer 3 = 2,050; layer 4 = 2,230 (kg/m^3).

Manga and Wang 2007). The threshold at which liquefaction occurs has also been investigated using laboratory experiments. For example, oscillatory deformation experiments of water-saturated granular matter and simultaneous pore-water pressure measurements have been conducted to constrain the strain needed for liquefaction (e.g., Vucetic 1994). Experiments for the situation in which a permeability barrier exists have also been conducted to study the formation process of water film and sand boils. In these studies, a one-dimensional tube test, to which an instant shock is applied (Kokusho 1999; Kokusho and Kojima 2002), or horizontal shake tables (Kokusho 1999; Yamaguchi et al. 2008) were used. Experiments more closely simulating natural conditions have also shown that a variety of structures can form as a consequence of shaking (Moretti et al. 1999).

On the other hand, the response of a dry granular matter under vertical shaking has intrigued physicists for a long time. Experiments have revealed instabilities such as surface waves, the Brazil nut effect, and granular convection (Duran 2000). Detailed parameter studies have shown that a variety of instabilities can occur, depending on the combination of acceleration and frequency of shaking (Burtally et al. 2002). Recently, shaking conditions required for granular convection have been investigated (Hejmandy et al. 2012). Experiments for liquid-saturated cases have also been conducted and have revealed that a number of novel phenomena can occur (e.g., Schleier-Smith and Stone 2001; Voth et al. 2002; Leaper et al. 2005; Clement et al. 2009, 2010).

Here, we consider the situation observed in Figure 1 and attempt to explain how the observed structures may have formed, and consider constraining the required shaking conditions. In order to answer these questions, we conduct a series of experiments in which a water-immersed granular medium with a permeability barrier is shaken vertically under a range of accelerations and frequencies.

Methods

Figure 2 shows the experimental setup. We use a styrol case with a width of 99.4 mm, a height of $H = 107.6$ mm, and a thickness of 22.0 mm, which is filled with a mixture of glass beads and water. We add 0.25 ml of surfactant to the water so that the glass beads do not form clusters. The lowermost 33.8 ± 0.5 mm of the cell consists of a glass beads layer comprised of two size-graded layers. The upper layer (thickness 9.8 ± 0.3 mm) consists of fine particles (white) with a diameter of $d = 0.05 \pm 0.01$ mm, a density of $\rho = 2,500$ kg/m^3, and a volumetric particle packing fraction of $\phi = 0.60 \pm 0.02$. The lower layer (thickness 24.0 ± 0.4 mm) consists of coarse particles (red) with $d = 0.22 \pm 0.04$ mm, $\rho = 2,500$ kg/m^3, and $\phi = 0.49 \pm 0.01$. The thicknesses of the upper and lower layers scaled by the particle diameters are $N_{\mathrm{upper}} \sim 200$ and $N_{\mathrm{lower}} \sim 110$, respectively. The values of ϕ for the upper and lower layers are smaller than $\phi \sim 0.64$ for dense random close packing of spheres (Mavko et al. 1998), indicating that the particles are loosely packed. After shaking for 5 s, compaction occurs and the packing fraction of the glass bead layer as a whole increases from $\phi = 0.52 \pm 0.01$ at $t = 0$ s to $\phi = 0.54 \pm 0.02$ at $t = 9.9$ s.

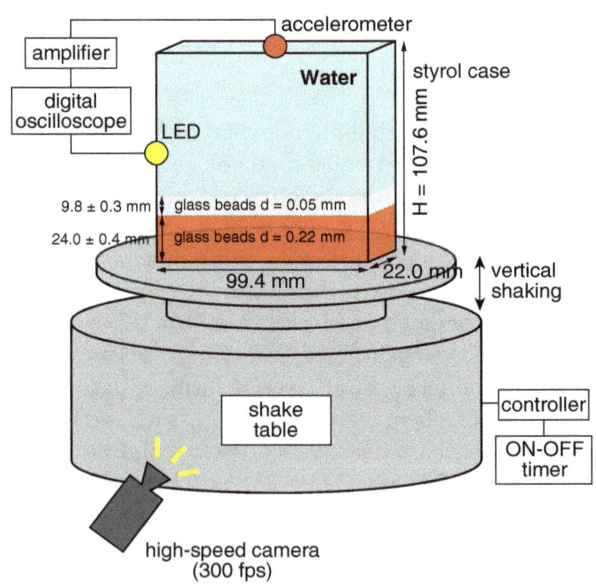

Figure 2 Experimental setup. Initial condition is such that the bottom granular layer is size-graded. The upper and lower granular layers consist of fine (diameter $d = 0.05$ mm) white particles, and coarse ($d = 0.22$ mm) red particles, respectively.

The above particle size ratio is chosen so that it is sufficiently large to result in an efficient size grading from settling, but not so large that the small ($d = 0.05$ mm) particles do not fit into the pore space of the large ($d = 0.22$ mm) particles. Particle settling velocity V_s can be estimated from the Stokes settling formula

$$V_s = \frac{\Delta\rho g d^2}{18\eta} \quad (1)$$

where $\Delta\rho$ is the particle-water density difference, g is the gravitational acceleration, and η is the viscosity of water. An efficient size grading occurs for a mixture of 0.05 mm and 0.22 mm particles, which can be understood as follows. The approximate fraction F of the small particles that can be included in the lower layer (consisting primarily of large particles) can be estimated using the time scale for a particle to settle $\sim H/V_s$. Assuming that the particles are initially homogeneously distributed, the fraction F of the small particles that have settled during the time needed for all of the large particles to settle is $F \sim (H/V_{s(la)})/(H/V_{s(sm)}) \simeq 0.06$, where the subscripts sm and la indicate small and large particles, respectively. A small $F(\ll 1)$ value indicates an efficient size grading. We also note that, since permeability scales as $\propto d^2$ (e.g., Andreotti et al. 2013), this particle size ratio results in a permeability ratio of 19, and the upper layer becomes a low permeability layer.

The experimental procedure is as follows. We first thoroughly shake the cell by hand and allow the particles to settle for 300 s, which is 5.7 times longer than the settling time of the 0.05 mm particles ($H/V_s = 53$ s). The cell is then attached to a shake table (Big Wave, Asahi Seisakusyo, Tokyo, Japan) and is shaken vertically for 5 s. The vertical displacement z changes with time t as

$$z = A \sin(2\pi f t) \quad (2)$$

where A is the amplitude and f is the frequency. An accelerometer (352A24, PCB Piezotronics, New York, USA) is attached to the case, and the signal is amplified (482C, PCB Piezotronics, New York, USA) and recorded by a digital oscilloscope (ZR-RX70, Omron, Kyoto, Japan). We use a high-speed camera (EX-F1, Casio, Tokyo, Japan) at 300 fps to record the motion within the cell. An LED lamp is used to synchronize the acceleration and image data.

We calculate the acceleration from the voltage output of the accelerometer. Here, we calculate the $\sqrt{2}\times$ root-mean-square of the acceleration data to obtain the average peak acceleration. This gives the exact peak acceleration for a sinusoidal shaking.

We conduct experiments under a peak acceleration range of 1.4 to 78.3 m/s^2 and a frequency range of 10 to 5,000 Hz. For comparison, this frequency range extends beyond the typical high frequency response limit (approximately 10 Hz) of seismometers (Shearer 1999). Digital images are analyzed using MATLAB. The spatial resolution of the images is 0.118 ± 0.002 mm per pixel. We binarized the images for the time interval of $t = 0$ to 9.9 s to obtain the heights of the granular layer surface and the two-layer interface, and analyzed how these heights change with time.

Results

Growth of an instability

Figure 3 and Additional file 1 show a typical example of an experiment in which the instability develops at the two-layer interface and a flame structure forms. We start shaking the case at $t = 0$ s and find that the heights of the lower and upper layers decrease with time. Initially the two-layer interface remains flat. This indicates that the interstitial water is percolating upwards in a laterally uniform manner, and the granular medium is compacting. After $t \sim 0.5$ s, undulations start to develop. The photo at $t = 1.6$ s in Figure 3a shows the occurrence of sand boils at the surface of the granular layer, indicating an upward channelized flow. Figure 3b is a close-up image at $t = 1.6$ s, indicating the formation of an instability with a wavelength of $\lambda \sim 2.7$ mm. Since this instability forms at the permeability barrier, we interpret this as the onset of a Rayleigh-Taylor type instability, which occurred as a consequence of the formation of a thin buoyant liquefied layer at the two-layer interface. The amplitude of the instability increases with time (Figure 3d), and undulations merge to form plumes (Figure 3c). However, the amplitude growth is suppressed and the plumes do not fully penetrate through the upper layer, in contrast to the Rayleigh-Taylor type instability of superimposed viscous fluids (Whitehead and Luther 1975). Here, we note that the shape of the flame structure deviates from the sinusoidal shape and is characterized by upward pointing cusps.

Figure 4a shows how the heights of the granular layer surface and the two-layer interface evolve with time for the same experiment shown in Figure 3. Here, thin vertical sections of the images, which are separated by a time interval of 0.033 s, are aligned horizontally as a function of time. The figure shows that in the initial stage, both the heights of the granular layer surface and the two-layer interface decrease with time as a result of compaction. The instability starts to grow at $t \sim 0.5$ s, but the height of the two-layer interface starts to increase at a later time, $t \sim 1.8$ s. This is because the instability growth and compaction occur simultaneously, and the height increase of the two-layer interface does not commence until the former exceeds the latter. Figure 4b shows the growth of the interfacial amplitude (δz) with time which was calculated as follows. We define the height at $t = 0$ as the

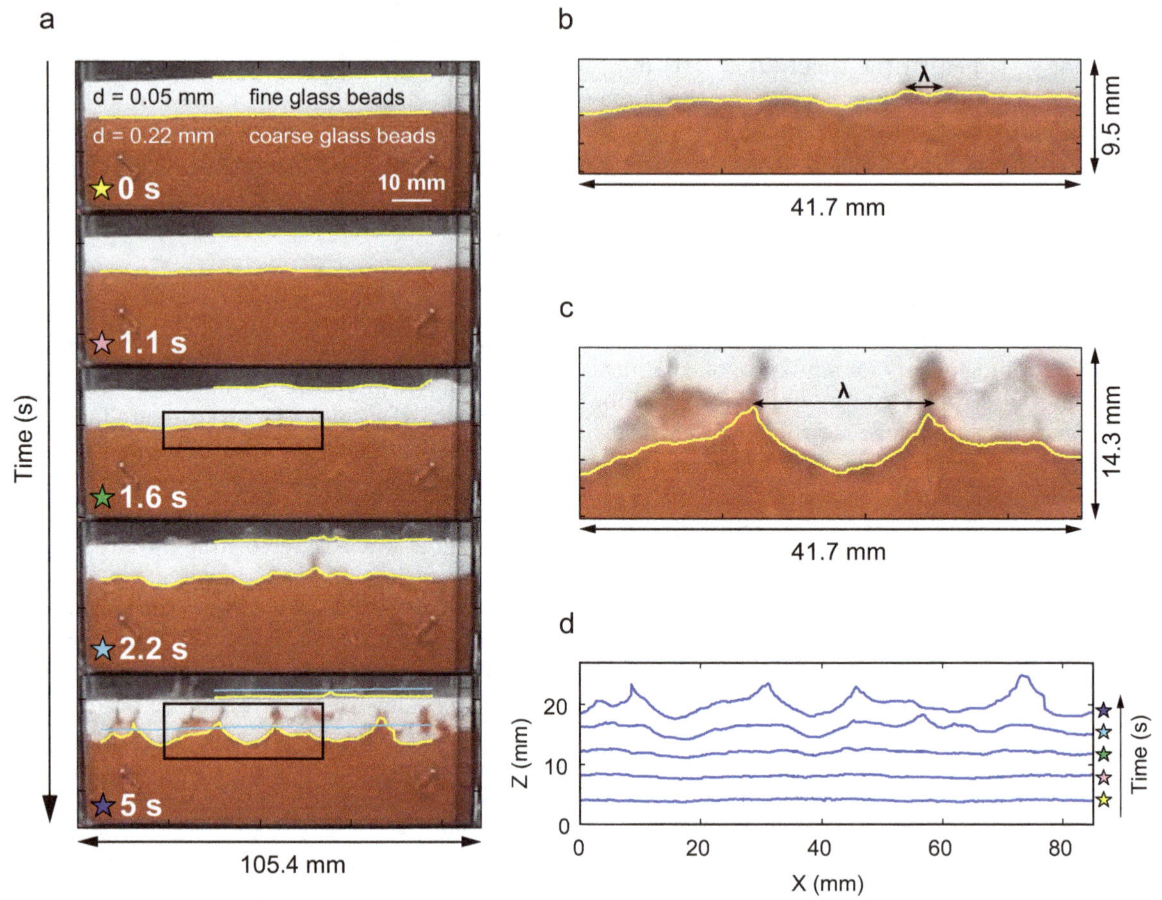

Figure 3 An experiment showing the formation of a flame structure. (a) The experimental cell is shaken for 5 s at an acceleration of 40.5 m/s^2, a frequency of 40 Hz, and an amplitude of 0.64 mm. The two yellow lines trace the surface of the granular layer and the two-layer interface. The two light blue reference lines at 5 s indicate these heights at $t = 0$ s. **(b)** A close-up of the two-layer interface at $t = 1.6$ s indicated by a rectangle in **(a)**. The measured wavelength of the instability is $\lambda = 3 \pm 1$ mm (average and standard deviation for three wavelengths). **(c)** Same as **(b)** but at $t = 5$ s with $\lambda = 21 \pm 6$ mm (average of three wavelengths). **(d)** Time evolution of the two-layer interface topography with time. The stars indicate the same timing as shown in **(a)**. Note that the amplitude and the wavelength increase with time.

reference and subtract it from each height data. We then subtract the linear fit to the data and define the amplitude by its standard deviation. The figure shows that the initial amplitude is $\delta z \simeq 0.068$ mm, which is intermediate between the particle sizes which comprise the upper and lower layers. The amplitude initially increases exponentially with time, but its growth stops at $t \sim 4.3$ s. The flame structure remains after we stop shaking at 5 s. Here, we define the relative amplitude $\delta z' = \delta z - \delta z_0$, where δz_0 is the amplitude at $t = 0.1$ s. Using $\delta z'$, we define the following stages I to III: stage I for $\delta z' < 0.1$ mm, stage II for $0.1 \leq \delta z' < 0.6$ mm, and stage III for $\delta z' \geq 0.6$ mm. The threshold of $\delta z' = 0.1$ mm is close to the spatial resolution limit of the images ($\simeq 0.12$ mm). For $\delta z' \geq 0.6$ mm, a fully developed flame structure forms. We consider these flame structures fully developed because a peak amplitude can be identified in the amplitude vs time plot during shaking, for most (14 out of 15) of the experiments which transit to stage III. Examples of such peaks are indicated by arrows in Figures 4b, 5b, and 6.

Acceleration and frequency dependence

We next show how the resulting instability changes when the peak acceleration and frequency of shaking are varied. Figure 5a and Additional file 2 show the acceleration dependence at a fixed frequency of 50 Hz. The figure shows that for a small acceleration of 2.1 m/s^2, the interstitial water percolates upwards, the granular layer compacts, and the instability ends in stage I. However, for the acceleration of 19.3 m/s^2, the instability evolves from stage I to stage II. Finally, for 40.5 m/s^2, the instability evolves through stages I and II, and ends in stage III. The growth of amplitudes for the three experiments shown in Figure 5a and two additional experiments at the same frequency is shown in Figure 5b. The figure shows that as peak acceleration increases, both the growth rate (initial

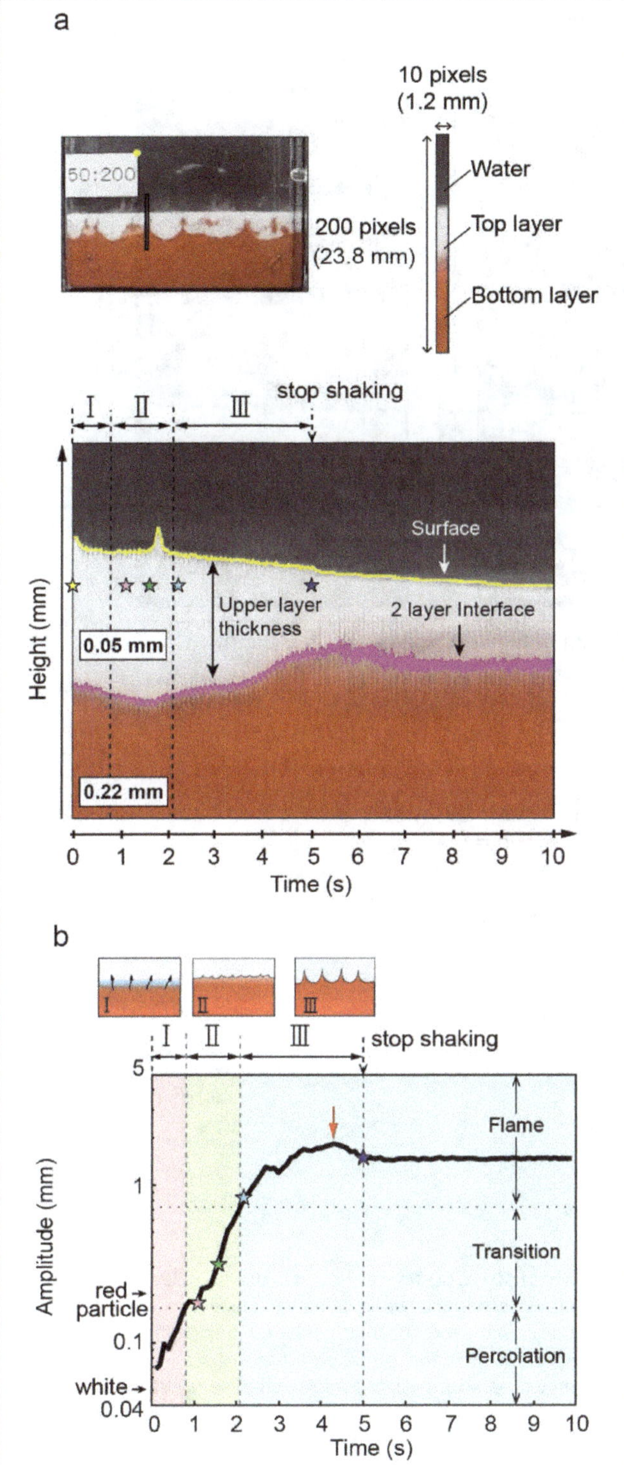

Figure 4 Time evolution of the height of the two-layer interface and the amplitude of the instability. (a) Growth of the interfacial heights with time for the same experiment shown in Figure 3. Here, a vertical section whose location is indicated in the left photo is cut from each image. The images are aligned horizontally at an interval of 0.033 s as a function of time. Yellow and pink lines trace the upper surface of the granular layer and the two-layer interface, respectively. I, II, and III indicate the three stages which are defined according to the relative amplitude $\delta z'$ (see text for the details). Stars correspond to the same timing as shown in Figure 3a. **(b)** Growth of the amplitude (δz) with time for the same experiment shown in **(a)**. The three stages are defined according to the relative amplitude ($\delta z'$) values of 0.1 and 0.6 mm. Diameters of the red and white particles which comprise the lower and upper layers, respectively, are indicated for comparison. The amplitude initially increases exponentially with time. At $t \sim 4.3$ s, the upper tip of the flame structure detaches and the amplitude slightly decreases. A red arrow indicates the peak amplitude. This amplitude remains after the shaking stops at 5 s.

slope of the curves in Figure 5b) and the amplitude of the instability increase.

Figure 6 shows the frequency dependence at a fixed acceleration of 5.1 ± 0.3 m/s^2, where the error is given by the standard deviation. Additional file 3 shows four selected examples from the experiments shown in Figure 6. Here, we note that at a constant acceleration, the amplitude (A) of shaking decreases with frequency (f) as $A \propto f^{-2}$. Figure 6 shows that at a frequency of 150 Hz, the instability growth is fastest and attains the largest amplitude.

We conducted a total of 73 experiments with different combinations of shaking acceleration and frequency, and classified the results into the following four regimes using the relative amplitude $\delta z'$ and the total compaction of the whole granular layer δh at $t = 9.9$ s (i.e., 4.9 s after the shaking stops): regime Ia (no change), when $\delta z' < 0.1$ mm and $\delta h < 0.1$ mm; regime Ib (percolation), when $\delta z' <$ 0.1 mm and $\delta h \geq 0.1$ mm; regime II (transition), when $0.1 \leq \delta z' < 0.6$ mm; and regime III (flame), when $\delta z' \geq$ 0.6 mm. Here, regimes I to III are defined using the same threshold values of $\delta z'$ used to define stages I to III, but at $t = 9.9$ s. δh is used to subdivide regime I into regimes Ia and Ib. For all experiments in regimes II to III, $\delta h \geq$ 0.1 mm, and therefore, we do not use δh to classify these regimes. For the amplitude data shown in Figures 5b and 6, the different resulting regimes are indicated by the line thicknesses.

In Figure 7a, we plot the classified regimes using different marker shapes and $\delta z'$ ranges using different marker sizes (colors), in the parameter space of acceleration and frequency. The figure shows that there is a critical acceleration above which the instability grows and regime Ib (percolation, in square markers) transits to regime II (transition, in triangle markers). In addition, we find

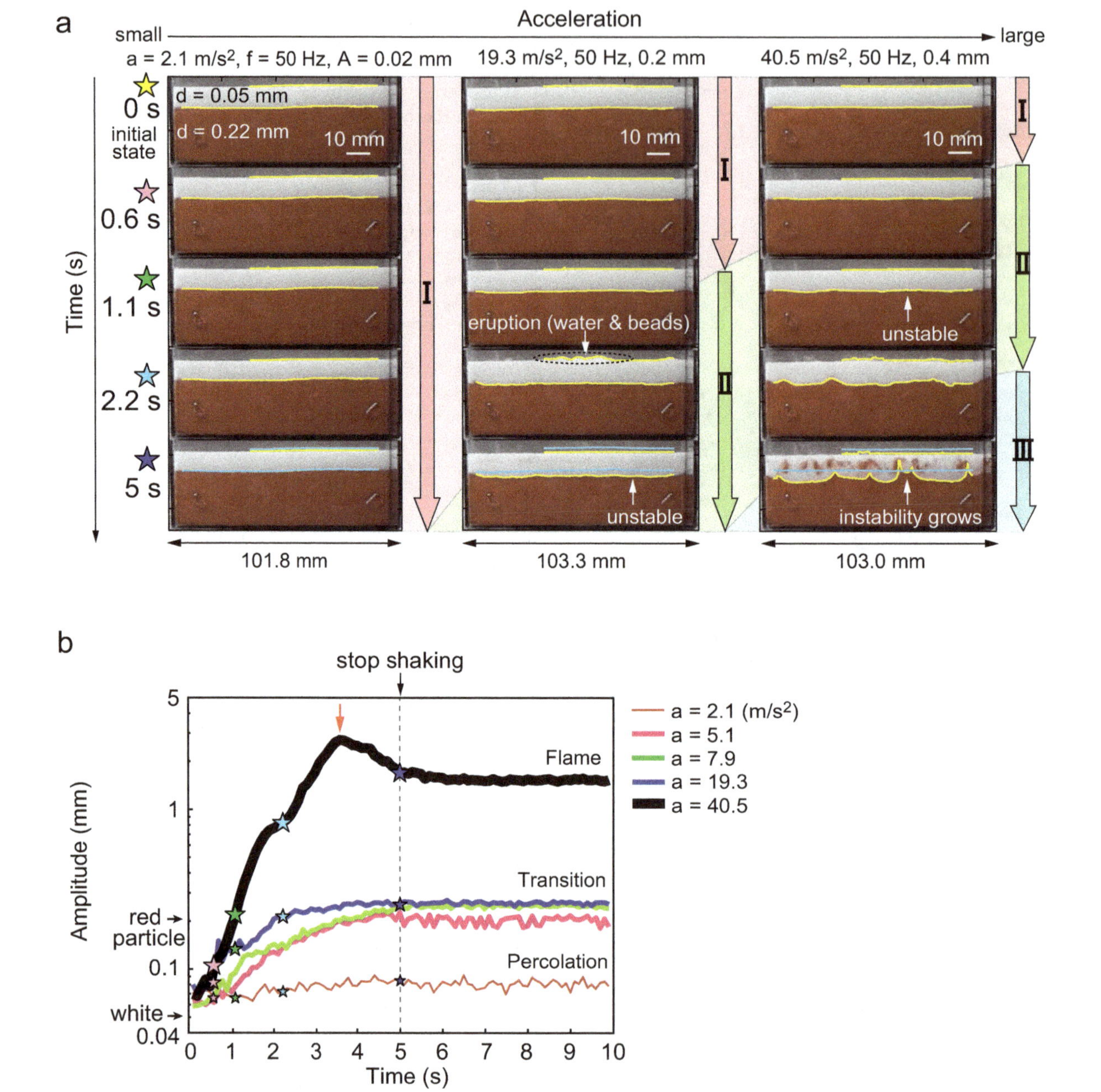

Figure 5 Acceleration dependence of the instability growth at a fixed frequency of 50 Hz. (a) Time evolution of the instability and the stage transitions at three different accelerations. Yellow lines trace the surface of the granular layer and the two-layer interface. Blue lines at $t = 5$ s indicate these heights at $t = 0$ s and are shown for reference. **(b)** Growth of amplitude δz with time. The thin, medium, and thick lines indicate the experiments in the percolation, transition, and flame regimes, respectively, and are classified using the relative amplitude $\delta z'$ at $t = 9.9$ s. A red arrow indicates a peak amplitude for the case in which a flame structure forms. Diameters of red and white particles which comprise the lower and upper layers, respectively, are indicated for comparison.

that the critical acceleration is frequency dependent and is minimum (3.4 m/s^2) at a frequency of approximately 100 Hz.

In Figure 7b, we show the same regime diagram using the dimensionless parameters. Here, we non-dimensionalize acceleration by the reduced gravity ($\Delta\rho g/\rho$) to obtain a dimensionless acceleration

$$\Gamma = \frac{\ddot{z}_{\text{peak}}}{\Delta\rho g/\rho} = \frac{A(2\pi f)^2}{\Delta\rho g/\rho}, \tag{3}$$

where $\ddot{z}_{\text{peak}} = A(2\pi f)^2$ is the peak acceleration. Similarly, we non-dimensionalize frequency by $V_s/d_{s(\text{sm})}$ to obtain a dimensionless frequency

$$f' = \frac{d}{V_{s(\text{sm})}} f. \tag{4}$$

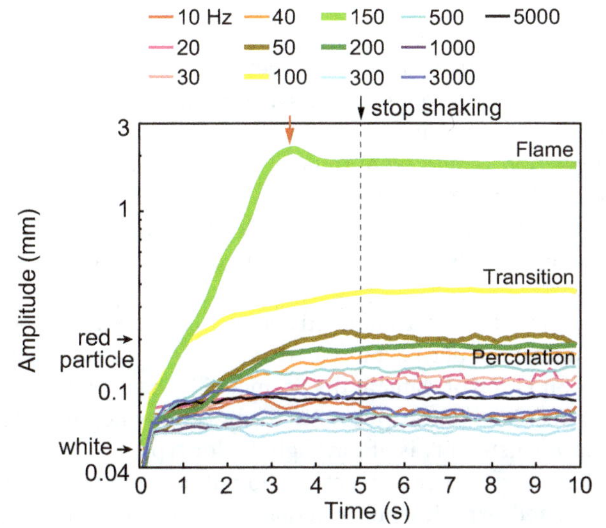

Figure 6 Frequency (f) dependence of amplitude (δz) growth in the range of f = 10-5,000 Hz at fixed acceleration of 5.1 ± 0.3 m/s^2. The thin, medium, and thick lines correspond to the percolation, transition, and flame regimes, which are classified using the relative amplitude $\delta z'$ at $t = 9.9$ s. The growth of the instability is fastest and attains the largest amplitude at an intermediate frequency of 150 Hz. A red arrow indicates a peak amplitude for the case in which a flame structure forms. Diameters of red and white particles which comprise the lower and upper layers, respectively, are indicated for comparison. Here, the amplitude data are smoothed (five-point running average) for the ease of visualization.

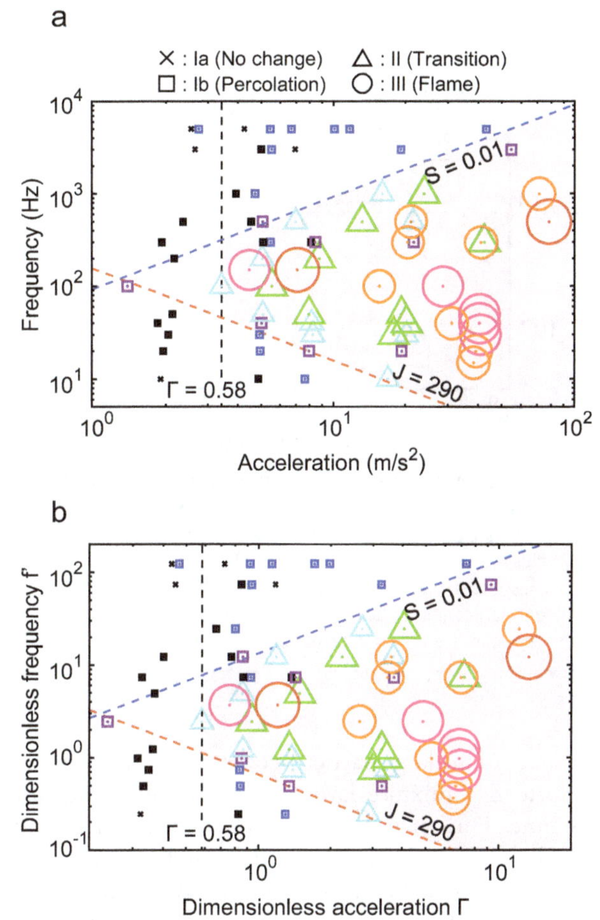

Figure 7 Regime diagram of the experiments. (a) Plotted in the parameter space of acceleration and frequency. The four regimes (in different marker shapes) are as follows: Ia, no change (crosses); Ib, percolation (squares); II, transition (triangles); and III, flame (circles). These are classified according to the relative amplitude $\delta z'$ (see text for details). The marker sizes (and colors) correspond to the magnitude of $\delta z'$ (in mm) at $t = 9.9$ s and are classified as follows. Squares and crosses in black $\delta z' < 0.01$. Squares in blue $0.01 \leq \delta z' < 0.05$ and in purple $0.05 \leq \delta z' < 0.1$. Triangles in light blue $0.1 \leq \delta z' < 0.15$ and in green $0.15 \leq \delta z' < 0.6$. Circles in orange $0.6 \leq \delta z' < 1.3$, in pink $1.3 \leq \delta z' < 2.0$, and in red $\delta z' \geq 2.0$. The black, blue, and red broken lines indicate the critical acceleration of 3.4 m/s^2 (corresponds to a dimensionless acceleration $\Gamma_c = 0.58$, where Γ is defined by Equation 3), the critical shaking strength $S_c = 0.01$ (corresponds to Equation 6 where S is defined in Equation 5), and the critical jerk strength $J_c = 290$ (corresponds to Equation 8 with J defined in Equation 7), respectively. The red colored domain indicates the conditions under which these three dimensionless parameters are supercritical (i.e., $\Gamma > \Gamma_c$, $S > S_c$, $J > J_c$). **(b)** Same as **(a)** but using dimensionless parameters Γ and f' (Equation 4).

Similar non-dimensionalizations have been used previously (e.g., Pak and Behringer 1993; Melo et al. 1995; Duran 2000; Schleier-Smith and Stone 2001; Burtally et al. 2002; Voth et al. 2002; Leaper et al. 2005; Eshuis et al. 2005, 2007). Here, we define the onset of instability when the relative amplitude is $\delta z' \geq 0.1$ mm (i.e., onset of the transitional regime). Figure 7b indicates that the minimum acceleration for the instability to occur is $\Gamma = 0.58$ at a frequency of $f' = 2.5$.

In Figure 8, we plot the relative amplitudes $\delta z'$ at $t = 9.9$ s, which were used to classify the regimes, as a function of acceleration $\ddot{z}_{\text{peak}}$ for five selected frequencies in the range of 40 to 200 Hz. The figure shows that the transition from percolation to flame regime is continuous. In particular, for the 100 Hz data, we obtain $\delta z' \propto (\ddot{z}_{\text{peak}})^{1.04}$, with an exponent close to a linear dependence.

Effects of other parameters

In addition to the shaking conditions, there are many other changeable parameters in our experiments. Here, we briefly describe the effects of other parameters which help to better understand the dynamics occurring in our experiments. Here, we define the experiments for the parameters shown in Figure 2 as the reference case.

First, we describe the results of two dry experiments (i.e., without water) and otherwise the same as the reference case. We find that the results are strikingly different. Most importantly, flame structure does not form in the dry experiments. At $\ddot{z}_{\text{peak}}$ = 20.7 m/s^2 and f = 50 Hz, the small particles comprising the upper layer settle into the

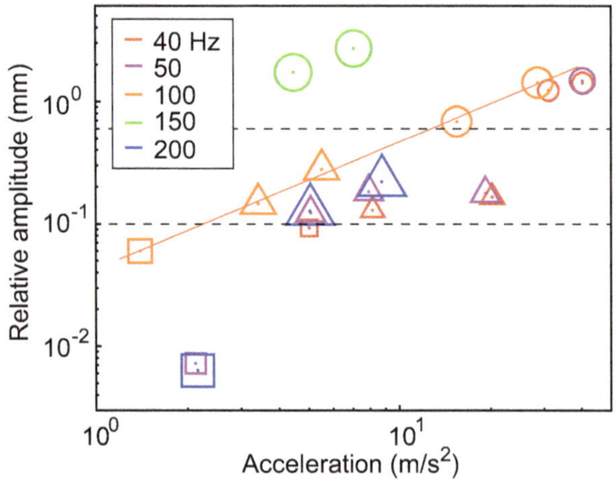

Figure 8 Acceleration dependence of the relative amplitude $\delta z'$ at $t = 9.9$ s. Here, the data for five different frequencies in the range of 40 to 200 Hz are plotted with different sizes (colors). Squares, triangles, and circles indicate regimes Ib (percolation), II (transition), and III (flame), respectively. Horizontal broken lines at $\delta z' = 0.1$ and 0.6 mm indicate the threshold values of regimes II and III, respectively. A power-law fit to the 100 Hz data ($\delta z' = 0.042(\ddot{z}_{peak})^{1.04}$) is also shown.

pore space of the large particles comprising the lower layer and formed a mixture layer whose thickness increased with time. Furthermore, the granular layer as a whole gradually tilts. At $\ddot{z}_{peak} = 39.6$ m/s^2 and $f = 50$ Hz, in addition to tilting, a large-scale whole layer granular convection with a length scale comparable to the case width occurs.

Second, we describe the results of experiments using a wider case (width 194 mm, height 104 mm, thickness 26 mm) and otherwise the same as the reference case. From a series of experiments at fixed accelerations of $\ddot{z}_{peak} = 5.0$ and 8.8 m/s^2 and different frequencies, we find that the final amplitude reached a maximum at a frequency of approximately 100 Hz, the same as the reference case. We also find a similar frequency dependence of the critical acceleration and a similar wavelength of the instability of $\lambda \sim 10$ mm, indicating that the effects of width dimension are not evident.

Third, we describe the results of different layer thicknesses. When there is no permeability barrier (i.e., one layer consisting of a particle size $d = 0.22$ mm with a thickness 21.9 mm), neither sand boil nor flame structure formation occurs. We also compare three experiments with the same thickness 33.8 ± 2.0 mm of the granular layer but with different thickness fractions of the upper layer: 0.19, 0.28, and 0.78. For $\ddot{z}_{peak} = 40.0 \pm 0.8$ m/s^2 and $f = 50$ Hz, for all three cases, a similar flame structure forms. However, we find that the relative amplitude at $t = 9.9$ s decreases by 35% as the upper layer becomes thicker.

Discussion

The origin of the flame structure

We consider that the flame structure in our experiments forms as a consequence of Rayleigh-Taylor type instability and not from surface instability, resonance, or granular convection. There seem to be at least three requirements for Rayleigh-Taylor instability to occur: the presence of a permeability barrier, a large fluid viscosity, and granular matter being dense. We explain each of these in detail below.

First, whenever a flame structure developed, it always formed at the permeability barrier. This indicates that the accumulation of water at the permeability barrier is necessary for the flame structure to form, from which we infer that the instability is of Rayleigh-Taylor type.

Second, granular convection did not occur in the water-immersed experiments, whereas it did occur in the dry experiments. On the other hand, in the dry experiments, the small particles comprising the upper layer settled downwards and the flame structure did not form. Here, we evaluate the importance of viscosity by comparing the terminal velocity given by the Stokes velocity V_s (Equation 1) and the free fall velocity $V_f = \sqrt{2(\Delta\rho/\rho)gd}$ where the particle diameter scale d is taken as the height scale. The ratio of these two velocity scales is the Stokes number $St = V_s/V_f$ (e.g., Andreotti et al. 2013). If $St \ll 1$, viscous drag is large such that the terminal velocity is attained within a particle diameter scale (viscous regime), and vice versa for $St \gg 1$ (free-fall regime). If we use the particle size d of the upper layer ($d = 0.05$ mm), for the water-immersed case, we obtain $St \sim 0.08 < 1$. In contrast for the dry case, we obtain $St \sim 6 > 1$. This suggests that the suppression of particle motions by viscosity played a role for the accumulation of water at the permeability barrier.

Third, in our experiments, the particles were initially closely packed, and even when they were shaken, we did not observe any standing surface waves indicating that the particles remained close to a close packed state. Even for the dry experiments, granular matter did not detach from the bottom of the case or bounce (e.g., Eshuis et al. 2007). The thickness of the whole layer scaled by the particle size is $N \sim 270$. A large value of N results in a larger frictional dissipation (Eshuis et al. 2007), which causes the granular matter to remain dense even when shaken.

An interpretation of a regime diagram

Here, we consider the origin of the critical acceleration and its frequency dependence. Our experiments indicate that the instability occurs when $\Gamma > \Gamma_c = 0.58$, which we indicate by vertical lines in Figure 7a,b. The critical value $\Gamma_c = 0.58$ can be understood in terms of the force balance. The inertial stress of a granular layer is $\sim \rho h \ddot{z}_{peak} = \rho A h (2\pi f)^2$ where h is the thickness scale.

Frictional stress is $\sim \mu \Delta\rho g h$, where μ is the friction coefficient. μ estimated from the angle repose θ_r of a pile of glass beads is $\mu = \tan\theta_r \sim 0.4$ to 0.5 (Samadani and Kudrolli 2001; Higashi and Sumita 2009). It follows that the ratio of the inertial stress to the frictional stress is $\sim \Gamma/\mu$, where we used the definition for Γ (Equation 3). Inertial stress must exceed the frictional stress to displace a particle from which we expect $\Gamma > \mu$ is required for instability to occur. Our experimental result of $\Gamma > 0.58$ for instability to occur is consistent with $\Gamma > \mu \sim 0.4$ to 0.5.

Next, we consider the frequency dependence. Here, we show that the high frequency limit can be understood in terms of energy balance, which is the integrated form of the force balance equation. For liquefaction to occur, the energy of shaking must be sufficiently large so that a particle can be lifted against gravity. If we use the particle diameter d for the height scale, the ratio of the energy of shaking to the work needed to lift a particle against gravity is expressed by the dimensionless shaking strength S (Pak and Behringer 1993; Eshuis et al. 2005, 2007):

$$S = \frac{A\ddot{z}_{\mathrm{peak}}}{\Delta\rho g d/\rho} = \frac{A^2(2\pi f)^2}{\Delta\rho g d/\rho} = \left(\frac{1}{\Delta\rho g d/\rho}\right)\left(\frac{\ddot{z}_{\mathrm{peak}}}{2\pi f}\right)^2. \tag{5}$$

A criterion can then be defined as $S > S_c$ where S_c is the critical value, which can be rewritten using Equation 5 as

$$f < \left(\frac{\rho}{\Delta\rho g d S_c}\right)^{1/2} \cdot \left(\frac{\ddot{z}_{\mathrm{peak}}}{2\pi}\right). \tag{6}$$

We use Equation 6 and draw a critical line in Figure 7a,b for $S_c = 0.01$, which separates regimes Ib and II fairly well. We note however that the high frequency limit may alternatively be interpreted in terms of critical shaking velocity v_c (Hejmandy et al. 2012), since the critical velocity line also scales as $f \propto \ddot{z}_{\mathrm{peak}}/v_c$.

The low frequency limit can be understood in terms of jerk, which is a time derivative of acceleration. In the limit of very low frequency, the acceleration can be approximated as constant. We propose that the change of acceleration (i.e., $d^3z/dt^3 \propto A(2\pi f)^3$) is important for liquefaction because it causes relative motion between the particles. On the other hand, in the absence of shaking, the characteristic scale for the change of acceleration of a particle settling in a viscous fluid is $\sim V_s^3/d^2$. We use this scale to non-dimensionalize jerk and define the jerk strength J

$$J = \left(\frac{d^3z}{dt^3}\right)_{\mathrm{peak}}\left(\frac{d^2}{V_s^3}\right) = \frac{A(2\pi f)^3}{V_s^3/d^2} = \left(\frac{2\pi f}{V_s^3/d^2}\right)\ddot{z}_{\mathrm{peak}}. \tag{7}$$

A criterion $J > J_c$ can then be rewritten as

$$f > \left(\frac{J_c}{2\pi}\frac{V_s^3}{d^2}\right)(\ddot{z}_{\mathrm{peak}})^{-1}. \tag{8}$$

In Figure 7a,b, we draw a line for $J_c = 290$, which separates regimes Ib and II fairly well. The conditions under which these three dimensionless parameters are supercritical, i.e., $\Gamma > \Gamma_c$, $S > S_c$, and $J > J_c$, correspond to the colored domain bounded by the three lines in Figure 7a,b.

Finally, we derive the frequency of approximately 100 Hz at which the critical acceleration is minimum. From combining the conditions $S_c \sim 0.01$, $J_c \sim 290$, and $\Gamma \sim 0.22$, which is the value of Γ at the intersection of the critical S, J lines (see Figure 7b), we obtain $f' = f \cdot d/V_s \sim 3$, which corresponds to approximately 100 Hz. We also obtain $A \sim 0.05d$ which is the amplitude of shaking relative to the particle size comprising the upper layer under this condition.

Comparison with the sedimentary rocks

Our experimental parameters (particle size pair, densities, layer thickness ratios) were not chosen to closely match or scale those of the sedimentary rocks in Figure 1. Therefore, our regime diagram (Figure 7) cannot be directly applied to these outcrops. If one wishes to constrain the exact shaking condition required for the flame structure in these outcrops to form, a similar set of experiments using the same particles from these outcrops need to be conducted. Nevertheless, we consider that the basic features of the regime diagram will remain unchanged.

Here, we point out several features which are common with the flame structures in Figure 1 and in Figure 3. First is that they both have upward pointing cusps. Similar cusps have been previously observed in Rayleigh-Taylor instability of a liquid-immersed granular medium (e.g., Michioka and Sumita 2005; Shibano et al. 2012). In addition, for both cases, the preserved flame structures do not fully penetrate through the upper layer. Second is that in the experiments, even when liquefaction occurs, only the uppermost part of the lower layer seems to be fluidized. We may approximate the situation in which water accumulates beneath the permeability barrier as a thin low viscosity layer with a thickness δ underlying a thick high viscosity layer. Here, we use the results of linear stability analysis (Whitehead and Luther 1975), from which we can constrain δ from the wavelength of instability λ as $\delta < \lambda/\pi$. It follows that for $\lambda \sim 21$ mm, $\delta < 7$ mm, implying that only the uppermost part of the coarse lower layer is fluidized. Similarly, for the outcrops, using the wavelength of $\lambda \sim 40$ mm, we obtain $\delta < 13$ mm, suggesting that only the uppermost one fifth or less of the tuff layer (layer 2 in Figure 1) became fluidized. Field observations indicate that the thickness of the tuff layer varies in the range of 20 to 80 mm, whereas the wavelength is found to be around $\lambda \sim 40$ mm. The constancy of the wavelength suggests that the thickness of the fluidized layer is also approximately constant, regardless of the thickness of the tuff layer.

Conclusions

We conducted a series of experiments in which a two-layered water-immersed granular medium, where the upper layer forms a permeability barrier, is shaken vertically at different combinations of accelerations and frequencies. We find that above a critical acceleration, the instability develops at the two-layer interface and grows. For a sufficiently large acceleration, a flame structure which is similar to those observed in sedimentary rocks forms. We also find that the critical acceleration is frequency dependent and is minimum at approximately 100 Hz. In other words, there is an optimum frequency band in which the flame structure most easily develops. These results were interpreted by combined conditions of inertial stress, energy, and jerk of shaking exceeding their critical values. Although further work is needed to clarify how the optimum frequency band depends on the parameters such as fluid viscosity and particle size, our experiments suggest that the occurrence of a flame structure may be used to constrain the shaking conditions when these structures formed.

Additional files

Additional file 1: Movie 1 in QuickTime. An example of an experiment in which a flame structure forms at an acceleration of 40.5 m/s^2, a frequency of 40 Hz, and an amplitude of 0.64 mm, replayed at $\times$0.2 speed. This is the same experiment which was shown in Figures 3 and 4.

Additional file 2: Movie 2 in QuickTime. Acceleration dependence at a fixed frequency of 50 Hz replayed at $\times$0.4 speed. These are four selected examples from those shown in Figure 5b. Accelerations in m/s^2 (and the resulting regime) are 2.1 (Ib: percolation), 7.9 (II: transition), 19.3 (II: transition), and 40.5 (III: flame), respectively.

Additional file 3: Movie 3 in QuickTime. Frequency dependence at a fixed acceleration of 5.0 $\pm$ 0.4 m/s^2 replayed at $\times$0.4 speed. These are four selected examples from those shown in Figure 6b. Frequencies in Hz (and resulting regime) are 10 (Ib: percolation), 100 (II: transition), 150 (III: flame), and 3,000 (Ib: percolation), respectively. Note that the flame structure forms at the intermediate frequency of 150 Hz.

Competing interests

The authors declare that they have no competing interests.

Authors' contributions

NY made the apparatus, planned and conducted the experiments, and analyzed and interpreted the results. IS contributed to setting up the apparatus and interpreting the results. Both authors contributed to the preparation of the manuscript. Both authors read and approved the final manuscript.

Acknowledgements

We thank Y. Shibano and A. Namiki for discussions and technical advice, and the two anonymous reviewers for helpful comments on the manuscript. This work was supported by KAKENHI (22109505, 24244073, 24510246).

References

Allen JRL (1985) Principles of physical sedimentology. Blackburn Press, Caldwell, New Jersey

Andreotti B, Forterre Y, Pouliquen O (2013) Granular media. Cambridge University Press, Cambridge

Burtally N, King PJ, Swift MR (2002) Spontaneous air-driven separation in vertically vibrated fine granular mixtures. Science 295:1877–1879. doi:10.1126/science.1066850

Clement CP, Pacheco-Martinez HA, Swift MR, King PJ (2009) Partition instability in water-immersed granular systems. Phys Rev E 80:011311

Clement CP, Pacheco-Martinez HA, Swift MR, King PJ (2010) The water-enhanced Brazil nut effect. EPL 91:54001. doi:10.1209/0295-5075/91/54401

Duran J (2000) Sands, powders, and grains: an introduction to the physics of granular materials. Springer, New York

Eshuis P, van der Weele K, van der Meer D, Lohse D (2005) Granular Leidenfrost effect: experiment and theory of floating particle clusters. Phys Rev Lett 95:258001

Eshuis P, van der Weele K, van der Meer D, Bos R, Lohse D (2007) Phase diagram of vertically shaken granular matter. Phys Fluids 19:123301

Hejmandy P, Bandyopadhyay R, Sabhapandit S, Dhar A (2012) Scaling behavior in the convection-driven Brazil nut effect. Phys Rev E 86:050301

Higashi N, Sumita I (2009) Experiments on granular rheology: effects of particle size and fluid viscosity. J Geophys Res 114:04413. doi:10.1029/2008JB005999

Kokusho T (1999) Water film in liquefied sand and its effect on lateral spread. J Geotech Geoenviron Eng 125:817–826

Kokusho T, Kojima T (2002) Mechanism for postliquefaction water film generation in layered sand. J Geotech Geoenviron Eng 128:129–137

Leaper MC, Smith AJ, Swift MR, King PJ, Webster HE, Miles NJ, Kingman SW (2005) The behaviour of water-immersed glass-bronze particulate systems under vertical vibration. Granular Matter 7:57–67. doi:10.1007/s10035-004-0185-7

Manga M, Brodsky E (2006) Seismic triggering of eruptions in the far field: volcanoes and geysers. Ann Rev Earth Planet Sci 34: 263–291. doi:10.1146/annrev.earth.34.031405.125125

Manga M, Wang C (2007) Earthquake hydrology. In: Kanamori H (ed) Treatise on geophysics, vol 4, 1st edn. Elsevier, Amsterdam, pp 293–320

Mavko G, Mukerji T, Dvorkin J (1998) The rock physics handbook. Cambridge University Press, Cambridge

Melo F, Umbanhowar PB, Swinney H (1995) Hexagons, kinks, and disorder in oscillated granular layers. Phys Rev Lett 75:3838–3384

Michioka H, Sumita I (2005) Rayleigh-Taylor instability of a particle packed viscous fluid. Geophys Res Lett 32:L03309. doi:10.1029/2004GL021827

Montenat C, Barrier P, d'Estevou PO, Hibsch C (2007) Seismites: an attempt at critical analysis and classification. Sedimentary Geol 196:5–30

Moretti M, Alfaro P, Caselles O, Canas JA (1999) Modelling seismites with a digital shaking table. Tectonophysics 304:369–383

Oehler DZ, Allen CC (2010) Evidence for pervasive mud volcanism in Acidalia Planitia, Mars. Icarus 208:636–657

Pak HK, Behringer RP (1993) Surface waves in vertically vibrated granular materials. Phys Rev Lett 71:1832–1835

Pondrelli M, Rossi AP, Ori GG, van Gasselt S, Ceramicola C (2011) Mud volcanoes in the geologic record of Mars: the case of Firsoff crater. Earth Planet Sci Lett 304:511–519

Shearer PM (1999) Introduction to seismology. Cambridge University Press, Cambridge

Samadani A, Kudrolli A (2001) Angle of repose and segregation in cohesive granular matter. Phys Rev E 64:051301

Shibano Y, Namiki A, Sumita I (2012) Experiments on upward migration of a liquid-rich layer in a granular medium: implications for a crystalline magma chamber. Geochem Geophys Geosyst 13:Q03007. doi:10.1029/2011GC003994

Schleier-Smith JM, Stone HA (2001) Convection, heaping, and cracking in vertically vibrated granular slurries. Phys Rev Lett 86:3016–3019
Vucetic M (1994) Cyclic threshold shear strains in soils. J Geotech Eng 120:2208–2228
Voth GA, Bigger B, Buckley MR, Losert W, Brenner MP, Stone HA, Gollub JP (2002) Order clusters and dynamical states of particles in a vibrated fluid. Phys Rev Lett 88:234301
Whitehead JA, Luther DS (1975) Dynamics of laboratory diapir and plume models. J Geophys Res 80:705–717
Yamaguchi A, Yoshida N, Tobita Y (2008) Estimation of thickness of liquefied layer from sand boiling pattern. J Japan Soc Civil Eng Div C 64:79–89

2

Three-dimensional mapping and kinematic characterization of mass transport deposits along the outer Kumano Basin and Nankai accretionary wedge, southwest Japan

Jason Lackey[1], Gregory Moore[1*] and Michael Strasser[2]

Abstract

Three-dimensional (3D) seismic data from the southern Kumano Basin of southwest Japan image a nested series of moderately sized mass transport deposits (MTDs) that slid from a slope along the seaward side of the forearc basin. The deposits are dated to be approximately 0.3 to 0.9 Ma. These MTDs are likely linked to the movement along a prominent out-of-sequence thrust (OOST) fault, regionally steeper slopes that would have existed during deposition, and shifts in sedimentation over the past 0.9 Ma. The spatial resolution provided by the 3D seismic data permits the identification of kinematic characteristics and the internal geometries of the MTDs which total over 2.8 km^3 in volume and cover more than 59 km^2 of the seafloor at various stratigraphic levels. Each MTD is well imaged and exhibits various kinematic indicators while most of the basal glide planes and original headwall scarps above the deposits have been partially or fully eroded by subsequent MTDs. There are at least seven individual deposits that range in volume from 0.005 to 1.16 km^3, in area from 0.2 to 21.8 km^2; have runouts between 0.55 and 7.9 km, and generally translate downslope from the SE to NW. Basal, internal, and top surface kinematic indicators, such as grooves, thrust and fold systems, and pressure ridges, show that these MTDs originate from a prominent slide scar recognized in the high-resolution regional bathymetry. This, combined with a regionally shifting depocenter and faulting related to the earthquake cycle, points to regional tectonic activity as being the most likely failure trigger for these nested landslides.

Keywords: Mass transport deposit, 3D, Landslide, Nankai, Accretionary, Subduction, Kinematic

Introduction

Subaqueous landslides occur in many different environments all over the world including both active and passive margins, along the flanks of volcanic islands, and river deltas (Prior and Coleman 1978; Moore et al. 1989; Lee 2009) and can occur on a great variety of slopes (ranging from < 0.01° to vertical (Prior and Coleman 1978)). These slides can be devastating to seafloor infrastructure via high-speed turbidity currents and to coastal communities in the form of tsunami (Bondevik et al. 1997; Bardet et al. 2003; Satake 2012). Characterization by their size, shape, and kinematics allows us to better understand how they initiate and translate downslope to the point at which they arrest (Ward and Day 2002).

The movement of the failed mass from the source of failure to the point of arrestment can be analyzed and described using kinematic indicators present in either seismic data (Prior et al. 1984; Bøe et al. 2000; Gee et al. 2005, 2006) or outcrops (Farrell 1984; Lucente and Pini 2003; Matheus et al. 2017). In a general sense, kinematics describes the motion of bodies without the consideration of the body's mass or forces that caused its motion. Here, we follow Bull et al. (2009) in defining "kinematic indicator" as a geologic structure or feature which records information related to the type and direction of motion at the time of emplacement. This information is of great use in helping us understand the motions involved with slide initiation, evolution during translation downslope, and cessation. Additionally, because of the complex nature of

* Correspondence: gmoore@hawaii.edu
[1]Department of Earth Science, University of Hawai'i, 1680 East-West Rd, Honolulu, HI 96822, USA
Full list of author information is available at the end of the article

submarine landsliding, many variably active processes are likely involved in each mass movement. This leads to a litany of potential classifications and terminology. Here, we chose to simplify the classification of deposits from various mass movements identified in the data as simply "mass transport deposits" (MTDs).

This study represents a unique opportunity to analyze several Quaternary MTDs via a 3D seismic reflection survey in the Kumano Basin along the Nankai accretionary prism. Some of these MTDs were previously identified by Moore and Strasser (2016); however, no kinematic analysis was performed nor was a reconstruction of the original slope attempted. Because they are fully imaged in the 3D survey, a complete analysis of kinematic indicators and morphology is possible. A kinematic analysis of these MTDs should provide greater insight into the depositional environment in which sliding occurred and determine a possible explanation as to the origin of a prominent seafloor scar (Lackey et al. 2018).

Regional setting and study area

The Kumano Basin is within the Nankai Trough subduction zone where the Philippine Sea Plate is presently subducting under the Amur Plate at a rate of 4 to 6 cm yr^{-1} (Seno 1989). The NW oblique subduction has led to the creation of an accretionary prism spanning roughly 700 km, with the overall oblique slip being apportioned along a large right-lateral strike-slip fault, known as the Median Tectonic Line, landward of the prism (Fitch 1972; Taira 2001). The accreted sediment is primarily terrigenous detritus from Honshu Arc, Japan, and hemipelagic deposits on the Philippine Sea Plate (Underwood and Moore 2012). The upper part of the hemipelagic section from the Philippine Sea Plate is accreted while the lower part is subducted beyond the inner prism (e.g., Kinoshita et al. 2009; Kimura et al. 2011; Strasser et al. 2011; Strasser et al. 2012; Underwood and Moore 2012; Moore et al. 2014). The inner prism is overlain by the Kumano Forearc Basin which has formed over approximately the past 1.95 to 2.0 Ma behind the outer ridge that is believed to be the result of the movement along the megasplay fault (Gulick et al. 2010; Moore et al. 2015). The seismic recurrence interval of roughly 100 to 200 years for large earthquakes has been well established for Nankai (Ando 1975). The rapid growth of the accretionary prism is an important factor for large, repeated earthquakes in this subduction zone setting (Ruff and Kanamori 1980).

The present study area is within the Kumano Forearc Basin just landward of the outer ridge of the accretionary prism (Fig. 1). The 3D seismic data used for this study covers roughly 275 km^2 (Moore and Strasser 2016) and is located entirely within the transition zone of the accretionary prism between 35 and 50 km from the deformation front (Kimura et al. 2007). Seafloor depths vary from 1700 m seaward of the study area to 2100 m landward. Notable characteristics include numerous faults in both the 3D data and bathymetry, temporally sequential landslide scars, and an absence of surficial landslide deposits (e.g., Strasser et al. 2009; Moore et al. 2013; Lackey et al. 2018).

Methods/Experimental

To identify MTDs and kinematic indicators, we interpreted several prominent horizons throughout our 3D seismic survey, including basal shear surfaces (BSS), tops of MTDs, and various depositional layers. Amplitude and coherency data are then displayed on these horizons to highlight geomorphic structures and kinematic indicators with the goal of reconstructing the various MTDs to their original positions. To validate the conclusions drawn from the data, we compare our results to other subaqueous landslide kinematic studies (e.g., Frey Martinez et al. 2005, 2006; Gee et al. 2005, 2006; Bull et al. 2009; Moore and Strasser 2016) that utilize 3D seismic reflection data.

3D volume

The full 3D survey covers roughly 12 km × 56 km extending from the Kumano Basin seaward to the frontal thrust in the dip direction (Fig. 1). The acquisition occurred aboard the M/V *Nordic Explorer* during April to May 2006 under contract by Petroleum Geo-Services (PGS). Two sound source arrays (totaling 51 L or 3090 $in.^3$) were utilized and fired alternately at 37.5 m shot intervals. The survey used four 4500 m long receiver cables spaced 150 m apart with 360 receiver groups at 12.5 m spacing. This geometry yielded eight source-receiver common midpoint (CMP) lines per sail line at 37.5 m spacing (Moore et al. 2009). The interval between the lines and cross lines of the resulting data set is 18.75 m and 12.5 m, respectively. The vertical resolution is ~ 5–7 m near the seafloor, degrading to ~ 10–20 m at depths near 1400 m below seafloor (mbsf; Moore et al. 2009). Standard pre-processing to reduce noise preceded pre-stack depth migration (PSDM) and produced a clear seismic image in depth (Moore et al. 2009). For the seaward part of our area, we were able to use a reprocessed version of the 3D data set that produced much higher-resolution images of the sedimentary section (Shiraishi et al. 2018).

Bathymetry data

High-resolution bathymetry data analyzed for surficial MTDs has been acquired during R/V *Sonne* Expedition 251 in October 2016 (Strasser et al. 2017). The R/V *Sonne* is equipped with an EM 122 KONGSBERG multibeam echosounder (MBES) operated at 12 kHz. The transducers have a nominal opening of 0.5° in

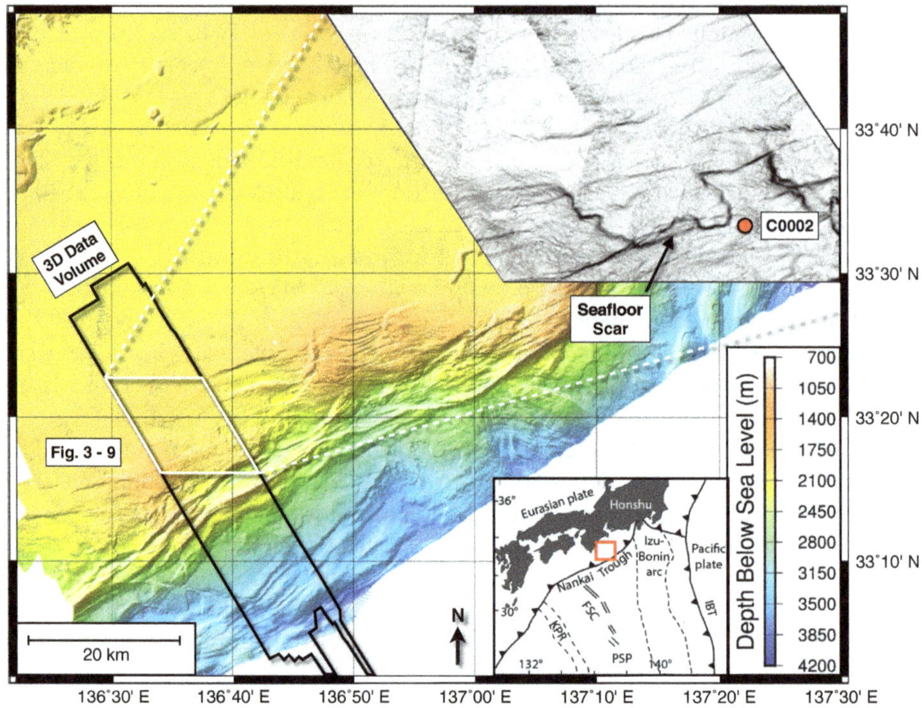

Fig. 1 Study area. High-resolution bathymetry of the Kumano Basin acquired by R/V *Sonne* during Cruise SO251 in 2016 (Strasser et al. 2017). The trapezoidal inset shows a slope gradient map of a prominent seafloor scar from Lackey et al. (2018) and the location of IODP drill site C0002

along-track direction and 1° in across-track direction. The MBES recorded 433 individual beams across track within a swath of 120°. Actual sound velocity profiles were recorded with the ships CTD and inserted as the basis for optimized performance. Data were processed with software MB-System (Caress and Chayes 1996).

Core data

Coring of the Kumano Basin sediments was accomplished on Integrated Ocean Drilling Program (IODP) Expeditions 315 (Kinoshita et al. 2009) and 338 (Strasser et al. 2014) (Fig. 1). The cored interval consists of hemipelagic mud and numerous thin interbeds of normally graded silt, sandy silt, and sand with local layers of volcanic ash (Underwood and Moore 2012). For a more detailed analysis of the core data, see Underwood and Moore (2012). These data demonstrate that nearly all the basin fills accumulated within the last 1.6 Ma.

MTD recognition and kinematic indicators

MTDs are typically identified by their internal chaotic reflection character or semitransparent to transparent seismic character. Once a MTD is identified, other seismic indicators (e.g., hummocky reflections, truncations of stratigraphic layers, folds, and thrusts) are used to characterize the remainder of the MTD. For the purposes of this study, we subdivide each MTD into three domains based on which kinematic indicators are most likely to occur in a typical "tripartite" MTD anatomy (Martinsen 1994; Lastras et al. 2002): the headwall, translational, and toe domains (Fig. 2). While there is almost always an overlap between these domains, it is important to have a clear delineation between them to properly study the kinematic indicators in the data.

Headwall domain

The two prominent kinematic indicators in this region are headwall scarps and extensional ridges and blocks that represent the extensional, upslope portion of the MTD (Bull et al. 2009). Crown cracks may also develop upslope of the primary headwall scarp (Fig. 2).

Translational domain

The translational domain is defined as the main body of failed material that has moved downslope. It is subdivided into four sections based on their differing physical parameters: lateral margins, basal shear surface, internal MTD body, and top MTD surface (Martinsen 1994; Frey Martinez et al. 2005; Bull et al. 2009).

Toe domain

Characterized by an overall convex-downslope morphology, the toe domain represents the region of downslope termination of the main MTD body and is characterized by thrust and fold systems and pressure ridges (Prior et al. 1984; Frey Martinez et al. 2005).

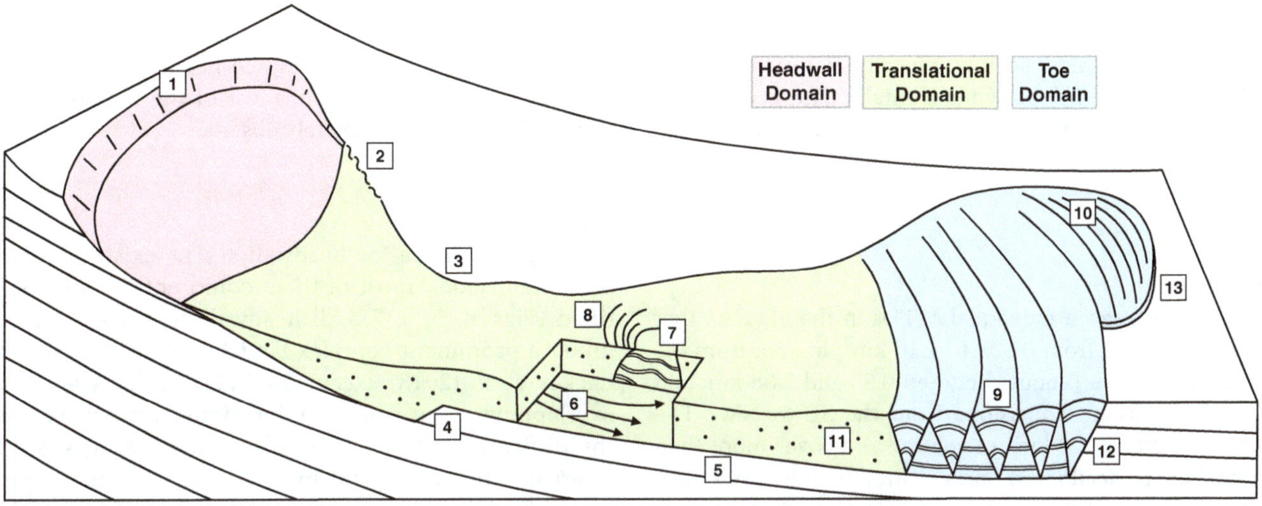

Fig. 2 Key geometric and geologic criteria for kinematic indicator recognition. 1—headwall scarp. The direction of initial movement roughly perpendicular to the headwall propagation. 2—en-echelon sigmoidal segments. Strike-slip affected MTD material differentiates lateral margin from headwall scarp. 3—lateral margins. Delimits strike-parallel extent of MTD and constrains gross general transport direction. 4—ramp. Location(s) where the BSS jumps to a higher stratigraphic level. 5—flat. Location(s) where the BSS is parallel with local stratigraphy. 6—grooves. Implies debris flow processes with a translation parallel to their trend. 7—translated block. Often align long-axis downslope. 8—secondary flow fabric. Translation direction indicated by the trend of flow parallel banding. 9—pop-up blocks. Typically hundreds of meters high, translation perpendicular to the alignment of ridges. 10—pressure ridges. Typically tens of meters high, translation perpendicular to the strike of thrust faults. 11—MTD matrix. 12—frontally confined toe. 13—frontally emergent toe. BSS—basal shear surface. MTD—mass transport deposit. Modified after Bull et al. (2009)

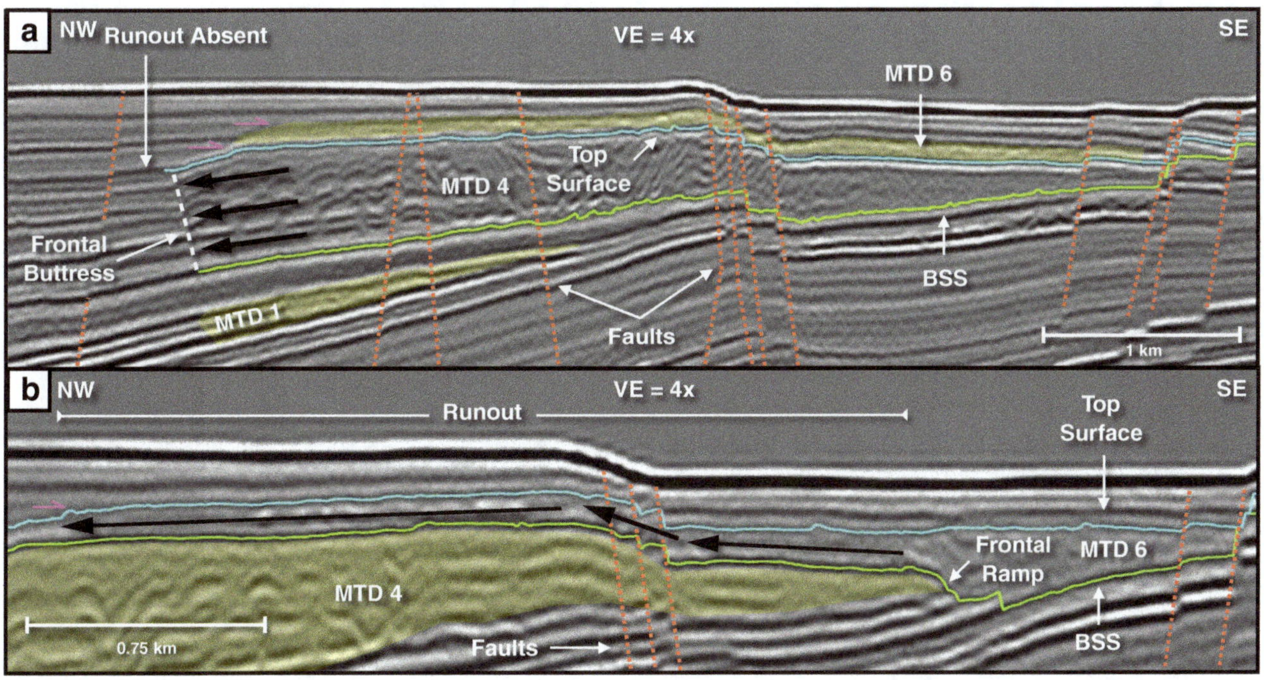

Fig. 3 Frontally confined versus frontally emergent MTDs. Black arrows indicate primary translation direction. Purple arrows indicate onlapping relationships. BSS—basal shear surface. **a** Frontally confined landslide. The arbitrary seismic line across MTD 4 (see Fig. 6 for location). Note how the landslide mass is buttressed against undisturbed strata. **b** Frontally emergent landslide. Seismic Inline 2352 across MTD 6 (see Fig. 7 for location). Note the runout of the landslide mass beyond the frontal ramp

Frontally confined MTDs are those where the toe domain exhibits impressive thrust and fold systems that are buttressed against stratigraphically equivalent undisturbed strata downslope while frontally emergent MTDs are able to ramp up from the original basal shear surface and translate freely across the seafloor (Frey Martinez et al. 2006) (Fig. 3).

Results

We identified seven individual MTDs in the 3D data that range in volume from 0.005 to 1.16 km^3, in area from 0.2 to 21.8 km^2, have runouts between 0.55 and 7.88 km, and generally translate downslope from the SE to NW. The MTDs total over 2.8 km^3 in volume and cover more than 59 km^2 of the seafloor at various stratigraphic levels. The MTDs have been designated as MTD 1 through MTD 7 based upon their relative temporal occurrences with MTD 1 being the oldest. Measurements for all MTDs are summarized in Table 1. Additionally, three surficial MTDs were identified in high-resolution bathymetry that were not fully resolvable in the 3D seismic data.

Headwall domain

Kinematic indicators for the headwall domain were difficult to identify in the data. While headwall scarps were partially or fully identified in most of the MTDs, extensional blocks were completely absent.

Headwall scarps

Although evidence for headwall scarps exists for most of the landslides, most of them could not be fully identified (Figs. 4, 5, 6, 7). All headwall scarps are located under a prominent complex landslide scar identified by Lackey et al. (2018) except for MTD 2. Because of the overlapping nature of their locations, we believe that the difficulty in identifying full headwall scarps is due to erosion by subsequent mass failures at higher stratigraphic levels. This is especially true for MTDs 1, 3, and 6 as there are no discernable headwall scarps in the data leading us to conclude that they have been fully eroded (Figs. 4b, 5b, 7b). Additionally, only a portion of the headwall scarp was identifiable for MTD 4 (Fig. 6b).

Table 1 Physical MTD parameters and identified kinematic indicators

MTD	Stratigraphic sequence and age	Volume (km^3)	Area (km^2)	Average thickness (m)	Runout (km)	Translation direction	Headwall domain	Translational domain	Toe domain (confined/emergent)
1	Kumano 4 0.9 Ma to 0.3–0.44 Ma	0.999	21.8	45.8	7.884	S ➔ N	None	LM: scarps, en-echelon sigmoidal segments BSS: flats, grooves IMB: translated block TMS: second-order fabric	Confined: thrusts
2	Kumano 4 0.9 Ma to 0.3–0.44 Ma	0.03	1.5	20	1.729	SE ➔ NW	Headwall	LM: scarps BSS: flats IMB: none TMS: none	Emergent: pressure ridges
3	Kumano 3 0.3–0.44 Ma to Present	0.358*	7.73*	46.3	3.238	SE ➔ NW	None	LM: scarps, en-echelon sigmoidal segments BSS: ramps, flats IMB: translated blocks TMS: none	Confined: thrusts
4	Kumano 3 0.3–0.44 Ma to Present	1.155	14.15	81.62	5.741	SE ➔ NW	Headwall	LM: scarps, en-echelon sigmoidal segments BSS: ramps, flats IMB: translated Blocks TMS: none	Mostly confined: thrusts
5	Kumano 3–2 0.3–0.44 Ma to Present	0.005	0.2	25	0.552	S ➔ N	Headwall	LM: scarps BSS: ramps, flats IMB: none TMS: none	Mostly confined: thrusts, pressure ridges
6	Kumano 2–1 0.3–0.44 Ma to Present	0.297	13.1	22.7	4.136	S ➔ N	None	LM: scarps BSS: ramps, flats IMB: none TMS: none	Emergent: none
7	Kumano 2–1 0.3–0.44 Ma to Present	0.031	1.2	25.8	1.508	SW ➔ NE	Headwall	LM: scarps BSS: flats IMB: none TMS: none	Emergent: pressure ridges

LM lateral margins, *BSS* basal shear surface, *IMB* internal MTD body, *TMS* top MTD surface
*This is a pre-erosional estimate based on the assumption that 1/3 of the MTD was eroded by MTD 4

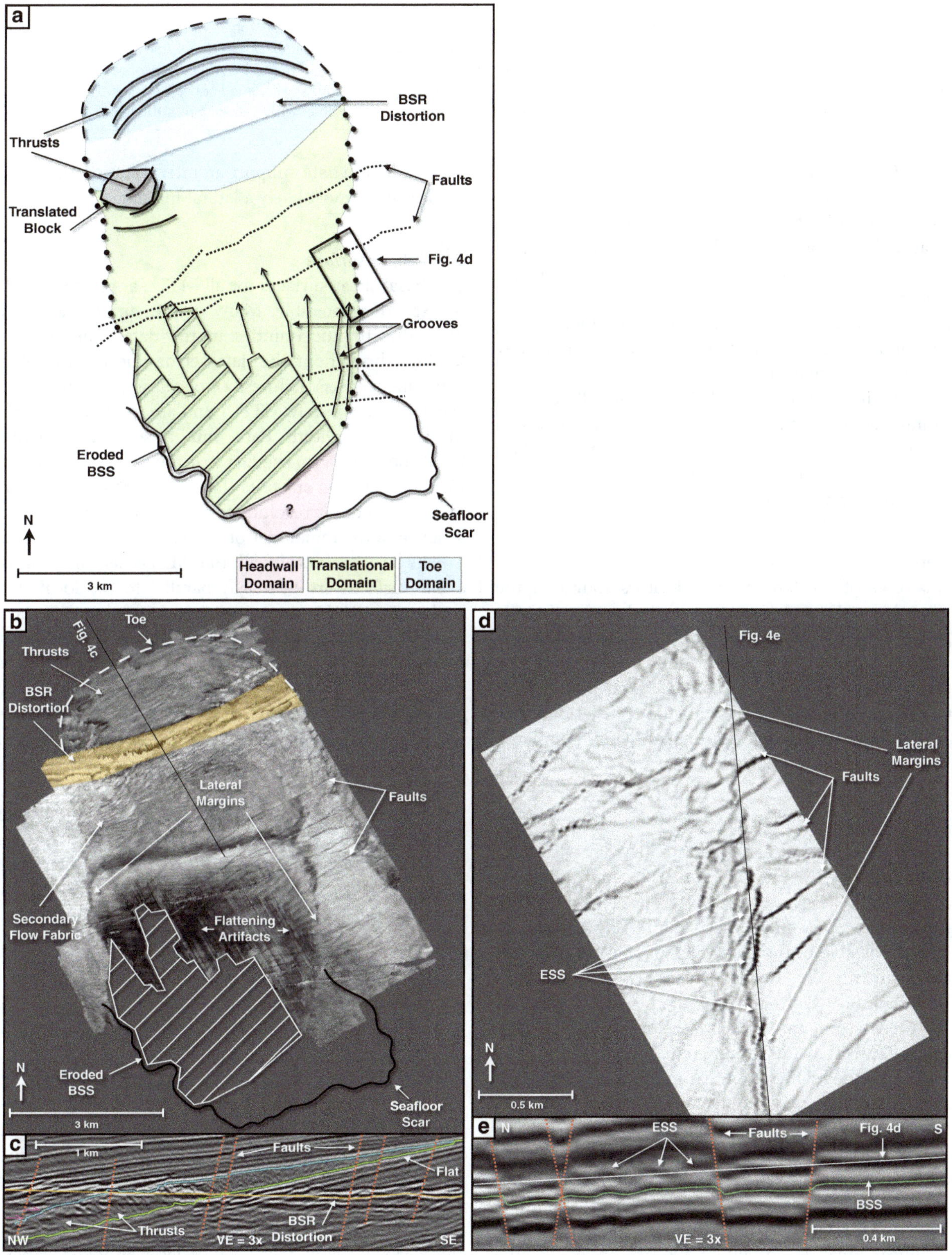

Fig. 4 (See legend on next page.)

(See figure on previous page.)
Fig. 4 MTD 1 interpretations. The primary direction of translation is S ➔ N. BSR—bottom-simulating reflector; BSS—basal shear surface; ESS—en-echelon sigmoidal segments. **a** Schematic depiction of the domains and main kinematic features within MTD 1. **b** Seismic amplitude map of structurally flattened BSS. **c** Inline seismic cross section 2480 transecting the translational and toe domains. Note that BSR reflection crosscuts the MTDs and disrupts the stratal continuity. Purple arrows indicate onlapping relationships. **d** Structurally flattened coherency slice showing a lateral margin of MTD 1 (location shown in **a**). **e** Arbitrary seismic cross section along the lateral margin of MTD 1. Location of the line is shown in **d**

Extensional ridges and blocks
No extensional blocks were identified for any of the MTDs. The most likely reasons for their absence are that the headwall regions were fully evacuated of material, subsequent erosion by stratigraphically higher failures (as with the headwall scarps), or insufficient resolution of the data. All headwall regions appear to be void of any failure material, indicating that any extensional blocks that may have formed continued to translate downslope and become blocks in the translational domain or fully disintegrated into the MTD matrix. Because we believe many of the headwall scarps were eroded by subsequent mass failures, it is also likely that any extensional blocks that may have remained immediately after failure were also eroded.

Translational domain
The majority of the kinematic indicators found in the data are located in the translational domain. Lateral margins, ramps, and flats are well represented in the data. We believe this domain to be so well represented due to its containing the bulk of the failed material for each MTD and because almost all the translational domains were well preserved and imaged when compared to the headwall and toe domains.

Lateral margins
All MTDs have lateral margin scarps that are visible in the data. Some of the margins are easily interpreted for the entire MTD (Figs. 4b, 6b) while others have either been eroded by subsequent mass failure (Fig. 5b) or are not fully resolvable in the data (Fig. 7b). En-echelon sigmoidal segments are imaged in MTDs 1, 3, and 4 (Figs. 4d, 5d, 6d). Only portions of the lateral margins exhibit these segments, and in each instance, the segments are only 50 to 100 m in length.

There was added difficulty in the accurate identification of lateral margins as the strata are offset by numerous faults throughout the 3D volume (Moore et al. 2013). To overcome this, seismic amplitude and coherency are displayed onto interpreted horizons that are then structurally flattened (Fig. 8). When viewing an internal amplitude slice using MTD 1's structurally flattened basal shear surface, a linear break in amplitude appears upslope of the main body of failed material (Fig. 4b). The orientation of this linear feature is consistent with the MTD's other kinematic indicators such as grooves found in the basal shear surface. These data support an interpretation that this linear feature is most likely a lateral margin.

Basal shear surface
The basal shear surfaces for all MTDs are mapped to their fullest extent possible. Many of the surfaces are incomplete due to uncertainties in the data from ubiquitous faulting throughout the data set or due to erosion from subsequent mass failures. Grooves appear in the basal shear surface of only MTD 1 and are made visible only by displaying amplitude data onto a structurally flattened basal shear surface. The fact that they do not show in seismic cross section suggests that these features are near the limit of the data's resolution. This could also explain their absence in the remainder of the MTDs.

Flats are present in all the MTDs as all their basal shear surfaces are primarily parallel to the local stratigraphy. Ramps, however, are only located in MTDs 3, 4, 5, and 6 (Figs. 5c, 6c, and 7c) and all trend perpendicular to their MTD's primary flow direction. The absence of parallel trending ramps, by definition, means that slots are also absent from the data (Bull et al. 2009). Additionally, there are no instances of the basal shear surface ramping down and cutting into the underlying stratigraphy, only ramping up to higher stratigraphic levels.

None of the MTD basal shear surfaces exhibit signs of striations. As these features are caused by outrunner blocks (of which there was none identified), their absence is not a surprise. It is possible, however, that these features exist, especially for the MTDs that are partially or fully emergent and are simply unresolvable in the data.

Internal MTD body
Translated blocks were identified in MTDs 1, 3, and 4 (Figs. 4a, 5a, 6a). Their absence in the remaining MTDs likely has a connection with their average thickness (Table 1). Because each stratigraphic horizon is roughly equal to the vertical resolution of the data, a block would need to be at least 20–28 m (× 4 the vertical resolution) in height to be easily distinguishable from the surrounding strata. The average thickness of MTDs 2, 5, 6, and 7 is 20–26 m. Therefore, any blocks within the failed mass of these MTDs could have been easily overlooked as belonging to the local stratigraphy. This may also explain the absence of remnant and outrunner blocks as neither of these block types were identified for any of the MTDs.

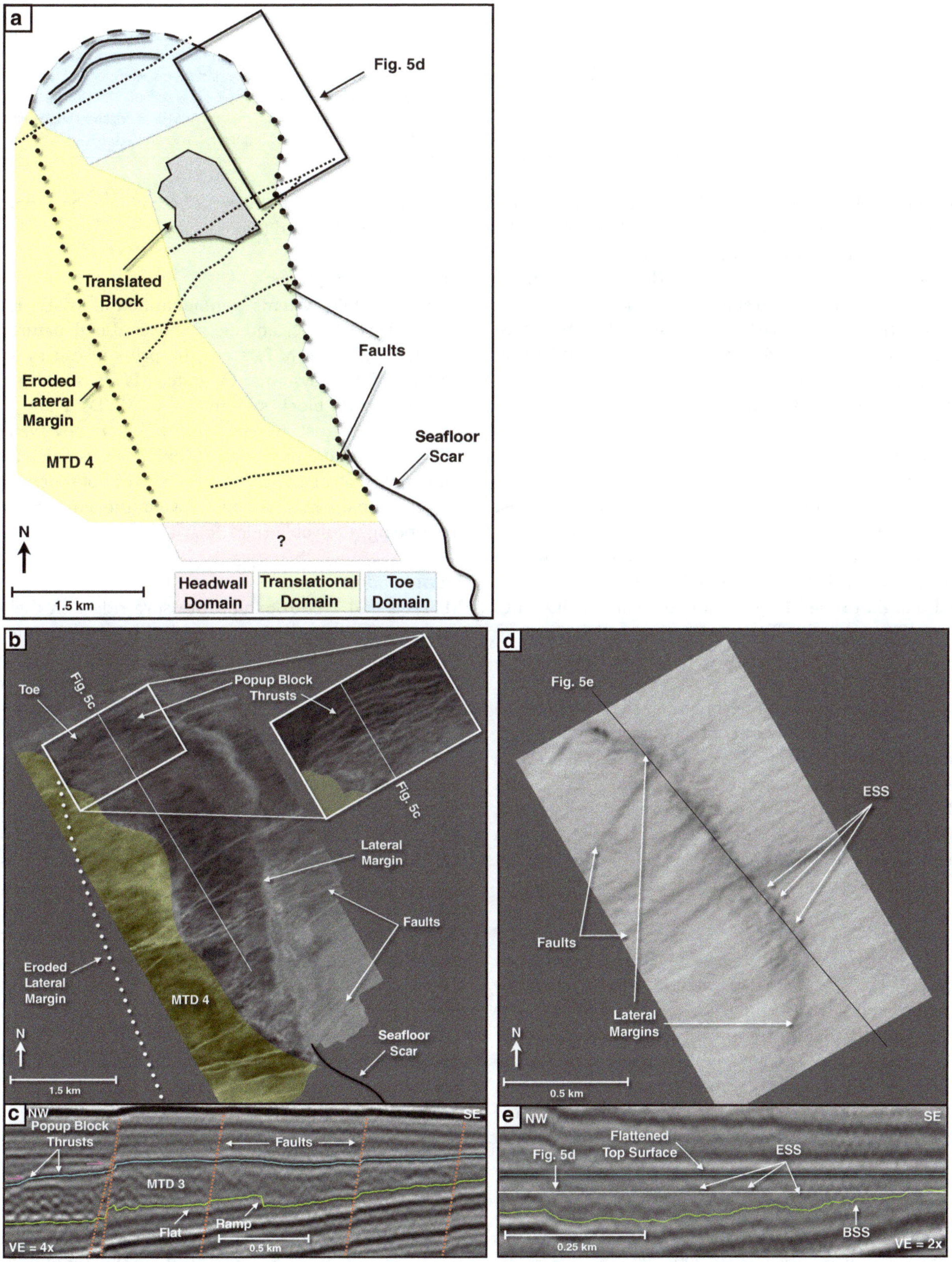

Fig. 5 (See legend on next page.)

(See figure on previous page.)
Fig. 5 MTD 3 interpretations. The primary direction of translation is SE → NW. BSS—basal shear surface; ESS—en-echelon sigmoidal segments. **a** Schematic depiction of the domains and main kinematic features within MTD 3. The question mark indicates an undefined headwall domain due to erosion. The large yellow area indicates the location of MTD 4. **b** Seismic amplitude map of the structurally flattened top surface of MTD 3 (blue line in **c**). Inset is a structurally flattened portion of the BSS showing thrusts in the toe domain. **c** Inline seismic cross section 2492 transecting the translational and toe domains (location shown in **b**). **d** Structurally flattened coherency slice showing a lateral margin of MTD 3 (location shown in **a**). Note the en-echelon sigmoidal segments. **e** Arbitrary seismic cross section along the lateral margin of MTD 3 (location shown in **d**)

Outside of the toe domain's thrust and fold systems, folds were not identified in any of the MTDs. Given the large degrees of deformation and the runouts for these deposits, it is unlikely that folding did not take place during emplacement. Therefore, the most likely explanation for their absence is that they are below the resolution of the data as folding has been identified in sub-seismic scale failures in outcrop (Farrell 1984; Lucente and Pini 2003; Matheus et al. 2017).

Top MTD surface

Only MTD 1 exhibited a kinematic indicator from the top surface. A second-order flow fabric was identified above and around a translated block (Fig. 4b). The likely explanation for this fabric centers around the MTD's translated block. It likely became buttressed against the lateral margin of the MTD, causing the failed material to be forced around the flow obstruction which leads to thrusting within the failed mass and a secondary fabric in the top surface (Fig. 4b).

There were no instances of longitudinal shear zones in any of the MTDs. The majority of these types of kinematic indicators have been found on much larger MTDs in other places around the world (Masson et al. 1993; Gee et al. 2005; Gafeira et al. 2007). Therefore, we believe that the MTDs in this study are either too small to exhibit such differences in flow speed within the failed mass or that this indicator is simply not resolved in the data.

Toe domain

The toe domain is identified on all MTDs. However, although the toe of MTD 6 was identifiable, it lacked any kinematic indicators within the domain (Fig. 7). The most likely explanation for this absence is that the toe domain is not thick enough to either contain or resolve any of the indicators. Also unique to MTD 6 were a hummocky terrain and turbidite located beyond the toe domain. Their existence is likely due to the amount of material scoured from MTD 4 during the emplacement of MTD 6 (eroded BSS; Fig. 6).

The toe domain of MTD 1 also proved difficult to map. A strong bottom-simulating reflector (BSR) caused by gas hydrate accumulations is imaged 200–400 mbsf across the entire area of interest within the 3D seismic volume (Miyakawa et al. 2014). This led to inaccuracies in mapping the toe domain of MTD 1, causing kinematic indicators that would otherwise be visible to be either partially or fully obscured (Fig. 4).

Thrust and fold systems

Thrust and fold systems were identified in MTDs 1, 3, 4, and 5 (Figs. 3c, 5c, and 6c). The confined nature and thickness of these MTDs are the primary causes of the thrust and fold systems. Of note, MTDs 3 and 4 both display pop-up block systems while MTDs 1 and 5 exhibit simple thrust systems. It is possible, however, that MTD 1 also contains pop-up blocks as its thickness is similar to that of MTD 3. The BSR that obscures MTD 1's toe domain may simply prevent the pop-up blocks from being resolvable (Fig. 3c).

Pressure ridges

MTDs 2 and 7 both exhibit pressure ridges in their toe domains. Because thrusts are typically found in conjunction with pressure ridges in outcrop (Masson et al. 1993; Gee et al. 2005; Gafeira et al. 2007), it is likely that some sort of thrust system exists in both MTDs and are simply too small to resolve in the 3D data.

Surficial MTDs

The seafloor scar identified by Lackey et al. (2018) appears to contain at least three surficial MTDs that could not be fully identified in the 3D data (Fig. 9b). These MTDs are clearly visible as seafloor features in high-resolution bathymetry (Strasser et al. 2017). Their absence from the 3D data is likely due to their relatively small scale and their recent occurrence. Because they are recent failures, they have yet to be buried and will, therefore, be poorly imaged in 3D seismic data. While kinematic indicators from these surficial MTDs cannot be identified in the 3D data, they are still important in understanding the evolution of the seafloor with respect to this study.

Discussion

The MTDs identified in this study exhibit numerous kinematic indicators that enable us to determine the spatial and relative temporal origin of each failure. We have reconstructed each of the failures to their original positions and, using the local stratigraphy, temporally sequenced each failure in the most probable order (Fig. 10). Each of the MTDs (with the notable exception of MTD 2) appears to originate stratigraphically below the

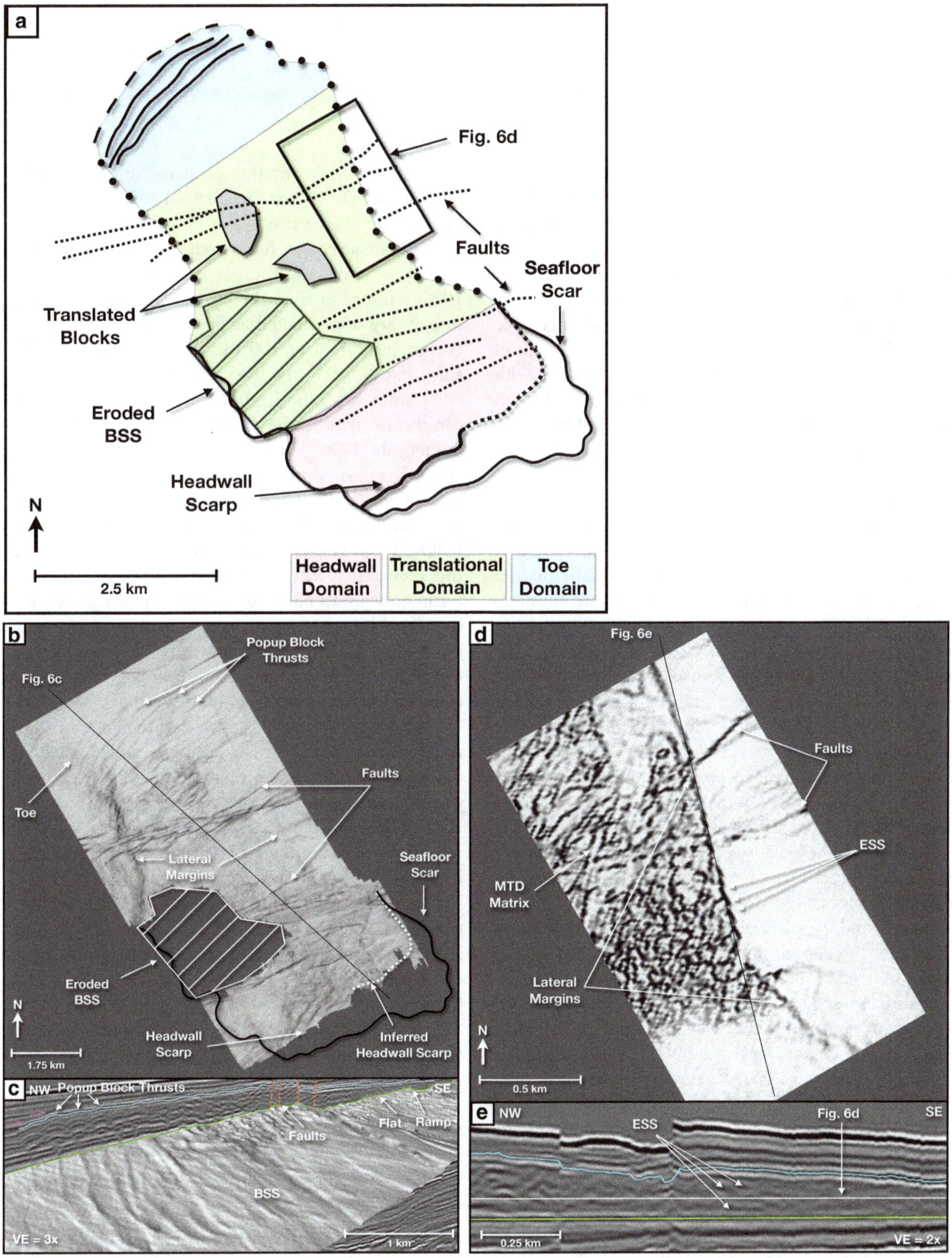

Fig. 6 (See legend on next page.)

(See figure on previous page.)
Fig. 6 MTD 4 interpretations. The primary direction of translation is SE ➔ NW. BSS—basal shear surface; ESS—en-echelon sigmoidal segments. **a** Schematic depiction of the domains and main kinematic features within MTD 4. **b** Coherency map of structurally flattened BSS. **c** Arbitrary seismic cross section transecting MTD 4 from the headwall domain to the toe domain (location shown in **b**). **d** Structurally flattened coherency slice showing a lateral margin of MTD 4 (location is shown in **a**). Note the en-echelon sigmoidal segments. **e** Arbitrary seismic cross section along the lateral margin of MTD 4 (location of the line shown in **d**)

seafloor scar identified by Lackey et al. (2018). These, combined with the three surficial deposits that could not be identified in the 3D data, fully explain the geometry of the scar.

We interpret the scar's location to be associated with the local depositional and tectonic environment over the past 0.9 Ma. All MTDs are located within the stratigraphic sequences Kumano 4–Kumano 1 identified by Gulick et al. (2010). Kumano 4 is the oldest of these sequences, with deposition beginning ~ 0.9 Ma. During this time, motion along a prominent out of sequence thrust (OOST) was reactivated, potentially by the continued subduction of a seamount on the Philippine Sea Plate (Kimura et al. 2011; Moore et al. 2013), causing the outer forearc to continue tilting landward as it had during the deposition of the Kumano 8 sequence ~ 1.2 Ma (Moore et al. 2015). A series of normal faults that are spatially correlated with a SW uplift penetrates through the Kumano 5 and Kumano 4 sequences but does not penetrate the Kumano 4 sequence boundary (K4) (Fig. 11b). This indicates that the uplift took place rapidly and ceased by ~ 0.3–0.44 Ma (Moore et al. 2015). This rapid uplift and resultant slope steepening coupled with seismic loading could explain the occurrence of MTDs 1 and 2 since they are fully contained within the Kumano 4 sequence and are spatially correlated. Of note, there is a lack of MTDs in stratigraphic sequences older than the Kumano 4 sequence. We believe this to be because of the hiatus in tilting (a key preconditioning factor) between the deposition of Kumano sequences 5–7 (Moore et al. 2015).

MTDs 3–7 and the surficial MTDs occur within or above the Kumano 3 sequence, making them < 0.3–0.44 Ma in age and indicating potentially different causal mechanisms. Kumano sequences 3–1 thicken in the landward direction from < 50 m in the seaward region to > 600 m toward the basin center (Moore et al. 2015). Isopach maps created by Moore et al. (2015) show a general widening and NW shift of the basin depocenter over time (Fig. 12). This shift has led to a decrease in deposition around the seafloor scar since 0.3–0.44 Ma and is the most likely explanation for the general decrease in the size of the MTDs over time. We infer this because, as sediment deposition decreases with time, there will be less material available to slide. Additionally, the edge of the forearc basin ceased tilting by the deposition of K4 as evidenced by the onlapping relationships of the Kumano 4–1 sequences (Gulick et al. 2010). Because neither local steepening nor high rates of sedimentation occurred during the emplacement of MTDs 3–7 or the surficial MTDs, another causal mechanism likely exists.

The Nankai region has a well-documented and regular history of small and great earthquake activity (Ando 1975; Kimura et al. 2011). Great earthquakes along the margin are documented as far back as 648 AD with a recurrence interval of 100 to 200 years (Ando 1975). If a splay fault, such as the Nankai megasplay, ruptures co-seismically, amplification of seismic ground shaking can occur within the hanging wall (Abrahamson and Somerville 1996). This has been suggested for Nankai and other splay fault systems (e.g., Plafker 1972; Tanioka and Satake 2001; Baba et al. 2006), thereby representing a plausible causal link between seismic loading and slope failure. However, as noted by Moore and Strasser (2016), the recurrence interval for landsliding in the Kumano Basin appears to be about 0.05 to 0.1 Ma, which is much less frequent than the recurrence interval for Nankai's great earthquakes. Kremer et al. (2017) suggested that this difference could be due to slopes being preconditioned to fail because of climate forcing during interglacial periods. However, they were unable to resolve the relative contributions of tectonic versus climate preconditioning factors, and Urlaub et al. (2013) showed that the ages of landslides on passive margins can be described by a temporally random Poisson distribution and therefore do not show a positive correlation between the frequency of major slope failures on passive margins and sea level changes during the past 0.18 Ma. Another possibility is that the extent of updip rupture propagation along the megasplay fault is not always sufficient to cause slope failure. Ikari et al. (2011) found that the triggering of slope failure for a statically stable slope by a M_w 6–8 earthquake requires the megasplay to rupture within 12 km of the slope.

Shear strengthening (Locat et al. 2002; Lee et al. 2004) could also be an explanation as to why failure is less frequent than the great earthquake recurrence interval, especially for M_w 6–8 earthquakes (Ikari et al. 2011). An instantaneous increase in pore pressure is induced on saturated, fine-grained sediment during an earthquake. This temporarily weakens the sediment. However, if it is not weakened to the point of failure, the pore pressure diffuses over time permitting the sediment to become over-consolidated. Therefore, an earthquake that does not cause slope failure could lead to an increase in the

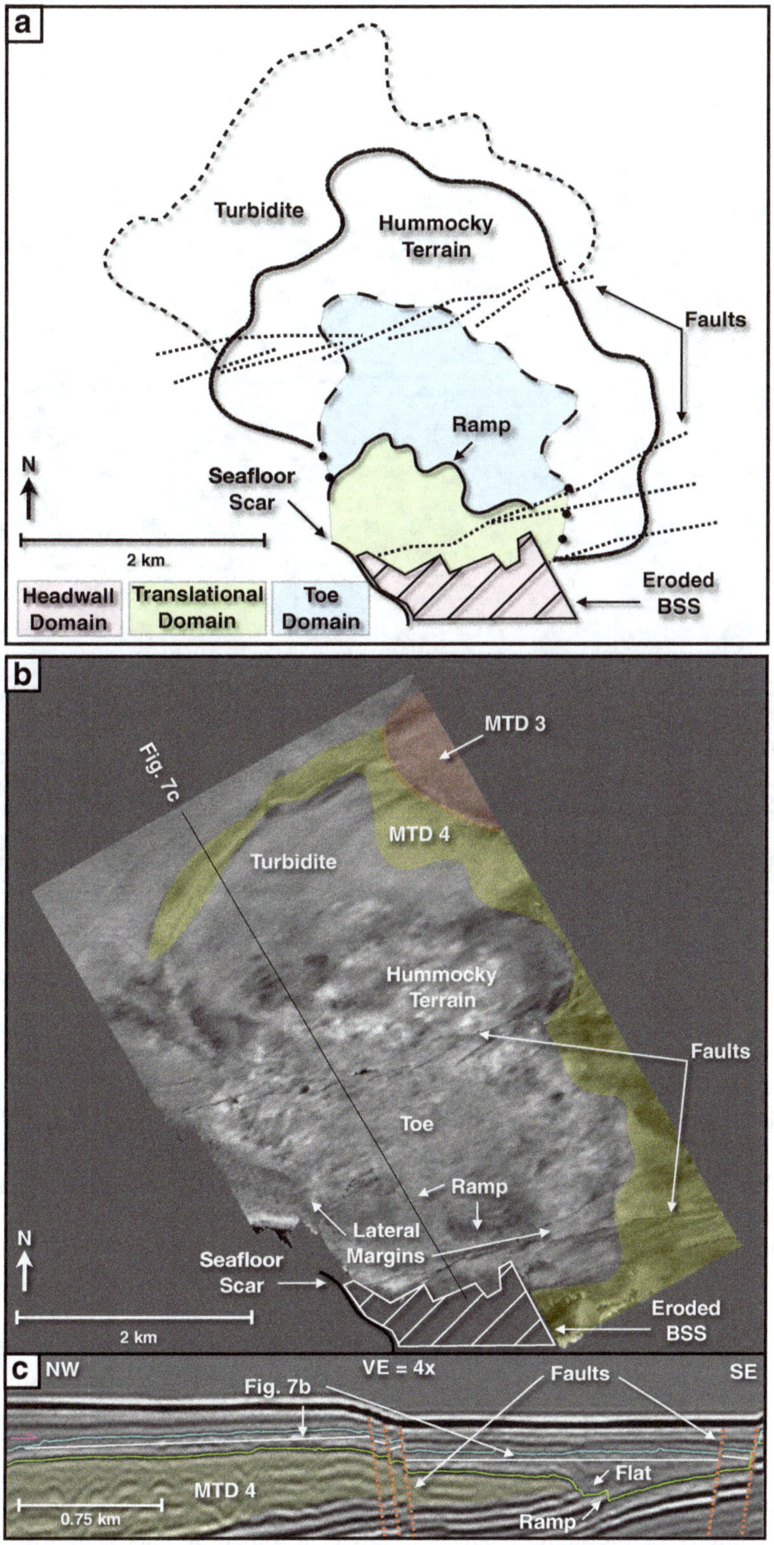

Fig. 7 MTD 6 interpretations. The primary direction of translation is S ➔ N. BSS—basal shear surface. **a** Schematic depiction of the domains and main kinematic features within MTD 6. Distribution of hummocky terrain and a turbidite are also identified. **b** Structurally flattened amplitude slice through MTD 6. Inline seismic cross section 2352 transecting the translational and toe domains of the MTD. Location shown in **b**. **c** Inline seismic cross section 2352 transecting all three domains (location shown in Fig 7b). Note the hummocky reflections in the NW half of the cross section

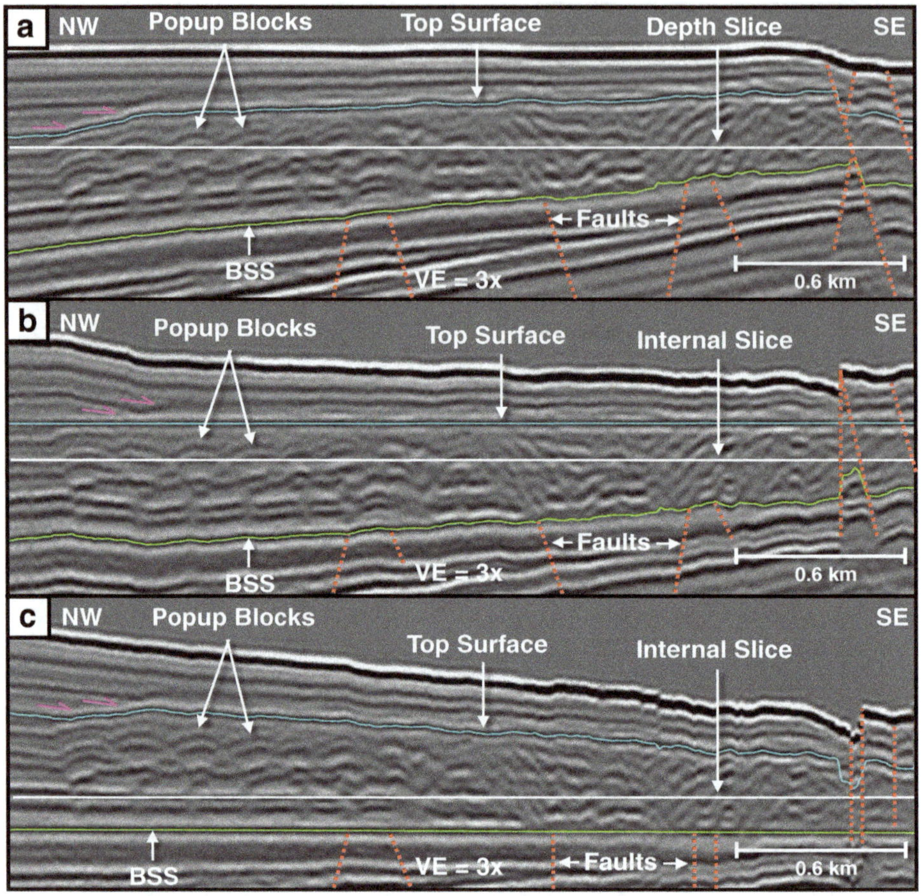

Fig. 8 Structural flattening of interpreted horizons, see Fig. 11a for arbitrary line location. The primary direction of translation is SE → NW. BSS—basal shear surface. Purple arrows indicate onlapping relationships. **a** Arbitrary seismic line of MTD 4 toe domain. Note how the depth slice (white line) intersects the pop-up blocks at various stratigraphic levels. **b** Structurally flattened top surface. Note how the flattened image permits an internal slice (white line) to cross the top surface of each pop-up block. **c** Structurally flattened BSS. Note how the flattened image permits an internal slice (white line) to cross the pop-up blocks at the same stratigraphic level near the BSS

slope's shear strength that increases its resistance to future failures (DeVore and Sawyer 2016). A resistance to future failure implies that either an even larger earthquake would be required to initiate sliding or another causal mechanism is ultimately at work such as the activation of a fault.

Three normal fault populations exist in the study area. The youngest population (phase 3; Sacks et al. (2013); Moore et al. (2013)) tends to cut the upper strata of the forearc basin and manifest as scarps on the seafloor which have been identified in this study. Sacks et al. (2013) proposed a mechanism, via the earthquake cycle, by which the differing orientation of these fault populations and the arcuate geometry of the faults associated with our study originated. During the late inter-seismic period, faults perpendicular to the trench can undergo slip, while faults oriented parallel to the trench undergo slip during post-seismic extension. Additionally, Moore et al. (2013) showed that these faults are likely short-lived rather than regional phenomenon as they lack growth structures. These two conclusions support the idea that these faults occur in conjunction with the earthquake cycle.

A landslide does not occur during each seismic event, possibly in part to the effects of shear strengthening. However, landsliding can occur because of the change in the regional stress regime during the post-seismic phase of the earthquake cycle leading to favorable normal faulting conditions that strike parallel to the trench axis. Because all of the MTDs in this study (except for MTDs 1 and 2, likely caused by local over steepening a seismic loading) appear to have headwall scarps either on or near a trench parallel fault within the youngest of Sacks et al.'s (2013) populations, we propose that the primary final trigger mechanism for these nested MTDs is failure as a result of faulting during the post-seismic relaxation of the accretionary prism or massive megathrust earthquakes along the megasplay fault.

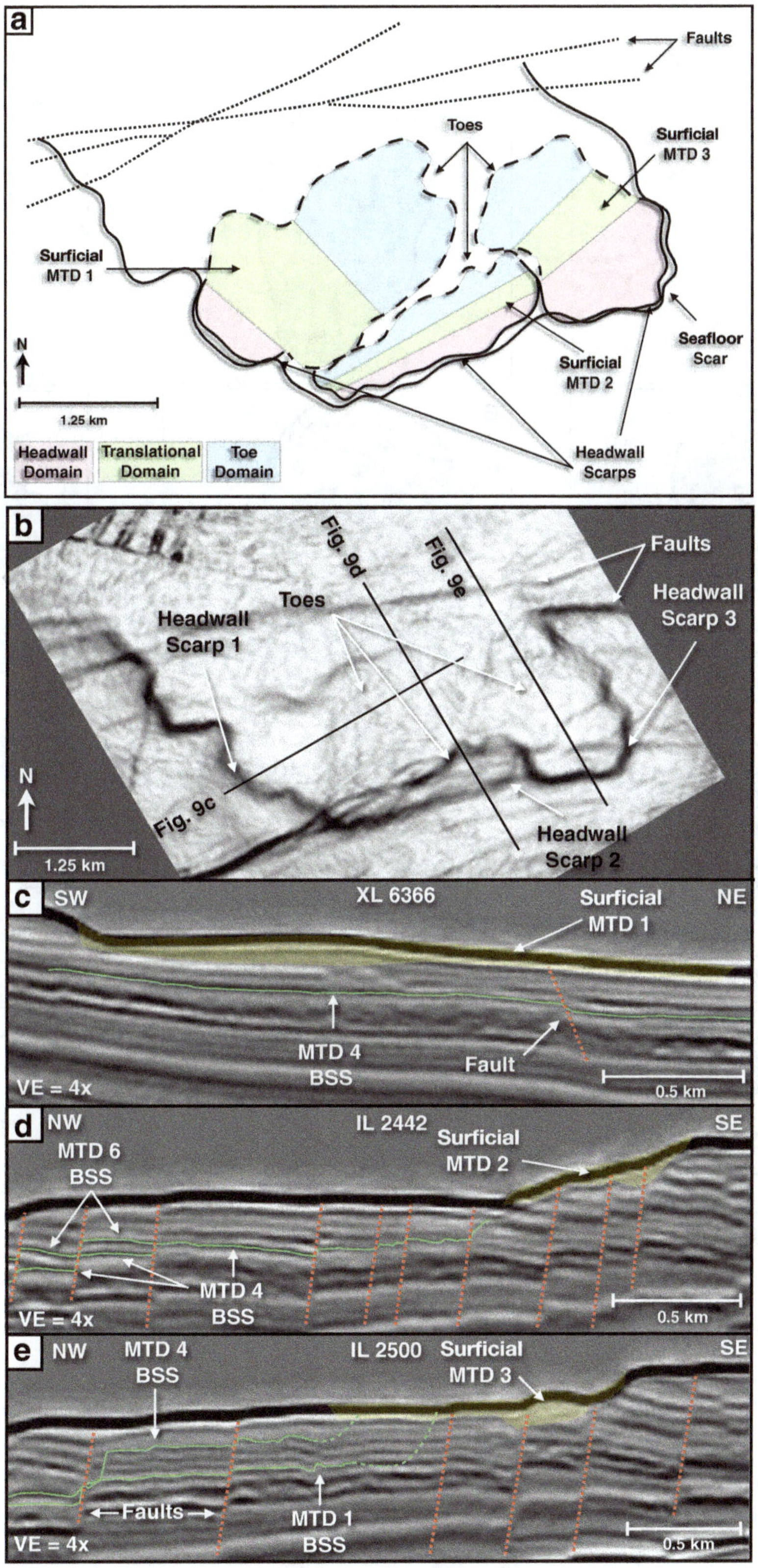

Fig. 9 Surficial MTDs. **a** Schematic depiction of three surficial MTDs. **b** Bathymetric image of surficial landslide scar showing the location of three surficial landslides. **c**–**e** Seismic cross sections across the surficial MTDs in the 3D seismic data. Green lines are BSSs of older landslides. Red dotted lines are normal faults

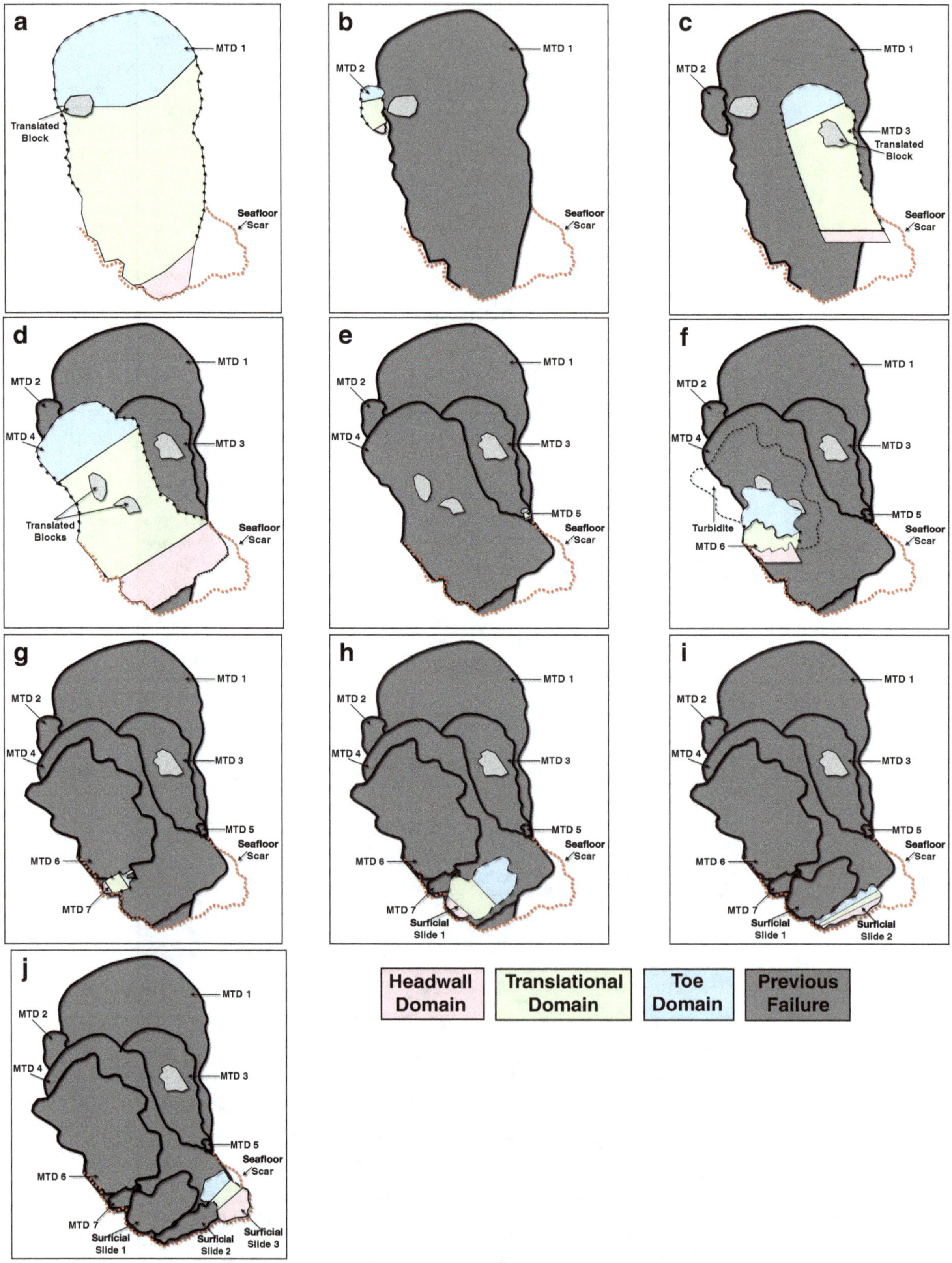

Fig. 10 (See legend on next page.)

(See figure on previous page.)
Fig. 10 Schematic reconstruction of the seafloor scar. Based on the kinematic indicators and geometry of each MTD. **a** MTD 1 occurs and begins the process toward creating the seafloor scar. **b** MTD 2 occurs because of rapid uplift to the SW (Fig. 11a). **c** MTD 3 emplaces on top of MTD 1 after a ~ 0.46 Ma hiatus in landsliding. **d** MTD 4 occurs soon after MTD 3 and erodes the W margin and at least 1/3 of MTD 3 as well as a portion of the BSS of MTD 1. **e** MTD 5 is possibly a result of lateral margin failure of either MTD 3 or 4. **f** MTD 6 erodes a portion of MTD 4's BSS and deposits a turbidite on its top surface. **g** MTD 7 erodes a portion of MTD 6's basal shear surface and creates the W boundary of the seafloor scar. **h–j** Three surficial landslides finish shaping the current seafloor scar

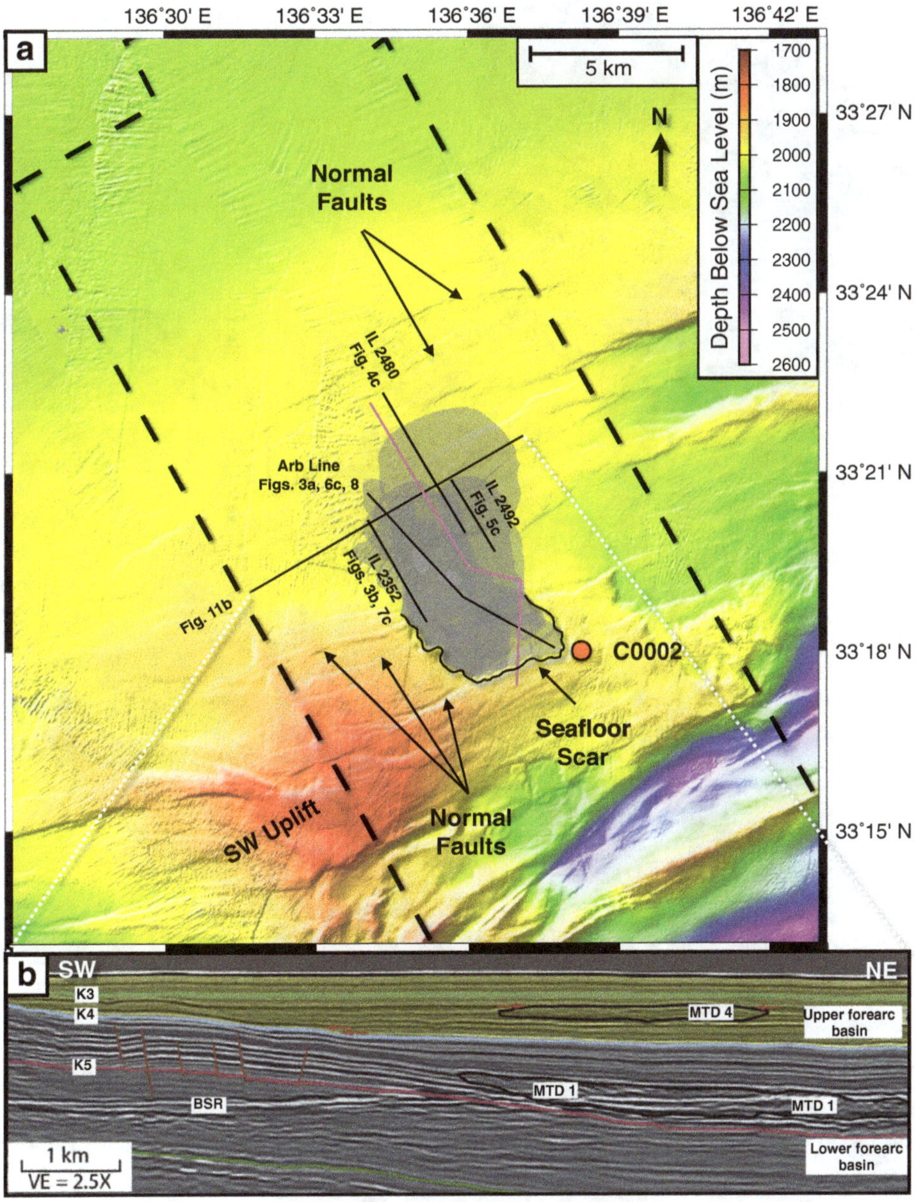

Fig. 11 a High-resolution bathymetry showing SW uplift in relation to the seafloor scar. The gray area represents all the identified MTDs. The purple line indicates the location of the Abstract figure. **b** Seismic cross section showing the relative stratigraphic positions of MTD 1 in the lower forearc basin stratigraphy and MTD 4 within the upper forearc basin stratigraphy. K3, K4, and K5 are upper boundaries of the Kumano Basin seismic sequences, based on toplap, downlap, onlap, or angular relations as defined by Gulick et al. (2010). BSR—bottom-simulating reflector. Modified from Moore et al. (2015)

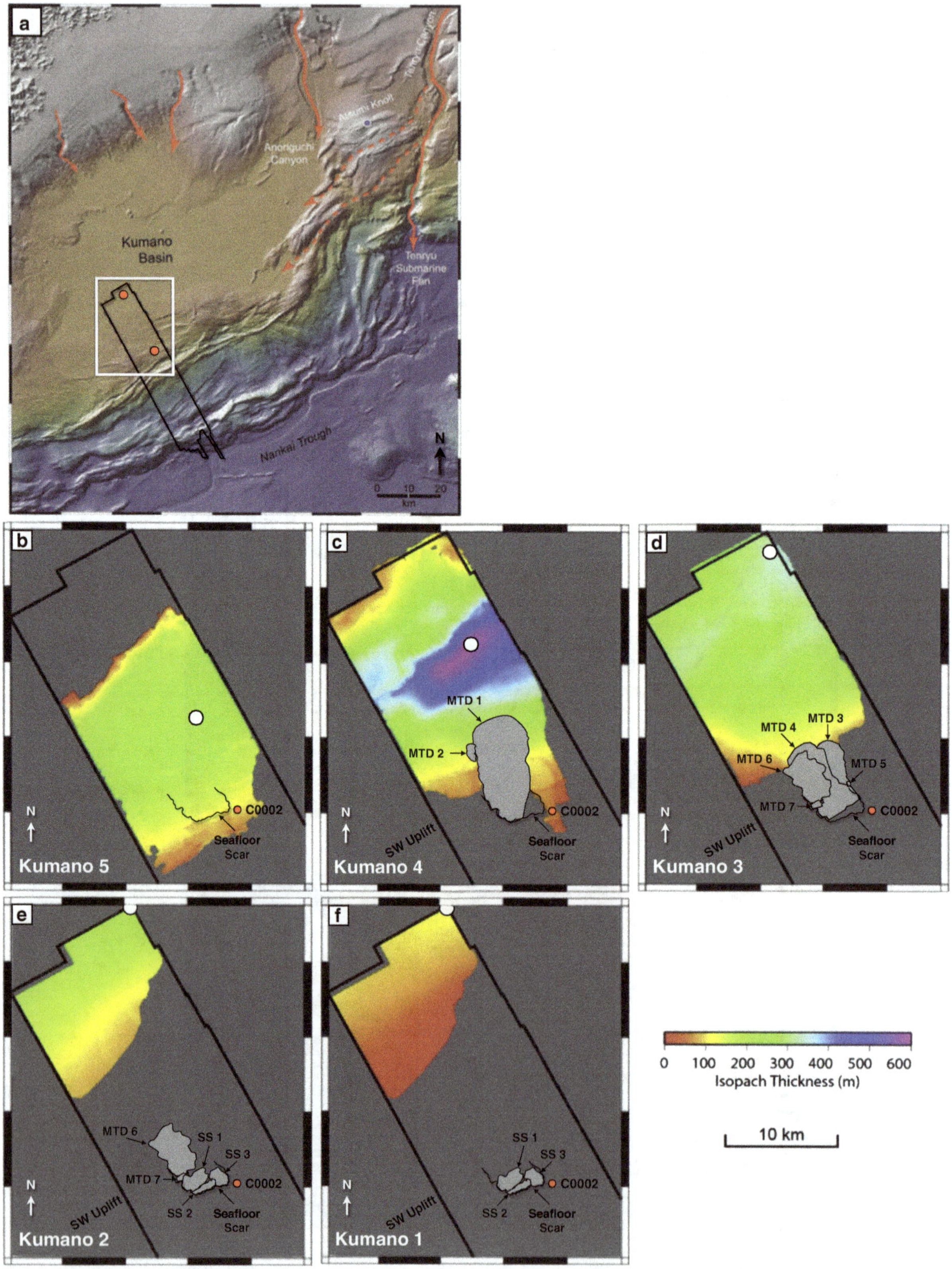

Fig. 12 (See legend on next page.)

(See figure on previous page.)
Fig. 12 Kumano Basin sedimentation through time. **a** Map showing the primary sediment inputs to the Kumano Basin. Solid lines show present-day pathways; dashed lines show possible older pathways. Note the bathymetric high to the NE of the 3D seismic volume that prevents modern sedimentation near IODP site C0002. Red dots are IODP drill sites. White box shows the location of isopach maps. **b–f** Isopach maps of Kumano Basin stratigraphic sequences. Sequence Kumano 1 is the youngest sequence; Kumano 5 is the oldest. White dots show the thickest point of the basin during each isopach interval. Identified MTDs that occur within each sequence are shown for reference. The pattern of deposition changes once the SW uplift appears during the Kumano 4 sequence. Additionally, a lack of sedimentation at the headwalls during Kumano sequences 3–1 helps to explain the reduction in the size of MTDs 5–7 and the surficial MTDs. Because their relative timing cannot be determined from the data, MTDs 6 and 7 are shown in both the Kumano 3 and 2 maps and the surficial MTDs are shown both in the Kumano 2 and Kumano 1 maps (**d–f**). SS—surficial slide. Modified from Moore et al. (2015)

Conclusions

A series of nested MTDs is imaged using both 3D seismic data and high-resolution bathymetry in the outer forearc of the Kumano Basin. The nested nature of the observed MTDs and the stratigraphic sequences in which they occur show how sedimentation of the Kumano Basin and tectonic activity of the Nankai accretionary prism affect local mass wasting. We use kinematic indicators from the observed MTDs to reconstruct a prominent seafloor scar. Local slope over-steepening from a rapid uplift SW of the study area possibly coupled with seismic loading caused the initiation of landsliding in the study area ~ 0.9 Ma. Subsequent failures likely occurred because of faulting caused by post-seismic relaxation of the accretionary prism and seismic activity from 0.3–0.44 Ma to present.

Abbreviations

3D: Three-dimensional; BSR: Bottom-simulating reflector; CMP: Common midpoint gather; IODP: International Ocean Drilling Program; MBES: Multibeam echosounder; MTD: Mass transport deposit; OOST: Out-of-sequence thrust; PGS: Petroleum Geo-Services; PSDM: Pre-stack depth migration

Acknowledgements

The new bathymetric compilation was made possible by the hard work of the SO251 Hydroacoustic Group (C. dos Santos Ferreira, K. Bachmann, K. Lange, J. Moernaut). We thank two anonymous reviewers for the comments that greatly improved the manuscript and Paradigm Geophysical and Blue Marble Geographics for making their software available to us. SOEST Contribution # 10345.

Funding

This work was supported by grants from the National Science Foundation, the US Science Support Program, the German Ministry of Science and Education (BMBF) through funding (grant 03G0251A) of the RV *Sonne* cruise SO251 that acquired the bathymetric data, and the Austrian Science Foundation (grant P29678-N28 to M.S). The 3D data reprocessing was supported by the Japan Society for the Promotion of Science (JSPS) through a KAKENHI Grant-in-Aid for Scientific Research S (JP15H05717) and by the Ministry of Education, Culture, Sports, Science, and Technology (MEXT) through management expense grants at the Research and Development Center for Ocean Drilling Science (ODS) and the Center for Deep Earth Exploration (CDEX) of Japan Agency for Marine-Earth Science and Technology (JAMSTEC).

Authors' contributions

GM proposed the topic and supervised the research. JL carried out the experimental study and analyzed and interpreted the data. MS proposed and directed the bathymetry data acquisition program and collaborated in the construction of the manuscript. All authors read and approved the final manuscript.

Competing interests

The authors declare that they have no competing interest.

Author details

[1]Department of Earth Science, University of Hawai'i, 1680 East-West Rd, Honolulu, HI 96822, USA. [2]Institute of Geology, University of Innsbruck, Innrain 52f, 6020 Innsbruck, Austria.

References

Abrahamson NA, Somerville PG (1996) Effects of the hanging wall and footwall on ground motions recorded during the Northridge earthquake. Bull Seismol Soc Am 86:S93–S99

Ando M (1975) Source mechanisms and tectonic significance of historical earthquakes along the Nankai Trough, Japan. Tectonophysics 27:119–140

Baba T, Cummins PR, Hori T, Kaneda Y (2006) High precision slip distribution of the 1944 Tonankai earthquake inferred from tsunami waveforms: possible slip on a splay fault. Tectonophysics 426:119–134. https://doi.org/10.1016/j.tecto.2006.02.015

Bardet JP, Synolakis CE, Davies HL, Imamura F, Okal EA (2003) Landslide tsunamis: recent findings and research directions. In: Bardet JP, Imamura F, Synolakis CE, Okal EA, Davies HL (eds) Landslide tsunamis: recent findings and research directions. Pageoph Topical Volumes. Birkhäuser, Basel, pp 1793–1809. https://doi.org/10.1007/978-3-0348-7995-8_1

Bøe R, Hovland M, Instanes A, Rise L, Vasshus S (2000) Submarine slide scars and mass movements in Karmsundetand Skudenedfjorden, southwestern Norway: morphology and evolution. Mar Geol 167:147–165

Bondevik S, Kaland PE, Svendsen JI, Johnsen G, Mangerud J (1997) The Storegga tsunami along the Norwegian coast, its age and runup. Boreas 26:29–53

Bull S, Cartwright J, Huuse M (2009) A review of kinematic indicators from mass-transport complexes using 3D seismic data. Mar Pet Geol 26(7):1132–1151. https://doi.org/10.1016/j.marpetgeo.2008.09.011

Caress DW, Chayes DN (1996) Improved processing of Hydrosweep DS multibeam data on the R/V Maurice Ewing. Mar Geophys Res 18:631–650

DeVore JR, Sawyer DE (2016) Shear strength of Siliciclastic sediments from passive and active margins (0–100 m below seafloor): insights into seismic strengthening. In: Lamarche G et al (eds) Submarine mass movements and their consequences. Advances in natural and technological hazards research, vol 41. Springer, Cham, pp 173–180. https://doi.org/10.1007/978-3-319-20979-1_17

Farrell SG (1984) A dislocation model applied to slump structures, Ainsa Basin, South Central Pyrenees. J Struct Geol 6:727–736

Fitch TJ (1972) Plate convergence, transcurrent faults, and internal deformation adjacent to Southeast Asia and Western Pacific. J Geophys Res (23):4432–4460

Frey Martinez J, Cartwright J, Hall B (2005) 3D seismic interpretation of slump complexes: examples from the continental margin of Israel. Basin Res 17:83–108

Frey Martinez J, Cartwright J, James D (2006) Frontally confined versus frontally emergent submarine landslides: a 3D seismic characterisation. Mar Pet Geol 23:585–604

Gafeira J, Bulat J, Evan D (2007) The southern flank of the Storegga Slide: imaging and geomorphological analyses using 3D seismic. In: Lykousis V, Sakellariou D, Locat J (eds) Submarine mass movements and their consequences. Springer, Dordrecht, pp 57–66

Gee MJR, Gawthorpe RL, Friedmann JS (2005) Giant striations at the base of a submarine landslide. Mar Geol 214:287–294. https://doi.org/10.1016/j.margeo.2004.09.003

Gee MJR, Gawthorpe RL, Friedmann SJ (2006) Triggering and evolution of a giant landslide, offshore Angola revealed by 3D seismic stratigraphy and geomorphology. J Sediment Res 76:9–19

Gulick SPS, Bangs NLB, Moore GF, Ashi J, Martin KM, Sawyer DS, Tobin HJ, Kuramoto S, Taira A (2010) Rapid forearc basin uplift and megasplay fault development from 3D seismic images of Nankai Margin off Kii Peninsula, Japan. Earth Planet Sci Lett 300(1–2):55–62. https://doi.org/10.1016/j.epsl.2010.09.034

Ikari MJ, Strasser M, Saffer DM, Kopf AJ (2011) Submarine landslide potential near the megasplay fault at the Nankai subduction zone. Earth Planet Sci Lett 312: 453–462 https://doi.org/10.1016/j.epsl.2011.10.024

Kimura G, Kitamura Y, Hashimoto Y, Yamaguchi A, Shibata T, Ujiie K, Sy O (2007) Transition of accretionary wedge structures around the up-dip limit of the seismogenic subduction zone. Earth Planet Sci Lett 255:471–484. https://doi.org/10.1016/j.epsl.2007.01.005

Kimura G, Moore GF, Strasser M, Screaton E, Curewitz D, Streiff C, Tobin H (2011) Spatial and temporal evolution of the megasplay fault in the Nankai Trough. Geochemistry Geophysics Geosystems 12. https://doi.org/10.1029/2010gc003335

Kinoshita M, Tobin H, Ashi J, Kimura G, Lallemant S, Screaton EJ, Curewitz D, Masago H, Moe KT, expedition scientists (2009) Proc. IODP, 314/315/316. https://doi.org/10.2204/iodp.proc.314315316.2009

Kremer K, Usman MO, Satoguchi Y, Nagahashi Y, Vadakkepuliyambatta S, Panieri G, Strasser M (2017) Possible climate preconditioning on submarine landslides along a convergent margin, Nankai Trough (NE Pacific). Prog Earth Planet Sci 4:20. https://doi.org/10.1186/s40645-017-0134-9

Lackey JK, Moore GF, Strasser M, Kopf A, Ferreira CS (2018) Spatial and temporal cross-cutting relationships between fault structures and slope failures along the outer Kumano Basin and Nankai accretionary wedge, southwest Japan. In: Submarine Mass Movements and their Consequences. Advances in Natural and Technological Hazards Research in press

Lastras G, Canals M, Hughes-Clarke JE, Moreno A, De Batist M, Masson DG, Cochonat P (2002) Seafloor imagery from the BIG'95 debris flow, western Mediterranean. Geology 30:871–874

Lee HJ (2009) Timing of occurrence of large submarine landslides on the Atlantic Ocean margin. Mar Geol 264:53–64

Lee HJ, Locat J, Boulanger E, Konrad JM (2004) Seismic strengthening, a conditioning factor influencing submarine lanslide development. Paper presented at the 57th Canadian Geotechnical Conference; 5th Joint CGS/IAH-CNC Conference, Quebec City, Quebec, pp 24–27

Locat J, Lee H, Kayen R, Israel K, Savoie M-C, Boulanger E (2002) Shear strength development with burial in Eel River margin slope sediments. Mar Georesour Geotechnol 20(2):111–135. https://doi.org/10.1080/03608860290051_831

Lucente CC, Pini GA (2003) Anatomy and emplacement mechanism of a large submarine slide within a Miocene foredeep in the northern Apennines, Italy: a field perspective. Am J Sci 303:565–602

Martinsen OJ (1994) Mass movements. In: Maltman A (ed) The geological deformation of sediments. Chapman and Hall, London, pp 127–165

Masson DG, Hugget QJ, Brunsden D (1993) The surface texture of the Saharan debris flow deposit and some speculation on submarine debris flow processes. Sedimentology 40:583–598

Matheus S, Sobiesiak G, Alsop I, Kneller B, Milana JP (2017) Sub-seismic scale folding and thrusting within an exposed mass transport deposit: a case study from NW Argentina. J Struct Geol 96:176–191. https://doi.org/10.1016/j.jsg.2017.01.006

Miyakawa A, Saito S, Yamada Y, Tomaru H, Kinoshita M, Tsuji T (2014) Gas hydrate saturation at site C0002, IODP expeditions 314 and 315, in the Kumano Basin, Nankai Trough. Island Arc 23:142–156

Moore GF, Boston BB, Sacks AF, Saffer DM (2013) Analysis of normal fault populations in the Kumano Forearc Basin, Nankai Trough, Japan: 1. Multiple orientations and generations of faults from 3-D coherency mapping. Geochem Geophys Geosyst 14:1989–2002. https://doi.org/10.1002/ggge.20119

Moore GF, Boston BB, Strasser M, Underwood MB, Ratliff RA (2015) Evolution of tectonosedimentary systems in the Kumano Basin, Nankai Trough forearc. Mar Petrol Geol 67:604–616. https://doi.org/10.1016/j.marpetgeo.2015.05.032

Moore GF, Kanagawa K, Strasser M, Dugan B, Maeda L, Toczko S, Exp 338 Sci. Party (2014) IODP Expedition 338: NanTroSEIZE Stage 3: NanTroSEIZE plate boundary deep riser 2. Sci Dril 17:1–12. https://doi.org/10.5194/sd-17-1-2014

Moore GF, Park JO, Bangs NL, Gulick SP, Tobin HJ, Nakamura Y, Saito S, Tsuji T, Yoro T, Tanaka H, Uraki S, Kido Y, Sanada Y, Kuramoto S, Taira A (2009) Structural and seismic stratigraphic framework of the NanTroSEIZE Stage 1 transect. Proc. IODP, 314/315/316, pp 1–46. https://doi.org/10.2204/iodp.proc.314315316.102.2009

Moore GF, Strasser M (2016) Large mass transport deposits in Kumano Basin, Nankai Trough, Japan. In: Lamarche G et al (eds) Submarine mass movements and their consequences advances in natural and technological hazards research, vol 41: 371–379. Springer, Cham. https://doi.org/10.1007/978-3-319-20979-1_37

Moore JG, Clague DA, Holcomb RT, Lipman PW, Normark WR, Torresan ME (1989) Prodigious submarine landslides on the Hawaiian ridge. J Geophys Res Solid Earth Planets 94:17465–17484

Plafker G (1972) Alaskan earthquake of 1964 and Chilean earthquake of 1960: implications for arc tectonics. J Geophys Res 77:901–925

Prior DB, Bornhold BD, Johns MW (1984) Depositional characteristics of a submarine debris flow. J Geol 92:707–727

Prior DB, Coleman JM (1978) Disintegrating retrogressive landslides on very-low-angle subaqueous slopes, Mississippi Delta. Mar Geotechnol 3(1):37–60

Ruff L, Kanamori H (1980) Seismicity and the subduction process. Phys Earth Planet Inter 23(3):40–252

Sacks AF, Saffer DM, Fisher DM (2013) Analysis of normal fault populations in the Kumano Forearc Basin, Nankai Trough, Japan: 2. Principal axes of stress and strain from inverson of fault orientations. Geochem Geophys Geosyst 14:1973–1988

Satake K (2012) Tsunamis generated by submarine landslides. In: Yamada Y (ed) Submarine mass movements and their consequences. Advances in natural and technological hazards research, vol 31. Springer, Dordrecht/Heidelberg/London/New York, pp 475–484. https://doi.org/10.1007/978-94-007-2162-3_42

Seno T (1989) Philippine Sea Plate kinematics. Mod Geol 14:87–97

Shiraishi K, Moore GF, Yamada Y, Kinoshita M, Kimura G (2018) Seismogenic zone structures revealed by improved 3D seismic images in the Nankai Trough off Kumano. Prog Earth Planet Sci

Strasser M, Dugan B, Kanagawa K, Moore GF, Toczko S, Maeda L, Expedition 338 Scientists (2014) Proc. IODP, 338: Yokohama (Integrated Ocean Drilling Program). https://doi.org/10.2204/iodp.proc.338.2014

Strasser M, Henry P, Kanamatsu T, Thu MK, Moore GF and scientists (2012) Scientific drilling of mass-transport deposits in the Nankai accretionary wedge: first results from IODP Exp 333. In: Yamada Y, et al (eds) Submarine mass movements and their consequences. Advances in natural and technological hazards research, vol 31, pp671–681. Springer, Dordrecht. doi: https://doi.org/10.1007/978-94-007-2162-3-60

Strasser M, Kopf A, Abegg FW, Asada M, Bachmann AK, Cuno P, Dos Santos Ferreira C, Fleischmann T, Fujiwara T, Hatakeyama E, Heesemann BR, Hillman JIT, Hoehne M, Huusmann H, Ikari M, Ikehara K, Jaeger FD, Kanamatsu T, Kang M, Kaul NE, Kioka A, Koelling M, Lange K, Luebben N, Matthiessen T, Mchugh CM, Meier A, Menapace W, Mochizuki K, Moernaut J, et al (2017) REPORT AND PRELIMINARY RESULTS OF R/V SONNE CRUISE SO251: Extreme events Archived in the GEological Record of Japan's Subduction margins (EAGER-Japan). Berichte, MARUM – Zentrum für Marine Umweltwissenschaften, Fachbereich Geowissenschaften, Universität Bremen

Strasser M, Moore G, Kimura G, Kopf A, Underwood M, Guo J, Screaton E (2011) Slumping and mass-transport deposition in the Nankai forearc: evidence from IODP drilling and 3-D reflection seismic data. Geochem Geophys Geosyst 12:Q0AD13. https://doi.org/10.1029/2010GC003431

Strasser M, Moore GF, Kimura G, Kitamura Y, Kopf AJ, Lallemant S, Park JO, Screaton EJ, Su X, Underwood MB, Zhao X (2009) Origin and evolution of a splay fault in the Nankai accretionary wedge. Nat Geosci 2:648–652

Taira A (2001) Tectonic evolution of the Japanese island arc system. Annu Rev Earth Planet Sci 29:109–134

Tanioka Y, Satake K (2001) Coseismic slip distribution of the 1946 Nankai earthquake and aseismic slips caused by the earthquake. Earth, Planets and Space 53:235–241

Underwood MB, Moore GF (2012) Evolution of sedimentary environments in the subduction zone of southwest Japan: recent results from the NanTroSEIZE Kumano transect. In: Busby CJ, Azor AP (eds) Tectonics of sedimentary basins: recent advances. Wiley-Blackwell, New York, pp 310–326

Urlaub M, Talling PJ, Masson DG (2013) Timing and frequency of large submarine landslides: implications for understanding triggers and future geohazard. Quat Sci Rev 72(0):63–82. https://doi.org/10.1016/j.quascirev. 2013.04.020

Ward SN, Day S (2002) Suboceanic landslides. 2002 Yearbook of science and technology, pp 249–352

3

Cenozoic sedimentary records of climate-tectonic coupling in the Western Himalaya

Peter D. Clift[1,2]

Abstract

Sedimentary archives in the Himalayan foreland basin and Indus submarine fan provide the most detailed records of how changing monsoon strength may have affected erosion and the development of tectonic structures in the western Himalaya during the Neogene. Muscovite Ar-Ar ages show that fast exhumation of the Greater Himalaya was earlier in the west (20–35 Ma) than in the central Himalaya (10–25 Ma), probably driven by progressive lithospheric slab tearing that caused both rock uplift and an intensified summer monsoon to start earlier in the western Himalaya. Rates of exhumation reduced after ~ 17 Ma as uplift also slowed, apparently unrelated to climate change. However, further reduction at 6–8 Ma coincided with a time of weakening summer rains, as shown by carbon isotope data from the foreland and hematite-goethite records from ODP Site 730 on the Oman margin. Coming at a time of stronger winds, this drying was linked to global climatic cooling. Weakening of Late Miocene monsoon rains coincided with a southward migration of the ITCZ and faster erosion of the Lesser rather than Greater Himalaya. Unroofing of the Inner Lesser Himalaya, starting after 9 Ma, followed as a consequence of duplex formation enabled by focused erosion, although widespread exposure of that range was delayed until after 6 Ma. Inner Lesser Himalayan exposure, combined with unroofing of the Nanga Parbat Massif, caused a decrease in average ε_{Nd} values of Indus River sediment after 5 Ma, rather than large-scale drainage capture.

Keywords: Monsoon, Climate-tectonic, Himalaya, Erosion, Exhumation, Arabian Sea, Foreland basin, Provenance, Geochemistry, Cenozoic, Weathering

Introduction

The building of high topography following the collision of India with Eurasia sometime in the early Cenozoic is proposed to have played a fundamental role in strengthening the Asian monsoon in South and East Asia (Kutzbach et al. 1993; Molnar et al. 1993). In turn, the monsoon climate is believed to feed back on the development of the mountains via surface processes and to partially control the tectonic evolution of the fold belts that have developed since the time of initial India-Eurasia collision (Clift et al. 2008; Whipple 2009; Willett 1999). In doing so, these interactions between the solid Earth and the atmosphere represent one of the world's great natural laboratories for studying the interplay of climate and tectonics over long periods of geological time. In particular, the Western Himalaya are a classic example of this interaction, despite being on the NW margin of the monsoon rainfall, which is centered instead around the Bay of Bengal. In fact, because of their location on the edge of the monsoon region, the Western Himalaya are more sensitive to changes in rainfall as monsoon precipitation waxed and waned. This area also has the advantage of having relatively well-documented, long-duration, marine paleoceanographic records (Prell et al. 1992; Kroon et al. 1991; Gupta et al. 2015), which can be compared readily with contemporaneous continental records.

In this review, I evaluate the evidence linking the climatic development of SW Asia to the processes of mountain building in the adjacent ranges of the Himalaya and Karakoram (Fig. 1), exploiting the sedimentary record preserved in the terrestrial and marine basins. This allows both the processes of erosion, as well as the environmental conditions that have affected the region since ~ 25 Ma to be compared. I focus on the Miocene to present, because this is the time period over which we have more complete records of terrestrial erosion and environmental conditions

Correspondence: pclift@lsu.edu
[1]Department of Geology and Geophysics, Louisiana State University, Baton Rouge, LA 70803, USA
[2]School of Geography Science, Nanjing Normal University, Nanjing 210023, China

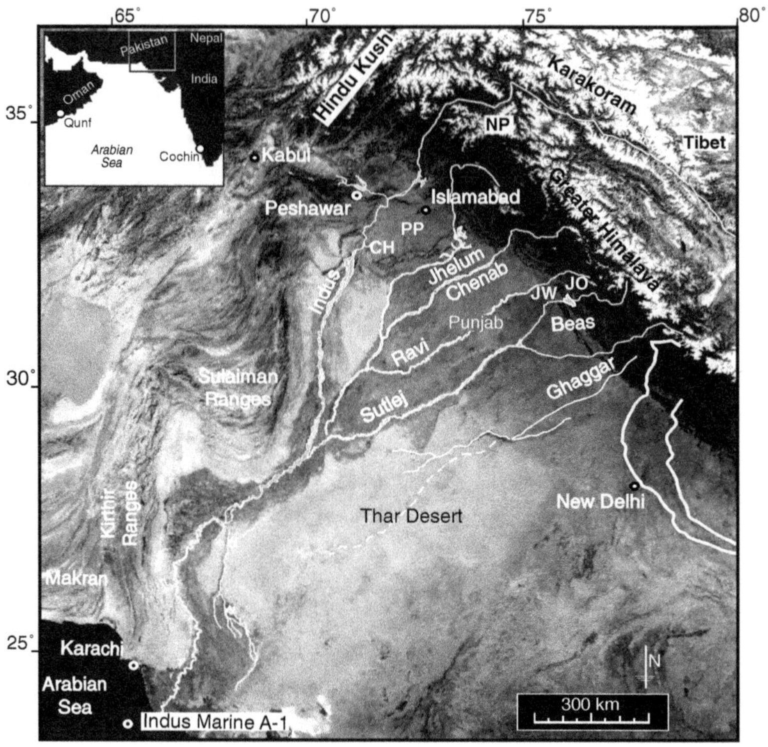

Fig. 1 Satellite image of the Indus Basin from Google Earth. Map shows the major tributaries and trunk stream of the Indus, together with the main ranges discussed in this study. PP—Potwar Plateau; CH—Chinji; NP—Nanga Parbat; JW—Jawalamukhi; JO—Jogindernagar

(Burbank et al. 1996; Najman 2006). It is also the time period for which new matching marine records are now available from scientific ocean drilling in the Arabian Sea (Pandey et al. 2015). Although climate-tectonic interactions may go back much further into the Paleogene, presently, there are no suitable sedimentary sequences in SW Asia that allow testing of whether the monsoon is really much older than the Miocene and whether the onset of mountain building immediately after India-Eurasia collision was linked to early Eocene intensification of summer rainfall, as has been proposed in some parts of South and East Asia (Licht et al. 2014).

The Indus River Basin

Much of the evidence constraining erosion and environmental conditions is preserved in the fill of the sedimentary basins of SW Asia, which have been supplied by the erosional load carried by the Indus River and its various tributaries (Fig. 1). This drainage system is the only major system in the western Himalaya (with the exception of the Oxus/Amu Darya flowing north from the Pamir) and comprises a trunk stream that originates in western Tibet that then flows westward along the Neotethyan Indus Suture Zone before turning south around the western syntaxis (Nanga Parbat) and heading to the Arabian Sea where it has supplied the material to construct the second largest sediment mass in the modern oceans, the Indus Submarine Fan (Fig. 2) (Clift et al. 2001; Naini and Kolla 1982; Kolla and Coumes 1987). The Indus Fan stretches >1500 km from the coast of Pakistan into the Indian Ocean and comprises $\sim 4.5 \times 10^6$ km^3 of sediment, around a third of the size of the neighboring Bengal Fan.

On the eastern side of the drainage, the Indus River (average trunk discharge 3.91×10^6 m^3/s) (Alizai et al. 2011) is supplied by discharge from four large tributaries, the Jhellum (average discharge 9.9×10^5 m^3/s), Chenab (1.1×10^6 m^3/s), Ravi (2.6×10^5 m^3/s), and Sutlej (4.0×10^5 m^3/s) (Fig. 1) that largely derive their sediment from the Greater and Lesser Himalaya before flowing across the plains of the Punjab to join the mainstream about half way on its journey from the mountain front to the sea (Inam et al. 2007). In doing so, these rivers also fill the foreland basin that borders the Himalaya on their southern flank. A smaller set of flexural basins borders the Indus River to the west where the oblique collision between the Indian Plate and Eurasia, represented by the Chaman Fault, has generated SE-propagating thrust belts in the Kirthar and Sulaiman Ranges (Fig. 1) (Roddaz et al. 2011). There is little sediment supplied from the west into the Indus River because of the generally arid conditions. The major exception to

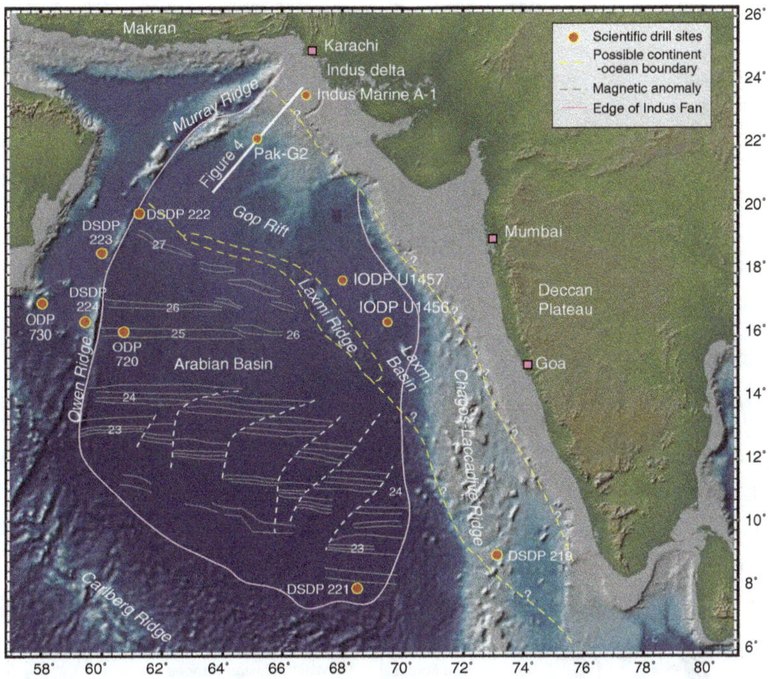

Fig. 2 Shaded bathymetric map of the Arabian Sea showing the extent of the Indus Fan (Kolla and Coumes 1987), the location of scientific and industrial boreholes discussed in this paper, and the main tectonic elements that bound the basin, especially the marine magnetic anomalies of Miles and Roest (1993). Base map is from GeoMapApp

this pattern is the Kabul River that brings sediment from the Hindu Kush.

Reconstructing erosion

If we are to assess the degree to which climate or solid Earth tectonic forces control the development of the Himalaya, then we need to derive independent records of how erosion and climate have evolved over long periods of geological time, in order to compare with the structural evolution of the mountains. This is most effectively done through detailed provenance studies that attempt to reconstruct how patterns and rates of erosion have changed through time, in order to determine whether changes in these processes correlate with climatic events. This is easier to do in the Indus River basin compared to many areas because of the compositional diversity of the source terranes. Figure 3 shows the different ranges and tectonic blocks that comprise the Western Himalaya and the associated ranges in the Karakoram and Hindu Kush. Each of these units has their own distinctive geological history and compositional characteristics that allow their influence on the net flux of sediment to the Arabian Sea to be resolved. Because of differences in the timing of tectonic or magmatic events in each block, the sediments eroded from each tends to have unique characteristics in terms of either geochemistry or the age of mineral cooling through a variety of different closure temperatures (Hodges 2003).

Source terrains

The Lesser, Greater, and Tethyan Himalaya comprise rock units that were previously part of a Greater India prior to its collision with Eurasia (Garzanti et al. 1987; Ali and Aitchison 2005). These units have some similarities in terms of their evolution prior to collision, although in general the rocks of the Lesser Himalaya tend to represent older crust (DeCelles et al. 2000; Gehrels et al. 2011), which has been more recently exhumed to the surface than the high mountains of the Greater Himalaya. Those in turn have been metamorphosed at higher temperatures and pressures compared to the lower ranges to the south (Searle 1996; Hodges and Silverberg 1988; Searle et al. 2006). The Tethyan Himalaya represent weakly metamorphosed sedimentary rocks whose origin was similar to the Greater Himalaya in representing the former sedimentary cover to the northern margin of Greater India (Robertson and Degnan 1994; Zhang et al. 2012; Garzanti 1993). To the north of these mountains lies the primitive mafic rocks of the Kohistan magmatic arc (Bignold et al. 2006; Khan et al. 1997), believed to have been formed in an intra-oceanic subduction setting, which is the primary geological unit within the Indus Suture Zone. In turn, the Kohistan Arc is juxtaposed along its northern edge

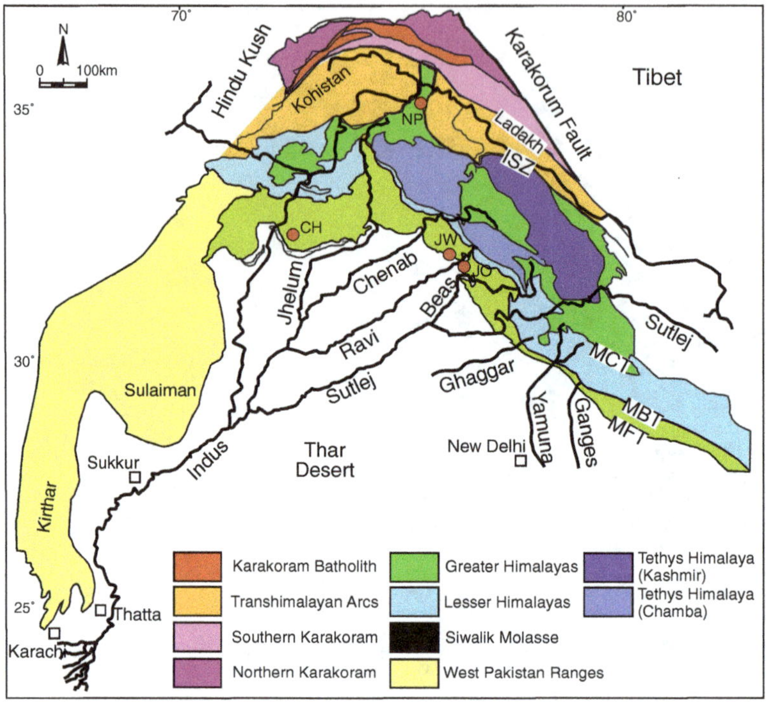

Fig. 3 Geological map of the study area showing the major tectonic units that are eroded by the Indus River and its tributaries. Map is modified after Garzanti et al. (2005). Rivers as shown in thick black lines. ISZ = Indus Suture Zone, MCT = Main Central Thrust, MBT = Main Boundary Thrust and MFT = Main Frontal Thrust. Sample locations are shown as filled circles. Thick black line shows major Himalayan tributaries. CH—Chinji; NP—Nanga Parbat; JW—Jawalamukhi; JO—Jogindernagar

against the Karakoram separated by the Shyok oceanic suture zone (Robertson and Collins 2002; Dunlap and Wysoczanski 2002). The Karakoram themselves represent a Mesozoic continental magmatic arc that preceded the closure of this northern Tethyan ocean but continued activity long after closure of the oceanic basin (Schärer et al. 1990; Searle et al. 1989; Fraser et al. 2001). Equivalent rocks in the Hindu Kush have similarities with the Karakoram but are generally not as strongly metamorphosed and have not experienced the recent high exhumation rates (Hildebrand et al. 2000), which are partly linked to activity along the Karakoram Fault starting in the middle Miocene (Phillips et al. 2004).

Another distinctive terrain within the Western Himalaya is the Nanga Parbat Massif that forms the linchpin of the Western Syntaxis (Fig. 1). This is well known as being a region of exceptionally fast rock uplift and exhumation, making it the source of young detritus into the Indus River (Zeitler et al. 1989; Treloar et al. 2000). However, earlier work on this system suggests that Nanga Parbat is less important as a source of sediment into the Indus than its Eastern equivalent at Namche Barwe (Lee et al. 2003; Clift et al. 2002b). In summary, the various ranges within the Indus basin have unique characteristics that allow their contribution to the erosional flux to be constrained if sufficient numbers of provenance proxies can be applied. Typically a single proxy is less effective at resolving erosional flux from a single tectonic block.

Marine sedimentary records

The vast majority (~ 66%) of the sediment eroded from the Himalaya since the Eocene lies in the Arabian Sea (Clift et al. 2001), even if some of the sediment was diverted to the west, into the Makran before the uplift of the Murray Ridge (Critelli et al. 1990). Although the region has been surveyed by seismic profiles over the thickest part of the depocenter, the bulk of the submarine fan has not yet been covered by representative seismic profiles. More importantly, the fan stratigraphy has not been penetrated to its total depth anywhere except the most distal locations in the southern part of the Arabian basin, for example at Deep Sea Drilling Project (DSDP) Site 221 (Whitmarsh et al. 1974). Most recently, International Ocean Discovery Program (IODP) recovered two sections from the Laxmi Basin, offshore Western India, IODP Sites U1456 and U1457 (Pandey et al. 2015). Although these provide relatively long records extending to ~ 10.8 Ma, they are unable to provide an erosional record of the Western

Himalaya back into the Paleogene. This is because of erosion of the fan in Laxmi Basin by the Nataraja Slide (mass transport deposit) that was deposited into the basin at around 10.8 Ma (Calvès et al. 2015). Furthermore, the IODP drill sites are unable to provide age control even to the younger part of the section across the proximal fan because they are separated from the main Arabian Basin by the Laxmi Ridge, preventing correlation of the dated reflectors from these sites towards the main sediment mass offshore the Indus River mouth.

Close to the modern river mouth around 11 km of sediment has been identified overlying igneous basement and carbonate sedimentary rocks (Clift et al. 2002a; Naini and Kolla 1982). The age of sedimentation here is only loosely controlled by industrial drilling sites both under the continental margin at Indus Marine A-1 (Shuaib and Tariq Shuaib 1999), as well as in deeper water at PAK-G2 (Fig. 4), which penetrated into Eocene carbonates related to the early stages breakup of the Arabian Sea. Although a detailed erosion budget is not possible at the present time, it is possible to date the onset of the construction of channel-levee complexes using the seismic and the industrial drilling data. This indicates that the first channel-levee complexes were being built during the Early Miocene and suggest a significant increase in sediment flux to the basin starting around that time (Clift et al. 2001; Droz and Bellaiche 1991). Nonetheless, several kilometers of sediment predate these features, indicating the longevity of the Indus River as the conveyor of sediment from the Western Himalaya since before the Early Miocene.

Sedimentology and mineralogy of the Himalayan foreland basin

To date most of the detailed reconstructions of erosion in the Western Himalaya have come from studies onshore, principally in the Himalayan foreland basin. As India has continued to move towards Eurasia, the plate has been flexed down by the weight of the subducting slab, as well as by the mountains resulting in a basin approximately 5–6 km deep (Raiverman et al. 1983). Sections of sediment from the basin have been progressively accreted into the mountains as the thrust front migrated southwards so that they are preserved both in the Sub-Himalaya and the Siwalik hills that form the first significant topography north of the Main Frontal Thrust (MFT). Figure 5 shows a generalized stratigraphic section of the sedimentary sequence. About half of the cumulative thickness is comprised of the Siwalik Group, which is a coarsening-upward sequence of fluvial rocks that date back to around 13 Ma (Burbank et al. 1996). These are principally dated through paleomagnetic methods coupled with radiometric dating of occasional volcanic ash beds that allow the depositional ages to be constrained (Ojha et al. 2000; Tauxe and Opdyke 1982; Johnson et al. 1985; Meigs et al. 1995). The Siwalik Group is divided into three different formations, with the Upper Siwalik being dominated by conglomerates of alluvial fan sedimentary facies (Rao 1993).

Although the coarsening-upward character might suggest an increasing intensity of erosion, this need not necessarily be the case because of the fact that any given point in a foreland basin will progressively approach the mountain front through time, as the Indian plate is underthrust, resulting in a long-term coarsening upwards even with constant erosional flux (DeCelles and

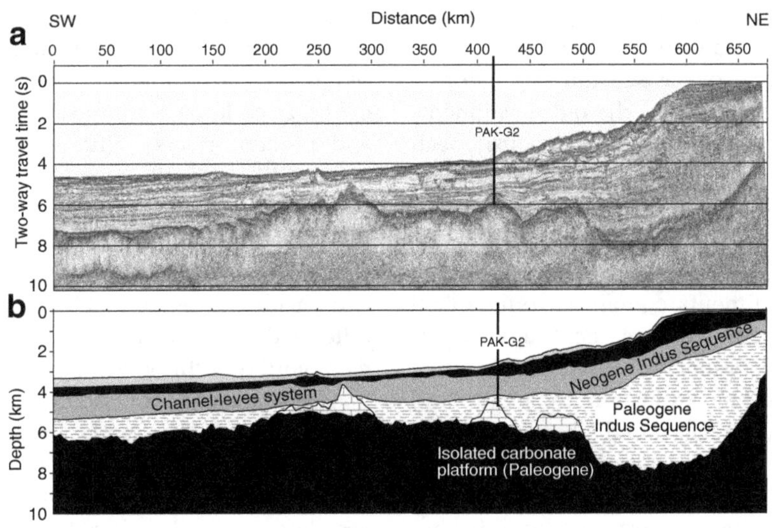

Fig. 4 NE-SW oriented seismic line TEPP207 through well PAK-G2, showing the original and interpreted sections. Modified from Calvès et al. (2008). See Fig. 2 for location

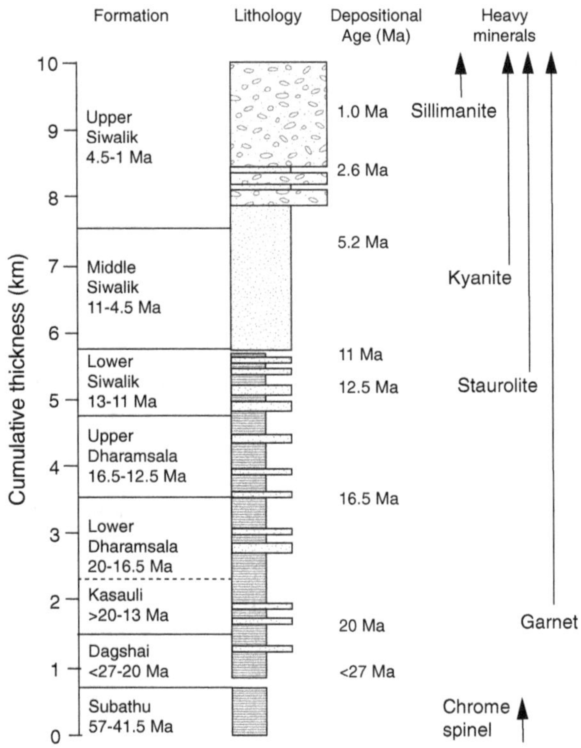

Fig. 5 Simplified stratigraphy of the western Himalayan foreland basin derived from the work of Jain et al. (2009), showing the general coarsening up-section and the appearance of a variety of higher grade metamorphic minerals that might be associated with the unroofing of the Greater and Inner Lesser Himalaya

Giles 1996). Those formations predating the Siwalik Group can be divided into two groups separated by a major unconformity of debatable duration but largely spanning the Oligocene. At one extreme, the hiatus spans from 40 to 22 Ma (Najman 2006; Jain et al. 2009), while others argue for a break of ≤ 3 m.y. or even less (Bera et al. 2008; Raiverman and Raman 1971). The muddy and carbonate rocks of the Subathu Formation at the base of the foreland section overlie older sedimentary rocks of the Indian Craton and have variously been interpreted as the products of sedimentation into an early foreland basin (Najman et al. 2001; Najman et al. 1997) or possibly into a back bulge location (i.e., on the distal side of the flexural forebulge) (DeCelles et al. 1998a). The Subathu sediments contain chrome spinel grains that were likely eroded from the Indus Suture Zone, presumably from ophiolites or magmatic units similar to those seen in Kohistan (Najman and Garzanti 2000). The units deposited above the long Oligocene (22–40 Ma) unconformity have a variety of names (Dagshai, Kasauli, Dharamsala, Murree and Kamlial Formations in the western foreland basin), but essentially comprise increasingly sandy fluvial sedimentary rcoks.

Studies of heavy minerals in these rocks show the appearance of garnet starting at about 20 Ma (Najman and Garzanti 2000; Sinha 1970), shortly after the onset of thrusting along the Main Central Thrust at ~ 24 Ma (Stephenson et al. 2001; Catlos et al. 2001), which is at least partly responsible for the exposure of the Greater Himalayan Sequence, one of the possible sources. It is however highly debatable whether the garnet grains in these sandstones were actually derived from these units because garnet is not unique to the Greater Himalaya. It is noteworthy that going up-section a series of increasingly high-grade metamorphic minerals were progressively input into the foreland basin. Staurolite appears in the Lower Siwalik Formation, after ~ 13 Ma (Fig. 5), while kyanite makes its initial appearance in the Upper Miocene, with sillimanite being the most recent addition within the Pleistocene Upper Siwalik Formation (Najman and Garzanti 2000). Sillimanite is formed at higher temperatures than kyanite, leading to the idea that this up-section evolution in detrital mineralogy reflects progressive unroofing of higher temperature source rocks. Quite what those sources might be is not immediately clear because the crystalline Inner Lesser Himalaya also comprise high-grade metamorphic rocks, similar in grade to some part of the Greater Himalaya (Célérier et al., 2009a, b). The presence of Miocene leucogranites in the Greater Himalaya does, however, provide a contrast to the generally lower temperature rocks of the sandwiching Lesser and Tethyan Himalaya (Searle and Fryer 1986; Vidal et al. 1982), and their erosion might provide a key indicator of Greater Himalayan erosion.

Isotopic constraints on evolving provenance

Contrasting crustal ages and petrogenesis has resulted in major differences in isotopic character between the different ranges of the Himalaya. In particular, Nd isotopes have been used to assess the origin of clastic sediments in the foreland basin and Indus Fan because of the large known compositional ranges in the sources and modern rivers (Clift et al. 2002b; Najman et al. 2009; Clift and Blusztajn 2005; Ahmad et al. 2000; Whittington et al. 1999). Furthermore, the Nd isotope ratio ($^{143}Nd/^{144}Nd$) of clastic sediment is not affected by transport or chemical weathering/diagenesis (Goldstein and Jacobsen 1988). Nd isotope composition in part reflects the average age of the crust being eroded. The data plotted in this study are provided in Table 1.

Here, I consider two isotope ratios ($^{143}Nd/^{144}Nd$) from the Himalayan foreland, one at Chinji in Pakistan (Chirouze et al. 2015), close to the location where the Indus River crosses the MFT, and another at Jawalamukhi which lies close to the Beas River (Fig. 1) (Najman et al. 2009). Each is expected to record the long-term evolution of these local river systems and thus the erosion of that part of the

Table 1 Synthesis of published Nd isotope data used in this study, together with a reference to the original sources from which they were derived

Sample	Age	$^{143}Nd/^{144}Nd$	± 2σ	Depositional Age (Ma)	εNd	Source
224-10R-1, 140 cm	Middle Eocene	0.512027	5	37.0	− 11.9	Clift and Blusztajn (2005)
224-10R-2, 49 cm	Middle Eocene	0.512357	5	37.5	− 5.5	Clift and Blusztajn (2005)
224-11R-2, 100 cm	Early Eocene	0.512159	5	50.0	− 9.3	Clift and Blusztajn (2005)
224-7R-1, 35 cm	Upper Oligocene	0.512071	5	26.5	− 11.1	Clift and Blusztajn (2005)
224-7R-CC, 0 cm	Upper Oligocene	0.512039	5	27.0	−11.7	Clift and Blusztajn (2005)
224-8R-2, 49 cm	Middle Oligocene	0.512049	5	28.0	− 11.5	Clift and Blusztajn (2005)
720A-3H-2, 110 cm	Pleistocene	0.511921	5	0.6	− 14.0	Clift and Blusztajn (2005)
720A-3H-2, 14 cm	Pleistocene	0.511996	5	0.6	− 12.5	Clift and Blusztajn (2005)
731A-40X-2, 103 cm	Lower Miocene	0.512051	5	20.0	− 11.5	Clift and Blusztajn (2005)
731C-13 W-1, 50 cm	Lower Oligocene	0.512066	5	29.0	− 11.2	Clift and Blusztajn (2005)
731C-19R-1, 50 cm	Lower Oligocene?	0.512037	9	31.0	− 11.7	Clift and Blusztajn (2005)
731C-1R-1, 50 cm	Lower Miocene	0.512003	5	16.0	− 12.4	Clift and Blusztajn (2005)
731C-21R-1, 50 cm	Middle Eocene (?)	0.512021	6	38.0	− 12.0	Clift and Blusztajn (2005)
731C-24R-1, 50 cm	Middle Eocene (?)	0.512001	10	40.0	− 12.4	Clift and Blusztajn (2005)
731C-2R-1, 50 cm	Lower Miocene	0.512071	5	21.5	− 11.1	Clift and Blusztajn (2005)
731C-7W-1, 50 cm	Upper Oligocene	0.512049	5	25.0	− 11.5	Clift and Blusztajn (2005)
IL3 0 – Chinji	Middle Miocene	0.512334	7	13.8	− 5.8	Chirouze et al. (2015)
IL4 – Chinji	Middle Miocene	0.512434	5	13.7	− 3.8	Chirouze et al. (2015)
IL5 – Chinji	Middle Miocene	0.512258	6	13.4	− 7.3	Chirouze et al. (2015)
IL6 – Chinji	Middle Miocene	0.512237	7	12.3	− 7.7	Chirouze et al. (2015)
IL7 – Chinji	Middle Miocene	0.512302	8	12.0	− 6.4	Chirouze et al. (2015)
IL7 – Chinji	Middle Miocene	0.512329	6	12.0	− 5.9	Chirouze et al. (2015)
IL8 – Chinji	Middle Miocene	0.512278	9	11.8	− 6.9	Chirouze et al. (2015)
IL9 – Chinji	Middle Miocene	0.512379	5	10.0	− 4.9	Chirouze et al. (2015)
Indus Marine A1, 1380 ft	Pliocene	0.512097	13	2.9	− 10.6	Clift and Blusztajn (2005)
Indus Marine A1, 1620 ft	Pliocene	0.512106	6	3.6	− 10.4	Clift and Blusztajn (2005)
Indus Marine A1, 2200 ft	Pliocene	0.512145	7	5.2	− 9.6	Clift and Blusztajn (2005)
Indus Marine A1, 3180 ft	Upper Miocene	0.512131	4	6.9	− 9.9	Clift and Blusztajn (2005)
Indus Marine A1, 4180 ft	Upper Miocene	0.512135	5	8.7	− 9.8	Clift and Blusztajn (2005)
Indus Marine A1, 4940 ft	Upper Miocene	0.512127	16	10.0	− 10.0	Clift and Blusztajn (2005)
Indus Marine A1, 5920 ft	Middle Miocene	0.512131	5	11.7	− 9.9	Clift and Blusztajn (2005)
Indus Marine A1, 6890 ft	Middle Miocene	0.512133	5	13.2	− 9.9	Clift and Blusztajn (2005)
Indus Marine A1, 7190 ft	Middle Miocene	0.512132	11	13.7	− 9.9	Clift and Blusztajn (2005)
Indus Marine A1, 7820 ft	Middle Miocene	0.512104	8	14.6	− 10.4	Clift and Blusztajn (2005)
Indus Marine A1, 8140 ft	Middle Miocene	0.512137	6	15.1	− 9.8	Clift and Blusztajn (2005)
Indus Marine A1, 8650 ft	Middle Miocene	0.512124	8	15.9	− 10.0	Clift and Blusztajn (2005)
Indus Marine A1, 9170 ft	Middle Miocene	0.512117	7	16.8	− 10.2	Clift and Blusztajn (2005)
JAM-Sr2	Late Paleocene	0.512273	9	57.5	− 7.0	Zhuang et al. (2015)
JW-03-1A-clast2	Jawalamukhi clast	0.511772	8	10.6	− 12.6	Najman et al. (2009)
JW-03-1A-clast3	Jawalamukhi clast	0.511762	8	10.6	− 12.3	Najman et al. (2009)
JW-03-1A-clast4	Jawalamukhi clast	0.511525	8	10.6	− 15.6	Najman et al. (2009)
JW-03-4A-clast1	Jawalamukhi clast	0.511847	10	10.9	− 10.6	Najman et al. (2009)
JW-03-4A-clast2	Jawalamukhi clast	0.512333	8	10.9	− 7.5	Najman et al. (2009)

Table 1 Synthesis of published Nd isotope data used in this study, together with a reference to the original sources from which they were derived *(Continued)*

Sample	Age	$^{143}Nd/^{144}Nd$	± 2σ	Depositional Age (Ma)	εNd	Source
JW03-3A	Jawalamukhi clast	0.511849	9	11.0	− 9.9	Najman et al. (2009)
JW03-3A	Jawalamukhi clast	0.511642	9	11.0	− 13.8	Najman et al. (2009)
JW03-3A	Jawalamukhi clast	0.511847	11	11.0	− 10.0	Najman et al. (2009)
JW03-3A-	Jawalamukhi clast	0.511801	9	11.0	− 10.5	Najman et al. (2009)
JW03-3B	Jawalamukhi clast	0.511864	7	11.3	− 8.4	Najman et al. (2009)
JW97-17B	Jawalamukhi clast	0.511570	8	8.8	− 15.0	Najman et al. (2009)
JW97-17B	Jawalamukhi clast	0.511309	9	8.8	− 20.0	Najman et al. (2009)
JW97-17B	Jawalamukhi clast	0.511857	5	8.8	− 10.3	Najman et al. (2009)
JW97-18B	Jawalamukhi clast	0.511331	5	8.3	− 19.5	Najman et al. (2009)
JW97-19A	Jawalamukhi mudstone	0.511735	8	4.5	− 17.6	Najman et al. (2009)
JW97-19B	Jawalamukhi clast	0.511391	5	5.2	− 19.1	Najman et al. (2009)
JW97-20B	Jawalamukhi clast	0.511303	6	5.6	− 19.8	Najman et al. (2009)
JW97-21B	Jawalamukhi mudstone	0.511647	8	5.5	− 19.3	Najman et al. (2009)
JW97-24B	Jawalamukhi mudstone	0.511881	9	9.5	− 14.8	Najman et al. (2009)
JW97-28A	Jawalamukhi mudstone	0.511956	9	8.5	− 13.3	Najman et al. (2009)
JW97-31A	Jawalamukhi mudstone	0.511875	8	10.5	− 14.9	Najman et al. (2009)
JW97-34C	Jawalamukhi clast	0.511331	9	7.7	− 19.5	Najman et al. (2009)
JW97-34C	Jawalamukhi clast	0.511300	13	7.7	− 20.0	Najman et al. (2009)
JW97-34C	Jawalamukhi clast	0.512273	9	7.7	− 2.8	Najman et al. (2009)
JW97-36B	Jawalamukhi clast	0.512088	9	7.3	− 7.8	Najman et al. (2009)
JW97-36B	Jawalamukhi clast	0.512192	11	7.3	− 7.7	Najman et al. (2009)
JW97-36B	Jawalamukhi clast	0.511873	11	7.3	− 9.2	Najman et al. (2009)
JW97-39B	Jawalamukhi mudstone	0.511897	11	6.5	− 14.5	Najman et al. (2009)
JW97-40A	Jawalamukhi mudstone	0.511909	5	11.5	− 14.2	Najman et al. (2009)
JW97-45B	Jawalamukhi mudstone	0.511861	7	12.5	− 15.2	Najman et al. (2009)
RKT-10-GC1	Early to Middle Paleocene	0.512622	7	61.6	− 0.2	Zhuang et al. (2015)
RKT-Sr1	Early Paleocene	0.512639	6	62.8	0.2	Zhuang et al. (2015)
SHN-Sr4	Early Eocene	0.512246	11	54.0	− 7.5	Zhuang et al. (2015)
SWN-Sr01	Early Paleocene	0.512648	6	64.9	0.4	Zhuang et al. (2015)
SWN-Sr04	Early Paleocene	0.512743	17	63.8	2.2	Zhuang et al. (2015)
SWN-Sr06	Campanian-Maastrichtian	0.512670	7	70.0	0.8	Zhuang et al. (2015)
SWN-Sr08	Middle Paleocene	0.512438	7	61.0	− 3.7	Zhuang et al. (2015)
SWN-Sr09	Middle Paleocene	0.512412	9	60.4	− 4.3	Zhuang et al. (2015)
SWN-Sr10	Middle Paleocene	0.512499	10	59.8	− 2.6	Zhuang et al. (2015)
SWN-Sr12	Middle Paleocene	0.512375	6	59.2	− 5.0	Zhuang et al. (2015)
SWN-Sr14	Early Miocene	0.511981	6	17.7	− 12.7	Zhuang et al. (2015)
SWN-Sr16	Early Miocene	0.511970	8	16.1	− 12.9	Zhuang et al. (2015)
SWN-Sr17	Late Early to Middle Miocene	0.512062	16	14.1	− 11.1	Zhuang et al. (2015)
SWN-Sr18	Late Early to Middle Miocene	0.512100	10	13.4	− 10.3	Zhuang et al. (2015)
SWN-Sr19	late early to middle Miocene	0.512009	11	12.5	− 12.1	Zhuang et al. (2015)
SWN-Sr20	Late early to middle Miocene	0.512096	5	11.6	− 10.4	Zhuang et al. (2015)
SWN-Sr21	Middle to Late Miocene	0.512137	9	8.4	− 9.6	Zhuang et al. (2015)
SWN-Sr22	Middle to Late Miocene	0.511961	7	6.9	− 13.1	Zhuang et al. (2015)

Table 1 Synthesis of published Nd isotope data used in this study, together with a reference to the original sources from which they were derived *(Continued)*

Sample	Age	$^{143}Nd/^{144}Nd$	± 2σ	Depositional Age (Ma)	εNd	Source
SWN-Sr23	Late Miocene	0.512151	8	5.3	− 9.3	Zhuang et al. (2015)
SWN-Sr24	Pliocene-Pleistocene	0.512016	13	2.7	− 12.0	Zhuang et al. (2015)
SWN-Sr26	Pliocene-Pleistocene	0.512011	8	0.5	− 12.1	Zhuang et al. (2015)
SWN-Sr27	Early Oligocene	0.511942	6	31.2	− 13.4	Zhuang et al. (2015)
SWN-Sr31	Early Miocene	0.511909	7	23.0	− 14.1	Zhuang et al. (2015)
SWN-Sr32	Early Miocene	0.511925	7	21.3	− 13.8	Zhuang et al. (2015)
SWN-Sr35	Early Miocene	0.512012	7	19.5	− 12.1	Zhuang et al. (2015)
SWN-Sr40	Late Early to Middle Miocene	0.512000	6	15.0	− 12.3	Zhuang et al. (2015)
SWN-Sr41	Middle to Late Miocene	0.512171	10	9.9	− 9.0	Zhuang et al. (2015)
SWN-Sr42	Early Paleocene	0.512670	6	66.0	0.8	Zhuang et al. (2015)

Himalaya upstream of that point. Paleomagnetic age constraints indicate that the Chinji section spans the period 8–17 Ma (Johnson et al. 1985), while the Jawalamukhi section captures the period 5–13 Ma (Meigs et al. 1995) (Fig. 6). As well as these proximal records, I consider Nd isotope records from the Kirthar and Sulaiman Ranges in the lower reaches of the Indus (Zhuang et al. 2015), as well as from offshore derived from samples from industrial well

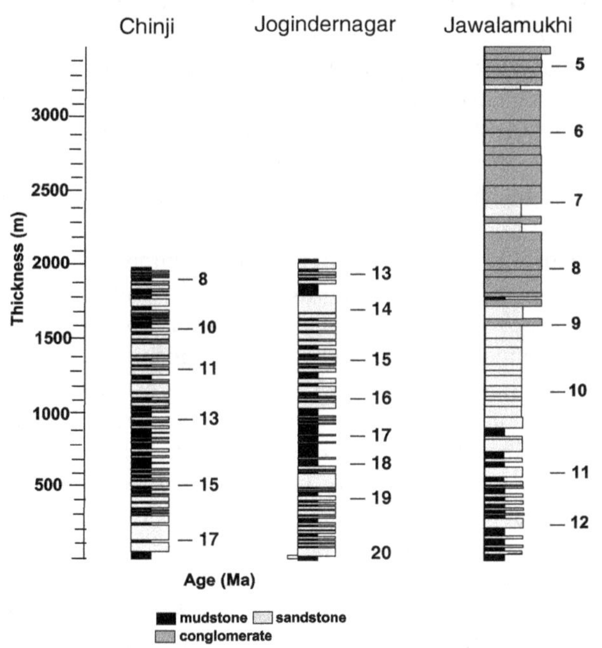

Fig. 6 Stratigraphic columns of three of the key foreland sections discussed in this paper, showing thickness, lithology and depositional ages. Chinji data are from Johnson et al. (1985), Jognidernagar data are from White et al. (2002), while Jawalamukhi data are from Najman et al. (2009). Note that the only well-studied section post-dating 8 Ma in the western Himalaya is at Jawalamukhi. None the sections extend older than 20 Ma meaning that they cannot be used to constrain changing erosion during the onset of rapid Greater Himalayan exhumation or the proposed intensification of the monsoon after ~ 24 Ma

Indus Marine A-1 and a variety of Ocean Drilling Program (ODP) and Deep Sea Drilling Project (DSDP) sites on the Indus Fan and Murray Ridge (Clift and Blusztajn 2005).

Figure 7 shows how the Nd isotope ratio ($^{143}Nd/^{144}Nd$) of the clastic sediment has evolved through time since ~ 25 Ma. In this figure, I use the ε_{Nd} notation to describe the $^{143}Nd/^{144}Nd$ by normalizing these values to Chondrite Uniform Reservoir (CHUR) (DePaolo and Wasserburg 1976). The Kirthar Ranges and offshore data generally plot close to one another, as might be expected, and show a gradual increase from ε_{Nd} values of − 14 at 25 Ma to around − 9 by 10 Ma. Subsequently, ε_{Nd} values fell, especially after 5 Ma reaching values close to − 15 in the modern river (Clift and Blusztajn 2005). In general, higher ε_{Nd} values are associated with more erosion from the Karakoram and other rocks of the suture zone, with Himalayan erosion resulting in lower ε_{Nd} values. Thus, the records from the lower reaches of the paleo-Indus River system indicate increasing flux from the Karakoram before 10 Ma and then more Himalayan influence after that time.

Clift and Blusztajn (2005) interpreted the sharp change to lower ε_{Nd} values after 5 Ma to represent major drainage capture of the four Punjabi tributaries (Jhellum, Ravi, Chenab and Sutlje) into the Indus basin after that time, but more recent work at Chinji by Chirouze et al. (2015) now suggests that the change in ε_{Nd} values at the delta is probably linked to changes in the composition of the trunk river. The Chinji section sediments show ε_{Nd} values of around − 5 at 10–15 Ma, but the modern river close to Chinji has an ε_{Nd} value of − 11 (Fig. 7). This change was attributed by Chirouze et al. (2015) to reflect the start of erosion of the Nanga Parbat Massif after 5 Ma because Nanga Parbat is known to have very negative ε_{Nd} values (Whittington et al. 1999).

Nd isotope data from Jawalamukhi tells a different story concerning erosion of the Himalaya during the Neogene. As might be expected, given the location of

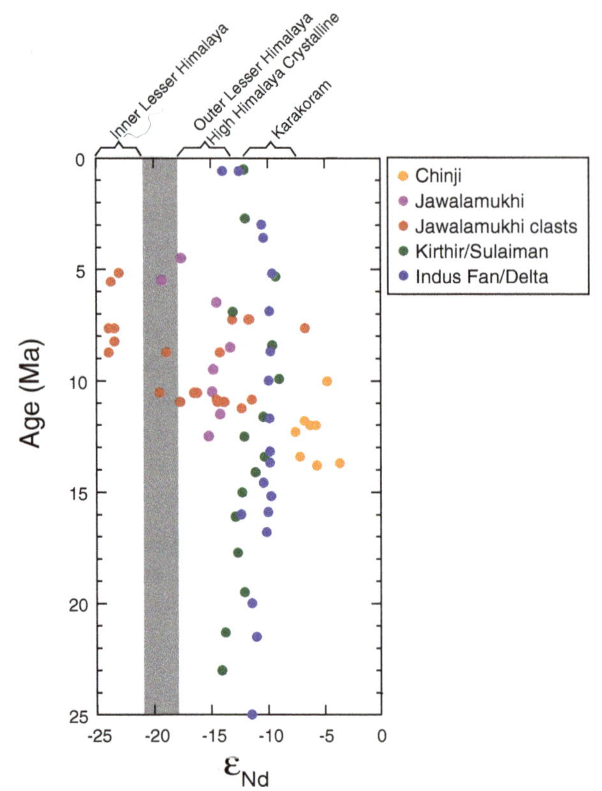

Fig. 7 Compilation of clastic sediment Nd isotope character from the Indus Basin since 25 Ma. Data from the Indus Shelf (Clift and Blusztajn 2005) and from the lower reaches in the Kirthir and Sulaiman Ranges (Zhuang et al. 2015) show a long term slow trend to more positive ε_{Nd} values followed by a drift to more negative ε_{Nd} values since ~ 5 Ma. In contrast, data from Chinji near the trunk Indus has consistently more positive ε_{Nd} reflecting the primitive sources in the suture zone (Chirouze et al. 2015). At Jawalamukhi in the eastern part of the drainage, ε_{Nd} values are more negative than the average. Extremely negative conglomerate clasts appear after ~ 9 Ma indicating erosion from the Inner Lesser Himalaya (Najman et al. 2009). Gray shading shows the demarcation between the Inner Lesser Himalaya and the Greater Himalayan Crystallines, Outer Lesser Himalaya and Karakoram

the section adjacent to Himalayan sources but far from the primitive sediment supply seen in the trunk river, ε_{Nd} values in those deposits are the most negative of any in the compiled dataset, reflective of their Himalayan sources (Najman et al. 2009). Prior to 9 Ma, both mudstones and conglomerate clasts in these sedimentary rocks show ε_{Nd} values close to − 15, consistent with erosion from the Greater Himalaya or from the Outer Lesser Himalaya. However, after 9 Ma conglomerate clasts with very negative ε_{Nd} values are found, falling to values close to − 25, which only readily correlate with the Inner Lesser Himalaya (Najman et al. 2009). As a result, Najman et al. (2009) concluded that the Inner Lesser Himalaya were only exposed at the surface after 9 Ma. Nonetheless, it is interesting to note that the isotopic composition of siltstones in the sequence continued to show ε_{Nd} values of around − 15, at least until around 6 Ma indicating that the total volume of material coming from the Inner Lesser Himalaya did not dominate over the Greater Himalayan material seen before that time. A shift to more negative ε_{Nd} values in the fine-grained sediment after around 6 Ma is however suggestive of an additional change starting at that time which can be further understood using thermochronologic evidence presented in the next section.

Thermochronology evidence for evolving erosion

Because the exhumation history of different source ranges changes across the Himalaya, the cooling ages of different minerals can be used to constrain where sediment sources. In particular, the application of Ar-Ar single grain muscovite mica dating has proven effective at resolving provenance within the foreland basin, as well as the Bengal Fan (Copeland and Harrison 1990). This is because the Greater Himalaya in NW India show rapid cooling starting in the Oligocene, largely ending in the Early Miocene (Stephenson et al. 2001; Walker et al. 1999), whereas the Inner Lesser Himalaya were not so deeply buried and were exposed to the surface more recently (Najman et al. 2009; Célérier et al., 2009a, b). Muscovite Ar-Ar ages are reset at temperatures at around 425 °C, consistent with burial in excess of around 15 km (Harrison et al. 2009). Because this is a relatively high temperature, these minerals are typically not reset during later burial in the foreland basin. Two sets of Ar-Ar detrital data have been collected from sections at Jawalamukhi (Najman et al. 2009) and Jogindernagar (White et al. 2002), which are relatively close to one another (Fig. 1) and which together provide a snapshot of erosion patterns and exhumation rates in that part of the Himalayan front since around 21 Ma.

Figure 8 shows the range of cooling ages for single mica grains compared with their depositional ages. A dashed line provides an indication of how rapid exhumation was by representing unity, i.e., when the two ages are within error of one another. Points plotting very close to this line are indicative of extremely rapid exhumation, such that grains can cool from 425 °C, and are then exhumed to the surface and deposited so quickly that the duration of grain erosion and transport is too short to be measured. The plot shows that at any given time there is typically a range of mica ages, some extending to significant ages of hundreds of m.y. or more. This pattern requires that both fast and slow cooling sources were contributing to the basin at any given particular time. It is, however, noteworthy that before around 17 Ma there is always a significant population of grains plotting very close to the unity line. This requires significant sediment flux from fast exhuming source areas. The fact that these Ar-Ar ages are

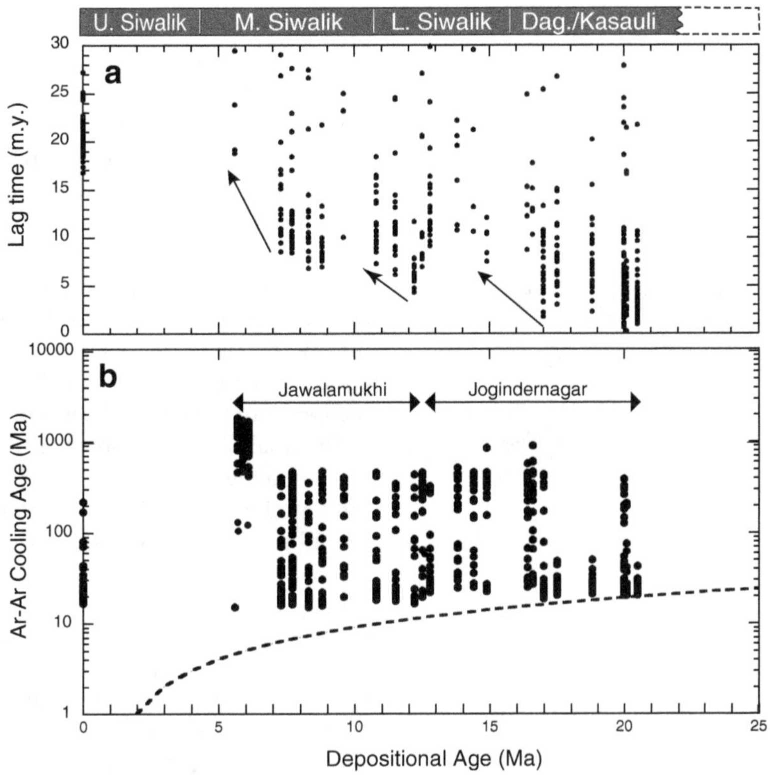

Fig. 8 Comparison of Ar-Ar muscovite cooling ages and depositional ages from the Jawalamukhi section of Najman et al. (2009) and the Jogindernagar section of White et al. (2002). Note that increase in lag of fastest cooled grains after ~ 17 Ma and ~ 8 Ma. **a** Lag times between cooling and sedimentation. **b** The cooling and deposition ages with the dashed line indicating zero lag time

Cenozoic indicates that those grains are derived from the Greater Himalaya or equivalent high-grade terranes exposed early on during the motion on the MCT. After 17 Ma, it is clear that there is a more discernable gap so that even the fastest cooled grains show a significant lag between cooling and deposition. Even though these young grains probably came from localized sources, this change was interpreted to indicate a slowing of exhumation rates in even these erosional "hotspots" after that time.

Figure 8a shows a close-up of the lag time of the fastest cooling grains for these data. This plot reinforces the conclusion that prior to 17 Ma at least some parts of the mountains were exhuming at very high rates but slowed after that time. It is also noteworthy that the lag time sharply increased after 6 Ma, with some more moderate slowing visible after 12 and 8 Ma as well. Najman et al. (2009) argued that a population with Paleozoic dates represented erosion from granites of that age, which have intruded the Crystalline Lesser Himalaya so that their appearance and dominance after 6 Ma were taken to indicate an exposure of that unit, at least in the watershed of these particular sections, corresponding now to the Beas River, close to Jawalamukhi. It is however unknown whether this change in erosion pattern and rate was a regional feature or related only to this particular area, as the section only preserves local rivers. Focused erosion along river valleys is well known to drive enhanced exhumation along these corridors (Montgomery and Stolar 2006; Simpson 2004), but this need not mean regional exposure of such units outside such restricted areas.

At a more regional scale, the examination of the Western Himalaya may be compared with other parts of the orogen by looking at the range of Ar-Ar cooling mica ages, as recently synthesized by Webb et al. (2017). Figure 9 shows the muscovite Ar-Ar ages seen across the entire foreland basin and which have been color-coded to show the longitude of sedimentation. This plot was designed so that the sediments at any one place might largely reflect the cooling history of the adjacent source ranges. What is clear is that the ages from the far western parts of the foreland basin have some of the oldest Ar-Ar ages of any sediment found in the Himalaya. They particularly contrast with those in the central Nepal Himalaya at around 83–86° East, which tend to cluster at ages of 15–20 Ma, compared to 25–35 Ma for muscovite Ar-Ar in northwest India (Fig. 9b). Such a distribution reveals a large-scale asymmetry where rapid cooling began earlier in the Indus Basin and later in the central Himalaya.

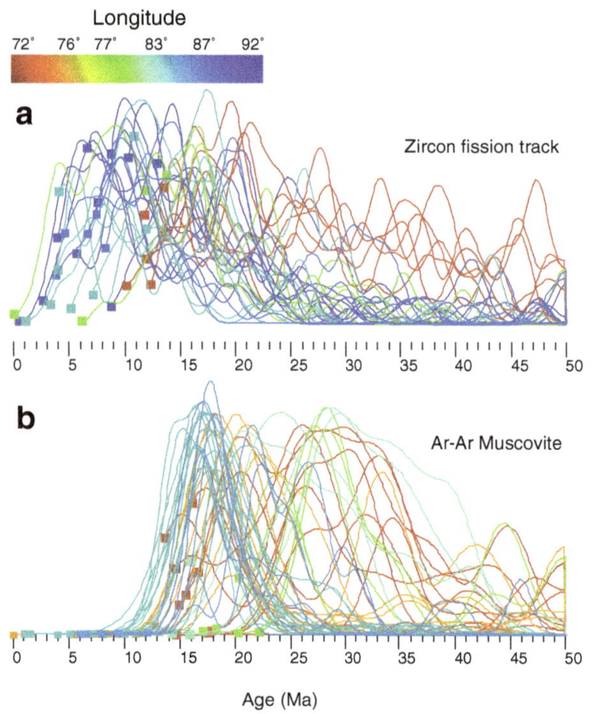

Fig. 9 Compilation of detrital thermochronology results from the Himalayan foreland basin, modified from Webb et al. (2017). Detrital thermochronology involves sampling sedimentary materials and acquiring cooling ages from detrital components, in order to constrain the cooling history of the sediment source regions. In parts **a** and **b**, $^{40}Ar/^{39}Ar$ muscovite and zircon fission track zircon results are plotted, respectively, for dates younger than 50 Ma. $^{40}Ar/^{39}Ar$ muscovite ages date the cooling of muscovite crystals below 425 ± 25 °C (Harrison et al. 2009), whereas fission track zircon ages date the cooling of zircon crystals below 240 ± 30 °C (Hurford 1986; Bernet and Garver 2005). The data are shown using the Kernel Density Estimation (KDE) methodology, which plots the detrital dates as a set of Gaussian distributions (Vermeesch 2012). This approach allows the age ranges and abundances of different detrital age populations to be compared: peaks in the curves represent peaks in the detrital age populations. For these plots, the population for a single sample is shown as a curve, sample longitude is keyed to a color spectrum, and the depositional age is shown via the squares at the young (left) terminations of the curve. These data provide an approximation of the cooling experienced by adjacent Himalayan regions. Signals in the Himalayan foreland basin can be complicated by river sediment transport along the range trend, since not all river systems transport sediment perpendicularly away from the mountains and thus might not only represent cooling and exhumation over the limited extent of the range immediately adjacent to the sampling location. A general trend appears in these data: peaks in the cooling age populations appear to young to the east from 25 to 20 Ma to 10–8 Ma. This is consistent with an eastwards migrating pulse of hinterland cooling and erosion during this period

The same general pattern was also shown in data from zircon fission track analysis (Fig. 9a), which dates the time at which detrital zircon grains cooled below around 200 °C (Hurford 1986; Tagami et al. 1990). Again, there are significant numbers of grains with old cooling ages extending back until 50 Ma, but ranging up until around 20 Ma in the Western foreland basin. In contrast, the central-eastern part of the foreland shows an onset to rapid cooling beginning in much younger times, generally after 20 Ma. Because these are detrital grains and not bedrock, we can be confident that this pattern is not a preservation issue linked to the removal of old material from the metamorphic rocks in the Greater Himalaya themselves but represents a real difference in the erosion history along the mountain chain. Because erosion tends to remove the older bedrock, the ages of the metamorphic and igneous rocks that now form the Greater Himalaya only tell us about the unroofing history of those particular rocks but do not help us constrain earlier erosion. This is only preserved in the sediments of the foreland basin or submarine fan.

The analysis by Webb et al. (2017) demonstrates that rapid cooling and erosion had began perhaps as early as the Late Eocene, and was certainly well underway during the Oligocene (30–35 Ma) within the Indus catchment. There was a period of rapid cooling around 20 Ma, shortly after the onset of motion on the MCT, with a corresponding drop off in rates shortly after that. As visible in Fig. 8a it seems that the basement sources began to cool more slowly after ~ 17 Ma. Webb et al. (2017) relate this burst of early cooling to a progressive breakoff of the subducting Indian lithospheric slab driven by a tear in the slab that propagated from the western syntaxis towards the east. This breakoff terminated crustal thickening and provoked a rebound of the Greater Himalaya that were then uplifted and incised rapidly, driving a wave of exhumation.

Changing rates of Himalayan erosion

Further detail about how erosion of the Western Himalaya differs from other parts of the range can be seen when considering studies from the central Himalaya. Bernet et al. (2006) conducted a study of the Siwalik Group in Western-Central Nepal where they applied zircon fission track dating to reconstruct the cooling history of that particular part of the Greater and Lesser Himalaya since 16 Ma. That study revealed that there were two groups of zircon fission track ages that remained essentially the same throughout the entire section and which pointed to relatively stable sources that continued to provide sediment to the basin at least since 16 Ma (Fig. 10). In contrast, after 11 Ma, a third minority age population appeared. This evolved to younger and younger ages as the sediment itself became younger. This implies the presence of a source that was rapidly uplifting and exhuming. Figure 10 demonstrates that this group had an approximate lag between cooling and sedimentation of 4 m.y. after 11 Ma. The younger of the stable fission track age populations centered around 16 Ma likely represents the Greater Himalaya (Bernet et al. 2006). The two younger sets of grains were interpreted to be derived from both the Greater and Lesser Himalaya, but

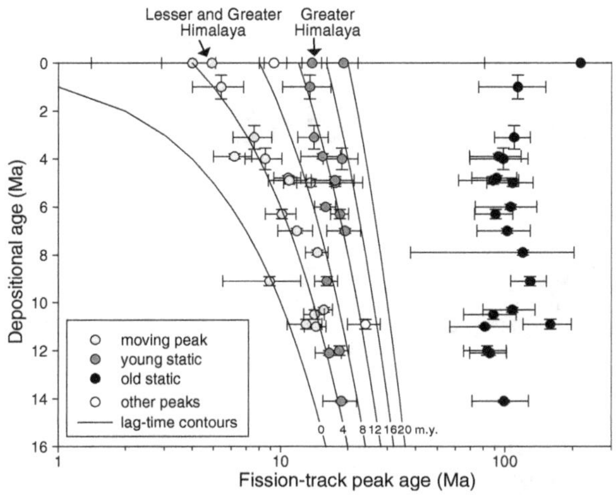

Fig. 10 Lag time plot of detrital zircon fission track peak ages versus depositional ages of Siwalik Group sedimentary rocks from Nepal and modern river sediment from Bernet et al. (2006). Peak ages are given with two σ errors. Old static peak refers to fission track age populations ~ 100 Ma and that do not change up-section. Young static peak refers to fission track age populations ~ 20 Ma and that do not change up-section. The young moving peak of fission track ages that appears after 11 Ma maintains a constant lag time of ~ 4 m.y. but constitutes a smaller fraction of the total sediment mass

particularly from the Lesser Himalaya, which became progressively more important as the Miocene progressed.

What is interesting to note in the study of Bernet et al. (2006) is that the rate of exhumation does not appear to speed or slow as time progressed towards the present day. This contrasts with the result from the muscovite Ar-Ar work at Jawalamukhi and Jogindernagar that implied a slowing of exhumation (Najman et al. 2009; White et al. 2002). Thus, either the micas and zircons were derived from different sources, or the exhumation history of the central Himalaya differs substantially from that seen in the Western Himalaya, potentially linked to the climatic gradient between them.

A note of caution is warranted however after considering what sort of record a slice of foreland strata might reflect. If those sediments were deposited close to the mountain front than they might reflect sediment being derived directly from the immediately adjacent ranges, whereas if the strata were deposited more centrally within the foreland they might show more evidence of material from a dominant axial river that is supplied by sediment from a broad region. This interplay between local and more regional influences was highlighted by Najman et al. (2009) in the case of the Jawalamukhi section where they inferred local dominance after 6 Ma due to the disappearance of flux from the Greater Himalaya. They reasonably presumed that the Greater Himalaya did not entirely stop eroding at that time and that the change in provenance in the preserved record was mostly linked to drainage reorganization at the local scale. Such local complexities are always an issue when dealing with proximal foreland sedimentary deposits. Although foreland sections will continue to be important, such complexity does highlight the additional need for a basin-wide view, such as provided by the integrated records from the Indian Ocean submarine fans.

Erosion and structural models of the Himalaya

Our understanding of how the Himalaya may have eroded based on the sedimentary records can now be compared with reconstructions concerning the orogenic architecture based on structural analysis. For example, Fig. 11 shows a reconstruction of the last 5.4 Ma derived from the work of Webb (2013). In this model, the Tethyan Himalaya essentially buried other structural units prior to 5.4 Ma and it is only after that time that they begin to come to the surface as a result of duplexing. The reconstruction at 1.9 Ma would appear to offer the first opportunity for rocks of the Greater Himalaya and of the Crystalline Lesser Himalaya to be exposed to the surface. This prediction would not necessarily preclude the changes in the rates of erosion demonstrated by the thermochronology data (e.g., Fig. 8). The structural reconstruction appears to indicate exposure of Inner Lesser Himalayan rocks a little later than would be inferred based on the Ar-Ar mica data from Najman et al. (2009) who showed the shift in provenance at about 6 Ma. However, the presence of conglomerate clasts from the Inner Lesser Himalaya seen at Jawalamukhi after 9 Ma would be much earlier than anticipated from the Webb (2013) structural model, but might reflect enhanced erosion along major river valleys (Simpson 2004; Montgomery and Stolar 2006) and reflect along-strike variability. It should be noted that other quite different structural and exhumation models have been proposed for the Himalayan foreland basin further east in Nepal. DeCelles et al. (1998b) used petrographic data from the pre-Siwalik Dumri Formation in Nepal to argue for erosion of sedimentary and low-grade metasedimentary rocks in the Tethyan Himalaya during the Early Miocene (~ 16–20 Ma). The presence of plagioclase grains in Dumri Formation sandstones suggested exposure of crystalline rocks of the Greater Himalaya at that time.

Certainly, the relative stability of Nd isotope composition until about 6 Ma in the siltstones of Jawalamukhi would also be consistent with relatively limited exposure of the Inner Lesser Himalaya until around that time. The Jawalamukhi isotope record is thus close to being in agreement with the structural reconstruction shown in Fig. 11. It is also noteworthy that the relatively late exposure of Inner Lesser Himalayan rocks predicted by the structural model would have served to input significant amounts of

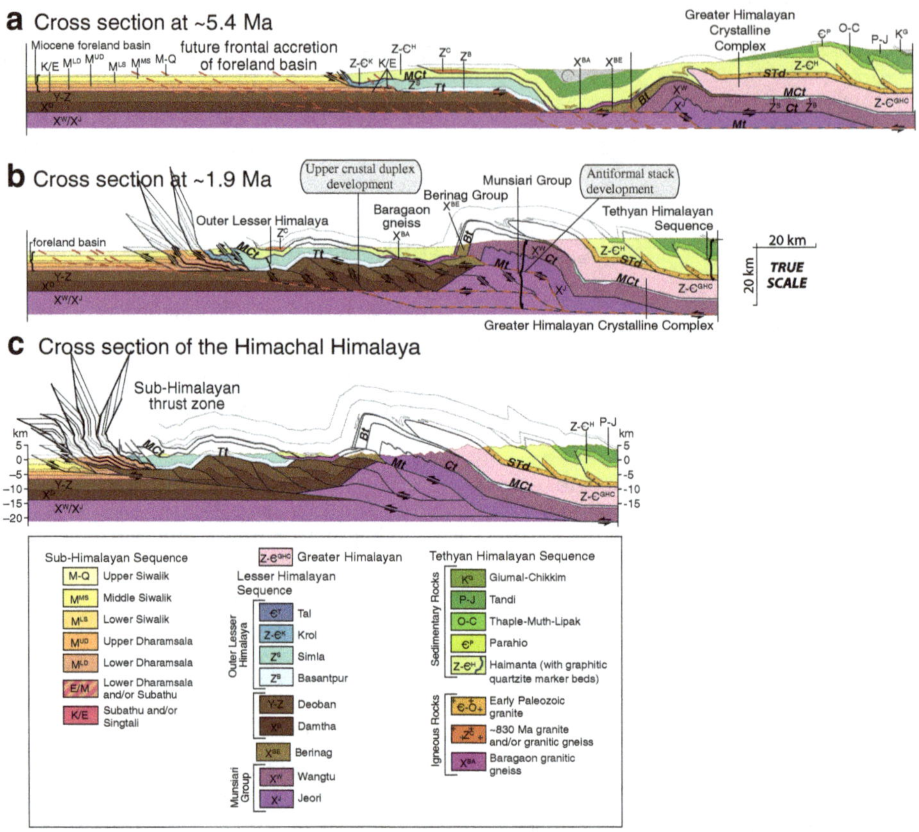

Fig. 11 Restorations of the Western Indian Himalaya by Webb (2013) at **a** ~5.4 Ma, **b** ~1.9 Ma and **c** the present day. Note that the crystalline Inner Lesser Himalaya and the Greater Himalaya are only predicted to breach the surface in this region after 5.4 Ma

relatively low ε_{Nd} material into the basin after 5.4 Ma. This provides an alternative mechanism to that suggested by Chirouze et al. (2015) to explain the basin-wide shift to more negative ε_{Nd} values after 5 Ma. Although the uplift and exposure of Nanga Parbat may have played a part in this process, it is possible that erosion of the volumetrically more important Lesser Himalaya were also driving the bulk Indus composition towards more negative ε_{Nd} values.

The Webb (2013) structural model is moreover consistent with the relatively recent appearance of both kyanite and sillimanite in the foreland basin as illustrated by Fig. 5, starting around 5 Ma at least in the Western Himalaya. Although the MCT must have been active much earlier, it appears that the final phase of erosion that brought kyanite and sillimanite-bearing sources to the surface may have been much more recent and not related either to motion along that fault, or to the slab break-off event that characterized the Oligocene to mid Miocene. Again, this may be linked to faster erosion caused by climatic variability. The apparent discrepancy between the Nepal reconstruction by DeCelles et al. (1998b) that requires earlier unroofing may indicate that one of these studies is in error or that there are major along strike differences in unroofing history. The much drier climate in the Indus basin might have slowed erosion and delayed Greater Himalayan unroofing compared to what is known in Nepal.

Erosion and climatic evolution

When trying to decide whether solid Earth tectonic forces or climate and surface processes control the structural evolution and exposure history of mountains, it is important to have an independent climate record to compare with erosional reconstruction in order to test for any linkages. There have been many attempts to try and reconstruct the evolution of the Asian monsoon over long periods of geological time, not all of which agree with one another. Kroon et al. (1991) looked at the varying abundance of *Globigerina bulloides*, a planktonic foraminifera that blooms offshore Oman during the summer monsoon because of the enhanced upwelling of nutrient-rich waters driven by the winds. These workers highlighted the time since 8 Ma as being one of intensified upwelling and therefore of stronger summer monsoon activity. Moreover, this reconstruction was broadly consistent with other evidence from the stable carbon isotope character of pedogenic carbonates within the Pakistani part of the foreland basin, i.e., the Potwar Plateau just to the east of Chinji. Quade et al. (1989)

showed a shift to more isotopically heavy carbon isotopes ($\delta^{13}C$) after ~7 Ma. This implied a shift from vegetation dominated by C_3 plant types, (i.e., mostly trees) to more C_4 domination (i.e., largely grassland). Although Quade et al. (1989) initially interpreted this to indicate a strengthening of the monsoon at that time, this was later changed to indicate a drying of the climate in the Indus floodplain after 7 Ma (Quade and Cerling 1995). The same general trend is also seen in NW India, where the first significant development of C_4 plants was dated at 7 Ma and with dominant C_4 growth after 5 Ma (Singh et al. 2011). The scatter to lower $\delta^{13}C$ values in NW India compared to Pakistan after 5 Ma may reflect the moderately wetter conditions seen going east along the Himalayan front and which favor C_3 flora (Fig. 12b). Oxygen isotopes have helped to confirm this interpretation of the vegetation changes. Dettman et al. (2001) provided oxygen isotope profiles through fossil freshwater bivalve shells and mammal teeth that help to constrain total rainfall and seasonality through the Late Miocene. These data showed the presence of a monsoon since at least 10.7 Ma but also demonstrated that the

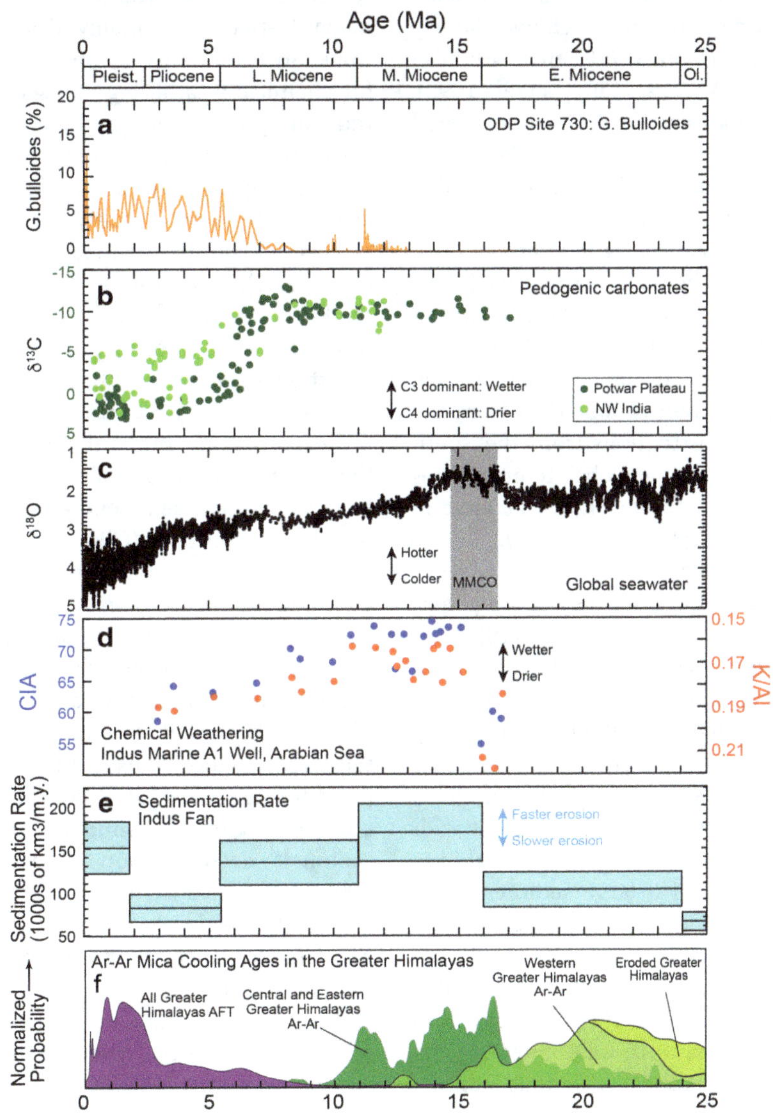

Fig. 12 Comparison of climate, erosion and exhumation proxies in the Himalaya. **a** Abundance of *G. bulloides* at ODP Site 730 on the Oman margin as a proxy for summer monsoon wind strength (Gupta et al. 2015). **b** Carbon isotope character of pedogenic carbonate in Pakistan as an indicator of dominant vegetation in the Potwar Plateau of Pakistan (Quade et al. 1989), and NW India (Singh et al. 2011). **c** Global d18O as a proxy for seawater temperatures and/or ice volume from Zachos et al. (2001). **d** Degree of chemical alteration of sediments on the Indus continental shelf at Indus Marine A-1 as measured by K/Al and CIA (Clift et al. 2008), **e** Rates of sediment supply to the Arabian Sea calculated from regional seismic (Clift 2006). **f** Exhumation rates of the Greater Himalaya tracked by bedrock Ar-Ar dating (Clift et al. 2008) and foreland basin sediment (Szulc et al. 2006)

$\delta^{18}O$ of wet-season rainfall was significantly more negative before 7.5 Ma than after that time. Assuming that there was not a big change in source water compositions at that time this shift in $\delta^{18}O$ would imply increasing aridity after 7.5 Ma.

Such interpretations beg the question as to why the climate would dry at a time when summer monsoon driven upwelling in the Indian Ocean appeared to be intensifying. It should be remembered that the upwelling proxy does not record summer rain, but rather is linked to the wind strength, which might potentially be decoupled from one another in the context of long-term global cooling during the Late Miocene. This is despite the fact that in the present day, coastal upwelling is largely modulated by summer monsoon winds that are also linked to precipitation in SW Asia (Curry et al. 1992). Chemical weathering records from the Indus continental margin show long-term decrease in the degree of alteration especially after around 10 Ma (Clift et al. 2008) that might indicate weaker monsoon over that timescale (Fig. 12d). Because humidity as well as temperature is a key control over the rates of chemical weathering (West et al. 2005), Clift et al. (2008) argued that the decrease in the degree of alteration seen in Indus River delta sediments reflects a drying of the climate after 10 Ma, broadly consistent with the carbon isotope data from the foreland basin. However, reconstructions from the South China Sea now suggest that falling global temperatures are the primary control on weathering rates, at least in southern China (Wan et al. 2012; Clift et al. 2014).

Although the age of drying/cooling inferred from weathering records appears to precede the 8 Ma upwelling increase documented by Kroon et al. (1991), it should be noted that more recent attempts to look at the abundance of *G. bulloides* have resulted in a reevaluation of the record. Although upwelling strengthened after 8 Ma, it is noteworthy that the initial increase began around 13 Ma (Gupta et al. 2015) (Fig. 12a). This older age for intensified wind strength is consistent with new results from scientific ocean drilling in the Maldives that showed the development of a large sediment drift link to prevailing monsoon winds, again starting at around 12.9 Ma (Betzler et al. 2016).

A new record for environmental evolution during this critical transition around 8 Ma is presented here based on new spectral color data scanned from ODP Site 730, the same site analyzed for *G. bulloides*. This analysis is based on the recognition that hematite and goethite are associated with characteristic wavelengths in the spectrum of the sediment (Deaton and Balsam 1991; Balsam and Deaton 1991). The ratio of these two wavelengths can be used to estimate the relative intensity of these two minerals (Giosan et al. 2002). This is significant because hematite is generally associated with hotter, drier, and more seasonal conditions, whereas goethite is more commonly recognized as a proxy for wetter, colder environments (Schwertmann 1971). Such an approach has been used effectively before in the South China Sea (Zhang et al. 2007; Clift 2006) as a means for tracking continental humidity and thus monsoon strength.

Figure 13 shows the hematite/goethite record from 10.7 to 8.5 Ma. Given the location of the core, this is presumed to reflect environmental conditions in Eastern Arabia, which is the source of the clastic sediment deposited at this drill site. Although there is a period between 9.9 and 9.5 Ma when hematite is relatively scarce, the overall trend after 9.5 Ma is towards more hematite-rich, presumably drier, more seasonal conditions. Such a record is consistent with the other proxies for chemical weathering and sediment supply that favors reducing precipitation and colder conditions, linked to a weakening summer monsoon into the Late Miocene. Exactly why the summer monsoon might be weakening at that time is not entirely clear, but could be related to the overall cooling of global climate at that time (Zachos et al. 2001), since times of cool dry conditions globally tend to correlate with times of weaker summer monsoons (Gupta et al. 2003; Clemens et al. 2010).

Although there has been some debate in the past concerning whether wetter climates might result in more or less erosion (Burbank et al. 1993), studies of the Quaternary evolution of the Indus and Ganges deltas have shown that sediment supply increased during times of strong monsoon (Clift and Giosan 2014; Goodbred and Kuehl 2000) when landsliding is also known to be common in the source areas (Bookhagen et al. 2005). This particular form of mountain erosion is well recognized as being particularly efficient at generating large volumes of sediment (Liu and Li 2015; Hovius and Stark 2006). It is thus interesting to note that the supply of sediment to the Indus delta and fan was estimated to peak during the middle Miocene, based on the interpretation of regional seismic reflection data coupled with sediment decompaction calculations (Fig. 12e) (Clift 2006). That reconstruction suggested that sediment delivery rates to the Arabian Sea dropped after the Middle Miocene to reach a minimum in the Pliocene, before rebounding to higher values during the Pleistocene. High rates of sediment delivery during the Middle Miocene coincided with times of stronger chemical weathering and suggest wet, more erosive climatic conditions at that time, at least in the Western Himalaya. Falling rates of sediment delivery and reduced erosion would then correlate with the drying climate in the Late Miocene, consistent with the pedogenic carbonate stable isotope data.

When considering the entire Himalayan range times of strong precipitation and rapid erosion broadly

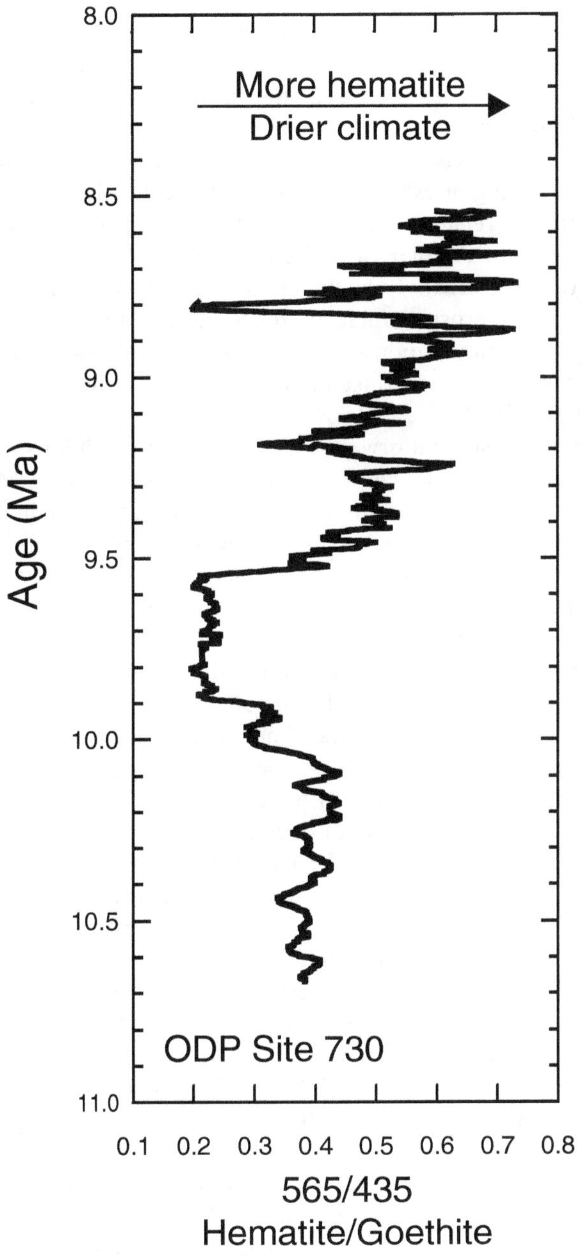

Fig. 13 Relative abundance of hematite versus goethite at ODP Site 730 on the Oman margin implies steady drying after 9.5 Ma following a wet period from 9.9 to 9.5 Ma. See Fig. 2 for location

correlate with episodes of fastest exhumation in the central and in Eastern Himalaya (Clift et al. 2008), but it is noteworthy that the time of fastest exhumation is oldest in the West (Webb et al. 2017). It is possible that this earlier phase of erosion relates to the initial phases of monsoon intensification whose timing is not well defined but appears to strengthen around 23 Ma based on data from the South China Sea (Clift et al. 2008). Unfortunately, no matching record from that time yet exists in the Arabian Sea. Certainly, detrital thermochronology from the foreland basin indicate that rapid erosion of the Western Greater Himalaya had begun during the Oligocene (Fig. 9) and that this would be consistent with an earlier initiation of the South Asian monsoon, as well as potentially with slab break-off (Webb et al. 2017).

Given that the climate is drier in the Western Himalaya compared to the eastern and central ranges, slower erosion in that former area would result in older fission track cooling ages being preserved in the bedrock sources. In contrast, faster erosion under the wetter climate in the East has stripped away bedrock with the older ages, which are only preserved in the Greater Himalayan crystallines exposed in the west. Alternatively, if the slab breakoff mechanism proposed by Webb et al. (2017) is correct, then the difference in fission track cooling ages may simply reflect the propagation of the tear and uplift from west to east during the Miocene.

Climate-tectonic linkages

Because the intensity of the South Asian monsoon has been modeled to be largely dependent on the height of the Himalayan barrier (Boos and Kuang 2010) rather than on the height and extent of the Tibetan Plateau, as previously favored (Kutzbach et al. 1993; Molnar et al. 1993), rebound of the Greater Himalaya caused by slab breakoff would have raised the topographic barrier along the southern edge of the Tibetan Plateau and driven a progressive increase in South Asian monsoon intensity starting shortly after ~ 30 Ma. Evidence for tearing of the Indian slab comes from a variety of sources. Seismic tomographic images of the mantle below India show a seismically fast region interpreted as detached Indian lithosphere (Replumaz et al. 2010). The distance between the detached slab and the Indian plate now underthrusting the plateau reduces from west to east leading Replumaz et al. (2010) to propose a lateral migration of slab detachment from the western Himalaya to the central Himalaya. These authors further estimated this tearing to have occurred from 25 to 15 Ma. This seismic evidence is further supported by an east-to-west younging of alkaline and adakitic magmatic rocks that are dated at ~ 17–19 Ma at the western end of the Himalaya, 25–30 Ma in the East and ~ 15–8 Ma in the east-central Himalaya has been interpreted as a product of lateral migration of slab detachment (Zhang et al. 2014; Pan et al. 2012).

In this scenario, the earlier onset of rapid erosion in the Western Greater Himalaya would be linked not only to uplift of the range caused by slab breakoff, but by a synchronous intensification of summer monsoon rains in that region driven by the increased height of the topographic barrier (Webb et al. 2017). Evidence for the height of the barrier is however rather thin. Hydrogen isotope data from minerals in shear zones around Mount Everest indicating that the barrier in that location was > 5 km in the late Early Miocene (Gébelin et al. 2013), but no similar dataset is known from those mountains in the Indus catchment. Likewise, stable oxygen isotope data, which is sensitive to paleoaltitudes, from Miocene-Pliocene carbonate sediments in southern Nepal has been used to demonstrate altitudes of the southern Tibetan Plateau close to the present data before 11 Ma (Garzione et al. 2000). Similar data would argue for altitudes > 4 km in the central Tibetan Plateau as long ago as 35 Ma (Rowley and Currie 2006). Not only do these data not constrain elevations in the western Himalaya but also they largely relate to the plateau and not the Himalayan barrier.

Unfortunately, a sedimentary record of environmental conditions and erosion during the Oligocene is mostly not present in the Himalayan foreland basin because of the long-duration unconformity that separates the Subathu Formation from the Dagshai Formation (Najman 2006). Although it has been suggested that this gap reflects a forebulge unconformity (DeCelles et al. 1998a), more recently, it has been suggested that this hiatus is instead the product of rebound of the foreland basin caused by faster erosion in the mountains, which reduced the size of the orogenic load, thus allowing the foreland basin to partially invert during the Early Miocene (Clift and VanLaningham 2010). A similar erosion-driven inversion has been proposed from the Himalayas during the recent glacial episodes (Burbank 1992). In this model the great foreland unconformity is another reflection, together with the rapid exhumation of the Greater Himalaya, of an intensifying summer monsoon. However, if the slab breakoff model is valid, then this might also result in a flexural uplift of the foreland basin and the generation of a regional unconformity as the tear propagated from West into the central Himalaya. It is possible that both mechanisms may be playing a role in the development of this break in the record. Either way, the only sedimentary record that spans this critical period must be in the Arabian Sea and this has yet to be recovered by scientific ocean drilling. Until this is remedied, it is unknown when South Asian climate first became strongly monsoonal and if that is linked to faster erosion.

Given the incompleteness of the record, is it possible to say anything meaningful about the role of solid Earth tectonic processes compared to climatically modulated surface processes in controlling the erosion of the Western Himalaya? Slowing of erosion after around 17 Ma does not obviously correlate with any known climatic transition although it does shortly follow the warmest conditions of the Mid-Miocene Climatic Optimum (MMCO), which is expected to be a period of strong tropical rainfall (You et al. 2009; Wan et al. 2009). Instead, slowing exhumation after the Mid Miocene may reflect slowing uplift related to earlier slab breakoff, which triggered the initial phase of rapid exhumation seen to peak in the Western Himalaya at 24–17 Ma (Figs. 8b and 12). The Ar-Ar mica data points to a slowing of exhumation after around 12 Ma (Fig. 8b), which correlates well to the strengthening wind conditions in the Arabian Sea (Betzler et al. 2016; Gupta et al. 2015) (Fig. 12a). It is possible that the Late Miocene slowing of erosion might be related to an initial weakening of summer monsoon rain linked to global cooling that occurs at and just before that time (Zachos et al. 2001) (Fig. 12c).

A more straightforward link to climatic conditions can be seen in the slowing of exhumation rates after ~ 7 Ma. This is the time when the Ar-Ar data indicate reduced cooling rates and is when environmental conditions in the Indus floodplain became drier (Quade and Cerling 1995), with lower degrees of chemical weathering and reduced sediment flux to the ocean (Clift et al. 2008). What is less clear is why reduced erosion rates and a weaker monsoon would result in the exhumation of the Inner Lesser Himalaya. This may be the simple culmination of a long-term process by which these rocks are being progressively unroofed through time. Although the monsoon is weaker in the Late Miocene than during the Middle Miocene, it was still active and would have eventually exposed Lesser Himalayan units as a result of the rock uplift driven by the structural imbrication then occurring at depth. This view implies that exposure would have been largely driven by solid Earth tectonic processes. Alternatively weakening monsoon might have resulted in a southward migration of the band of most intense rainfall (i.e., the Intertropical Convergence Zone; ITCZ), which would have been pushed further north onto the edge at the Tibetan Plateau during times of strong summer monsoon. A review by Armstrong and Allen (2011) has proposed that the ITCZ has migrate southward during the Neogene. This model was largely based on the position of the smectite–illite transition that marks the boundary between wind-blown clay derived from Asia and that from Central and South America in the central Pacific Ocean (Lyle et al. 2002). In the western Pacific, a less dramatic migration southward was noted on the basis of the composition of manganese crusts that are dependent on biologic and detrital fluxes (Kim et al. 2006). As the monsoon weakened, this band of intense rainfall would have migrated south and became positioned over the region where the Inner Lesser Himalaya are now exposed. Analog modeling of the formation of duplexes in mountain belts suggests that these are preferentially formed beneath areas where surface processes are driving intensified erosion (Malavieille 2010). Formation of a duplex in the Inner

Lesser Himalaya after 5.4 Ma is indicative of intensified erosion in this region, linked to migration of the climatic bands associated with the summer monsoon.

The weakening of the summer monsoon during the Late Miocene does however appear to be largely linked to global cooling and not to changes in the height of the Himalayan topographic barrier, which was inferred to control monsoon strength during the initial intensification during the later Oligocene to early Miocene (Webb et al. 2017). In that model, exhumation of the Lesser Himalaya is largely driven by changes in summer monsoon intensity and location rather than solid Earth processes, which tended to be more steady state. The Late Miocene to Pliocene exposure of the Lesser Himalaya and duplex formation can be linked to monsoon evolution driven by global climatic processes.

Conclusions

Sedimentary records from the Himalayan foreland and Arabian Sea show that changing environmental conditions linked to summer monsoon intensity played an important role in controlling the structural evolution of the Himalaya. Greater Himalaya exhumation started earlier in the western than the central Himalaya, possibly linked to lithospheric slab breakoff that drove rock uplift and intensified the summer monsoon by raising the topographic barrier of the Himalaya. This in turn further increased erosion rates that peaked in the Middle Miocene in the Indus Fan. A regional unconformity in the foreland basin in the Oligocene to Early Miocene reflects both uplift driven by slab breakoff and the strengthening monsoon that heightened erosion rates and reduced the orogenic load on the subducting plate. However, the presence of the unconformity means that deep sea records are required if we are to extend our reconstruction beyond 20 Ma and so better correlate climate and erosion with Greater Himalayan exhumation.

Ar-Ar ages from detrital micas in the foreland basin indicate that exhumation in the NW Himalaya slowed after 17 Ma, apparently unrelated to monsoon variability. Initial strengthening of monsoonal winds is now dated to start at 12.9 Ma but is not positively correlated with South Asian rainfall. Indeed, exhumation rates appear to slow after 12 Ma implying a drier, less erosive climate. Stronger drying of the climate is documented after ~ 10 Ma, as shown by carbon isotopes in pedogenic carbonates, increasing hematite/goethite values offshore Oman and falling sediment supply rates in the Indus Fan, caused the rainfall maxima to migrate south in the Himalaya. Focused erosion after that time allowed the Inner Lesser Himalaya to duplex and become initially exposed after 9 Ma. Wider exposure was delayed until ~ 6 Ma. Together with unroofing of the Nanga Parbat Massif, this process drove the average ε_{Nd} value of the Indus Delta to more negative values, removing the need for major drainage reorganization to account for this trend. Focused erosion encouraged thrust duplex formation in the Inner Lesser Himalaya, facilitating the exposure both of this unit and the Greater Himalaya after ~ 6 Ma. This explains the relatively late appearance of kyanite and sillimanite in sediments of the Middle and Upper Siwaliks respectively. Climate as well as solid Earth tectonic forces is seen to be crucial in the formation of the NW Himalaya.

Abbreviations

DSDP: Deep Sea Drilling Project; IODP: International Ocean Discovery Program; ITCZ: Intertropical Convergence Zone; MCT: Main Central Thrust; MFT: Main Frontal Thrust; MMCO: Mid-Miocene Climatic Optimum; ODP: Ocean Drilling Program

Acknowledgements

PC gratefully acknowledges the travel grant from Japan Geoscience Union to attend the JpGU-AGU joint meeting 2017 held at Makuhari, Chiba, Japan, and the Charles T. McCord Chair in Petroleum Geology for other research expenses.

Funding

PC was supported by the Charles T. McCord Jr. Chair in Petroleum Geology.

Author's information

PC is the Charles T. McCord Jr. Professor of Petroleum Geology and Dr. Henry V. Howe Distinguished Professor in Geology and Geophysics at Louisiana State University. He has worked on the geological evolution and erosion history of the Western Himalaya for 20 years focusing on the Indus River system. PC works both on and offshore, most recently as co-chief scientist on IODP Expedition 355 in the Arabian Sea. Prior to 2012, he was Kilgour Professor of Geology at the University of Aberdeen, Scotland.

Competing interests

The author declares that he has no competing interests.

References

Ahmad T, Harris N, Bickle M, Chapman H, Bunbury J, Prince C (2000) Isotopic constraints on the structural relationships between the Lesser Himalayan series and the high Himalayan crystalline series, Garhwal Himalaya. Geol Soc Am Bull 112(3):467–477

Ali JR, Aitchison JC (2005) Greater India. Earth Sci Rev 72:169–188

Alizai A, Carter A, Clift PD, VanLaningham S, Williams JC, Kumar R (2011) Sediment provenance, reworking and transport processes in the Indus River by U-Pb dating of detrital zircon grains. Glob Planet Chang 76:33–55. doi:10.1016/j.gloplacha.2010.11.008

Armstrong HA, Allen MB (2011) Shifts in the Intertropical Convergence Zone, Himalayan exhumation, and late Cenozoic climate. Geology 39(1):11–14. doi:10.1130/G31005.1

Balsam WL, Deaton BC (1991) Sediment dispersal in the Atlantic Ocean: evaluation by visible light spectra. Rev Aquat Sci 4:411–447

Bera MK, Sarkar A, Chakraborty PP, Loyal RS, Sanyal P (2008) Marine to continental transition in Himalayan foreland. Geol Soc Am Bull. doi:10.1130/B26265.1

Bernet M, Garver JI (2005) Fission-track analysis of detrital zircon. Rev Min Geochem 58:205–238

Bernet M, van der Beek P, Pik R, Huyghe P, Mugnier J-L, Labrin E, Szulc AG (2006) Miocene to recent exhumation of the central Himalaya determined from combined detrital zircon fission-track and U/Pb analysis of Siwalik sediments, western Nepal. Basin Res 18:393–412. doi:10.1111/j.1365-2117.2006.00303

Betzler C, Eberli GP, Kroon D, Wright JD, Swart PK, Nath BN, Alvarez-Zarikian CA, Alonso-García M, Bialik OM, Blättler CL, Guo JA, Haffen S, Horozai S, Inoue M, Jovane L, Lanci L, Laya JC, Mee ALH, Lüdmann T, Nakakuni M, Niino K, Petruny LM, Pratiwi SD, Reijmer JJG, Reolid J, Slagle AL, Sloss CR, Su X, Yao Z,

Young JR (2016) The abrupt onset of the modern South Asian monsoon winds. Sci Rep:29838. doi:10.1038/srep29838
Bignold SA, Treloar PJ, Petford N (2006) Changing sources of magma generation beneath intra-oceanic island arcs: an insight from the juvenile Kohistan island arc, Pakistan Himalaya. Chem Geol 233(1-2):46–74
Bookhagen B, Thiede RC, Strecker MR (2005) Late Quaternary intensified monsoon phases control landscape evolution in the northwest Himalaya. Geology 33(2):149–152. https://doi.org/10.1130/G20982.1
Boos WR, Kuang Z (2010) Dominant control of the South Asian monsoon by orographic insulation versus plateau heating. Nature 463:218–222. doi:10.1038/nature08707
Burbank DW (1992) Causes of recent Himalayan uplift deduced from deposited patterns in the Ganges basin. Nature 357:680–683
Burbank DW, Beck RA, Mulder T (1996) The Himalayan foreland basin. In: Yin A, Harrison TM (eds) The tectonics of Asia. Cambridge University Press, New York, pp 149–188
Burbank DW, Derry LA, France-Lanord C (1993) Reduced Himalayan sediment production 8 Myr ago despite an intensified monsoon. Nature 364:48–50
Calvès G, Clift PD, Inam A (2008) Anomalous subsidence on the rifted volcanic margin of Pakistan: no influence from Deccan plume. Earth Planet Sci Lett 272:231–239
Calvès G, Huuse M, Clift PD, Brusset S (2015) Giant fossil mass wasting off the coast of West India: the Nataraja submarine slide. Earth Planet Sci Lett 432:265–272. doi:10.1016/j.epsl.2015.10.022
Catlos EJ, Harrison TM, Kohn MJ, Grove M, Ryerson FJ, Manning CE, Upreti BN (2001) Geochronologic and thermobarometric constraints on the evolution of the main central thrust, central Nepal Himalaya. J Geophys Res 106(B8): 16,177–116,204
Célérier J, Harrison TM, Beyssac O, Herman F, Dunlap WJ, Webb AAG (2009a) The Kumaun and Garwhal Lesser Himalaya, India: part 2. Thermal and deformation histories. Geol Soc Am Bull 121(9-10):1281–1297. doi:10.1130/B26343.1
Célérier J, Harrison TM, Webb AAG, Yin A (2009b) The Kumaun and Garwhal lesser Himalaya, India: part 1. Structure and stratigraphy. Geol Soc Am Bull 121(9-10): 1262–1280. doi:10.1130/B26344.1
Chirouze F, Huyghe P, Chauvel C, van der Beek P, Bernet M, Mugnier J-L (2015) Stable drainage pattern and variable exhumation in the Western Himalaya since the Middle Miocene. J Geol 123:1–20. doi:10.1086/679305
Clemens SC, Prell WL, Sun Y (2010) Orbital-scale timing and mechanisms driving Late Pleistocene Indo-Asian summer monsoons: reinterpreting cave speleothem $\partial^{18}O$. Paleoceanography 25 (PA4207). doi:10.1029/2010PA001926
Clift P, Gaedicke C, Edwards R, Lee J II, Hildebrand P, Amjad S, White RS, Schlueter H-U (2002a) The stratigraphic evolution of the Indus fan and the history of sedimentation in the Arabian Sea. Mar Geophys Res 23(3): 223–245
Clift PD (2006) Controls on the erosion of Cenozoic Asia and the flux of clastic sediment to the ocean. Earth Planet Sci Lett 241(3-4):571–580
Clift PD, Blusztajn JS (2005) Reorganization of the western Himalayan river system after five million years ago. Nature 438(7070):1001–1003
Clift PD, Giosan L (2014) Sediment fluxes and buffering in the post-glacial Indus Basin. Basin Res 26:369–386. doi:10.1111/bre.12038
Clift PD, Hodges K, Heslop D, Hannigan R, Hoang LV, Calves G (2008) Greater Himalayan exhumation triggered by Early Miocene monsoon intensification. Nat Geosci 1:875–880. doi:10.1038/ngeo351
Clift PD, Lee JI, Hildebrand P, Shimizu N, Layne GD, Blusztajn J, Blum JD, Garzanti E, Khan AA (2002b) Nd and Pb isotope variability in the Indus River system; implications for sediment provenance and crustal heterogeneity in the western Himalaya. Earth Planet Sci Lett 200(1-2):91–106. doi:10.1016/S0012-821X(02)00620-9
Clift PD, Shimizu N, Layne G, Gaedicke C, Schlüter HU, Clark MK, Amjad S (2001) Development of the Indus fan and its significance for the erosional history of the western Himalaya and Karakoram. Geol Soc Am Bull 113:1039–1051
Clift PD, VanLaningham S (2010) A climatic trigger for a major Oligo-Miocene unconformity in the Himalayan foreland basin. Tectonics 29 (TC5014). doi:10.1029/2010TC002711
Clift PD, Wan S, Blusztajn J (2014) Reconstructing chemical weathering, physical erosion and monsoon intensity since 25 Ma in the northern South China Sea: a review of competing proxies. Earth-Sci Rev 130:86–102. doi:10.1016/j.earscirev.2014.01.002
Copeland P, Harrison TM (1990) Episodic rapid uplift in the Himalaya revealed by 40Ar/39Ar analysis of detrital K-feldspar and muscovite, Bengal fan. Geology 18:354–357. doi:10.1130/0091-7613(1990)018
Critelli S, De Rosa R, Platt JP (1990) Sandstone detrital modes in the Makran accretionary wedge, Southwest Pakistan; implications for tectonic setting and long-distance turbidite transportation. Sediment Geol 68:241–260
Curry WB, Ostermann DR, Guptha MVS, Itekkot V (1992) Foraminiferal production and monsoonal upwelling in the Arabian Sea; evidence from sediment traps. In: Summerhayes CP, Prell WL, Emeis KC (eds) Upwelling systems; evolution since the early Miocene, Special publication, vol 64. Geological Society, London, pp 93–106
Deaton BC, Balsam WL (1991) Visible spectroscopy—a rapid method for determining hematite and goethite concentration in geological materials. J Sediment Petrol 61(4):628–632
DeCelles PG, Gehrels GE, Quade J, LaReau B, Spurlin M (2000) Tectonic implications of U-Pb zircon ages of the Himalayan orogenic belt in Nepal. Science 288(5465):497–499. doi:10.1126/science.288.5465.497
DeCelles PG, Gehrels GE, Quade J, Ojha TP (1998a) Eocene-early Miocene foreland basin development and the history of Himalayan thrusting, western and central Nepal. Tectonics 17(5):741–765
DeCelles PG, Gehrels GE, Quade J, Ojha TP, Kapp PA, Upreti BN (1998b) Neogene foreland basin deposits, erosional unroofing, and the kinematic history of the Himalayan fold-thrust belt, western Nepal. Geol Soc Am Bull 110(1):2–21
DeCelles PG, Giles KA (1996) Foreland basin systems. Basin Res 8(2):105–123
DePaolo DJ, Wasserburg GJ (1976) Inferences about magma sources and mantle structure from variations of $^{143}Nd/^{144}Nd$. Geophys Res Lett 3(12):743–746
Dettman DL, Kohn MJ, Quade J, Ryerson FJ, Ojha TP, Hamidullah S (2001) Seasonal stable isotope evidence for a strong Asian monsoon throughout the past 10.7 m.y. Geology 29(1):31–34
Droz L, Bellaiche G (1991) Seismic facies and geologic evolution of the central portion of the Indus fan. In: Weimer P, Link MH (eds) Seismic facies and sedimentary processes of submarine fans and turbidite systems. Springer Verlag, Berlin, pp 383–402
Dunlap WJ, Wysoczanski R (2002) Thermal evidence for Early Cretaceous metamorphism in the Shyok suture zone and age of the Khardung volcanic rocks, Ladakh, India. J Asian Earth Sci 20(5):481–490
Fraser JE, Searle MP, Parrish RR, Noble SR (2001) Chronology of deformation, metamorphism, and magmatism in the southern Karakoram Mountains. Geol Soc Am Bull 113(11):1443–1455
Garzanti E (1993) Sedimentary evolution and drowning of a passive margin shelf (Giumal Group; Zanskar Tethys Himalaya, India): palaeoenvironmental changes during final break-up of Gondwanaland. In: Treloar PJ, Searle MP (eds) Himalayan tectonics, Special publication, vol 74. Geological Society, London, pp 277–298. doi:10.1144/GSL.SP.1993.074.01.20
Garzanti E, Baud A, Mascle G (1987) Sedimentary record of the northward flight of India and its collision with Eurasia (Ladakh Himalaya, India). Geodin Acta 1(4/5):297–312
Garzanti E, Vezzoli G, Ando S, Paparella P, Clift PD (2005) Petrology of Indus River sands; a key to interpret erosion history of the western Himalayan syntaxis. Earth Planet Sci Lett 229(3-4):287–302. doi:10.1016/j.epsl.2004.11.008
Garzione CN, Dettman DL, Quade J, DeCelles PG, Butler RF (2000) High times on the Tibetan Plateau; paleoelevation of the Thakkhola Graben, Nepal. Geology 28(4):339–342
Gébelin A, Mulch A, Teyssier C, Jessup MJ, Law RD, Brunel M (2013) The Miocene elevation of Mount Everest. Geology 41(7):799–802. doi:10.1130/G34331.1
Gehrels GE, Kapp P, DeCelles P, Pullen A, Blakely R, Weislgel A, Ding L, Guynn J, Marin A, McQuarrie N, Yin A (2011) Detrital zircon geochronology of pre-Tertiary strata in the Tibetan-Himalayan orogen. Tectonics 30 (TC5016). doi:10.1029/2011TC002868
Giosan L, Flood RD, Grutzner J, Mudie P (2002) Paleoceanographic signi¢cance of sediment color on western North Atlantic drifts: II. Late Pliocene-Pleistocene sedimentation. Mar Geol 189:43–61
Goldstein SJ, Jacobsen SB (1988) Nd and Sr isotopic systematics of river water suspended material; implications for crustal evolution. Earth Planet Sci Lett 87(3):249–265
Goodbred SL, Kuehl SA (2000) Enormous Ganges-Brahmaputra sediment discharge during strengthened early Holocene monsoon. Geology 28(12): 1083–1086
Gupta AK, Anderson DM, Overpeck JT (2003) Abrupt changes in the Asian southwest monsoon during the Holocene and their links to the North Atlantic Ocean. Nature 421:354–356
Gupta AK, Yuvaraja A, Prakasam M, Clemens SC, Velu A (2015) Evolution of the South Asian monsoon wind system since the late Middle Miocene. Palaeogeogr Palaeoclimatol Palaeoecol 438:160–167. doi:10.1016/j.palaeo.2015.08.006

Harrison TM, Célérier J, Aikman A, Hermann J, Heizler M (2009) Diffusion of 40Ar in muscovite. Geochim Cosmochim Acta 73:1039–1051

Hildebrand PR, Searle MP, Shakirullah, van Heijst HJ (2000) Geological evolution of the Hindu Kush, NW Frontier Pakistan; active margin to continent-continent collision zone. In: Khan MA, Treloar PJ, Searle MP, Jan MQ (eds) Tectonics of the Nanga Parbat syntaxis and the western Himalaya, Special publication, vol 170. Geological Society, London, pp 277–293

Hodges K (2003) Geochronology and thermochronology in orogenic systems. In: Rudnick R (ed) The crust. Elsevier-Science, Amsterdam, pp 263–292

Hodges KV, Silverberg DS (1988) Thermal evolution of the Greater Himalaya, Garhwal, India. Tectonics 7(3):583–600

Hovius N, Stark CP (2006) Landslide-driven erosion and topographic evolution of active mountain belts. In: Evans SG, Mugnozza GS, Strom A, Hermanns RL (eds) Landslides from massive rock slope failure. Springer Netherlands, Dordrecht, pp 573–590. doi:10.1007/978-1-4020-4037-5_30

Hurford AJ (1986) Cooling and uplift patterns in the Lepontine Alps, South Central Switzerland and an age of vertical movement on the Isubric fault line. Contrib Mineral Petrol 92:413–427

Inam A, Clift PD, Giosan L, Tabrez AR, Tahir M, Rabbani MM, Danish M (2007) The geographic, geological and oceanographic setting of the Indus River. In: Gupta A (ed) Large rivers: geomorphology and management. Wiley, Chichester, pp 333–345

Jain AK, Lal N, Sulemani B, Awasthi AK, Singh S, Kumar R, Kumar D (2009) Detrital-zircon fission-track ages from the lower Cenozoic sediments, NW Himalayan foreland basin: clues for exhumation and denudation of the Himalaya during the India-Asia collision. Geol Soc Am Bull 121(3/4):519–535. doi:10.1130/B26304.1

Johnson NM, Stix J, Tauxe L, Cerveny PF, Tahirkheli RAK (1985) Palaeomagnetic chronology, fluvial processes and tectonic implications of the Siwalik deposits near Chinji Village, Pakistan. J Geol 93:27–40

Khan MA, Stern RJ, Gribble RF, Windley BF (1997) Geochemical and isotopic constraints on subduction polarity, magma sources, and palaeogeography of the Kohistan intra-oceanic arc, northern Pakistan Himalaya. J Geol Soc Lond 154:935–946

Kim J, Hyeong K, Jung HS, Moon JW, Kim KH, Lee I (2006) Southward shift of the Intertropical Convergence Zone in the western Pacific during the late Tertiary: evidence from ferromanganese crusts on seamounts west of the Marshall Islands. Paleoceanography 21 (Pa4218). doi: doi:10.1029/2006pa001291

Kolla V, Coumes F (1987) Morphology, internal structure, seismic stratigraphy, and sedimentation of Indus fan. AAPG Bull 71:650–677

Kroon D, Steens T, Troelstra SR (1991) Onset of monsoonal related upwelling in the western Arabian Sea as revealed by planktonic foraminifers. In: Prell W, Niitsuma N (eds) Proceedings of the ocean drilling program, scientific results, vol 117. Ocean Drilling Program, College Station, pp 257–263

Kutzbach JE, Prell WL, Ruddiman WF (1993) Sensitivity of Eurasian climate to surface uplift of the Tibetan Plateau. J Geol 101:177–190

Lee JI, Clift PD, Layne G, Blum J, Khan AA (2003) Sediment flux in the modern Indus River traced by the trace element composition of detrital amphibole grains. Sediment Geol 160:243–257. doi:10.1016/S0037-0738(02)00378-0

Licht A, van Cappelle M, Abels HA, Ladant J-B, Trabucho-Alexandre J, France-Lanord C, Donnadieu Y, Vandenberghe J, Rigaudier T, Lecuyer C, Terry D, Adriaens R, Boura A, Guo Z, Soe AN, Quade J, Dupont-Nivet G, Jaeger J-J (2014) Asian monsoons in a late Eocene greenhouse world. Nature 513:501–506. doi:10.1038/nature13704

Liu F, Li J (2015) Landslide erosion associated with the Wenchuan earthquake in the Minjiang River watershed: implication for landscape evolution of the Longmen Shan, eastern Tibetan Plateau. Nat Hazards 76(3):1911–1926. https://doi.org/10.1007/s11069-014-1575-8

Lyle MW, Wilson PA, Janecek TR, Shipboard party (2002) Leg 199 summary. In: Lyle MW, Wilson P, Janecek TR (eds) Proceedings of the ocean drilling program, initial reports, vol 199. Ocean Drilling Program, College Station, pp 1–89

Malavieille J (2010) Impact of erosion, sedimentation, and structural heritage on the structure and kinematics of orogenic wedges: analog models and case studies. GSA Today 20(1). doi:10.1130/GSATG48A.1

Meigs AJ, Burbank DW, Beck RA (1995) Middle-late Miocene (>10 Ma) formation of the main boundary thrust in the western Himalaya. Geology 23(5):423–426

Miles PR, Roest WR (1993) Earliest seafloor spreading magnetic anomalies in the north Arabian Sea and the ocean-continent transition. Geophys J Int 115:1025–1031

Molnar P, England P, Martinod J (1993) Mantle dynamics, uplift of the Tibetan Plateau, and the Indian monsoon. Rev Geophys 31(4):357–396

Montgomery DR, Stolar DB (2006) Reconsidering Himalayan river anticlines. Geomorphology 82:4–15. doi:10.1016/j.geomorph.2005.08.021

Naini BR, Kolla V (1982) Acoustic character and thickness of sediments of the Indus fan and the continental margin of western India. Mar Geol 47:181–195

Najman Y (2006) The detrital record of orogenesis: a review of approaches and techniques used in the Himalayan sedimentary basins. Earth-Sci Rev 74(1-2):1–72

Najman Y, Bickle M, Garzanti E, Pringle M, Barfod D, Brozovic N, Burbank D, Ando S (2009) Reconstructing the exhumation history of the Lesser Himalaya, NW India, from a multitechnique provenance study of the foreland basin Siwalik Group. Tectonics 28 (TC5018). doi:10.1029/2009TC002506

Najman Y, Garzanti E (2000) Reconstructing early Himalayan tectonic evolution and paleogeography from Tertiary foreland basin sedimentary rocks, northern India. Geol Soc Am Bull 112(3):435–449

Najman Y, Pringle M, Godin L, Oliver G (2001) Dating of the oldest continental sediments from the Himalayan foreland basin. Nature (London) 410(6825): 194–197

Najman YMR, Pringle MS, Johnson MRW, Robertson AHF, Wijbrans JR (1997) Laser 40Ar/39Ar dating of single detrital muscovite grains from early foreland-basin sedimentary deposits in India; implications for early Himalayan evolution. Geology 25(6):535–538

Ojha TP, Butler RF, Quade J, DeCelles PG, Richards D, Upreti BN (2000) Magnetic polarity stratigraphy of the Neogene Siwalik Group at Khutia Khola, far western Nepal. Geol Soc Am Bull 112(3):424–434

Pan F-B, Zhang H-F, Harris N, W-C X, Guo L (2012) Oligocene magmatism in the eastern margin of the east Himalayan syntaxis and its implication for the India-Asia post-collisional process. Lithos 154:181–192. doi:10.1016/j.lithos.2012.07.004

Pandey DK, Clift PD, Kulhanek DK, Expedition 355 Scientists (2015) Arabian Sea Monsoon. Int Ocean Disc Prog Prelim Rpt 355. doi:10.2204/iodp.pr.355.2015

Phillips RJ, Parrish RR, Searle MP (2004) Age constraints on ductile deformation and long-term slip rates along the Karakoram fault zone, Ladakh. Earth Planet Sci Lett 226:305–319

Prell WL, Murray DW, Clemens SC, Anderson DM (1992) Evolution and variability of the Indian Ocean summer monsoon: evidence from the western Arabian Sea drilling program. In: Duncan RA, Rea DK, Kidd RB, von Rad U, Weissel JK (eds) Synthesis of results from scientific drilling in the Indian Ocean, Geophysical monograph, vol 70. American Geophysical Union, Washington, DC, pp 447–469

Quade J, Cerling TE (1995) Expansion of C4 grasses in the late Miocene of northern Pakistan: evidence from stable isotopes in Paleosols. Palaeogeogr Palaeoclimatol Palaeoecol 115(1-4):91–116

Quade J, Cerling TE, Bowman JR (1989) Development of Asian monsoon revealed by marked ecological shift during the latest Miocene in northern Pakistan. Nature 342(6246):163–166

Raiverman V, Kunte SV, Mukherjea A (1983) Basin geometry, Cenozoic sedimentation and hydrocarbon prospects in northwestern Himalaya and Indo-Gangetic plains. Petrol Asia J 6:67–92

Raiverman V, Raman KS (1971) Facies relations in the Subathu sediments, Simla Hills, N.W. Himalaya, India. Geol Mag 108:329–341

Rao AR (1993) Magnetic-polarity stratigraphy of upper Siwalik of north-western Himalayan foothills. Curr Sci 64(11/12):863–873

Replumaz A, Negredo AM, Villaseñor A, Guillot S (2010) Indian continental subduction and slab break-off during tertiary collision. Terra Nova 22:290–296. doi:10.1111/j.1365-3121.2010.00945.x

Robertson AHF, Collins AS (2002) Shyok suture zone, N Pakistan; late Mesozoic-Tertiary evolution of a critical suture separating the oceanic Ladakh Arc from the Asian continental margin. J Asian Earth Sci 20(3):309–351

Robertson AHF, Degnan PJ (1994) The Dras arc complex: lithofacies and reconstruction of a late cretaceous oceanic volcanic arc in the Indus suture zone, Ladakh Himlaya. Sediment Geol 92:117–145

Roddaz M, Said A, Guillot S, Antoine PO, Montel JM, Martin F, Darrozes J (2011) Provenance of Cenozoic sedimentary rocks from the Sulaiman fold and thrust belt, Pakistan: implications for the palaeogeography of the Indus drainage system. J Geol Soc Lond 168:499–516. doi:10.1144/0016-76492010-100

Rowley DB, Currie BS (2006) Palaeo-altimetry of the late Eocene to Miocene Lunpola basin, central Tibet. Nature 439:677–681

Schärer U, Copeland P, Harrison TM, Searle MP (1990) Age, cooling history, and origin of post-collisional leucogranites in the Karakoram Batholith; a multi-system isotope study. J Geol 98(2):233–251

Schwertmann U (1971) Transformation of hematite to goethite in soils. Nature 232:624–625

Searle MP (1996) Cooling history, erosion, exhumation and kinematics of the Himalaya-Karakoram-Tibet orogenic belt. In: Yin A, Harrison TM (eds) The tectonic evolution of Asia. Cambridge University Press, Cambridge, pp 110–137

Searle MP, Fryer BJ (1986) Garnet, tourmaline and muscovite-bearing leucogranites, gneisses and migmatites of the higher Himalayas from Zanskar, Kulu, Lahoul and Kashmir. In: Coward MP, Ries AC (eds), Special Publications, vol 19. Geological Society, London, pp 185-201 doi:10.1144/GSL.SP.1986.019.01.10

Searle MP, Law RD, Jessup M, Simpson RL (2006) Crustal structure and evolution of the greater Himalaya in Nepal – South Tibet: implications for channel flow and ductile extrusion of the middle crust. In: Law RD, Searle MP, Godin L (eds) Channel flow, ductile extrusion and exhumation in continental collision zones, Special publication, vol 268. Geological Society, London, pp 355–378

Searle MP, Rex AJ, Tirrul R, Rex DC, Barnicoat A, Windley BF (1989) Metamorphic, magmatic and tectonic evolution of the central Karakoram in the Biafo-Baltoro-Hushe regions of north Pakistan. Geol Soc Am Spec Pap 232:47–73

Shuaib SM, Tariq Shuaib SM (1999) Geology and oil/Gass presence in the offshore Indus Basin of Pakistan. In: Meadows A, Meadows PS (eds) The Indus River: biodiversity, resources, humankind. Oxford University Press, Karachi, pp 249–265

Simpson G (2004) Role of river incision in enhancing deformation. Geology 32(4): 341–345. https://doi.org/10.1130/G20190.2

Singh S, Parkash B, Awasthi AK, Kumar S (2011) Late Miocene record of palaeovegetation from Siwalik palaeosols of the Ramnagar sub-basin, India. Curr Sci 100(2):213–222

Sinha RN (1970) Heavy mineral investigation of the Siwaliks of Mohand, district Saharanpur, Uttar Pradesh. J Geol Soc India 11:163–177

Stephenson BJ, Searle MP, Waters DJ, Rex DC (2001) Structure of the main central thrust zone and extrusion of the high Himalayan deep crustal wedge, Kishtwar-Zanskar Himalaya. J Geol Soc Lond 158(4):637–652

Szulc AG, Najman Y, Sinclair HD, Pringle M, Bickle M, Chapman H, Garzanti E, Ando S, Huyghe P, Mugnier J-L, Ojha T, DeCelles PG (2006) Tectonic evolution of the Himalaya constrained by detrital 40Ar/39Ar, Sm/Nd and petrographic data from the Siwalik foreland basin succession, SW Nepal. Basin Res 18(4):375–391

Tagami T, Ito H, Nishimura S (1990) Thermal annealing characteristics of spontaneous fi ssion tracks in zircon. Chem Geol 80:159–169

Tauxe L, Opdyke ND (1982) A time framework based on magnetostratigraphy of the Siwalik sediments of the Khaur arae, northern Pakistan. Palaeogeogr Palaeoclimatol Palaeoecol 37:43–61

Treloar PJ, Rex DC, Guise PG, Wheeler J, Hurford AJ, Carter A (2000) Geochronological constraints on the evolution of the Nanga Parbat syntaxis, Pakistan Himalaya. In: Khan MA, Treloar PJ, Searle MP, Jan MQ (eds) Tectonics of the Nanga Parbat syntaxis and the western Himalaya, Geological society special publication, vol 170. Geological Society, London, pp 137–162

Vermeesch P (2012) On the visualisation of detrital age distributions. Chem Geol 312–313:190–194. doi:10.1016/j.chemgeo.2012.04.021

Vidal P, Cocherie A, Le Fort P (1982) Geochemical investigations of the origin of the Manaslu leucogranite (Himalaya, Nepal). Geochim Cosmochim Acta 46: 2279–2292

Walker J, Martin MW, Bowring SA, Searle M, Waters DJ, Hodges K (1999) Metamorphism, melting, and extension: age constraints from the high Himalayan slab of southeast Zanskar and northwest Lahaul. J Geol 107:473–495

Wan S, Clift PD, Li A, Yu Z, Li T, Hu D (2012) Tectonic and climatic controls on long-term silicate weathering in Asia since 5 Ma. Geophys Res Lett 39(L15611). doi:10.1029/2012GL052377

Wan S, Kürschner WM, Clift PD, Li A, Li T (2009) Extreme weathering/erosion during the Miocene climatic optimum: evidence from sediment record in the South China Sea. Geophys Res Lett 36(L19706). doi:10.1029/2009GL040279

Webb AAG (2013) Preliminary palinspastic reconstruction of Cenozoic deformation across the Himachal Himalaya (northwestern India). Geosphere 9:572–587

Webb AAG, Guo H, Clift PD, Husson L, Müller T, Costantino D, Yin A, Xu Z, Cao H, Wang Q (2017) The Himalaya in 3D: slab dynamics controlled mountain building and monsoon intensification. Lithosphere. doi:10.1130/L636.1

West AJ, Galy A, Bickle MJ (2005) Tectonic and climatic controls on silicate weathering. Earth Planet Sci Lett 235:211–228. doi:10.1016/j.epsl.2005.03.020

Whipple KX (2009) The influence of climate on the tectonic evolution of mountain belts. Nat Geosci 2:1–8. doi:10.1038/ngeo413

White NM, Pringle M, Garzanti E, Bickle M, Najman Y, Chapman H, Friend P (2002) Constraints on the exhumation and erosion of the high Himalayan slab, NW India, from foreland basin deposits. Earth Planet Sci Lett 195:29–44

Whitmarsh RB, Weser OE, Ross DA, Shipboard Scientific Party (1974) Site 221. Initial reports of the Deep Sea drilling project, vol 23, pp 167–210

Whittington A, Foster G, Harris N, Vance D, Ayres M (1999) Lithostratigraphic correlations in the western Himalaya—an isotopic approach. Geology 27(7):585–588

Willett SD (1999) Orogeny and orography: the effects of erosion on the structure of mountain belts. J Geophys Res 104:28957–28981

You Y, Huber M, Müller RD, Poulsen CJ, Ribbe J (2009) Simulation of the Middle Miocene climate optimum. Geophys Res Lett 36(4). doi:10.1029/2008GL036571

Zachos J, Pagani M, Sloan L, Thomas E, Billups K (2001) Trends, rythms and abberations in global climate 65 Ma to present. Science 292:686–693

Zeitler PK, Sutter JF, Williams IS, Zartman RE, Tahirkheli RAK (1989) Geochronology and temperature history of the Nanga Parbat-Haramosh Massif, Pakistan. In: Malinconico LL, Lillie RJ (eds) Tectonics of the western Himalayas, Special Paper, vol 232. Geological Society of America, Boulder, pp 1–22

Zhang L-Y, Ducea MN, Ding L, Pullen A, Kapp P, Hoffman D (2014) Southern Tibetan Oligocene-Miocene adakites: a record of Indian slab tearing. Lithos 210-211:209–223. doi:10.1016/j.lithos.2014.09.029

Zhang Q, Willems H, Ding L, Gräfe K-U, Appel E (2012) Initial India-Asia continental collision and Foreland Basin evolution in the Tethyan Himalaya of Tibet: evidence from Stratigraphy and Paleontology. J Geol 120(2):175–189

Zhang YG, Jia J, Balsam WL, Liu L, Chen J (2007) High resolution hematite and goethite records from ODP 1143, South China Sea: co-evolution of monsoonal precipitation and El Niño over the past 600,000 years. Earth Planet Sci Lett 264(1-2):136–150

Zhuang G, Najman Y, Guillot S, Roddaz M, Antoine P-O, Métaise G, Carter A, Marivaux L, Solangig SH (2015) Constraints on the collision and the pre-collision tectonic configuration between India and Asia from detrital geochronology, thermochronology, and geochemistry studies in the lower Indus basin, Pakistan. Earth Planet Sci Lett 432:363–373. doi:10.1016/j.epsl.2015.10.026

4

Reconstruction of sediment-transport pathways on a modern microtidal coast by a new grain-size trend analysis method

Shota Yamashita[1], Hajime Naruse[2*] and Takeshi Nakajo[3]

Abstract

A new method for granulometric-parameter-based reconstruction of sediment-transport pathways is proposed and is termed P-GSTA (grain-size trend analysis using principal component analysis) herein. The main advantage of this method is its applicability to depositional environments involving mixed transport processes, for instance, fluvial, tidal, and wave-influenced environments. In the P-GSTA method, a linear function with all significant granulometric parameters that are summed with different weighting factors was used to infer sediment-transport direction (sediment flux pattern); the previous grain-size trend analysis (GSTA) methods considered only three parameters (mean grain size, sorting, and skewness) with equal weighting. This study chose six parameters (namely, median grain size, coefficient of variation, skewness, kurtosis, and mud and gravel log-ratios) for calculation. First, the zero values of mud and gravel fractions are replaced, and their log-ratios are defined. Then, all values are standardized. Thereafter, principal component analysis (PCA) is conducted to determine the weighting factor of each granulometric parameter. Each principal component is then interpreted, and the function best representing a sediment flux pattern is chosen from these components. Trend vectors are calculated, solely on the basis of a map interpolated from the scores of the chosen principal component, as the two-dimensional gradient of this value. The P-GSTA method proposed in this study was applied to a modern microtidal coast (tidal sand flat along the Kushida River Delta, central Japan). Sediment-transport pathways reconstructed by this method were consistent with observed sediment-transport patterns determined by field experiments using tracer sediments and geomorphologic observation; the results of the previous GSTA method were inconsistent with the observations. The proposed method also revealed additional minor depositional processes on the sand flat, namely, the deposition of fluvial-channel lags and muddy particles. Thus, this study demonstrates that the proposed P-GSTA method is a potentially powerful tool to reconstruct sediment-transport patterns even under mixed transport processes, where the estimation of the sediment-transport function is difficult.

Keywords: GSTA, Microtidal, Sediment grain size, Sediment transport, Tidal flat

Introduction

Depositional systems are consequences of sediment dispersal from their provenances. Progressive and selective transport of grains differentiates lithofacies deposited in different sedimentary environments (Swift et al. 2003). In particular, in coastal environments, sediment supply and spatial changes in dominant sediment transport processes (wave, fluvial, or tidal currents) play a major role in producing a unique pattern of shoreline geomorphology, offshore bathymetry, and sediment distribution patterns, in addition to the geomorphological geometry of the depositional system and the type of the depositional environment (Tanner 1962; Liu et al. 2000). The size-selective transport process of sediments results in downcurrent variation of the sediment grain-size distribution. Therefore, spatial variation in the grain size distribution of bottom sediments can be an important clue in reconstruction of sediment-transport pathways, which is important in coastal engineering and environmental protection.

* Correspondence: naruse@kueps.kyoto-u.ac.jp
[2]Division of Earth and Planetary Sciences, Graduate School of Science, Kyoto University, Kitashirakawa Oiwakecho, Sakyo-ku, Kyoto 606-8502, Japan
Full list of author information is available at the end of the article

Many researchers have attempted to identify spatial variations of granulometric parameters, which have been referred to as 'grain-size trends,' associated with net sediment-transport pathways (McCave 1978; McLaren 1981; Le Roux and Rojas 2007). This is because methods to reconstruct sediment-transport pathways from granulometric patterns are inexpensive and easily available compared to direct measurements of sediment movement. In addition, there is a possibility that ancient sediment-transport patterns can be reconstructed by applying these methods to subsurface sediments. Early studies suggested that sediments become finer in the direction of transport (e.g., Pettijohn and Ridge 1932; Self 1977), while other investigations have indicated the opposite trend (e.g., McCave 1978; Nordstrom 1981, 1989). In order to address this disparity, McLaren and Bowles (1985) discussed the combination of three parameters (mean grain size, sorting, and skewness) on the basis of flume experiments and statistical calculations. They concluded that sediments always become better sorted during transport, whereas the mean grain size and skewness vary with specific combination patterns (FB–: finer mean grain size, better sorting value, and greater negative skewness; or CB+: coarser mean grain size, better sorting value, and greater positive skewness).

Gao and Collins (1992) applied their theory to two-dimensional data of grain-size distribution in an inner bay environment. In their method, after comparing three granulometric parameters (mean, sorting, and skewness values) of adjacent points, vectors of unit length (100 m) are drawn between two points if they satisfy the conditions (FB– or CB+). These vectors are calculated from every point with respect to the eight immediately neighboring grid points. Summing the vectors at each sample point produces a single-trend vector of sediment transport pathways. This approach is called grain-size trend analysis (GSTA) and has been applied in a variety of depositional environments (e.g., Pedreros et al. 1996; Cheng et al. 2004).

It has been reported, however, that there may be serious discrepancies between the GSTA model and actual sediment-transport patterns (Asselman 1999; Masselink et al. 2008). This is probably because grain-size trends do not always follow the assumptions of McLaren and Bowles (1985), especially under mixed transport processes. Grain-size trends due to sediment transport can vary due to the depositional processes, environments, and sediment provenances. Deviation from the presumed trend of even a single parameter is sufficient to result in the GSTA not accurately reflecting the trend vectors. Indeed, Flemming (1988) disputed whether sorting value would always improve during transport. In addition, it seems problematic that the GSTA model considers only three equally weighted parameters, but completely ignores other parameters representing grain-size distribution, such as kurtosis and mud/gravel fractions. In order to address these problems, Rojas et al. (2000), Rojas (2003), and Rojas and Le Roux (2003) proposed a model that combines four grain-size parameters (mean grain size, sorting, skewness, and kurtosis) into a linear granulometric facies function with differential weighting. These weighting factors are modified incrementally and iteratively for comparison with actual sediment-transport directions measured from bedforms or visual observations, so that the error between the observed vector and the generated vector is minimized. This method can determine the optimum weights of granulometric parameters, but requires detailed observational data of existing sediment-transport patterns, and thus vitiates the abovementioned advantages of grain-size trend analysis.

To this end, the present study employs principal component analysis (PCA) (Davis 1986; Middleton 2000) for the purpose of treating multiple grain-size parameters with appropriate weightings. PCA is a multivariate technique for explaining the correlation between explanatory variables and automatically organizing these parameters into a few linear synthetic variables with different weights. It is reasonable to regard this spatial variation of grain-size parameters as resulting from the sediment-transport processes. Therefore, it can be expected that parameters that strongly reflect sediment transport are emphasized by PCA. Hereinafter, this new method of GSTA employing PCA is referred to as P-GSTA.

Previous GSTA methods also include problems in validation of the methods, as sediment-transport trends vary according to the time scale of the observation. Sediment-transport pathways of daily processes such as fair-weather waves can counteract those of monthly or yearly transport processes such as storms or slope failure events. Thus, the time scale of the field observation must be specified in order to validate the method for reconstructing sediment-transport pathways. The time scale of sediment-transport patterns estimated by any GSTA method is defined by the sampling interval (Gao and Collins 1992). Subsurface sediments in shallower sampling intervals can be entrained and transported by waves or currents, whereas deeper layers are "historical" sediments that were already fixed and are not entrained by the surface processes. Nevertheless, this problem has been given little attention.

To reconstruct the ongoing sediment-transport patterns, a sampling depth should be assigned to the thickness of the active layer of the bottom-surface sediments. The active layer is a concept proposed by Hirano (1971), indicating the uppermost layer of the bottom-surface sediments, which are frequently reworked and deposited by currents and wave processes. Thus, grain-size variation of bottom-surface sediments accompanied by sediment transport occurs within this layer (Hirano 1971). It should be noted that the thickness of the active layer varies with time under non-uniform flow conditions in natural environments; it generally becomes thicker in accordance with the time development of the probability distribution of burial and reworking as functions of depth below the bed surface (Ribberink 1987; Armanini 1995; Parker et al. 2000).

A microtidal sand flat along the Kushida River Delta, Ise Bay, central Japan, was chosen as the study area. This microtidal sand flat is a largely natural environment, lacking breakwaters or dams, and is therefore suitable for observation of sediment transport. The active-layer thickness and net sediment-transport directions over a specific time scale were then measured by tracer experiments and geomorphological observation. Finally, sampling for the GSTA method (the proposed P-GSTA and previous GSTA) was conducted according to the active-layer thickness, and the results were validated with net sediment-transport patterns recorded by field measurements.

The proposed P-GSTA method successfully reproduced the actual sediment-transport patterns, whereas the results of the previous GSTA model were inconsistent. In addition, application of the PCA model revealed details of previously unrecognized depositional processes on the sand flat. To summarize, this method is flexible because it requires no assumption on the sediment transportation processes and the influence of sediment transportation is estimated only from the characteristics of the obtained data. Thus, the method is effective under mixed transport processes, in which estimation of the sediment-transport function is difficult.

Methods/Experimental

Granulometric parameters used for P-GSTA method

On the basis of grain-size measurements of samples, each of the granulometric parameters (mean grain size μ, sorting σ, skewness S_k, and kurtosis k_t) is generated by the moment method (Folk 1966; Harrington 1967), as follows:

$$\mu = \sum pd \tag{1}$$

$$\sigma = \sqrt{\frac{\sum p(d-\mu)^2}{100}} \tag{2}$$

$$S_k = \frac{\sum p(d-\mu)^3}{100\sigma^3} \tag{3}$$

$$k_t = \frac{\sum p(d-\mu)^4}{100\sigma^4} \tag{4}$$

where d is the representative value of each grain-size class (in increments of 0.1ϕ) and p is the weight fraction. Physical characteristics of mud and gravel are evaluated as weight percentages (mud and gravel fractions: F_{mud} and F_{gravel}, respectively). The coefficient of variation (CV) is also employed to represent the degree of sediment sorting. CV is the dimensionless sorting value divided by mean value, calculated as follows:

$$\mathrm{CV} = \frac{\sigma}{\mu} \tag{5}$$

The P-GSTA method employs six parameters (median grain size d_{50}, CV, S_k, k_t, and fractions of mud and gravel F_{mud} and F_{gravel}) in a sediment-transport function. Here, the median grain size d_{50} is employed rather than mean grain size, because median grain size is mathematically independent of the other granulometric parameters, whereas mean grain size is used in the calculation of CV, S_k, and k_t. In addition, CV value is employed here instead of sorting value (standard deviation) σ. Compared to sorting value σ, CV value has the advantage of avoiding mathematical correlation with mean value. When comparing grain-size distributions whose mean values are nearly identical, sorting values are a good index of the dispersion of the grain-size distribution. However, sorting values cannot be used when comparing grain-size distributions with quite different means because sorting value (standard deviations) has an inevitable positive correlation with the mean value of the grain size distributions under consideration. In such cases, CV rather than sorting value needs to be used in general sediment evaluation.

Parametric interpolation by kriging

Kriging interpolation is employed to estimate spatial variations of granulometric parameters (Burgess and Webster 1980a, 1980b; Kohsaka 1999). All GSTA methods have a problem in that the estimated trend vectors are potentially affected by the sampling intervals. Asselman (1999) solved this problem using kriging to interpolate observational data sets (Burgess and Webster 1980a, 1980b; Kohsaka 1999). Kriging is an algorithm based on least squares, which is used to estimate the spatial variation of a real-valued function, and assumes that spatial variation can be estimated from a linear combination of measured values (Kohsaka 1999). Weighting coefficients are obtained on the basis of the spatial dependence of a variable, which can be represented by a semivariogram, i.e., the scatter diagram of covariance with respect to spatial distance. The weighting of the running average is then determined by the variogram model function, which is the function fitted to the semivariogram (Burgess and Webster 1980a). Theoretically, the covariance of spatial data increases with distance and becomes a steady value over some distance (Burgess and Webster 1980a, 1980b; Kohsaka 1999). This limited distance is the range within which the data indicate spatial dependence. The spatial dependencies of the data are also shown in the variances between the estimations, which provide a measure of uncertainty in the interpolation values.

Data preparation

Prior to statistical analysis, granulometric parameters should be converted into appropriate forms for analysis by conventional statistical techniques.

It is common to report grain-size fractions in percentages, but such data are not independent, because they must sum to 100%. Compositional data such as mud and gravel fractions are therefore difficult to use in statistical analyses as they are (Aitchison 1986, 2003; Aitchison and Egozcue 2005). Aitchison (1986, 2003) proposed, instead, to evaluate the relative magnitudes of constituents in comparison to other compositional variables and to use logarithms to simplify the computation of the variance and covariance. After taking logarithms, the resulting transformed values are free to vary over the entire range of real numbers. However, a zero value in the compositional data prohibits log-ratio transformation, because the transformed value is $-\infty$, and therefore cannot be included in the statistical analysis. In order to avoid this problem, a multiplicative approach (Martín-Fernández et al. 2003) replaces a zero value with a value that is sufficiently small, as follows:

$$\hat{F}_j = \begin{cases} \delta_j & (F_j = 0) \\ \left(1 - \dfrac{\sum_{k|F_k=0} \delta_k}{c}\right) F_j & (F_j > 0) \end{cases} \tag{6}$$

where F_j and $\hat{F}_j$ denote the original and substitute compositional values respectively; δ_j, the input value of F_j; and c, the sum-constraint constant (100%). The modification of nonzero values is multiplicative, as suggested by the simple replacement rule. Generally, δ_j is the maximum rounding error, but the range $\delta_y/5 \le \delta_j \le 2\delta_y$, where δ_y is the maximum rounding error, has been suggested to be reasonable (Aitchison 1986; Martín-Fernández et al. 2003). This nonparametric replacement is stable, i.e., independent of both data numbers and component numbers, and retains the original information of the compositional data (Aitchison 1986; Arai and Ohta 2006). In the present study, mud/gravel fractions were treated via these two transformational processes. First, zero values of mud/gravel fractions were replaced by the maximum rounding error (0.01%). Secondly, the dimensionless magnitudes of the mud and gravel fractions were defined as the mud and gravel log-ratios r_{mud} and r_{gravel}, as follows:

$$r_{\text{mud}} = \log\left(\frac{\hat{F}_{\text{mud}}}{100 - \hat{F}_{\text{mud}} - \hat{F}_{\text{gravel}}}\right) \tag{7}$$

$$r_{\text{gravel}} = \log\left(\frac{\hat{F}_{\text{gravel}}}{100 - \hat{F}_{\text{mud}} - \hat{F}_{\text{gravel}}}\right) \tag{8}$$

where $\hat{F}_{\text{mud}}$ and $\hat{F}_{\text{gravel}}$ are the substitute values for mud and gravel fractions (Eq. 6), respectively.

The units, mean values, and standard deviation of variables strongly influence the results of PCA. To avoid this problem, the following formula was used to standardize all data:

$$x_i^* = \frac{x_i - m_x}{s_x} \tag{9}$$

where m_x is mean and s_x is the standard deviation of each granulometric parameter x. This operation transforms the mean and standard deviation of each variable into 1 and 0, respectively.

Weighting (principal component analysis)

The new method for defining the linear granulometric function proposed in this study is based on principal component analysis (PCA), which is a technique for explaining the correlation between multiple variables and accounting for a large variance using only a few synthetic variables (principal components) (Davis 1986; Middleton 2000).

Let $x_1, x_2, \cdots, x_p$ be variables, and let $z_1, z_2, \cdots, z_m$ $(m \le p)$ be their respective weighted synthetic variables. These synthetic variables are defined as the principal components.

$$\begin{cases} z_1 = L_{11}x_1^* + L_{12}x_2^* + \ldots + L_{1p}x_p^* \\ \quad \vdots \\ z_i = L_{i1}x_1^* + L_{i2}x_2^* + \ldots + L_{ip}x_p^* \\ \quad \vdots \\ z_m = L_{m1}x_1^* + L_{m2}x_2^* + \ldots + L_{mp}x_p^* \end{cases}$$
$$L_{i1}^2 + L_{i2}^2 + \ldots + L_{ip}^2 = 1, \quad (i = 1, 2, \ldots, m) \tag{10}$$

Here, the matrix $(L_{i1}...L_{ip})$ is the eigenvector of the ith principal component. The m synthetic variables have the following two properties: (1) the correlation of any two synthetic variables is zero (orthogonal) and (2) the following inequality holds for the variances of synthetic variables $\text{Var}(z_i)$: $\text{Var}(z_1) \ge \text{Var}(z_2) \ge \ldots \text{Var}(z_m)$. The amplitude of the variance of principal components can be regarded as the amount of information in each principal component. The factor loading, which represents the relative importance of each variable in the principal component, is defined as the product of the root of the eigenvalue and the eigenvector. The absolute value of the factor loading is proportional to its significance.

Calculation of trend vectors of sediment transport by P-GSTA method

The procedures to obtain trend vectors of sediment transport by GSTA with principal component analysis (PCA) weighting (P-GSTA) are described herein. As discussed above, this method employs six granulometric parameters (median grain size, CV, skewness, kurtosis, and mud and gravel fractions). First, the zero values of mud and gravel fractions are replaced (Eq. 6), and mud and gravel log-ratios are defined (Eqs. 7 and 8). Then, all values are standardized (Eq. 9). After these preliminary procedures, PCA is conducted using these six transformed parameters (Eq. 10). Each principal component is then interpreted, and the first principal component is chosen as the sediment-transport function $T(x, y)$. The trend vectors $V(x, y)$ are then calculated, solely on the basis of the interpolated map of the scores of the chosen principal component as the two-dimensional gradient of this value, as follows:

$$V(x,y) = \frac{\partial T}{\partial x}\boldsymbol{i} + \frac{\partial T}{\partial y}\boldsymbol{j} \tag{11}$$

where $V(x, y)$ is an estimated trend vector of sediment transport, and $\boldsymbol{i}$ and $\boldsymbol{j}$ are unit vectors (magnitude: 100 m) in the x and y directions, respectively.

Application of P-GSTA method to microtidal sand flat

We applied our method to a microtidal sand flat located at the mouth of the Kushida River. The headwaters of the Kushida River are in the Daikou Mountains of Japan; the river flows eastward to the lowlands and then northward, to form a delta at Ise Bay (Fig. 1). The catchment area of the river is approximately 461 km^2. According to the Japanese Ministry of Land, Infrastructure, and Transportation (MLIT), the annual average fluvial discharge was 11.56 m^3/s in 2010 (http://www1.river.go.jp/). The delta is lobate or arcuate and is classified as fluvial-dominated or complex fluvial-wave-tide dominated on the basis of the classification of Galloway (1975).

The tidal range monitored on the left bank of the Kushida River (36° 36′ 0″ N, 136° 33′ 45″ E) is approximately 2 m during a spring tide (microtidal). Wave heights have been measured at the river mouth (34° 22′ 16″ N, 137° 07′ 40″ E) and in the enclosed section of Ise Bay (34° 55′ 12″ N, 136° 44′ 25″ E) (MLIT). Measurements by MLIT in 2002 at the river mouth showed that waves occur with a period of 5 to 10 s and are 1–2 m in height, increasing to 4–5 m in height during typhoons and winter storms (http://www.mlit.go.jp/kowan/nowphas/index.html). In the investigated area, waves mainly wash from the northeast throughout the year, but also occur from the northwest or north–northwest during winter seasons (Nakajo 2004). Tidal residual current is mainly from northwest to southeast in the southwest part

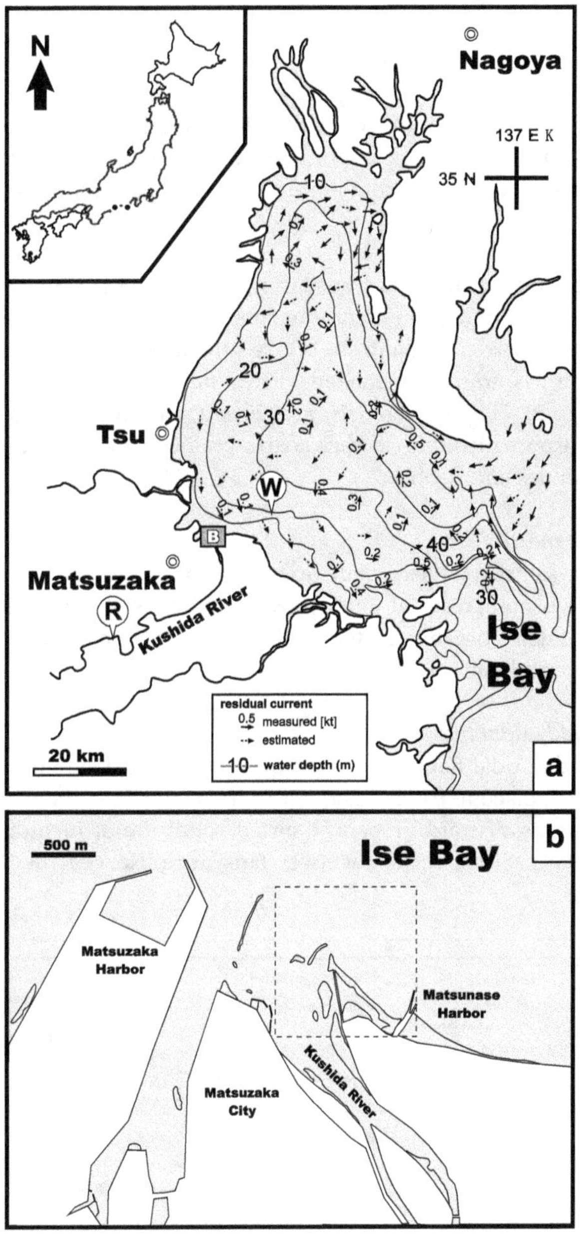

Fig. 1 Index map showing locality of the study area. The data of the residual current velocity (in **a**) are from the Coastal Oceanography Research Committee, The Oceanographic Society of Japan (1985). 'R' and 'W' (in **a**) are measurement localities of rainfall and water depth, respectively. Data are from the Japan Meteorological Agency (JMA) (see Fig. 10; http://www.data.jma.go.jp/obd/stats/etrn/index.php). Details of the geomorphological features of the study area (dotted line in **b**) are described in Fig. 2

of Ise Bay (Coastal Oceanography Research Committee, The Oceanographic Society of Japan 1985; Fig. 1). Typhoons struck this area less than five times per year from 2005 to 2011 (measured by the Japan Meteorological Agency [JMA]).

The geomorphology of the Kushida River Delta is classified into four components, namely, spit, salt marsh, muddy tidal flat, and tidal sand flat (Fig. 2).

Spit

The spit extends northwest from Matsunase Harbor and branches into several components. The spit is composed of well-sorted medium- to coarse-grained sand. Branching channels from the Kushida River flow behind the spit (Fig. 2). These channels are 20–30 cm in depth and transport medium- to very coarse-grained sand, including granules and pebbles.

Salt marsh

The salt marsh lies mainly on the landward side of the spit and is composed of muddy sediment and soil containing abundant organic material (Fig. 2). Sediment coarsens toward the main fluvial channel (Kushida River).

Muddy tidal flat

Muddy tidal flats are located on the landward side of the spit, adjacent to the salt marsh (Fig. 2). The sediment consists of mud or poorly sorted sandy mud, including gravels. Small-scale washover fans are observed on the muddy flat. Burrows of crustaceans are very common, and mud-flat sediment and coarse sediment of washover fans are intermixed by their benthic activities.

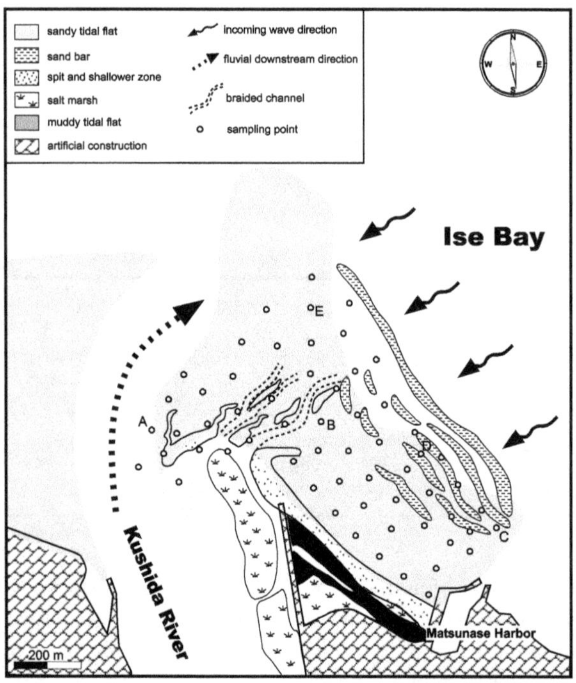

Fig. 2 Geomorphology of the study area and sampling localities for GSTA methods. Geomorphology is after fluvial flooding in July 2011

Tidal sand flat

The microtidal sand flat extends in front of the spit, on the right bank of the river mouth (Fig. 2). Its maximum area, during ebb spring tide, is approximately 0.4 km^2 (Fig. 2). The offshoreward development of the intertidal sand flat is most prominent around the river mouth and the southeast part of the study area (adjacent to *Matsunase Harbor*) (Fig. 2). The primary geomorphological features of the sand flat consist of sand bars and braided channels, and no tidal channels are observed. The sand bars are 60 to 100 cm in height and 50 to 100 m in wavelength (Fig. 2). They are arranged almost parallel to the shoreline and normal to the direction of incoming waves (Fig. 2). Breaking waves are found on the crests of these sand bars, and wave ripples are abundant on their surface. No significant migration of these bars was observed in the period from 2002 to 2011 (Nakajo 2004; Yamashita et al. 2009). Braided channels (approximately 20–30 cm deep) are observed intermittently on the sand flat, around the river mouth. These channels are temporary braided streams from the river. They are generally produced by floods and rapidly migrate and vanish. There are very few muddy particles on the sand flat, except for soon after a fluvial flooding event (Nakajo 2004; Yamashita et al. 2009; Yamashita et al. 2011). Sediment transport on the sand flat is mainly controlled by fluvial and wave activities (Yamashita et al. 2009).

Sampling and grain-size measurements

Bottom-surface sediments on the microtidal sand flat were collected during the ebb spring tides on 1 August 2011. A total of 60 sampling sites were located at horizontal intervals of approximately 100 m, both parallel and perpendicular to the spit, with extension to the river mouth and the northern margin of the sand flat (Fig. 2). Sediments were excavated by shovel in intervals, from the top of the bedforms to 8 cm below the surface, corresponding to the active layer thickness observed on 31 July 2011. Each sample (50–100 g weight) was homogenized before grain-size analysis.

The grain-size distribution of sand-sized particles in the samples was measured by the settling tube method (Gibbs 1974; Tucker 1988) after removing mud-sized ($> 4\phi$) and gravel-sized ($< -1\phi$) particles by sieving. Shell fragments were also removed before measurement. Grain-size distributions were calculated using S-Tube software (Friedman et al. 1992; Naruse 2005). The conversion method for the settling velocity of grains was proposed by Ferguson and Church (2004), using an empirical formula for natural sand.

Model validation by monitoring of actual sediment-transport patterns

Actual sediment-transport patterns on the tidal sand flat were monitored from 16 May to 31 July 2011, to enable validation of the GSTA models. Therefore, the sediment-transport patterns were examined in both normal (calm) conditions and in an episodic depositional event. The thickness of the active layer was also measured; the active layer was considered to be the mixed layer of tracer grains and natural sands.

The JMA reported rainfall measured in the upper reaches of the Kushida River and the water level of the river mouth (http://www.data.jma.go.jp/obd/stats/etrn/index.php; Fig. 3). There was moderate rainfall (< 15 mm/h) during the period from 16 May to 3 June. The water level indicates a usual spring-neap-spring cycle in this period (Fig. 3). In contrast, heavy rainfall was recorded on 8 July (> 45 mm/h) and 17–19 July (maximum: > 25 mm/h). In particular, unusually high water levels during ebb tides were monitored soon after the 17–19 July rainfall (18–20 July) (Fig. 3). The rainfall during this period was due to a typhoon (No. 6, MA-ON). Fluvial flooding events caused by the rainfalls during this period were reported through an enquiry survey.

Tracer experiments and geomorphological observations revealed net sediment-transport patterns on the tidal sand flat. Blue-colored sand was employed as tracer sediment, as it can be clearly identified from natural sand, both visually and by microscopic observations. The tracer sediments were well sorted and their mean grain size was ca. 1.5 ϕ. The tracer sediments were situated at five localities (points 1–5) on 16 May 2011 (spring tide) (Fig. 4). GPS and marking poles recorded the localities. At each locality, the bottom surface sediment was excavated approximately 0.3 (length) × 0.3 (breadth) × 0.2 (depth) m, and the hole was filled by the tracer sediments (approx. 25 kg). Two weeks after input of the tracer sediments (3 June 2011), bottom surface sediments were collected from four surrounding points that were located at a distance of 10 m from each tracer pit. Each sample was homogenized with sufficient stirring by hand in a transparent plastic case (8 × 15 cm basal area, and 3 cm in height). The basal sediment in this plastic case was scanned at high resolution (2400 dpi), and the number of colored sand grains in the basal area was counted to estimate the volume of surface sediments transported in each direction from the original tracer site. Sampling was also conducted 2 months after the tracer sediments were set (31 July), but no tracer grains were detected.

Mean transport direction of sediments $\overline{\theta}$ was estimated from observed colored sand grains, as follows:

$$\overline{\theta} = \arctan\frac{\sum \sin\theta_i}{\sum \cos\theta_i} \tag{12}$$

where θ_i is a direction from the tracer pit to the sampling locality at which the ith grain was detected. The mean resultant length (MRL) L_{MR} of the vector was employed as a proxy for concentration of sediment-transport directions, which was calculated as follows:

$$L_{MR} = \frac{\sqrt{\left(\sum \cos\theta_i\right)^2 + \left(\sum \sin\theta_i\right)^2}}{n} \tag{13}$$

where n is the number of detected colored grains. Furthermore, the thickness of the active layer of basal

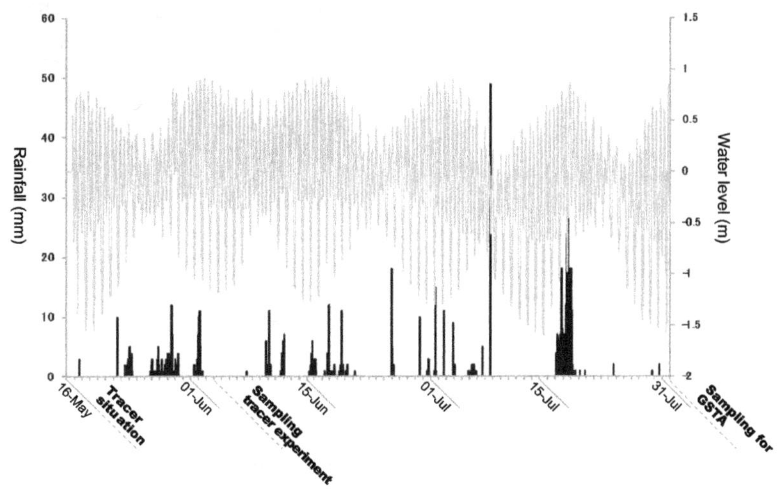

Fig. 3 The rainfall and water level of the Kushida River. The rainfall and the water level were measured respectively at the upper reaches ('R' in Fig. 1a) and at the river mouth ('W' in Fig. 1b) during a period from May 16 to July 31, 2011 (http://www.data.jma.go.jp/obd/stats/etrn/index.php)

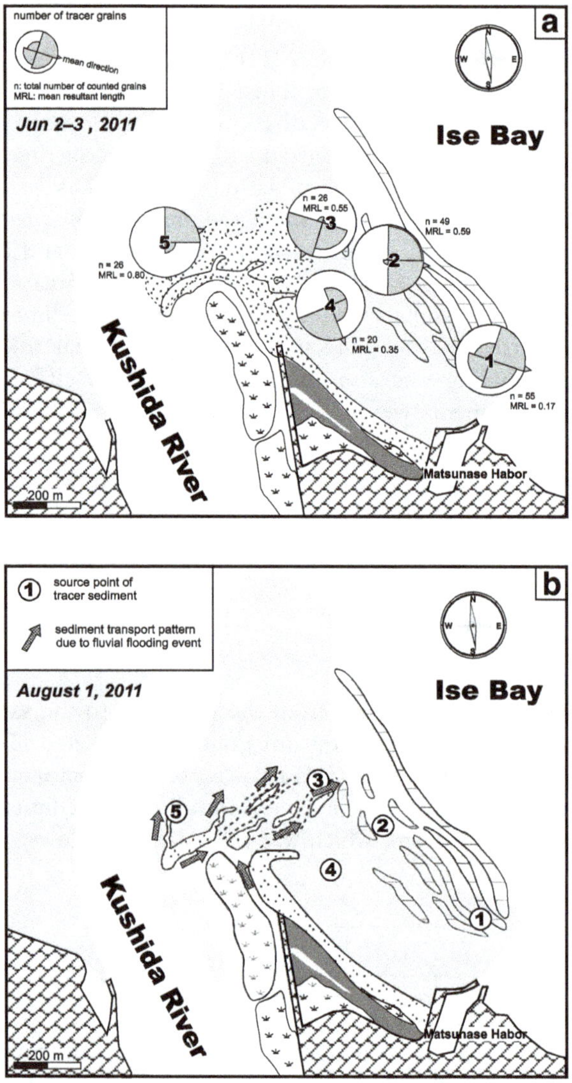

Fig. 4 **a** The result of the tracer experiment. Sampling was conducted on Jun 2–3, 2011. **b** The drastic geomorphological changes due to a fluvial flooding event in July 2011, and sediment-transport patterns suggested by the directions of channels and bedforms. Data was obtained on August 1, 2011

sediment was measured after excavating sediments at the source points (Fig. 5).

The sediment-transport patterns were also estimated on the basis of the observation of geomorphological development (Figs. 4 and 5) during the period from 16 May to 31 July 2011.

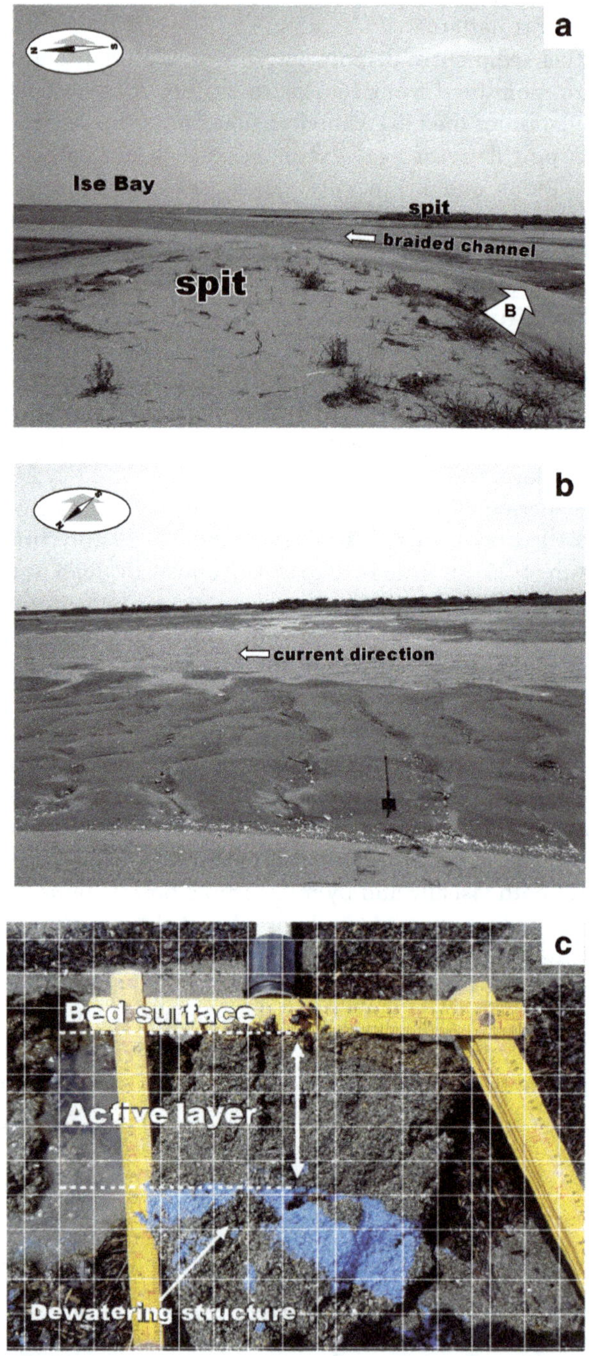

Fig. 5 **a** A photograph of a braided channel formed by the fluvial flooding event in July 2011. The braided channel cuts the spit and flows into the tidal sand flat. **b** Dunes in the braided channel. The left side of the photo is seaward. **c** The appearance of the active layer observed at point 3 soon after the fluvial flooding (August 1, 2011)

Results

Results of grain-size analysis

Spatial distributions of eight granulometric parameters (mean and median grain size, sorting, CV, skewness, kurtosis, and mud and gravel fractions) were calculated and were interpolated to a grid interval of 100 m by kriging (Figs. 6 and 7). Semivariograms were calculated from the measured data, and the variogram models were fitted to them by the least-squares method (Fig. 6). The estimation error is shown as a false-color image, and the

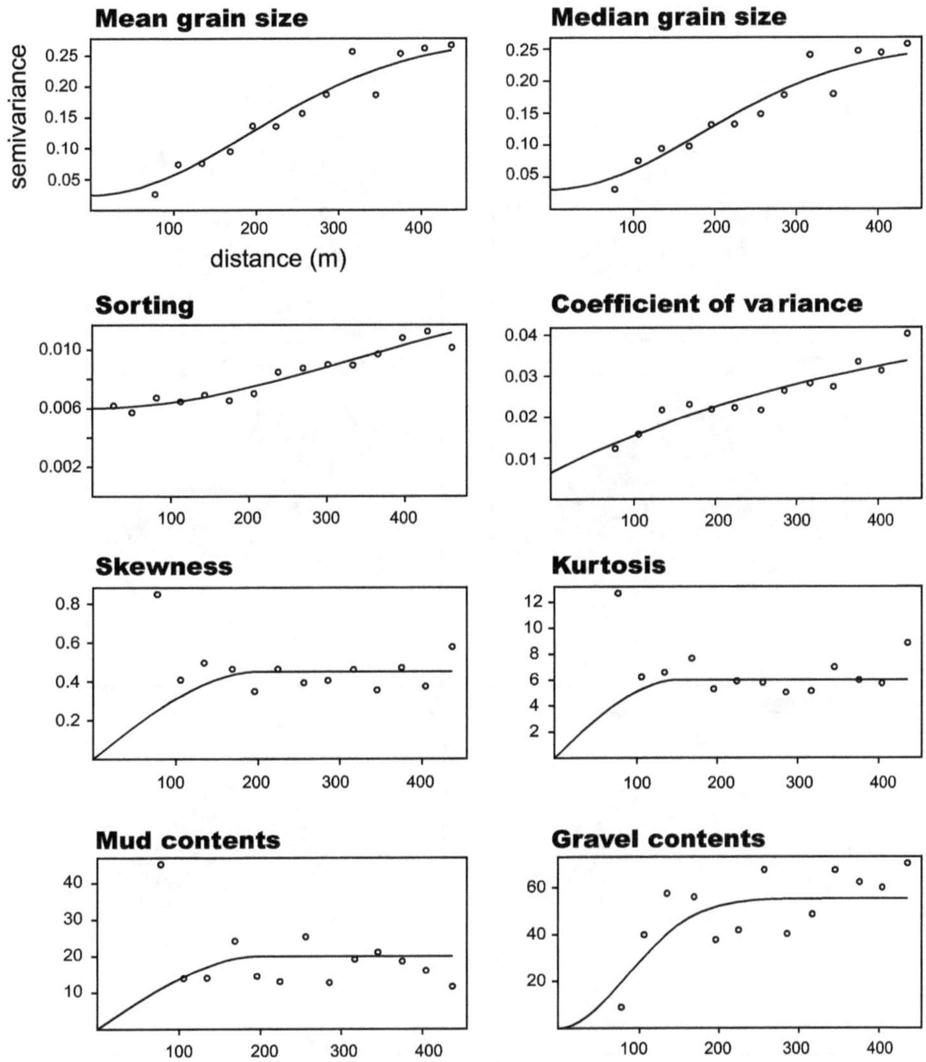

Fig. 6 Semivariograms and fitted variogram models of granulometric parameters. Variogram models were fitted by the least-squares method, and the model that produces the fewest residual errors was chosen from among the Gaussian, spherical, and exponential models. The spherical model was used for mean and median grain-size, sorting, and gravel fractions, the exponential model was used for the CV value, and the Gaussian model was used for skewness, kurtosis, and mud fractions

value of the parameter is shown by an overlying contour map (Fig. 7).

Mean and median grain size

The interpolated maps of the mean and median grain size show similar characteristics. The coarser sediments (< 1.0 ϕ) are distributed around the river mouth (Fig. 7). The grain size becomes finer in a radial pattern from the river mouth, and grain sizes finer than 1.50ϕ are found at the southeast part of the sand flat (Fig. 7). The variograms of the mean and median grain size reveal their high spatial dependency (Fig. 6). Therefore, they have relatively high accuracy (Fig. 7).

Sorting

The highest sorting values (> 0.56) are found in the central to southeast part of the sand flat (Fig. 7). The sorting value tends to decrease seaward, but a lower value (< 0.55) is also evident in the river mouth (Fig. 7). The semivariogram of the sorting value suggests relatively high spatial dependence (Fig. 6). However, the *y* intercept of the fitted curve for semivariance of the sorting value largely deviates from the origin (nugget effect), indicating that high measurement error exists (Fig. 6).

CV value

The map of the CV value indicates that higher values (> 0.7) are distributed in the river mouth. The

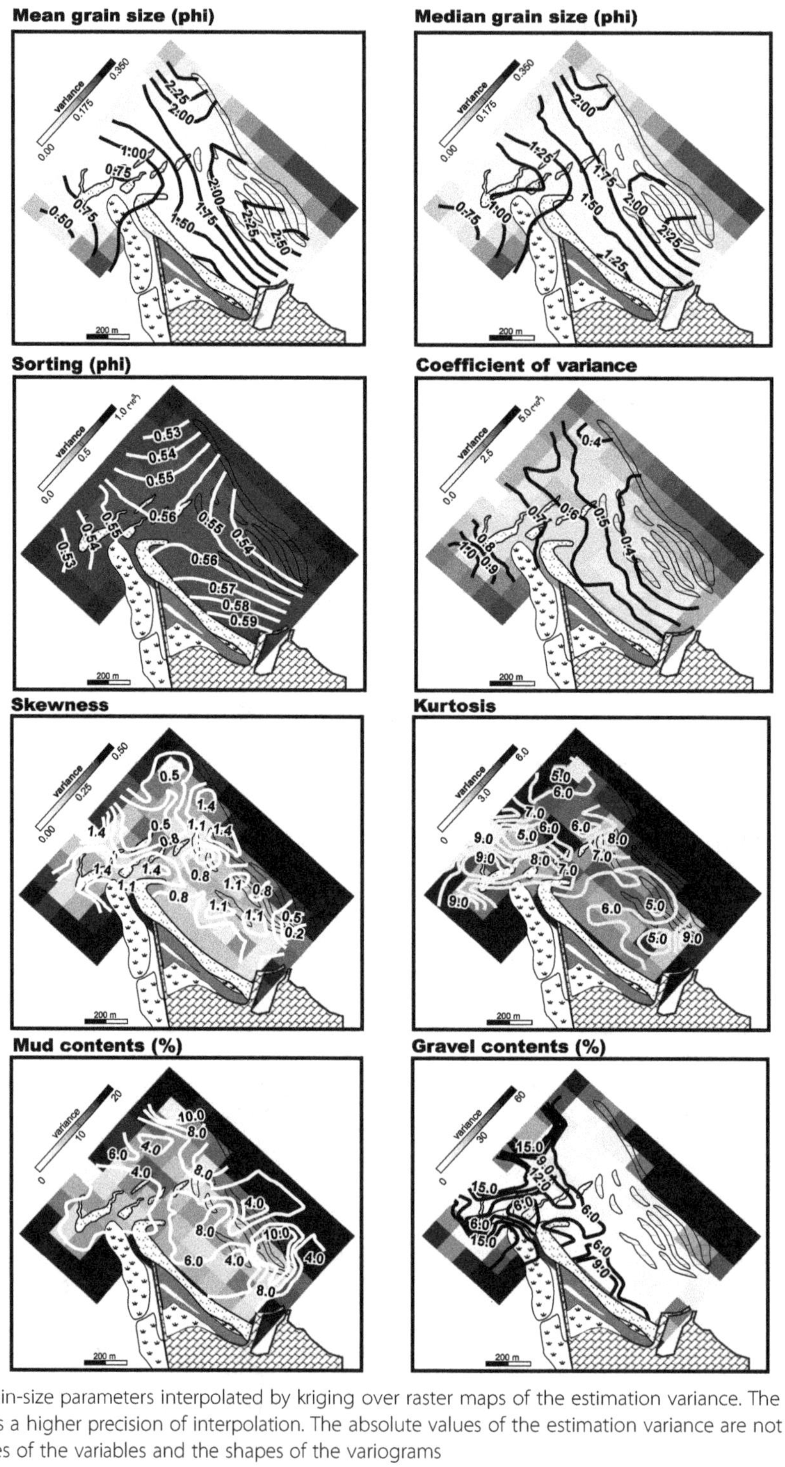

Fig. 7 Contour maps of grain-size parameters interpolated by kriging over raster maps of the estimation variance. The relatively lighter gray color in each raster map indicates a higher precision of interpolation. The absolute values of the estimation variance are not meaningful because these values depend on the values of the variables and the shapes of the variograms

CV value decreases seaward (Fig. 7). The variogram of the CV value shows high spatial dependency (Fig. 6); therefore, the accuracy of the CV value is relatively high.

Skewness

The higher skewness values (> 1.4) are mainly distributed in the river mouth and are also found in the northern part of the sand flat (Fig. 7). Skewness tends to decrease to the southeast (Fig. 7). The lowest value (< 0.2) is distributed primarily in the southeastern margin of the sand flat (Fig. 7). The accuracy of interpolation is low because of the anomalous value of the semivariogram in the range of 100 m (Fig. 6).

Kurtosis

The higher kurtosis values are mainly distributed in the river mouth (7.0–9.0), and the value tends to decrease in a seaward direction (Fig. 7). However, high values are also found in the southeastern margin of the sand flat. The accuracy of interpolation is relatively low because the variogram indicates anomalous values in the range of 100 m (Fig. 6).

Mud fractions

The spatial distribution of the mud fractions is quite random, and therefore, the accuracy of interpolation is very low (Fig. 6). Relatively high mud fractions (> 10.0%) are distributed in patches in the northeastern and southeastern parts of the sand flat (Fig. 7).

Gravel fractions

Gravels are abundant in the sediment around the river mouth (> 9.0%) and decrease seaward, becoming sparse in the southeast part of the sand flat (< 6.0%) (Fig. 7). The variogram shows high spatial dependency of the gravel fractions (Fig. 6); therefore, the accuracy of this value is relatively high (Fig. 7).

Results of principal component analysis of granulometric parameters

The contribution values, which indicate the percentages of the original information represented by each principal component of PC1 through PC3, are 56.77, 20.32, and 11.76, respectively (Table 1). This means that more than 80% of the original information is represented by PC1 through PC3. Therefore, these three principal components are interpreted further. Kriging interpolation was conducted in order to visualize the spatial distribution patterns of scores of PC1 through PC3, using the same method as for the original granulometric parameters (Figs. 8 and 9).

Table 1 Result of principle component analysis (PCA) for the samples from the Kushida River Delta. Factor loadings and contribution values are shown

	PC1	PC2	PC3	PC4	PC5	PC6
Median grain size	0.93	− 0.23	− 0.04	− 0.14	− 0.15	0.20
CV	− 0.92	0.23	0.16	− 0.05	0.21	0.17
Skewness	− 0.57	− 0.65	0.33	0.38	− 0.11	0.03
Kurtosis	− 0.58	− 0.66	− 0.02	− 0.47	0.01	− 0.04
Mud log-ratio	0.65	0.09	0.74	− 0.15	0.06	− 0.04
Gravel log-ratio	− 0.79	0.50	0.16	− 0.15	− 0.30	0.00
Contribution (%)	56.77	20.32	11.76	7.15	2.78	1.23

PC1: sediment-transport function

PC1 is negatively correlated with the gravel log-ratio (factor loading: − 0.79) and CV value (− 0.92) and is positively correlated with the median grain size (ϕ) (0.93) (Table 1). Correlations with skewness (− 0.57), kurtosis (− 0.58), and mud log-ratio (0.65) have relatively weak influences on this value (Table 1). In other words, the increase of the PC1 value indicates that the sediment becomes finer, better sorted, and less gravely (Fig. 10); this means a selective decrease in coarser materials from grain-size distribution (Fig. 10). The variogram and interpolated map of the PC1 scores indicate that the spatial variation of PC1 is high (Figs. 8 and 9). The lowest value of PC1 is found around the river mouth, and this value increases radially northeastward (Fig. 9). The scatter diagram of PC1 values against PC2 values indicates that the samples take wider values of PC1 (from − 4.5 to 4.0) than values of PC2 (from − 3.0 to 2.0); no notable clusters were formed (Fig. 11). Thus, it is interpreted that PC1 reflects the suite of size-selective sediment-transport processes resulting in fining-downcurrent trends with well-sorted grain-size distributions. The gradual decrease of coarser materials, which is represented by the increase of PC1, probably reflects the progressive lowering of bed shear stress of both fluvial and wave activities (Fig. 10). Settling velocity of finer materials is low, so that they have greater mobility as suspended loads. Therefore, the finest fraction of sediments is broadly dispersed downcurrent, producing the negative skewness of grain-size distribution. The relatively high spatial variation and the radial distribution pattern from the river mouth in PC1 strongly suggest that the PC1 value represents the sediment supply from the Kushida River and the dispersal by waves or tidal currents (Fig. 6). Therefore, the decreasing pattern of PC1 can be regarded as the sediment-transport function.

PC2: indicator of lag deposit of a fluvial flooding event

PC2 is negatively correlated to the skewness (− 0.65) and kurtosis value (− 0.66), and positively influenced by gravel log-ratio (0.55) (Table 1). Other granulometric parameters have only slight influences on PC2 (Table 1). In other words, the increase of the PC2 value indicates the more negatively skewed and platykurtic shape of grain-size distribution and gravely sediment (Fig. 10). The higher values of PC2 (1.2–2.0) cluster around the breached section of the spit and the newly formed braided channel on the sand flat (Fig. 9). This value decreases around the river mouth and offshore area (< 4.0). The spatial dependency and precision of interpolation is low because of the anomalous values of *semivariance* at approximately 100 m (Fig. 8). The exclusive distribution of higher PC2 values

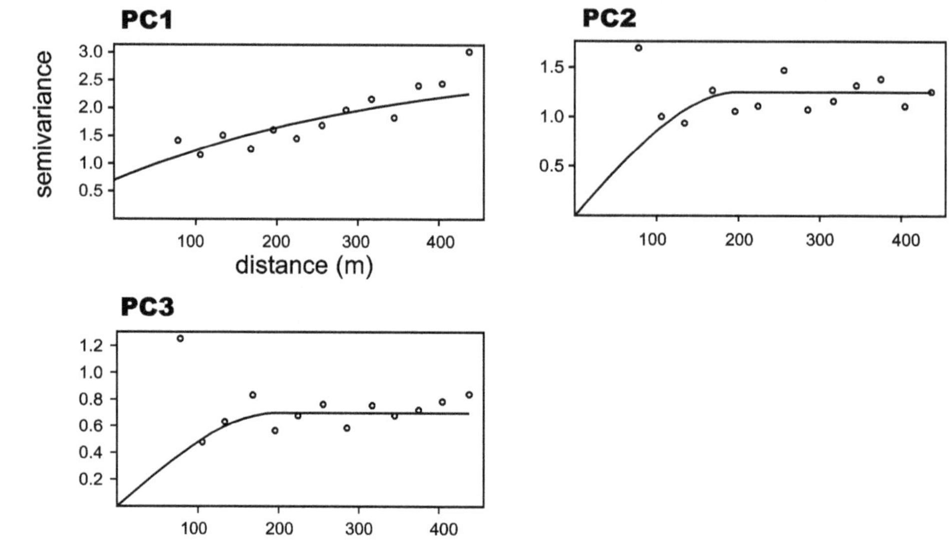

Fig. 8 Semivariograms and fitted variogram models of principal components (PC1–PC3). The variogram models were fitted by the least-squares method, and the model that produces the fewest residual errors was chosen from among the Gaussian, spherical, and exponential models. The exponential model was used for PC1 and the spherical model for PC2 and PC3

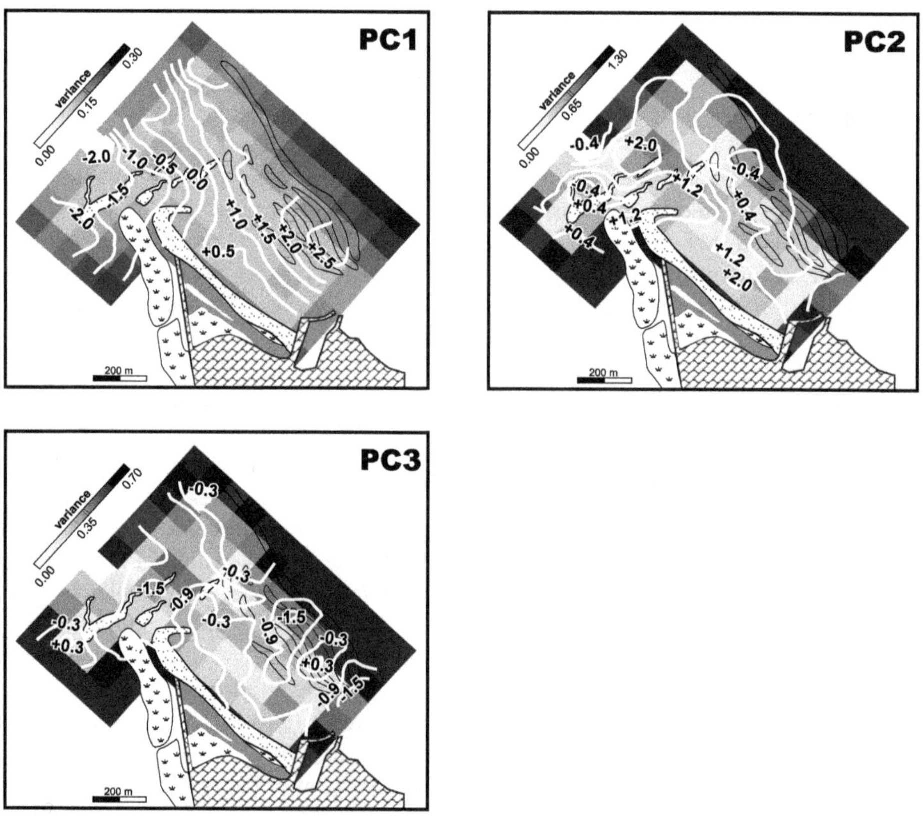

Fig. 9 Contour maps of the principal components interpolated by kriging over raster maps of estimation variance

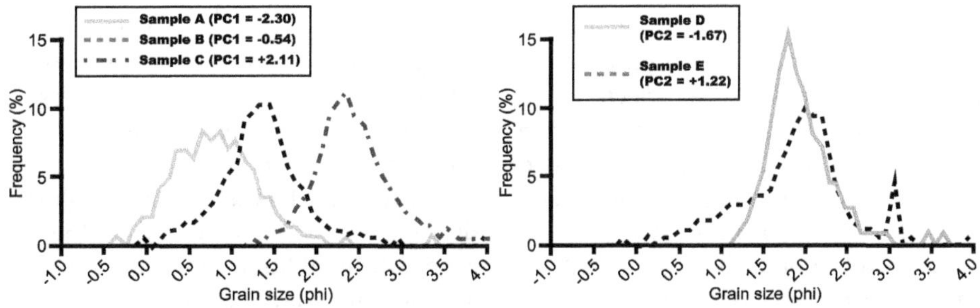

Fig. 10 Typical transformation pattern of the grain-size distribution curves represented by PC1 (left) and PC3 (right). Sampling locations of samples A–E are shown in Fig. 2. The values of PC1 through PC3 for each sample (A–E) are shown in the scatter diagram of Fig. 11

around the newly formed braided channel suggests a linkage to the episodic deposition by the fluvial flooding event during July 2011. The platykurtic and negatively skewed shape of grain-size distribution as well as the high gravel fraction indicates that the sediments largely consist of coarser fractions and this is in agreement with the direct observation of a high gravel fraction (Fig. 10). These coarser fractions can be interpreted as lag deposits in braided channels. Coarser sediments may have been episodically deposited during periods of high shear stress flow at times of fluvial flooding and remained through the subsequent reworking by wave or tidal currents.

PC3: indicator of muddy particles

PC3 is positively influenced by the mud log-ratio (0.74), and other granulometric parameters show limited influences (from – 0.02 to 0.33) (Table 1). This means that the increase in the PC3 score suggests muddier sediment. The variogram and interpolated map of PC3 show that the spatial dependence of PC3 is weak (Figs. 8 and 9). The spatial distribution of PC3 is very heterogeneous. Higher PC3 values (> – 0.3) are found in the northern and southeastern parts of the sand flat, and this characteristic distribution pattern is similar to the map of the mud content (Figs. 7 and 9). The range of variation in PC3 values (from – 2.0 to 2.5) is more limited than for PC1 (from – 4.5 to 4.0) (Fig. 11). Very poor correlation between mud log-ratio and other granulometric parameters of sand-sized particles indicates that muddy particles are deposited by quite different processes from those associated with sandy particles. As described above, the rapid deposition of muddy particles on the sand flat by a fluvial flood event (17–19 July 2011) is recognized, and these muddy particles preferentially remained in local depressions. The heterogeneity of the spatial distribution of PC3 might reflect a relic of mud deposition caused by this event and their local resistance to erosion.

Trend vectors obtained by P-GSTA and previous GSTA methods

Grain-size trend vectors were calculated by the P-GSTA method, in which the increase of the PC1 value is

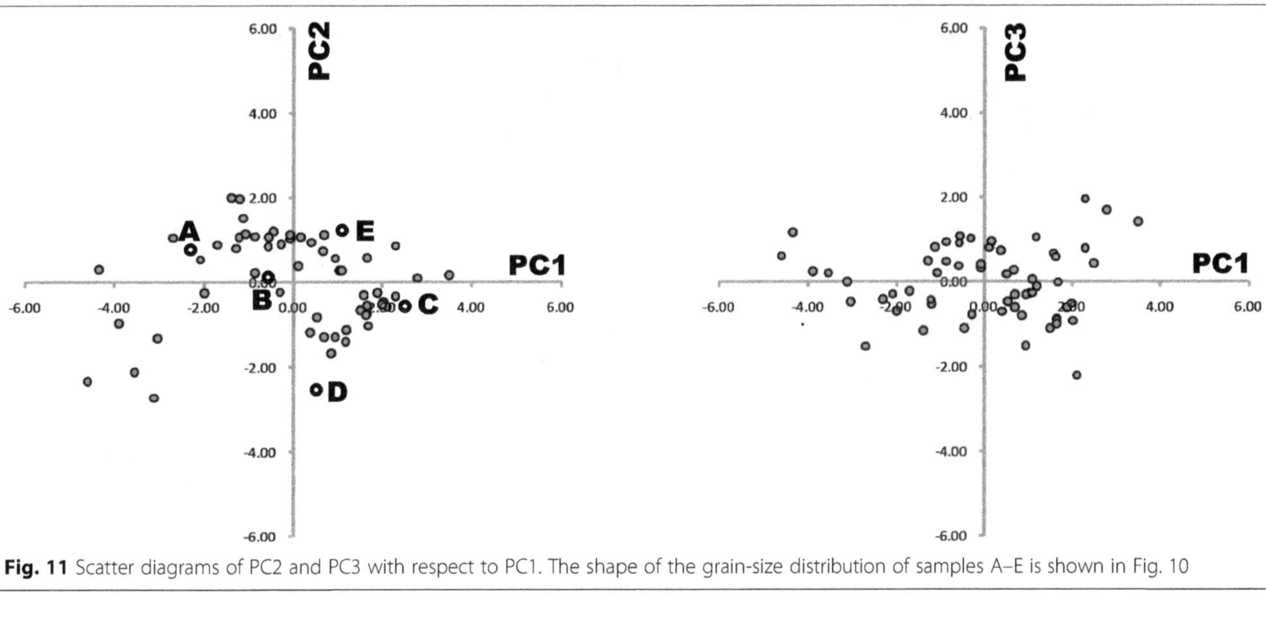

Fig. 11 Scatter diagrams of PC2 and PC3 with respect to PC1. The shape of the grain-size distribution of samples A–E is shown in Fig. 10

defined as a sediment-transport function (Fig. 12). Trend vectors directed to the north or northeast are dominant along the main fluvial channel and northern parts of the sand flat (Fig. 12). Northeastward-directed vectors are dominant around the breached section of the spit. In the central part of the sand flat, the vectors are directed to the southeast. In the southeastern regions, the trend vectors are directed toward the northeast (Fig. 12). To summarize, the trend vectors obtained by the P-GSTA method show radial dispersion of coastal sediment from the river mouth with a small influence from the shore-parallel current.

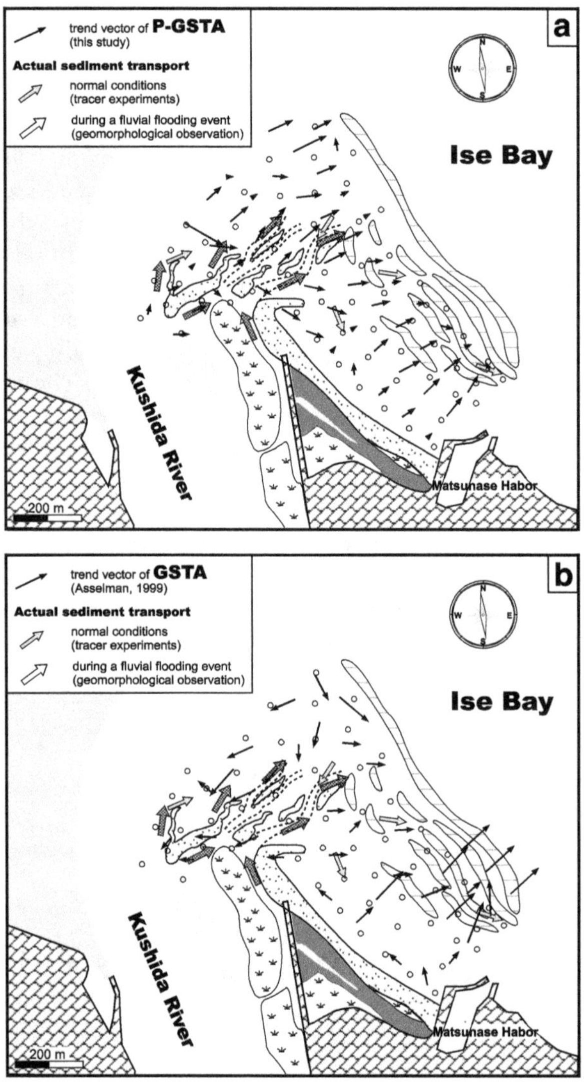

Fig. 12 Grain-size trend vectors. The actual sediment-transport directions during normal conditions are shown as the mean direction of tracer results (see Fig. 4a). Circles indicate the sampling points. **a** Grain-size trend vectors calculated by P-GSTA. **b** Grain-size trend vectors calculated by the previous GSTA method (Asselman 1999). Some of sampling points lack the trend vector because the GSTA method does not recognize the grain-size trend when no surrounded locations satisfy the conditions FB– or CB+

The previous GSTA model (Gao and Collins 1992; Asselman 1999) was also applied for the purposes of comparison. Trend vectors obtained by the previous GSTA method showed quite diverse and inconsistent estimates of sediment transport pathways (Fig. 12). Landward-directed vectors were dominant along the main fluvial channel, while offshore (southeastward) directed vectors were found at the central part of the sand flat. In addition, transport vectors were not defined around the southeastern region of the study area because neither FB– nor CB+ patterns were satisfied in this region.

Results of tracer experiments and topographic observations

The tracer experiments clarified the actual sediment-transport patterns during the period 16 May to 3 June (Fig. 4). Along the main fluvial channel, the transport pattern was directed to the northeast (point 5) (Fig. 7) and the degree of concentration (MRL) was relatively high (Fig. 7). In the central part of the sand flat (points 3 and 4), southward (landward) transport was dominant (Fig. 4) and the degree of concentration (MRL) was moderate (Fig. 4). In the offshore area (point 2), eastward sediment transport was dominant (Fig. 4) and the degree of concentration was moderate (Fig. 4). At the southeastern margin of the sand flat (point 1), sediment dispersed in various directions (Fig. 4); northeastward transport is dominant, according to the number of tracer grains, but the mean direction indicates east–southeastward transport (Fig. 4).

On 31 July 2011, no tracer grains were detected from the samples, probably because they had been highly dispersed. Instead, sediment-transport patterns were estimated on the basis of the observation of the geomorphological development derived from a fluvial flooding event of 17–19 July (Fig. 3). The most prominent feature was incision of the spit and formation of braided channels (Figs. 4 and 5). The spit was breached at two points around its folding point, 30–50 m in width (Figs. 4 and 5). At the breach locations, the spit was incised by shallow, braided channels (0.3–0.5 m in depth), which were connected to branched channels behind the spit (Figs. 4 and 5). These channels flowed into the tidal sand flat more than 300 m offshore (Figs. 4 and 5). Sand bodies were deposited on the sand flat 100 m off the breach point (Figs. 4 and 5). These sand bodies extended approximately 100 m northeast and were approximately 0.5 m in height (Fig. 4). They were composed of medium- to coarse-grained sands, including shell fragments, which were similar to sediments comprising the spit and the fluvial channels, and were clearly coarser than the sediments on the tidal sand flat. This probably means that these sand bodies were formed by sediments derived from the breached part of the spit and from sediments in the fluvial channels. Dunes of approximately 1.0 m wavelength and 0.15 m amplitude formed in the braided channels (Fig. 5). These dunes exhibited the northeastward (offshoreward) current directions around points 3 and 5 (Figs. 4 and 5). At the eastern part of the sand flat (around points 1 and 2), muddy deposits were thickly deposited in the troughs between sand bars. The average thickness of the muddy sediment was 1–3 cm, but it occasionally reached 5 cm. No other notable geomorphological developments were observed in this area.

The active layer was clearly shown as the mixed layer of tracer grains and natural sands, and the boundary with the underlying substratum was sharp (Fig. 5). The thickness of the active layer was ca. 8 cm at point 3 (Fig. 5). The thickness was not measured at other locations due to the loss of the marking poles.

Actual sediment-transport patterns on the sand flat during a summer season

As discussed above, sediment-transport patterns were measured during a summer season (May to July 2011), via tracer experiments and geomorphological observation. The results showed that fluvial discharge, mainly from the braided channels that accompanied the incision of the spit, supplied sediments to the sand flat; these drifted southward at the central part of the sand flat and were dispersed eastward or northeastward by wave activity during normal conditions.

The drastic geomorphological development associated with the 17–19 July fluvial flooding event revealed episodic sediment-transport processes around the river mouth. Fluvial sediments were supplied to the sand flat toward the northeast, mainly from braided channels incising the spit, rather than from the main fluvial channel (Fig. 4). The incision of the spit and fluvial sediment supply was also observed, in October 2009, by Yamashita et al. (2011), who reported the northeastward transport of fluvial sediment into the sand flat by a fluvial flooding event.

It is estimated that the fluvial sediments supplied to the sand flat were dispersed by wave or tidal currents during normal conditions. The eastward transport in a seaward area (points 1 and 2) roughly coincides with the direction of coastal (residual) currents (Coastal Oceanography Research Committee, The Oceanographic Society of Japan 1985; Figs. 1 and 4). The landward (southward) transport at points 3 and 4 can be interpreted as representing a depositional wave condition during summer season (Hayes 1972; Fig. 4).

Comparison with trend vectors obtained by P-GSTA and previous GSTA methods

Grain-size trend vectors calculated by the P-GSTA method showed reasonable similarity to the actual sediment-transport patterns described above (Fig. 12). A north to northeastward trend along the main fluvial channel and northern parts of the sand flat coincided with the net sediment-transport directions observed by the tracer experiment at point 5. Northeastward-directed vectors were dominant around the breached section of the spit, and coincided with the downstream direction of the braided channels formed by the fluvial flooding event of July 2011 (Fig. 12). Thus, the trend vector of P-GSTA at point 5 coincides with the sediment transport direction estimated from geomorphological observation, whereas it contradicts that derived from the tracer data. In the central part of the sand flat, the vectors were directed to the southeast and coincided approximately with the tracer result at point 4 (Fig. 12). In addition, the vectors coincided very well with the tracer result at point 2, in which they were directed eastward. In the southeastern regions, the trend vectors were directed toward the northeast (Fig. 12), which is in slight disagreement with the tracer data, indicating eastward transport (Fig. 12).

On the other hand, sediment trend vectors calculated by the previous GSTA method did not show reasonable net sediment-transport patterns compared to the results for P-GSTA (Fig. 12). Around point 2, trend vectors showed eastward transport patterns and coincided with the tracer results (Fig. 12). Southeastward vectors found at the central part of the sand flat (around point 4) were also comparable with the tracer results. However, landward-directed vectors along the main fluvial channel clearly conflict with the tracer data at point 5 (Fig. 12). In addition, transport vectors were not defined along the braided channels and in the vicinity of point 3.

Discussion

In contrast with the results obtained using the previous GSTA method, the grain-size trend analyzed using the P-GSTA method closely matches the actual sediment-transport estimated from the tracer experiments and geomorphological features observed soon after fluvial flooding events (Gao and Collins 1992; Asselman 1999; Fig. 12). In the GSTA method, trend vectors opposite to the fluvial downcurrent direction are produced by the lower sorting values in the river mouth (Figs. 7 and 12). The coarse sediment (lower ϕ value) in the river mouth results in a lower sorting value due to the inevitable correlation between mean grain size and sorting value, and therefore, sorting values cannot represent the actual degree of sediment sorting, as supposed by the previous methods. The previous GSTA method performs quite poorly, in that no trend vector is defined around the braided channels, despite direct observations of the sediment transport by a fluvial flooding event in July 2011 (Figs. 3, 4, 5, and 12). In contrast, sediment-transport patterns are better reconstructed via the P-GSTA method, especially around the river mouth and braided channels (Fig. 12). At point 3, trend vectors are parallel to the downcurrent direction of braided channels and thus represent sediment-transport direction by the fluvial flooding event, rather than during normal conditions indicated by tracer experiments (Fig. 12). This is probably because the sampling for P-GSTA was conducted after the flooding event while the tracer experiments were conducted before the event (Fig. 3).

The disparity between trend vectors calculated by the GSTA method and observational data shows that the assumption of McLaren and Bowles (1985) is not appropriate in this case, probably because the dominant sediment-transport processes are mixed. In contrast, the P-GSTA method can detect the unique grain-size trend in each depositional setting and is therefore effective even under mixed transport processes, in which the precise sediment-transport function is difficult to determine. Principal component analysis (PCA) can easily explore the characteristic changes of the grain-size distribution curves, which result from size-selective sediment transport in a specific depositional setting, by a linear combination of multiple grain-size parameters (Eq. 10). PCA can automatically weight multiple grain-size parameters and combine them into a linear function. In addition, the magnitude of the weighting is proportional to the spatial variation within each parameter (Middleton 2000). Therefore, the resultant linear combination of grain-size parameters, summarized by the first principal component (PC1), represents the predominant deformation pattern of the grain-size distribution curves in the sampling area. In other words, the first principal component can generally be regarded as the primary function of size-selective sediment-transport in which a specific grain-size distribution curve varies spatially. If some minor transport processes exist, it is expected that the deformation pattern of the grain-size curves affected by these processes can be identified by the subsequent principal components. In the present study, PC2 and PC3 recognized two minor depositional processes (channel-lag deposits and muddy deposits resulting from episodic flooding). Indeed, the trend vectors derived from GSTA show directions incompatible with observations in the region around the spit (Fig. 12). In this region, PC2 values are large (1.2–2.0), suggesting the influence of channel-lag deposits formed by washover processes (Fig. 9). It is notable that the trend vectors of P-GSTA show directions consistent with the observation of sediment transport (Figs. 5b and 12). On the other hand, PC3 shows smaller values (< -1.5) around the river mouth and rhythmic bars (Fig. 9), implying the presence of fine-grained flooding deposits. Again, the trend vectors derived from GSTA are incompatible with the directions of actual sediment transport (Figs. 5 and 12), which suggests the effectiveness of the P-GSTA method.

Although P-GSTA is a very flexible method, a number of precautions must be taken in applying this method. First, careful selection of the effective parameters that represent the shape of the grain-size curve is necessary. If too many irrelevant parameters are included, the results of the PCA will be meaningless. Conversely, parameters that have a mathematically inevitable correlation will be erroneously emphasized, according to their spatial variances. Parameters should preferably be dimensionless values that are as independent from each other as possible. In the present study, median grain size and CV value are employed instead of mean grain size and sorting value, which are inevitably correlated to other granulometric parameters. In addition, the consistency of the sediment-transport function should be considered in some cases. If more than two sediment-transport functions dominate in a depositional setting, PCA alone cannot detect them. To deal with these situations, before PCA, samples should be divided into several groups on the basis of the similarities of grain-size characteristics.

Furthermore, the applicability of all GSTA methods should be verified in various depositional environments. GSTA methods preferentially target depositional environments rather than erosional environments. The bottom-surface sediments sampled for GSTA methods should be grains being actively exchanged with grains in a bedload layer, not the 'historical layer' exposed by erosion. In addition, attention should be paid to topographical undulation. Masselink et al. (2008) introduced a case of failure of the previous GSTA method (Gao and Collins 1992) on a tidal shoreface and described this failure in connection with the geomorphological variations in shoreface bars and runnels. In the case of regions such as shoreface bars or runnels where topographic undulation is highly developed, the active-layer thickness varies considerably within a depositional environment. Thus, GSTA methods are preferentially applied in depositional environments where topographical undulation is relatively minor and deposition occurs throughout the surveyed region; otherwise, the active-layer thickness should be identified at all sampling locations.

Conclusions

1. Principal component analysis (PCA) revealed the grain-size trend in a modern microtidal sand flat along the Kushida River Delta, central Japan. PCA enables multiple grain-size parameters to be organized into a simple linear function with different weights on the basis of their spatial variation.
2. The grain-size trend analysis using the results of PCA (P-GSTA) revealed that sediment becomes

finer, better sorted, and less gravely through major transporting processes and reconstructed sediment-transport pathways in the microtidal sand flat.

3. Application of the P-GSTA method revealed the presence of local depositional processes on the sand flat: deposition of fluvial-channel lags and muddy particles.

Abbreviations

CV: Coefficient of variation; GSTA: Grain-size trend analysis; JMA: Japan Meteorological Agency; MLIT: Japanese Ministry of Land, Infrastructure, and Transportation; PC: Principal component; PCA: Principal component analysis; P-GSTA: Grain-size trend analysis using principal component analysis

Acknowledgements

The authors are sincerely grateful to Prof. Makoto Ito and Dr. Shuji Yoshida (Chiba University) for their scientific discussions. Dr. Tomoyuki Sato, Dr. Naohisa Nishida (AIST), and Dr. Go-ichiro Uramoto (Kochi Core Center) provided constructive comments and warm encouragement. The authors also thank Ms. Kazuno Arai, Mr. Yoshinori Nemoto, Ms. Naomi Noguchi, Mr. Hidemichi Konishi, Ms. Kozue Itomine, and Mr. Takuya Ishimaru (Chiba University) for their help during the fieldwork and discussions. Many members of the Sedimentology Research Group of Chiba University also provided aid and encouragement. Informational contributions were made by Google Inc., Mie Prefecture, the MLIT, the Japan Coast Guard, and the JMA. Full facilities for the tracer experiments were provided by the Chubu Area Development Bureau of the MLIT, the Fisheries Cooperative Association of Matsusaka, and the Matsusaka Building Office of Mie Prefecture. All the source codes and grain-size data used in this study are on the author's website (http://turbidite.secret.jp/P-GSTA/).

Funding

The present thesis was supported by research grants from Chiba University (Research Support Program 2009 for Young Scientists) and a Grant-in-Aid from the Japan Society for the Promotion of Science (JSPS) Fellows (22000605).

Authors' contributions

SY proposed the topic and conceived and designed the study. SY, HN, and TN carried out the field survey. SY analyzed the data, and HN and TN assisted in the interpretation. All authors read and approved the final manuscript.

Competing interests

The authors declare that they have no competing interests.

Author details

[1]JX Nippon Oil & Gas Exploration Corporation, JX building, 1-1-2, Ote-Machi, Chiyoda-Ku, Tokyo, Japan. [2]Division of Earth and Planetary Sciences, Graduate School of Science, Kyoto University, Kitashirakawa Oiwakecho, Sakyo-ku, Kyoto 606-8502, Japan. [3]Osaka Museum of Natural History, Nagai Park 1-23, Higashi-Sumiyoshi-ku, Osaka 546-0034, Japan.

References

Aitchison J (1986) The statistical analysis of compositional data. Chapman & Hall, London

Aitchison J (2003) The statistical analysis of compositional data. Blackburn Press, Hoboken

Aitchison J, Egozcue JJ (2005) Compositional data analysis: where are we and where should we be heading? Math Geol 37:829–850

Arai H, Ohta T (2006) Note on zero and missing values in compositional data. Chishitsugaku Zasshi 112:439–451

Armanini A (1995) Non-uniform sediment transport: dynamics of the active layer. J Hydraul Res 33:611–622

Asselman NEM (1999) Grain-size trends used to assess the effective discharge for floodplain sedimentation, river Waal, the Netherlands. J Sediment Res 69:51–61

Burgess TM, Webster R (1980a) Optimal interpolation and isarithmic mapping I. The semi-variogram and punctual kriging. Eur J Soil Sci 31:315–331

Burgess TM, Webster R (1980b) Optimal interpolation and isarithmic mapping. II. Block kriging. Eur J Soil Sci 31:505–524

Cheng P, Gao S, Bokuniewicz H (2004) Net sediment transport patterns over the Bohai strait based on grain size trend analysis. Estuar Coast Shelf Sci 60:203–212

Coastal Oceanography Research Committee, The Oceanographic Society of Japan (1985) Coastal oceanography of Japanese islands. Tokai University Press, Tokyo

Davis JC (1986) Statistical and data analysis in geology, 2nd edn. Wiley, Hoboken

Ferguson RI, Church M (2004) A simple universal equation for grain settling velocity. J Sediment Res 74:933–937

Flemming BW (1988) Process and pattern of sediment mixing in a microtidal coastal lagoon along the west coast of South Africa. In: de Boer PL, van Gelder A, Nio SD (eds) Tide-influenced sedimentary environments and facies. D. Reidel Publ. Co, Dordrecht

Folk RL (1966) A review of grain-size parameters. Sedimentology 6:344–359

Friedman GM, Sanders JE, Kopaska-Merkel DC (1992) Sediments: names, particles, bulk properties. In: Friedman GM, Sanders JE, Kopaska-Merkel DC (eds) Principles of sedimentary deposits. Macmillan Publishing Company, New York

Galloway WE (1975) Process framework for describing the morphologic and stratigraphic evolution of deltaic depositional systems. In: Broussard ML (ed) Deltas, models for exploration, Houston Geological Society

Gao S, Collins M (1992) Net sediment transport patterns inferred from grain-size trends, based upon definition of "transport vectors". Sediment Geol 81:47–60

Gibbs RJ (1974) A settling tube system for sand-sized analysis. J Sediment Petrol 44:583–588

Harrington RF (1967) Field computation by moment methods, 1st edn. The Macmillan Co., New York

Hayes MO (1972) Forms of sediment accumulation in the beach zone. Waves on beaches and resulting sediment transport. In: Meyer RE (ed) Waves on beaches and resulting sediment transport: proceedings of an advanced seminar conducted by the Mathematics Research Center, the University of Wisconsin, and the Coastal Engineering Research Center, U.S. Army, at Madison October 11–13, 1971. Academic Press, New York

Hirano M (1971) River-bed degradation with armoring. Proc Jpn Soc Civ Eng 195:55–65

Kohsaka H (1999) Kriging and its geographic applications. Bulle Nihon Univ Coll Hum Sci 34:27–35

Le Roux JP, Rojas EM (2007) Sediment transport patterns determined from grain size parameters: overview and state of the art. Sediment Geol 202:473–488

Liu JT, Huang J, Hsu RT, Chyan J (2000) The coastal depositional system of a small mountainous river: a perspective from grain-size distributions. Mar Geol 165:63–86

Martín-Fernández JA, Barceló-Vidal C, Pawlowsky-Glahn V (2003) Dealing with zeros and missing values in compositional data sets using nonparametric imputation. Math Geol 35:253–278

Masselink G, Buscombe D, Austin M, O'Hare T, Russell P (2008) Sediment trend models fail to reproduce small-scale sediment transport patterns on an intertidal beach. Sedimentology 55:667–687

McCave IN (1978) Grain-size trends and transport along beaches: example from eastern England. Mar Geol 28:43–51

McLaren P (1981) An interpretation of trends in grain size measures. J Sediment Res 51:611–624

McLaren P, Bowles D (1985) The effects of sediment transport on grain-size distributions. J Sediment Petrol 55:457–470

Middleton GV (2000) Data analysis in the Earth sciences using MATLAB. Prentice-Hall, Inc., Upper Saddle River

Nakajo T (2004) Geomorphology and sedimentation of microtidal flat in the Kushida River estuary, Ise Bay, central Japan. Bull Osaka Mus Nat Hist 58:69–78

Naruse H (2005) Usage and advantages of an application program 'STube' for settling tube grain-size analysis. Taisekigaku Kenkyu 62:55–61

Nordstrom KF (1981) Differences in grain size distribution with shoreline position in a spit environment. Northeast Geol 3:252–258

Nordstrom KF (1989) Downdrift coarsening of beach foreshore sediments at tidal inlets: an example from the coast of New Jersey. Earth Surf Process Landf 14:691–701

Parker G, Paola C, Leclair S (2000) Probabilistic Exner sediment continuity equation for mixtures with no active layer. J Hydraul Eng 126:818–826

Pedreros R, Howa HL, Michel D (1996) Application of grain size trend analysis for the determination of sediment transport pathways in intertidal areas. Mar Geol 135:35–49

Pettijohn FG, Ridge JD (1932) A textural variation series of beach sands from Cedar Point, Ohio. J Sediment Petrol 2:76–88

Ribberink JS (1987) Mathematical modeling of one-dimensional morphological changes in rivers with non-uniform sediment. Communications on hydraulic and geotechnical engineering of the Faculty of Civil Engineering, Delft University of Technology, report no. 87-2

Rojas EM (2003) Determinación de vectors de transpote, utilizando informacióGranulométrica: aplicación al delta tipo Gilbert, del rìo Pescado, Llago Llanquìhue, X Región, Chile. Masters Thesis, Universidad de Chile
Rojas EM, Le Roux JP (2003) Determinación de vectores de transporte, utilizando información granulométrica: Aplicación al delta tipo gilbert del río pescado, lago llanquíhue, X región, Chile. Proceedings X Congreso Geológico Chileno, Sociedad Geológica de Chile, Santiago (CD-ROM)
Rojas EM, Le Roux JP, Cisternas M (2000) Metodología y algoritmos que permiten determinar las direcciones de transporte de sedimentos utilizando parámetros granulométricos: Caso teórico y aplicación a una cuenca active. Abstr Vol IX Congreso Geol Chile 1:539–543
Self RP (1977) Longshore variation in beach sands, Nautla area, Veracruz, Mexico. J Sediment Petrol 47:1437–1443
Swift DJP, Parsons BS, Foyle A, Oertel GF (2003) Between beds and sequences: Stratigraphic organization at intermediate scales in the Quaternary of the Virginia coast, USA. Sedimentology 50:81–111
Tanner WF (1962) Geomorphology and the sediment transport system. Southeast Geol 4:113–126
Tucker ME (1988) Techniques in sedimentology. Blackwell, Oxford
Yamashita S, Nakajo T, Naruse H, Sato T (2009) The three-dimensional distribution of sedimentary facies and characteristics of sediment grain-size distribution in a sandy tidal flat along the Kushida River estuary, Ise Bay, central Japan. Sediment Geol 215:70–82
Yamashita S, Nakajo T, Nishida N, Naruse H (2011) Characteristics and depositional pattern of fluvial-flood deposits on the river mouth: case study of a fluvial-flooding event at 2009 in the Kushida River Delta, Ise Bay, central Japan. Taisekigaku Kenkyu 70:81–92

5

Role of the capillary fringe on the dilatation of a low porosity limestone submitted to unidirectional freezing

Celine Thomachot-Schneider*, Emilie Huby, Kevin Chalons, Xavier Drothière and Patricia Vazquez

Abstract

As a rule, usually only one face of the building stones of a monument is exposed to climatic variations. Penetration of temperature and humidity stress is unidirectional from the external surface while capillary absorption comes from the bottom part of the accumulation zones such as soil or cornice band. This configuration has to be taken into account in the setting of frost weathering tests in the laboratory. Formation of ice is accompanied by changed volumes and by movements of liquid water due to cryosuction. These phenomena lead to the dilatation of the porous network as assessed by strain gage monitoring. In this study, an experimental setting was developed to simulate the condition of a building stone submitted to frost and water supply by capillary absorption. Strain gages and thermocouples were attached to a limestone sample in order to assess thermal penetration and dilatation changes. Two types of saturation conditions were applied to the sample: water supply only by capillarity and water supply by capillarity after immersion during 48 h. Freeze-thaw cycles of 24-h freezing at − 15 °C and 24-h thawing at + 10 °C were performed. Results showed that deformation perpendicular to the freezing direction was negligible except for the fringe zone where it could reach 6×10^{-4}. In contrast, on both sides of the wet fringe, a deformation parallel to the freezing direction occurred, corresponding to an expansion upon freezing and a contraction upon thawing, with an intensity correlated in both cases to saturation. This original test also showed the importance of the capillary fringe in the frost action on building stones that is not taken into account in classical frost weathering tests. And it is to be noted that the strain gages allowed to measure immediate damage before it reached the inside of the stone and before it could be visually perceived.

Keywords: Frost weathering, Capillary fringe, Unidirectional freezing, Strain gages, Limestone

Introduction

Frost action is one of the major weathering factors of building stones. Freezing of the water inside the porous network causes dilatation as a result of a change in its volume (9%) as well as of the movements of liquid water following cryosuction. Freezing in a porous network is not uniform (Thomas 1938). Water can remain liquid at temperatures below zero, a process called "supercooling" (Chahal and Miller 1965; Prick 1995). In this case, the freezing temperature depends on the pore size; the smaller the pore, the lower the freezing temperature (Everett 1961). The saturation state of the porous network is thus a critical parameter in the stress caused by the movement of water while freezing. In the event of a solute in the water, the solution can be cooled below the freezing point of the corresponding pure liquid, this being due to the presence of salts and called the "freezing point depression" (Conde et al. 2018). The freezing temperature of the water decreases depending on the salt content up to a minimum value corresponding to the eutectic point.

In a porous network saturated by capillary absorption, it is only its free porosity that is taken up by water while its trapped porosity (mainly macropores) remains occupied by air (Hirschwald 1912). In this case, the changed volume of the ice is accommodated by the trapped porosity and the ice crystallizes

* Correspondence: celine.schneider@univ-reims.fr
GEGENAA EA 3795, University of Reims Champagne-Ardenne (URCA), Reims, France

without generating any pressure (Powers 1948; Everett 1961; Litvan 1978). Friction resistance to the water movements following cryosuction is the main cause leading to an expansion or contraction of the porous network and thus to cracking (theory of hydraulic pressures of Powers 1948 and Litvan 1978). A high freezing rate, a large amount of available water and a long distance to reach a trapped pore will intensify the generated pressure. These water movements are irreversible and the water attracted by cryosuction remains in the macropores even during thawing. In the event of a water supply, the repeated cycles of freeze-thaw will lead to the increase of the amount of water in the porous network (Mc Greevy and Whalley 1985; Thomachot and Jeannette 2002). As a result, capillary saturation will progressively change into total saturation.

n a porous network totally saturated under vacuum, ice first grows within macropores but then because all the porous network is occupied by water, it has no longer any space to go on growing. It exerts a pressure on unfrozen water (Powers 1948) and when macropores are totally frozen, extrusion of ice toward the finer pores occurs and leads to so-called capillary pressures (Everett 1961). These pressures are proportional to the difference between the macropore radius and the bottleneck of the capillary.

In a monument, frost has the particularity of entering the stone only through the exposed facade. The surrounding blocks and the milder indoor conditions protect the other faces. To assess the effect of freeze-thaw on stones, standard tests are commonly used or adapted in terms of temperature or freezing time. All these tests, even the customized ones, consist in freezing the samples after prior immersion in water and thawing them in room temperature water (EN. 12371 2010; Tan et al. 2011; Ruedrich et al. 2011). All the stone faces are exposed to damage, yet these tests do not reproduce freeze-thaw conditions in situ.

These standardized tests estimate the decay produced by the ice as causing loss of material (visible or by weight measurements) and decrease in tensile strength. In both cases, if the test itself is not destructive enough to produce material detachment, these evaluation methods are irrelevant. In addition, these tests cannot be performed outside laboratories and in real monuments with immovable materials.

So in this study, we tested an alternative solution to these constraints thanks to a non-destructive setting that allowed freezing to be produced only on one face of the stone and to the use of strain gages attached along the stone as a way of assessing microscopical structural damage. This setting was tested on a common Lutetian limestone building in the surroundings of Reims (France).

Methods/Experimental

Materials

In order to develop an experimental setting able to simulate the natural conditions of a building stone placed in elevation, a building stone sample was necessary. The Middle Lutetian limestone called Courville limestone was chosen for this study. It is a common building stone in Reims (France) and its surroundings (Fronteau et al. 2010). Actually, the famous Reims Cathedral or the St-Remi Basilica was originally built with Lutetian limestones (Turmel et al. 2014, Fig. 1a, b). One of the main weathering patterns observed on this building stone is the scaling of millimetric layers of stones placed in the capillary zone or in elevation, especially when placed as cornice band, column, or statues (ICOMOS, Fig. 1c). The angular parts of the blocks mainly suffer from scaling attributed to the alternation of wetting-drying cycles combined with the action of temperature variations. Nowadays, there are no more quarries. The last one located 20 km west of Reims was closed in 2005. The Courville limestone is a cream-colored packstone/wackestone presenting a heterogeneous micritic matrix, which was recrystallized into microsparite with mainly shells, miliolids, and *Orbitolites complanatus* formaminifera (Fig. 1d, e). In this study, the used Courville limestone belongs to the "Roche" bed, which is the less porous bed of the formation (Fronteau et al. 2002). This special type of Courville limestone was mainly used for the lower section of buildings. The porosity value of the studied block determined by mercury porosimetry is 11.2% with a mean pore radius of 0.05 μm corresponding to a microporous network with a unimodal distribution (Fig. 2a). Capillary kinetics is quite low with values of 790 g m^{-2} $h^{-1/2}$ for the coefficient relative to the

Fig. 1 **a** Location of Reims (France), **b** Reims Cathedral built with Lutetian limestone, **c** weathering patterns on a capillary zone of the Cathedral, **d** Courville limestone (Lutetian) (G. Fronteau), and **e** thin section of Courville limestone showing fossils of miliolid and *Orbitolites complanatus* inside a micrite (G. Fronteau)

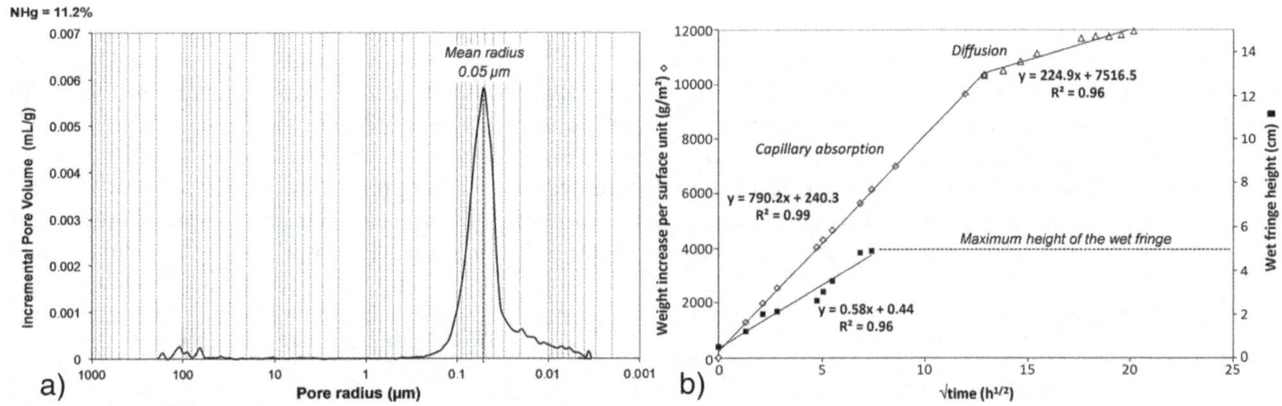

Fig. 2 **a** Pore distribution obtained by mercury injection porosimetry and **b** capillary kinetics of the Courville limestone "Banc Roche"

weight increase (first curve with diamond points in Fig. 2) and of 0.58 m $h^{1/2}$ for the coefficient relative to the migration of the weight fringe (Fig. 2b). Due to a weak capillary transfer explaining its use for the low sections of buildings or pavements, the wet fringe of a 15-cm high sample stopped at around a 6 cm height (Fig. 2b). Even if the visible wet fringe stops at a 6-cm height after 50 h, capillary absorption continues as shown by the first curve of the weight increase per surface unit. The break of slope, corresponding to the end of the capillary absorption strictly speaking, occurs after 170 h. The saturation of the porous network goes on by diffusion (curve with a weak slope).

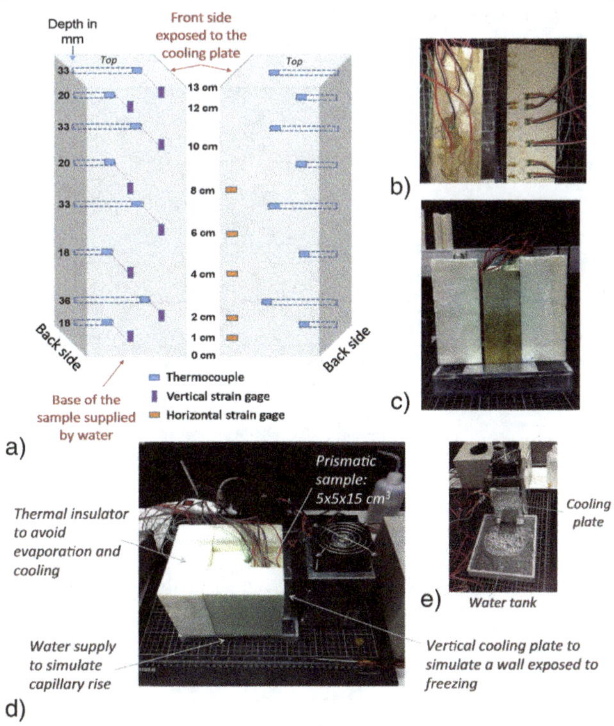

Fig. 3 **a** Diagram of the setting of strain gages and thermocouples on the sample, **b** view of the horizontal and vertical strain gages, **c** lateral view of the face in contact with the cooling plate, and **d**, **e** setting of the unidirectional and vertical frost experiment with water supply

Experimental setting

The aim of the setting was, as on the wall of the lower section of a building, to reproduce unidirectional freezing on the vertical face of a sample while keeping its other faces isolated, and with a capillary rise of water from the ground, supplied in it base.

To produce freezing, a cooling plate Tetech J061 (8 × 13 cm) of temperatures ranging from 80 to – 30 °C and a variation of 0.01 °C during the test was placed vertically in contact with one vertical face of the prismatic sample of 5 × 5 × 15 cm. The rest of the faces

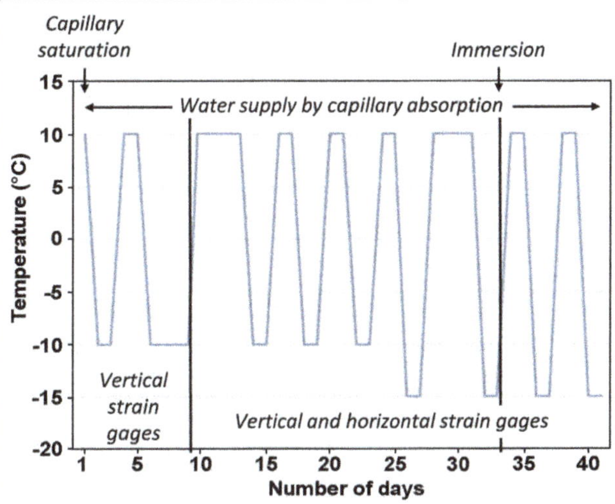

Fig. 4 Main conditions of the experiment

was sealed with a thermal insulator to avoid evaporation and cooling (Fig. 3c, d).

To assess frost action, strain gages from Kyowa, KFWSS-2N-120-C1-11L3M2R, with a length of 2 mm, a gage factor of 2.12, and a resistance of 120.0 ± 0.8 Ω, were set up at different heights of the sample with different orientations from the direction of freezing penetration. Strain gages measuring the vertical deformation were placed offset at 1 and 3 cm away from the face in contact with the cooling plate, at 1, 2, 4, 6, 8, 10, 12, and 13 cm away from the bottom of the sample. Testing was first performed in this setting, and subsequently horizontal strain gages, placed perpendicularly to the cooling plate, were added at heights of 2, 4, 6, and 8 cm on the opposite face, and 0.5 cm away from the surface in contact with the cooling plate (Fig. 3a–c).

To control the freezing penetration inside the sample, holes of 2 mm diameter were drilled at the same heights

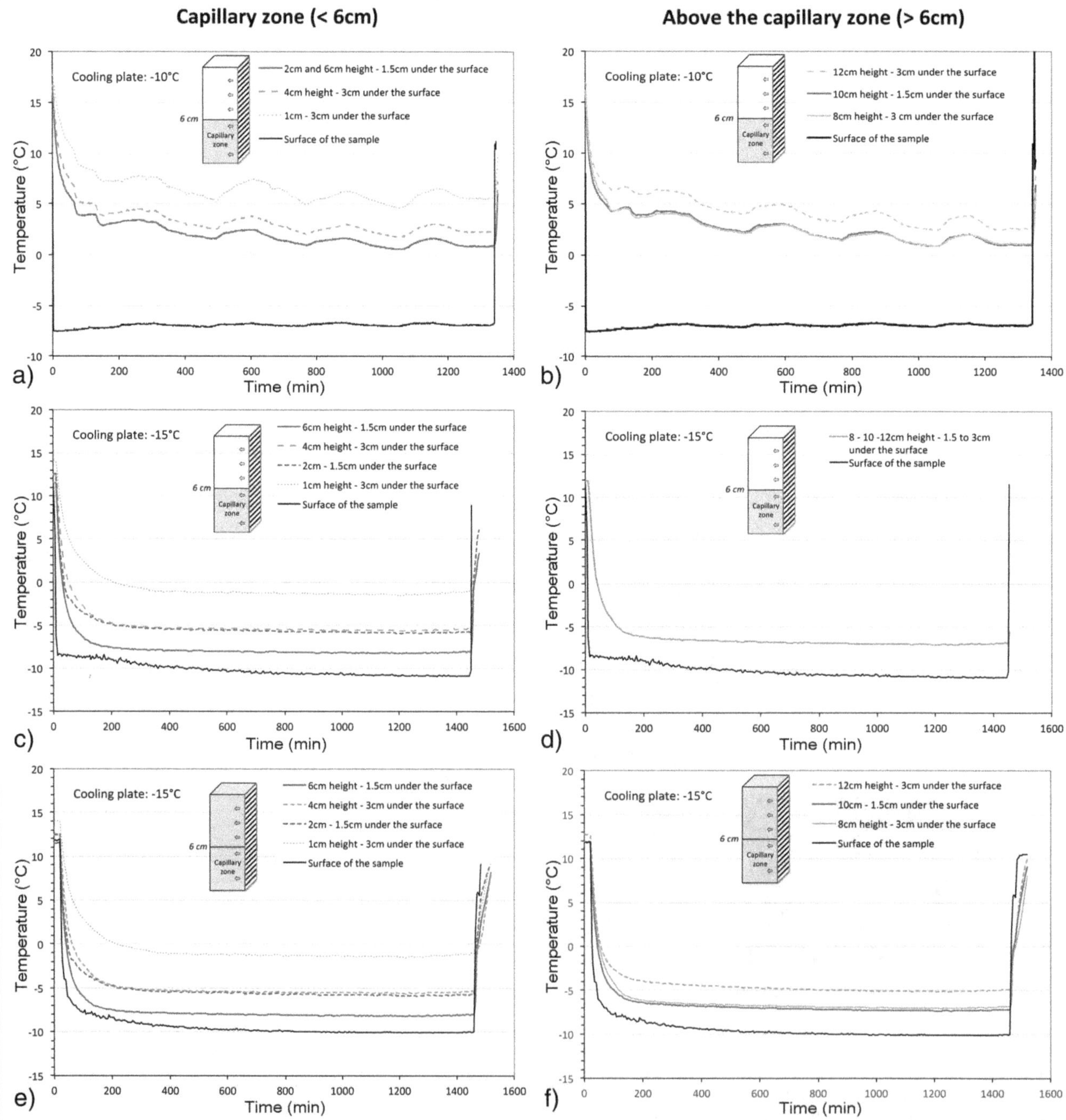

Fig. 5 Temperature kinetics inside the sample **a**, **c**, **e** in the capillary zone and **b**, **d**, **f** above the capillary zone

as the strain gages but from the face opposite the one submitted to the cooling plate. Type K thermocouples with an accuracy of 0.1 °C were placed within them at depths corresponding to the location of the strain gage for control of the penetration of freezing within the stone. Even though this was a partially destructive method, it was necessary to confirm whether freezing was produced within the stone and at which moment, in order to establish the best experimental protocol. The data from the strain gages and thermocouples were recorded with an Agilent Data Logger 3. The data were obtained every 5 min by the interface Benchlink.

Before experimenting, the thermal dilatation of the stone was recorded on the dry sample. The thermal expansion coefficient alpha expresses the expansion of a material as a function of temperature. The linear coefficient of thermal expansion for the investigated samples was calculated from the experimentally determined temperature and dilatation data (alpha = Delta L/L*Delta T: T is temperature and L is length). Temperature was set at + 40 °C and then down to – 15 °C to determine the average coefficient.

As regards freezing testing, two kinds of water saturation were tested: (i) partial saturation obtained by continuous capillary water supply, and (ii) partial saturation obtained by a previous 48-h total immersion. Freeze-thaw cycles were divided into 24 h freezing periods at – 10 °C for the first five cycles then at – 15 °C for the next five cycles and 24 h thawing periods at + 10 °C (Fig. 4). The whole experiment was repeated two times. During the whole time of the two tests, water was supplied to the sample by capillary absorption from a bottom water tank.

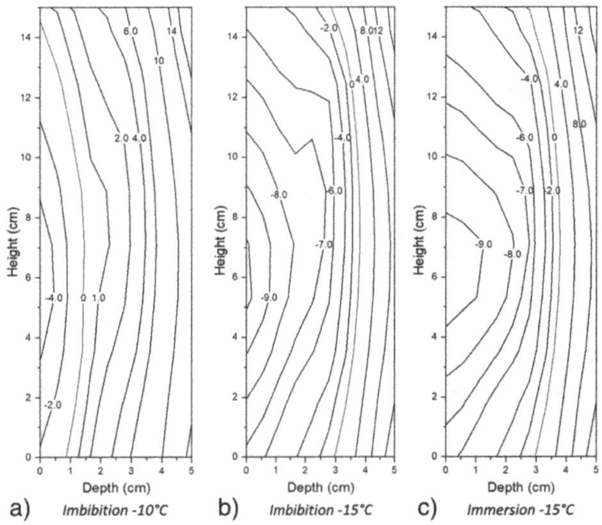

Fig. 6 Temperature mapping inside the sample after **a** capillary absorption and freezing at – 10 °C, **b** capillary absorption and freezing at – 15 °C, and **c** after immersion and freezing at – 15 °C

Results and discussion

For the first five cycles, the temperature of the cooling plate was set at – 10 °C. The temperature recorded at the surface of the sample was around – 6.7 °C (Fig. 5a, b). This temperature was enough to freeze only the surface of the sample and the first centimeter in depth, but freezing did not reach the inner part of the sample. Thermocouples placed at 1.5 to 3 cm in depth did not record temperatures lower than around 1 °C. Regarding the distribution of temperature extrapolated from the thermocouple data (Fig. 6a), the maximum for freezing was reached in the central part of the sample surface corresponding to a 4 to 8 cm height. This temperature distribution was similar in the other types of settings. This was due to a side effect of the cooling plate, the efficiency of which was maximum in the center. The time necessary to reach a stable temperature in the sample was between 2 and 3 h according to the thermal conductivity of the sample (Fig. 6).

For the next five cycles, the temperature was set at – 15 °C. This allowed to reach a temperature of – 10 °C at the surface of the sample and of 0 °C at around 3.5 cm in depth whatever the saturation of the sample (Figs. 5a, d–f and 6b, c). This difference in surface temperature of 3 °C between the two settings (– 10 and – 15 °C) led to a 2 cm change in frost penetration even if the kinetics of penetration was similar. The time necessary to reach a stable temperature inside the sample was also between 2 and 3 h. The gradient of temperature inside the sample was similar for both types of saturation (capillary and immersion). Due to its low porosity (11.2%) and low capillary saturation, the saturation of the Courville limestone by simple immersion was not enough to disturb significantly the heat conduction in the sample. In a porous network, a 0 °C temperature does not mean systematically that water is frozen. Water can remain in a liquid state at a temperature below 0 °C, this being called supercooling (Chahal and Miller 1965). Supercooling, and thus the freezing temperature of water inside a porous network, depends on the size of the pores: the smaller the pores, the lower the freezing temperature (Everett 1961; Blachère 1975). For example, Tourenq (1970) determined a freezing temperature around – 2 to – 4 °C for stones with main porosity larger than 0.03 μm while Fagerlund (1971) determined freezing temperature around 0 to – 3 °C for stones with main porosity ranging from 0.02 to 0.5 μm. The mean radius of the studied limestone is around 0.05 μm which means that freezing should occur at temperatures around – 2 °C. According to the gradient of temperature (Fig. 6), with a cooling plate set at – 10 °C, the freezing front, corresponding to the zone cooled at temperatures below – 2 °C, remained superficial (less than 0.5 cm in depth) (Fig. 6a). For the two other settings at – 15 °C with capillary absorption only and after immersion, the zone cooled under – 2 °C was deeper and corresponded to the first 3 cm.

The thermal dilatation of the stone was recorded on the dry sample at temperature ranging from room temperature to + 40 °C and then from room temperature to – 15 °C. It showed values respectively of + 1.2 mm/m and – 3.0 mm/m, which corresponded to thermal coefficients ranging from 1.0 to 2.0×10^{-6}/K. These values were of the same order of magnitude as those measured by other authors on limestone, sandstone, porphyry, tuff, or granite (Thomachot and Jeannette 2002; Ruedrich et al. 2011).

During the first part of the freezing test, the sample was partially saturated by capillary absorption. The weak capillary properties of this Lutetian bedding ("Banc Roche") due to its low porosity mainly made of microporosity, explained why the wet fringe on the 15 cm long prismatic sample did not reach a height of 6 cm or more. Differential dilatation behavior between the capillary zone and the upper dry zone was expected.

In the second part of testing when the sample was saturated by immersion during for 48 h, a more homogeneous deformation along the sample was expected even if the sample was still supplied by capillary absorption from the bottom.

Dilatation recorded by the vertical strain gages during the freeze-thaw cycles with the cooling plate set at – 10 °C was negligible. The sensors, placed along the sample, were

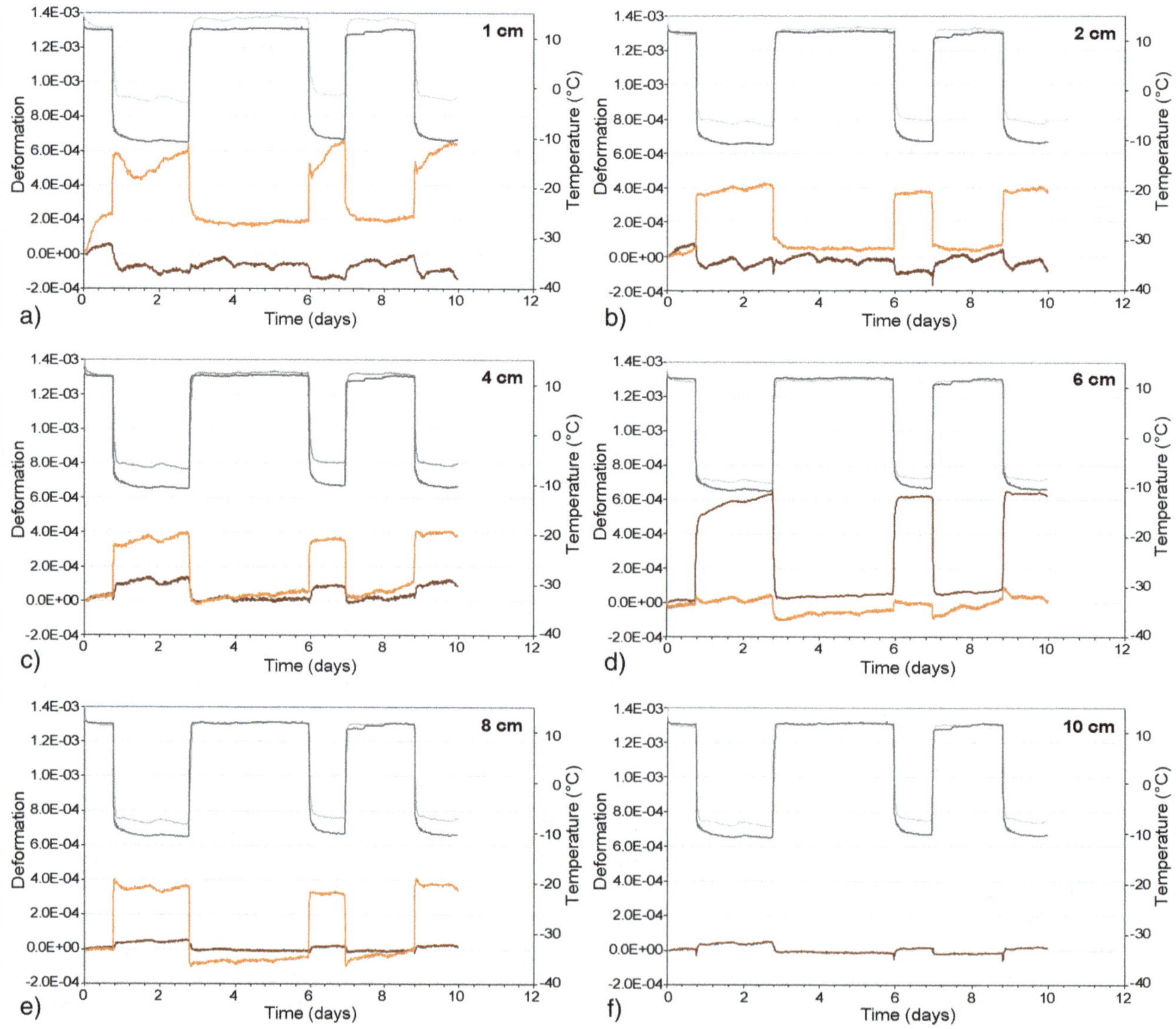

Fig. 7 Main horizontal and vertical dilatation behavior according to freeze-thaw cycles and corresponding temperature of the surface and in depth at the level of the strain gage at **a** 1 cm, **b** 2 cm, **c** 4 cm, **d** 6 cm, **e** 8 cm, and **f** 10 cm

too far away from the ice front located in the first millimeters of the surface. A significant expansion of 1.2×10^{-4} was recorded only by the strain gage at 6 cm height during freezing.

Results of freeze-thaw cycles with the cooling plate set at − 15 °C showed that whatever the saturation of the sample, the dilatation behavior was not homogeneous along the sample and that it could be divided into three parts corresponding to the capillary zonation: the capillary zone (0–6 cm, Fig. 7a–c), the fringe limit (6 cm, Fig. 7d), and the upper zone not soaked by capillary absorption during the test (> 6 cm, Fig. 7e, f).

In the capillary zone and the upper zone, the dilatometric behavior recorded was similar. The deformation perpendicular to the freezing direction (vertical strain gages) was negligible while the deformation parallel to the freezing direction corresponded to an expansion during freezing and a contraction during thawing. Values of expansion were close to 4×10^{-4} whatever the position of the strain gage. In the zone of the capillary fringe, the behavior was opposite: the deformation parallel to the freezing direction (horizontal strain gages) was negligible while the vertical deformation reached 6×10^{-4} expansion during freezing (Fig. 7b). During freezing, no contraction was recorded before expansion suggesting that no supercooling occurred.

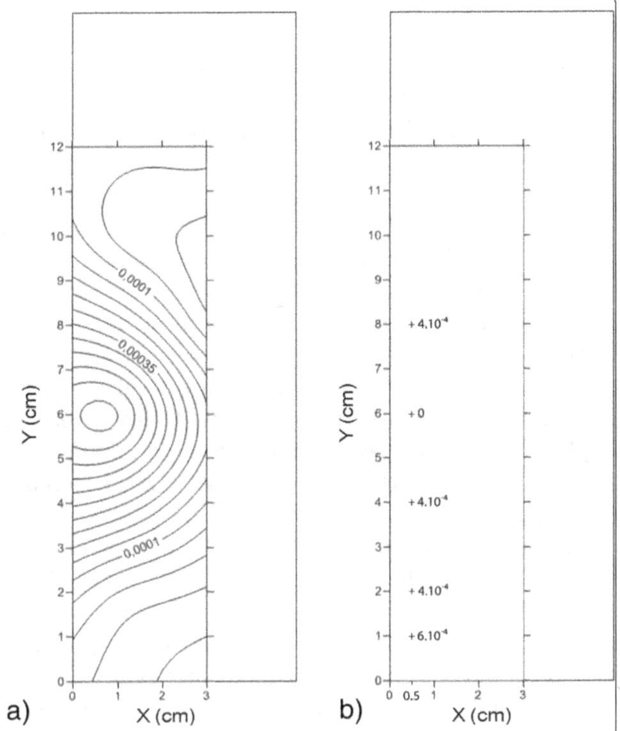

Fig. 8 Mapping of deformation inside the sample **a** vertically and **b** horizontally

The different dilatation behavior depending on the orientation of the freezing front showed that the deformation was mainly unidirectional (Fig. 8). Indeed, the deformation did not occur in all directions but was conditioned by the combination of the water flux around the water fringe and the freezing front. When a stone is submitted to a previous imbibition, the water is homogeneously distributed through the porous network, and when the temperature decreases sufficiently for water to freeze, the deformation of the stone will follow the penetration of the freezing front. A deformation perpendicular to the freezing front along the sample was thus expected: indeed, this corresponds to the expansion of ice when crystallizing and growing deeper in the porous network as associated to cryosuction of remained liquid water toward the ice front (Matsuoka 1990; Weiss 1992; Prick 1996; Thomachot et al. 2005). The presence of water supply during the experiment introduced another water flow perpendicular to the freezing front whose effect was increased by the 6 cm height threshold of the capillary fringe. In a capillary absorption, the repartition of water inside the stone is not homogeneous: the saturation is higher at the bottom of the sample supplied by water and at the level of the capillary fringe (Tournier 2001). Initially, the capillary fringe zone was more saturated than the other parts of the sample. Furthermore, the water supply during the freezing test enables cryosuction and leads to an increase of stone saturation (Mc Greevy and Whalley 1985). This water concentration interface inside the stone was enough to cause vertical expansion even if the freezing was superficial (cooling plate temperature at − 10 °C). With a stronger freezing (− 15 °C), the vertical expansion, perpendicular to the frost penetration, was maximum at the wet fringe zone due to the maximum water content and the water flux associated to capillary rise and cryosuction. A higher amount of ice crystallizing extruded toward less saturated zones led to a strain increase stronger than in less saturated parts. This phenomenon was accentuated by the fact that the cooling plate was more efficient in the central part that corresponds to the wet fringe. The penetration of the freezing front was faster and stronger in the part corresponding to the capillary fringe zone. Thus extrusion of ice during the freezing penetration caused dilatation on both sides of the wet fringe (upward and downward), causing vertical deformation.

When combining values of the different strain gages, it appears that deformation at the capillary fringe (6 cm) was opposite to the 2 and 10 cm one below and above the capillary fringe (4 and 8 cm), deformation was transitional with both directions of deformation (Fig. 8).

The lack of difference between dilatation after capillary absorption or immersion showed that the liquid within the pore space had no influence in the dilatation behavior of this limestone.

The opposite strains developed around the capillary fringe (Fig. 9) led to quite a lot of strains in this zone. A stone submitted to ice pressure would be subjected to an increase of porosity by cracking or widening of the pores. This implies that with a number of freeze-thaw cycles, the porosity of the fringe zone will increase as well as its saturation. If because of a modification of the porous network the maximum height of the wet fringe changes positions, the sensitive zone will increase with time. The zone submitted to a differential expansion will extend and consequently a larger part of the stone will be more prone not only to frost weathering but also to salt crystallization and drying impact inside the stone, and thus increase scaling as observed in monuments (Fig. 1c).

The stone tested is a homogeneous limestone in which the anisotropy of the capillary rise is weak to null. However, in the case of an anisotropic stone such as one of several limestones and sandstones, the damage produced by capillary frost weathering will be more evident. In elevations, stones are placed in a horizontal bedding while they are placed vertically when used for columns or rails. Capillary kinetics are usually higher when parallel to the bedding. The water movements associated to the freezing front will be easier and will cause less strain in the horizontal direction in blocks in elevations and in the vertical direction in columns or rails. The repartition and intensity of strain around the capillary fringe as shown in this study will change accordingly. Further experiments simulating rain with a water supply on the external surface would complete the display of strain caused by climatic changes in buildings.

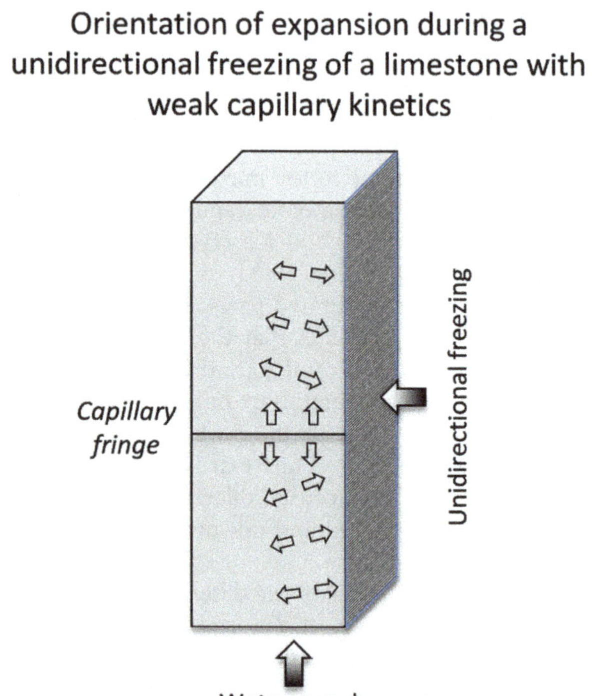

Fig. 9 Schematic distribution of expansion strains during unidirectional freezing of a sample supplied by water

Conclusions

Strain gages are a very accurate non-destructive technique that could be applied in real stone monuments. They measure the mechanical stress produced on their surface not only by ice but also by temperature changes or liquid water movements. The gages provide reliable measurement of the immediate damage before it reaches the inside of the stone and before its visual and irreversible degradation.

This original test shows the importance of the zone around the capillary fringe in the frost action on building stones. Around this zone, the directions of dilatation in the stone are opposite to the rest of the sample. This could lead to the growth of high mechanical stress on the fringe zone, making it more sensitive to frost action.

Abbreviation

GEGENAA: Groupe d'étude des géomatériaux naturels, anthropiques et archéologiques

Acknowledgements

CTS and PV gratefully acknowledge PEPS grant to attend the JpGU symposium 2017 held at Tokyo, Japan. KC gratefully acknowledges American Geoscience Union travel grant to attend the JpGU symposium 2017 held in Tokyo, Japan. Our thanks to A Thomachot for her editing assistance.

Funding

This work was supported by GEGENAA EA 3795, URCA, and Grand Reims through the Remitherm project.

Authors' contributions

CTS proposed the topic, conceived and designed the study. KC, EH, and XD carried out the experimental study. CTS, KC, and EH analyzed the data and helped in their interpretation. PV collaborated with the corresponding author in the construction of the manuscript. All authors read and approved the final manuscript.

Authors' information

CTS and PV are assistant professors at Reims University in a laboratory (GEGENAA EA-3795) involved in understanding and providing solutions to the weathering mechanisms of building materials. EH and KC are respectively PhD and Bachelor students who work on the thermal behavior of building stones. This study is the result of their work.

Competing interests

The authors declare that they have no competing interests.

References

Blachère JR (1975) Le gel de l'eau dans les matériaux poreux. VIème Congrès international F.F.E.N, Le Havre, pp 296–303

Chahal RS, Miller RD (1965) Supercooling of water in glass capillaries. Br J Appl Phys 16:231–239

Conde MM, Rovere M, Gallo P (2018) Molecular dynamics simulations of freezing-point depression of TIP4P/2005 water in solution with NaCl. J Mol Liq 261(2018):513–519
EN.12371 (2010) Natural stone test methods—determination of frost resistance. In: European standard
Everett DH (1961) The thermodynamics of frost damage to porous solids. Trans Faraday Soc 57:1541–1551
Fagerlund G (1971) Degré critique de saturation: un outil pour l'estimation de la résistance au gel des matériaux de construction. Matériaux et Constructions RILEM 4(23):271–283
Fronteau G, Moreau C, Thomachot-Schneider C, Barbin V (2010) Variability of some Lutetian building stones from the Paris Basin, from characterization to conservation. Eng Geol 115:158–166
Fronteau G, Pascal A, Barbin V (2002) Characterization of Courville stone (Lutetian, Marne, France). Relations between diagenesis and petrophysical properties, Géologie de la France 4:51–59
Hirschwald J (1912) Handbuch der bautechnischen Gesteinsprüfung, Berlin Borntraeger. Illustrated glossary on stone deterioration patterns, ICOMOS-ISCS. isbn:978-2-918086-00-0
Litvan G.G. (1978) Freeze-thaw durability of porous building materials, in durability of building materials and components, special technical publication STP 691. American Society for Testing and Materials 9:455–463
Matsuoka N (1990) Mechanisms of rock breakdown by frost action, an experimental approach. Cold Reg Sci Technol 17:253–270
Mc Greevy JP, Whalley WB (1985) Rock moisture content and frost weathering under natural and experimental conditions, a comparative discussion. Arct Alp Res 17(3):337–346
Powers TC (1948) The air requirement of frost-resistant concrete. Proc Highway Rese Board 29:184–211
Prick A (1995) Dilatometrical behaviour of porous calcareous rock samples subjected to freeze-thaw cycles. Catena 25:7–20
A. Prick (1996) La désagrégation mécanique des roches par le gel et l'haloclastie, Thèse de Sciences géographiques de l'Université de Liège
Ruedrich J, Kirchner D, Siegesmund S (2011) Physical weathering of building stones induced by freeze-thaw action: a laboratory long-term study. Environ Earth Sciences 63(7):1573–1586
Tan X, Chen W, Yang J, Cao J (2011) Laboratory investigations on the mechanical properties degradation of granite under freeze-thaw cycles. Cold Reg Sci Technol 68:130–138
Thomachot C, Jeannette D (2002) Evolution of the petrophysical properties of two types of alsatian sandstone subjected to simulated freeze thaw conditions. Geol Soc Spec Publ 205:19–32
Thomachot C, Matsuoka N, Kuchitsu N, Morii M (2005) Frost damage of bricks composing a railway tunnel monument in Central Japan: field monitoring and laboratory simulation. Nat Hazards Earth Syst Sci 5(4):465–476
W.N. Thomas (1938) Experiments on the freezing of certain buildings materials, Building Research, Technical Paper 17
C. Tourenq (1970) La gélivité des roches, application aux granulats, Rapport de Recherche N°6 LCPC
Tournier B (2001) Transferts par capillarité et évaporation dans des roches : rôles des structures poreuses, Thèse de doctorat de l'Université Louis Pasteur de Strasbourg, p 204
Turmel A, Fronteau G, Thomachot-Schneider C, Moreau C, Chalumeau L, Barbin V (2014) Stone uses in Reims cathedral: provenance, physical properties and restoration phases. Geol Soc Lond Spec Publ 391(1):17–30
Weiss G (1992) Die Eis und Salzkristallisation im Porenaum von Sandsteinen und ihre Auswirkungen auf das Gefüge unter besonderer Berücksichtigung gesteinsspezifischer Parameter. Münchner Geowissenschaftiche Abhandlungen B 9:118

6

Natural moissanite (SiC) – a low temperature mineral formed from highly fractionated ultra-reducing COH-fluids

Max W Schmidt[1*], Changgui Gao[1,2,4], Anastasia Golubkova[1], Arno Rohrbach[3] and James AD Connolly[1]

Abstract

Natural moissanite (SiC) is reported from dozens of localities, most commonly from ultramafic rocks where it may be associated with diamond and iron silicides. Yet, formation conditions of moissanite remain in the realm of speculation. The key property of SiC is its extremely reduced nature. We have experimentally equilibrated SiC with olivine and orthopyroxene at 1300-1700°C, 2 and 10 GPa, to determine the oxygen fugacity of the C + orthopyroxene = SiC + olivine + O_2 buffer (MOOC) and the equilibrium X_{Mg} of coexisting mantle silicates. The experiments resulted in olivine and orthopyroxene with X_{Mg} of 0.993-0.998 in equilibrium with SiC and iron silicides. Calculated oxygen fugacities are 5-6.5 log units below the iron-wustite (IW) buffer at 2-10 GPa. The experimental results concur with calculated phase relations for harzburgitic mantle under reducing conditions that include metal alloys, carbides and silicides. The extremely reducing character of SiC precludes coexistence with silicates with appreciable Fe^{2+}, and hence excludes equilibrium with mantle phases with typical X_{Mg}'s of ~0.9. Calculated Fe-Mg diffusion lengths reveal that SiC grains of 1 mm would react with the Fe-component of olivine to iron carbide or metal and orthopyroxene within <1 Ma at temperatures above 800°C. We thus conclude that SiC forms through a low-temperature process (<700-800°C) where equilibrium is only reached at the grain scale. The most plausible formation mechanism is a strong fractionation of a C-O-H fluid from metamorphosed sediments originally rich in organic material. Such a fluid is initially saturated with graphite or diamond and is slightly more reduced than the H_2O-C join. Fluid percolation in the mantle leads to H_2O-sequestration by crystallizing hydrous phases (most likely serpentine, brucite or phase A), and hence O_2-removal from the fluid causing its reduction and continuous graphite or diamond precipitation. A small, highly fractionated fluid fraction may then reach CH_4 and H_2 concentrations that allow SiC formation on grain boundaries without equilibration with the bulk rock on a larger scale. Such a mechanism is corroborated by the strongly negative $\delta^{13}C$ of moissanites (-20 to -37), consistent with reduced fluids originating from metamorphosed organic carbon.

Keywords: Moissanite; SiC; Ultra-reducing; Iron silicides and carbides; COH-fluid; Fluid-fractionation

Background

Moissanite is a refractory mineral stable at extremely reducing conditions with respect to typical Earth environments. Natural occurrences are manifold (Lyakhovich, 1980; Trumbull et al. 2009) but many fall into three categories: (i) peridotites, serpentinites (Xu et al., 2008) and podiform chromitites (Bai et al., 2000); (ii) kimberlites (Leung et al., 1990; Shiryaev et al., 2011); and (iii) metasomatic rocks (Lyakhovich, 1980; Di Pierro et al., 2003). In most cases, moissanite is recovered from heavy mineral separates with little textural and phase assemblage information available. Often Fe-silicides or metallic Si are attached to SiC grains (Trumbull et al. 2009), there are also diamonds which contain moissanite inclusions (Gurney 1989; Leung, 1990).

The genesis and crystallization conditions of natural SiC remains unclear, nevertheless, agreement exists about the extremely reduced nature of moissanite. This property has led to hypotheses such as moissanite being a remnant from a primordial ultra-reduced mantle or from the core-mantle boundary as discussed by Mathez et al. (1995). The extreme

* Correspondence: max.schmidt@erdw.ethz.ch
[1]Department of Earth Sciences, ETH, Zurich 8092, Switzerland
Full list of author information is available at the end of the article

oxygen fugacities required for moissanite formation have also led to propose extremely high temperatures and/or pressures of formation (Trumbull et al., 2009), in part motivated by a not uncommon association with diamonds, in part probably motivated by the high temperature synthesis practiced for industrial SiC. As conditions in the upper mantle would not allow moissanite to be stabilized at the predominant oxygen fugacities near the iron-wustite (IW) reference equilibrium (Ballhaus 1995, Woodland and Koch 2003, Frost and McCammon 2008), moissanite has been hypothesized to stem from the lower mantle (Mathez et al., 1995).

The most basic fact pertinent to moissanite genesis is that the quartz + C = SiC + O_2 equilibrium is at oxygen fugacities ≥5-7 log units below those of the iron-wustite buffer (Mathez et al. 1995; Ulmer et al., 1998). At such low oxygen fugacities, most Fe^{2+} should be reduced to metal. Consequently, coexisting mantle phases such as olivine and orthopyroxene should have unusually high X_{Mg} values.

Inspiration for the low temperature hypothesis which forms the starting point of this study stems from a moissanite location in a Cenozoic basalt from North China, where SiC appears in mantle xenoliths infiltrated by carbonatites (Gao and Liu, 2008); these composite xenoliths are in turn erupted by alkali basalts. Clinopyroxene, orthopyroxene and olivine in these xenoliths have average mantle X_{Mg} values near 0.90 and abundant interstitial calcite with no indication for a reduced environment. Nevertheless, moissanite and a panoply of metals (Fe, FeNiSi-alloy, Ti) and Fe-silicides occur in these xenoliths but are constrained to small voids. The close spatial occurrence of an assemblage (i.e. calcite + mantle silicates) requiring oxygen fugacities close to the CCO buffer (graphite/diamond-CO-CO_2, i.e.) with another assemblage requiring oxygen fugacities 10 log-units below this buffer motivates the formulation of a low-temperature hypothesis. In this model, the SiC in the voids would be precipitated from an ultra-reduced highly fractionated fluid that is not in equilibrium with the adjacent silicate and carbonate phases of the xenolith itself.

In this study, we demonstrate experimentally that olivine and orthopyroxene in equilibrium with SiC are almost purely magnesian, containing only 2-7‰ of their Fe-endmembers. Consequently, close spatial association of moissanite and silicates with X_{Mg} ≤0.9 demonstrates chemical disequilibrium and indicates temperatures lower than those allowing for diffusional equilibration. Secondly, we combine thermodynamic data for Si- and Fe-carbides, stoichiometric FeSi-compounds and FeSiC alloys with silicate data to compute P-T-f_{O2} phase relations, phase compositions and hence the stability of moissanite for an Fe-Mg-Si-O-C peridotitic bulk composition.

Methods

High pressure experiments

One set of experiments reacted San Carlos olivine (X_{Mg} = 0.90) and a natural orthopyroxene from the Urals (X_{Mg} = 0.91) with SiC at 2 GPa, 1300 and 1500°C; and at 10 GPa, 1500 and 1700°C (Table 1). The 2 GPa experiments were performed in an endloaded, 14 mm bore diameter piston cylinder employing a talc–pyrex–graphite-MgO assemblage. The 10 GPa experiments were performed in a multi anvil employing 18 mm Cr_2O_3-doped MgO-octahedra on a 11 mm truncation. The multi-anvil assemblage contained a zirconia insulator, a stepped $LaCrO_3$ furnace, inner MgO pieces, and a Mo ring and disc between the furnace and WC cube truncation faces. In both apparatus B-type PtRh thermocouples were used. In experiment G#1 we intended to monitor oxygen fugacity by using an Ir sensor (Woodland and O'Neill, 1997), but a ternary Fe-Si-Ir alloy resulted for which thermodynamic data are not available.

A second set of experiments intended to synthesize SiC first at low temperature from tetrakis-silane ((($(CH_3)_3Si)_4Si$) or stearic acid ($C_{18}H_{36}O_2$) and olivine + serpentine or talc + MgO at 500°C, 0.2 GPa. These experiments did not result in SiC or equilibrium assemblages. We then aimed at synthesizing SiC + magnesio-silicates at 1000-1600°C, 2-10 GPa employing talc + MgO and tetrakis-silane as starting materials. At 6 GPa, 1300°C we obtained SiC + olivine + opx, but the assemblage was unequilibrated with strong compositional zoning in the silicates and further experiments in this direction were abandoned.

Thermodynamic data

Thermodynamic databases suited for calculating phase equilibria in Earth sciences mostly contain only silicates and oxides. In order to perform calculations in mantle materials at ultra-reduced conditions, alloys, carbides and silicides were combined with silicate data from Holland and Powell (2011). The data sources employed were: Cementite (Fe_3C): calorimetric data from Gustafson (1985), volume data from Li et al. (2002), Wood et al. (2004); Fe_7C_3: Djurovic et al. (2011), Mookherjee et al. (2011), Nakajima et al. (2011), Moissanite (SiC): Lacaze and Sundman (1991), Miettinen (1998), Aleksandrov et al (1989), Li and Bradt (1987). The FeSiC-alloy and the iron silicides are based on the 1-atm phase diagram by Lacaze and Sundman (1991) and on the volume EoS of Brosh et al. (2009). Details of the adoption of the ternary metal, carbides and silicides are given in a forthcoming paper (Golubkova et al. 2014). Phase relations were computed by free energy minimization with the Perple_X software package (Connolly JAD 2009).

Table 1 Experimental phase compositions and calculated oxygen fugacity

	P [GPa]	T [°C]	t [h]	$X_{Mg}^{olivine}$	X_{Mg}^{opx}	Metal or silicide [wt%]					$\Delta logf_{O2}^{IW}$ SiC-C-ol-opx
						Fe	Ni	Si	Ir		
G#1	2	1300	48	---*	0.982(5)	8.1(26)	0.24(9)	59.4(43)	20.3(20)	+quartz	---
G#2	2	1500	15	0.997(1)	0.998(1)	77.4(13)	3.3(4)	18.0(4)	---	Fe_2Si	-6.4
G#3	10	1500	72	---*	0.995(1)	60.3(10)	2.7(8)	35.0(17)	---	FeSi	(-4.9)**
						46.6(21)	0.9(5)	49.5(16)	---	$FeSi_2$	
						30.8(6)	1.3(2)	64.4(20)	---	$FeSi_4$	
G#6	10	1700	54	0.993(2)	0.995(1)	75.6(11)	2.6(9)	20.8(4)	---	Fe_2Si	-5.6

*Rims or silicide-free exsolution zones of previously San Carlos olivine too small to measure by EMPA.
**From equilibrium (1), the FeSi is most likely formed metastably in the early stage of the experiment.

Results

Phase equilibria and phase compositions

Experiments at 2 GPa, 1500°C (G#2) and 10 GPa, 1700°C (G#6) resulted in equilibrated olivine + orthopyroxene + SiC plus iron silicides (Table 1, Figure 1). The experiment at 10 GPa, 1500°C (G#3) resulted in coarse opx + SiC plus three iron silicides (FeSi, $FeSi_2$, and $FeSi_4$) and olivine that did not develop equilibrium rims large enough for measurement. These charges did not contain any residual silicates from the starting material, but in both the 1500°C and 1700°C experiments at 10 GPa, most of the San Carlos olivine disintegrated to an almost Fe-free olivine containing many submicron iron silicide inclusions (Figure 1b). These inclusions are too small to determine their stoichiometry. The equilibrium X_{Mg} of olivine and opx in these experiments are 0.993-0.998 (Table 1). Initially, we added Ir metal as internal oxygen fugacity monitor (Woodland and O'Neill, 1997; Stagno and Frost, 2010), this experiment (G#1) at 1300°C, 2 GPa yielded opx (X_{Mg} = 0.982) and quartz. However, abundant residual starting material San Carlos olivine testifies for incomplete equilibration questioning whether the X_{Mg} of the opx of this experiment corresponds to equilibrium with SiC. We have thus only used experiments #2, #3 and #6 for oxygen fugacity calculations.

Oxygen fugacity (f_{O2}) in the experiments

For the experiments, which yielded iron silicides and equilibrium compositions of olivine and orthopyroxene, oxygen fugacities were calculated from equilibrium (1) and (2), employing the olivine and orthopyroxene activity formulations from O'Neill and Wall (1987).

$$1 \text{ ferrosilite} = 2\,\text{FeSi} + 3\,O_2 \quad (1)$$

$$1 \text{ fayalite} = 1\,Fe_2Si + 2\,O_2 \quad (2)$$

At equilibrium,

$$0 = 2G_{FeSi}^{P,T} + 3\,G_{O2}^{P,T} - G_{fs}^{P,T} - R\,T\,\ln\frac{a_{O2}^3}{a_{fs}^{opx}}, \quad (3)$$

FeSi being a pure phase and

$$0 = G_{Fe2Si}^{P,T} + 2\,G_{O2}^{P,T} - G_{fay}^{P,T} + R\,T\ln\frac{a_{O2}^2}{a_{fay}^{olivine}}, \quad (4)$$

Fe_2Si being a pure phase. This leads to

$$logf_{O2} = \left(-\frac{\left(2\,G_{FeSi}^{P,T} + 3\,G_{O2}^{P,T} - G_{fs}^{P,T}\right)}{2.3025\,R\,T} + loga_{fs}^{opx}\right)/3, \quad (5)$$

using the measured opx composition, $\log f_{O2}$ = – 11.2 (5) or $\Delta logf_{O2}^{IW} = -4.9$ results for G#3 at 10 GPa, 1500°C. For the Fe_2Si equilibrium (4)

$$\log f_{O2} = \left(-\frac{\left(G_{Fe2Si}^{P,T} + 2\,G_{O2}^{P,T} - G_{fay}^{P,T}\right)}{2.3025\,R\,T} + \log a_{fay}^{olivine}\right)/2, \quad (6)$$

measured olivine compositions yield $\log f_{O2}$ = – 13.3 (10a, 10b) or $\Delta logf_{O2}^{IW} = -4.6$ for G#2 at 2 GPa, 1500°C; and $\log f_{O2}$ = – 11.9 (4) or $\Delta logf_{O2}^{IW} = -5.8$ for G#6 at 10 GPa, 1700°C. For the two experiments G#2 and G#6, which resulted in equilibrated olivine and orthopyroxene, oxygen fugacities were also calculated from

$$1 \text{ ferrosilite} + 1 \text{ graphite/diamond} = 1 \text{ fayalite} + 1\,\text{SiC} + O_2 \quad (7)$$

At equilibrium

$$0 = 1\,G_{SiC}^{P,T} + 1\,G_{O2}^{P,T} + 1\,G_{fay}^{P,T} - 1\,G_{fs}^{P,T} - 1\,G_{graph/diam}^{P,T} + R\,T\ln\frac{a_{O2}\cdot a_{fay}^{olivine}}{a_{fs}^{opx}} \quad (8)$$

and

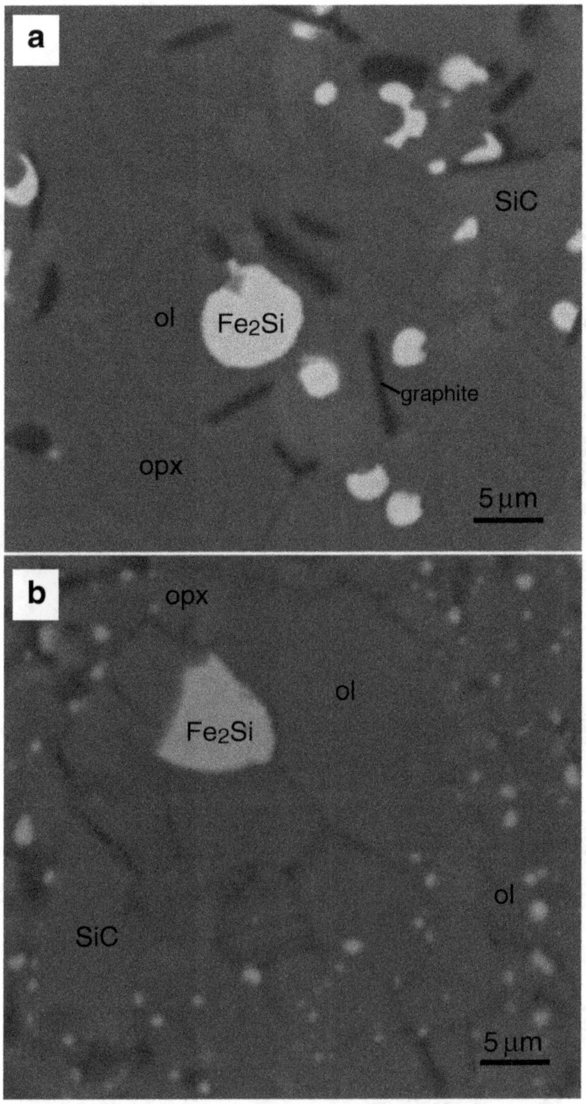

Figure 1 BSE images of run products from (a) experiment G#2 (1500°C, 2 GPa) and (b) experiment G#6 (1700°C, 10 GPa). Olivine and orthopyroxene have $X_{Mg} > 0.99$ and cannot be well distinguished from each other. Note the graphite flakes at 2 GPa. At 10 GPa diamonds result mostly in plug-outs during preparation and are hence rarely visible. Note that in **(b)** the original San Carlos olivine forms almost pure forsterite with many Fe-rich inclusions which stoichiometry cannot be determined due to their small size.

$$log f_{O2} = \frac{\left(G_{fs}^{P,T} + G_{graph/diam}^{P,T} - G_{fay}^{P,T} - G_{SiC}^{P,T} - G_{O2}^{P,T}\right)}{2.3025\ R\ T} - log a_{fay}^{olivine} + log a_{fs}^{opx}. \quad (9)$$

SiC and graphite or diamond being pure phases. The resulting oxygen fugacities are $\log f_{O2} = -15.1$. (6) or $\Delta log f_{O2}^{IW} = -6.4$. at 2 GPa, 1500°C (G#2), and $\log f_{O2} = -10.6$ (4) or $\Delta log f_{O2}^{IW} = -5.6$ at 10 GPa, 1700°C (G#2).

Calculation of T-f_{O2} sections

Reactions governing reduced phases in the mantle

To understand the succession of phase assemblages with decreasing oxygen fugacity, we calculated f_{O2}-temperature diagrams (Figure 2) and phase compositions (Figures 3 and 4) for a model harzburgite at 2 and 10 GPa. For reference, we also give the quartz-fayalite-magnetite (QFM), the graphite/diamond-CO_2-CO equilibrium (CCO), and the iron-wustite buffer (IW), even if these do not occur as reactions in the harzburgite phase diagram. The position of the iron-wustite buffer is calculated from Campbell et al. (2009), which yields about half a log-unit lower values than the formulation of O'Neill (1988).

In the carbon-bearing upper mantle, the iron-wustite equilibrium does not occur, its equivalent leading to the appearance of reduced Fe^0 are the iron carbide forming reactions

$$6\ \text{olivine} + 2\ \text{graphite} = 3\ \text{opx} + 2\ \text{cementite}\ (Fe_3C) + 3\ O_2 \quad (10a)$$

and

$$14\ \text{olivine} + 3\ \text{diamond} = 7\ \text{opx} + 2\ Fe_7C_3 + 7\ O_2 \quad (10b)$$

which are only stable at relatively low temperatures, i.e. 1140 to 1260°C at 2 to 10 GPa. At higher temperatures, carbides do not form but a metal alloy results from

$$2\ \text{olivine} + 2x\ \text{graphite/diamond} = 1\ \text{opx} + (2+2x)\ FeSiC^{alloy} + 1\ O_2 \quad (11)$$

where $FeSiC^{alloy}$ is a ternary metal solid solution. The stoichiometry of reaction (11) depends on the alloy composition. The molar fraction of C in the alloy ranges from 0.13 to 0.33 at 2-10 GPa, 1150-1700°C. At oxygen fugacities of initial metal formation, Si-fractions are <0.1 mol% in the alloy. For a graphite/diamond saturated mantle with a bulk X_{Mg} of 0.9, reactions (10a, 10b) and (11) occur at 2 GPa about one log-unit below the iron-wustite buffer and about one log-unit above IW at 10 GPa (Figure 2a,b). About 5-9 log-units below the carbide-forming reactions, increasing chemical potential of Si leads to the replacement of Fe_3C or Fe_7C_3 by the FeSiC-alloy then containing significant Si (Figure 5b,d). Silicon contents in the alloy increase with decreasing oxygen fugacity leading to a phase transition from a face centered to a body centered metal structure (fcc and bcc in Figure 5).

Silica carbide forms principally through a reaction involving moissanite, olivine, orthopyroxene and elemental carbon as graphite or diamond (moissanite-olivine-opx-carbon: MOOC)

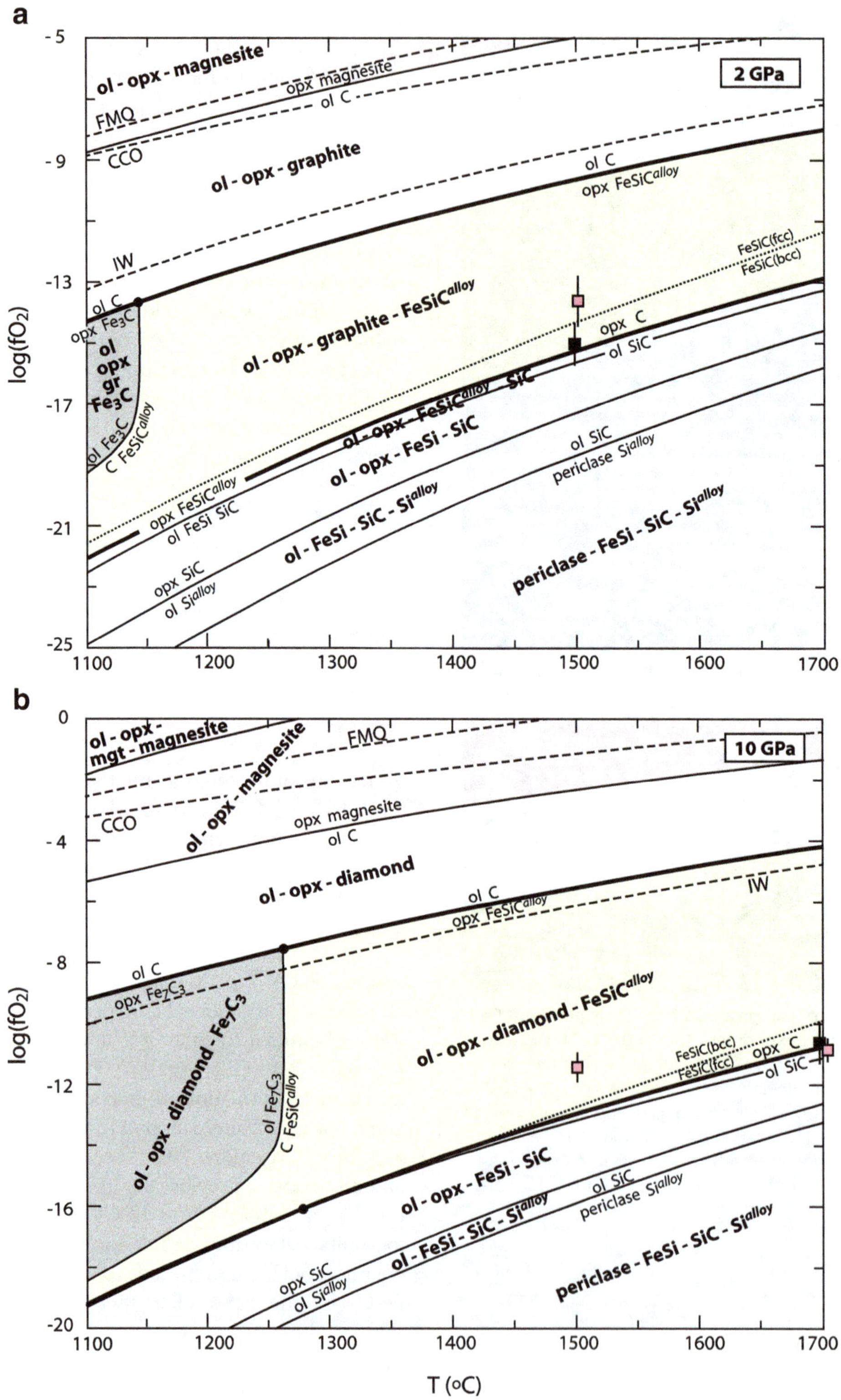

Figure 2 Calculated f_{O2}-temperature diagrams for carbon-bearing harzburgitic peridotite simplified in Mg-Fe-Si-C-O_2 at (a) 2 GPa and (b) 10 GPa. The saturation hierarchy at moderately oxidizing conditions is chosen as olivine > orthopyroxene > graphite/diamond > magnetite. The FMQ, CCO and IW buffers are given for reference. Squares are calculated oxygen fugacities for the experiments, red fill for f_{O2}'s calculated from the iron silicide involving equilibria (1) and (2), black fill for f_{O2}'s calculated from the SiC-involving reaction (7).

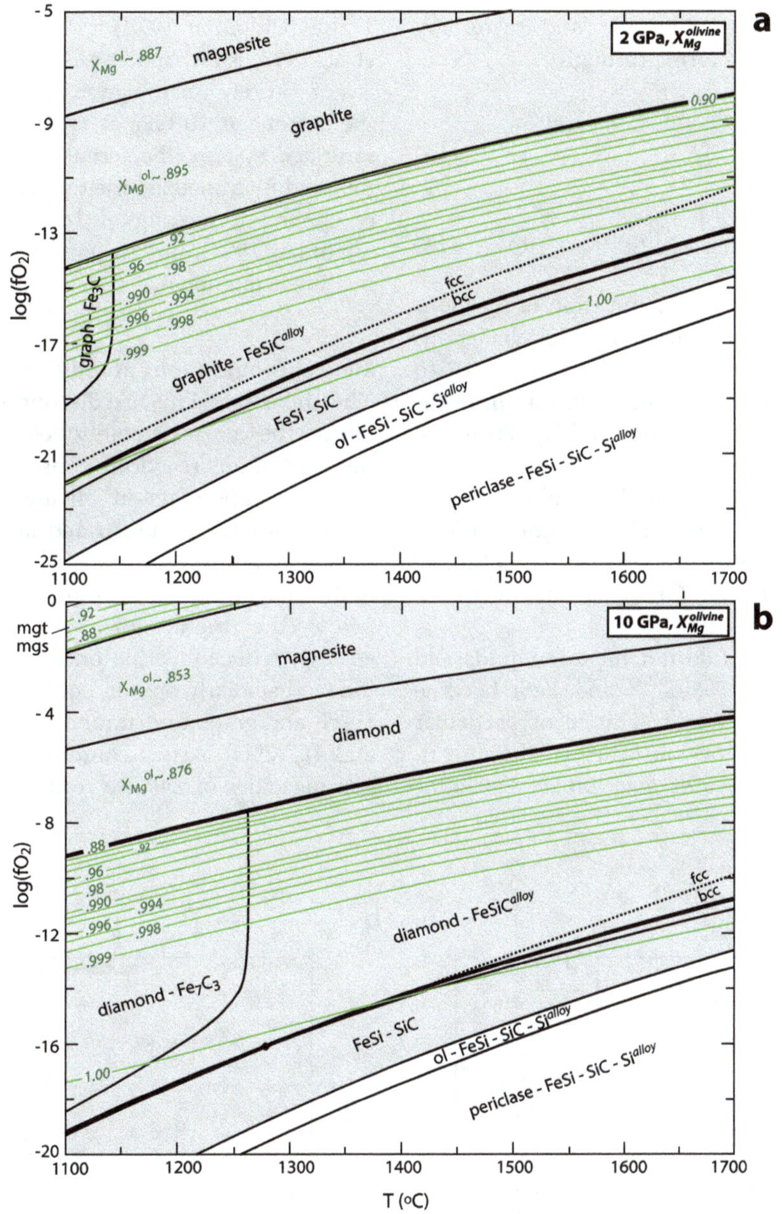

Figure 3 Isopleths of olivine composition in terms of X_{Mg} in the same harzburgitic bulk composition as in Figure 2 for (a) 2 GPa and (b) 10 GPa. Within two log-units of the metal forming reaction, $X_{Mg}^{olivine}$ increases to 0.99, when reaching the SiC-buffer, olivines are essentially Fe-free. $X_{Mg}^{olivine}$ is almost constant in the stability fields of magnesite or graphite/diamond. The grey field corresponds to olivine + orthopyroxene saturation and only additional phases are indicated.

$$1\text{ opx} + 1\text{ graphite/diamond} = 1\text{ olivine} + 1\text{ SiC} + 1\text{ O}_2 \qquad (12)$$

at 2 GPa at 7.7-4.8 log-units (1100-1700°C) below equilibria (10a, 10b) and (11). At 10 GPa this difference is 9.9-6.5 log units (1100-1700°C). This corresponds to a $\Delta\log f_{O2}$ of IW-9 to IW-6, the difference decreasing with temperature. Along a mantle adiabat, moissanite forms at IW-8.0 at 2 GPa and at IW-6.7 at 10 GPa. Reaction (12) was previously identified as being responsible for the appearance of moissanite in the upper mantle (Mathez et al. 1995; Ulmer et al. 1998).

Further reduction causes the FeSiC-alloy to be replaced by stoichiometric iron silicide FeSi. At 2 GPa this happens about ½ a log-unit below the SiC-buffer (12), while at 10 GPa, the FeSi-forming reaction intersects reaction (12) leading to an invariant point. At about 3 and 5 log-units below the SiC-forming reaction (12), opx and then olivine (at these conditions pure enstatite and forsterite, respectively) become unstable and a ternary metal alloy of almost pure Si forms through

$$1\ \text{opx} + 0.01\ \text{SiC} = 1\ \text{olivine} + 1.01\ \text{Si}^{\text{alloy}} + 1\ \text{O}_2 \quad (13)$$

$$1\ \text{olivine} + 0.01\ \text{SiC} = 1\ \text{periclase} + 1.01\ \text{Si}^{\text{alloy}} + 1\ \text{O}_2 \quad (14)$$

where the Si-alloy has 1 mol% C. Completion of reactions (13) and (14) would leave no oxidized Si. The calculations do not indicate any Fe in the Si metal and only 0.2 to 1.3% C. Reaction (13) is relevant in nature as pure Si-metal attached to SiC has been reported by Trumbull et al. (2009).

Summarizing, a harzburgitic mantle would have the following succession of reduced phases: graphite/diamond → (Fe_3C/Fe_7C_3) → Fe-rich FeSiC^{alloy} → SiC → FeSi → Si^{alloy}. At pressures ≥10 GPa, FeSi may form before moissanite.

Note that our calculations do not include liquids and that at high temperatures FeSiC^{alloy} and FeSi become metastable with respect to liquid. This is of particular relevance at oxygen fugacities just below the metal-forming reaction (11) where a *de facto* binary FeC-alloy is almost Si-free. The Fe-C eutectic is located at 1150°C, 1 atm (Chipman 1972) and 1210°C, 10 GPa (Hirayama et al. 1993; Rohrbach et al. 2014), although this location is not entirely unambiguous as Lord et al. (2009) report the eutectic at 10 GPa at 1420°C. In a graphite/diamond-saturated system, the ternary FeSiC-alloy would then be replaced by a metallic melt phase. FeSi has a temperature of congruent melting of 1410°C at 1 atm (Lacaze and Sundman 1991), but to our knowledge its high pressure melting is unknown.

Mineral compositions at reduced conditions

The above reactions are discontinuous reactions that delineate the various stability fields. For the silicate mantle the continuous reactions consuming the fayalite and ferrosilite components in olivine and orthopyroxene are more dramatic. As illustrated in Figures 3 and 4, X_{Mg} in olivine is almost invariant in each field of coexistence with magnesite or graphite/diamond. Immediately below the metal forming reaction, Fe^{2+} becomes strongly reduced, within only 2 log-units, the X_{Mg} of olivine (and similarly opx) in equilibrium with metal or Fe_3C/Fe_7C_3 and graphite/diamond increases to ~0.99 (Figures 3 and 4). When oxygen contents corresponding to the oxygen fugacities of SiC are reached, the equilibrium X_{Mg} in

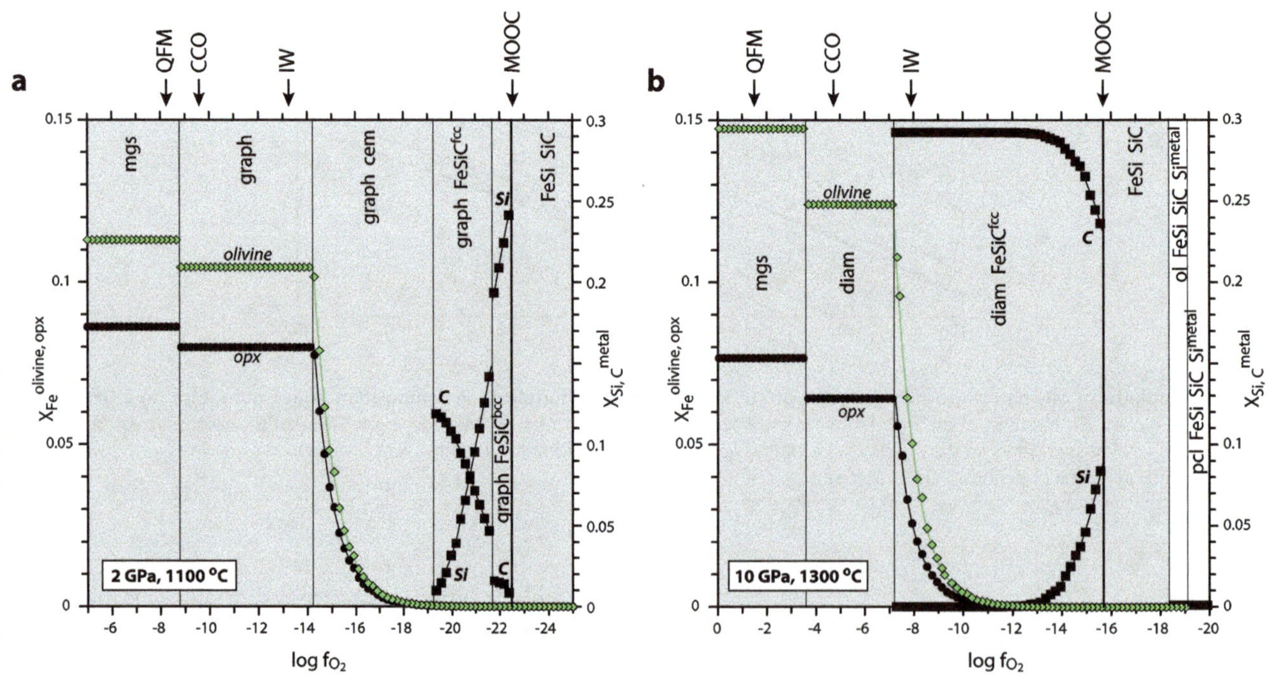

Figure 4 Isothermal composition – f_{O2} sections at (a) 2 GPa, 1100°C and (b) 10 GPa, 1300°C. At oxygen fugacities below the carbide- or metal-forming reaction, $X_{Fe}^{olivine}$ drops rapidly to per mill values. The grey field corresponds to olivine + orthopyroxene saturation and only additional phases are indicated. With decreasing oxygen fugacity the FeSiC-alloy becomes more and more Si-rich, leading to the fcc-bcc phase transition and then to iron silicide stability. Note that the Si-alloy formed orthopyroxene and olivine breakdown is essentially pure Si with 1 mol% C and almost no iron. The calculations are done at discrete f_{O2}-values represented by symbols; green: $X_{Mg}^{olivine}$, black: X_{Mg}^{opx}, red: C content in metal alloy, blue: Si-content in metal alloy.

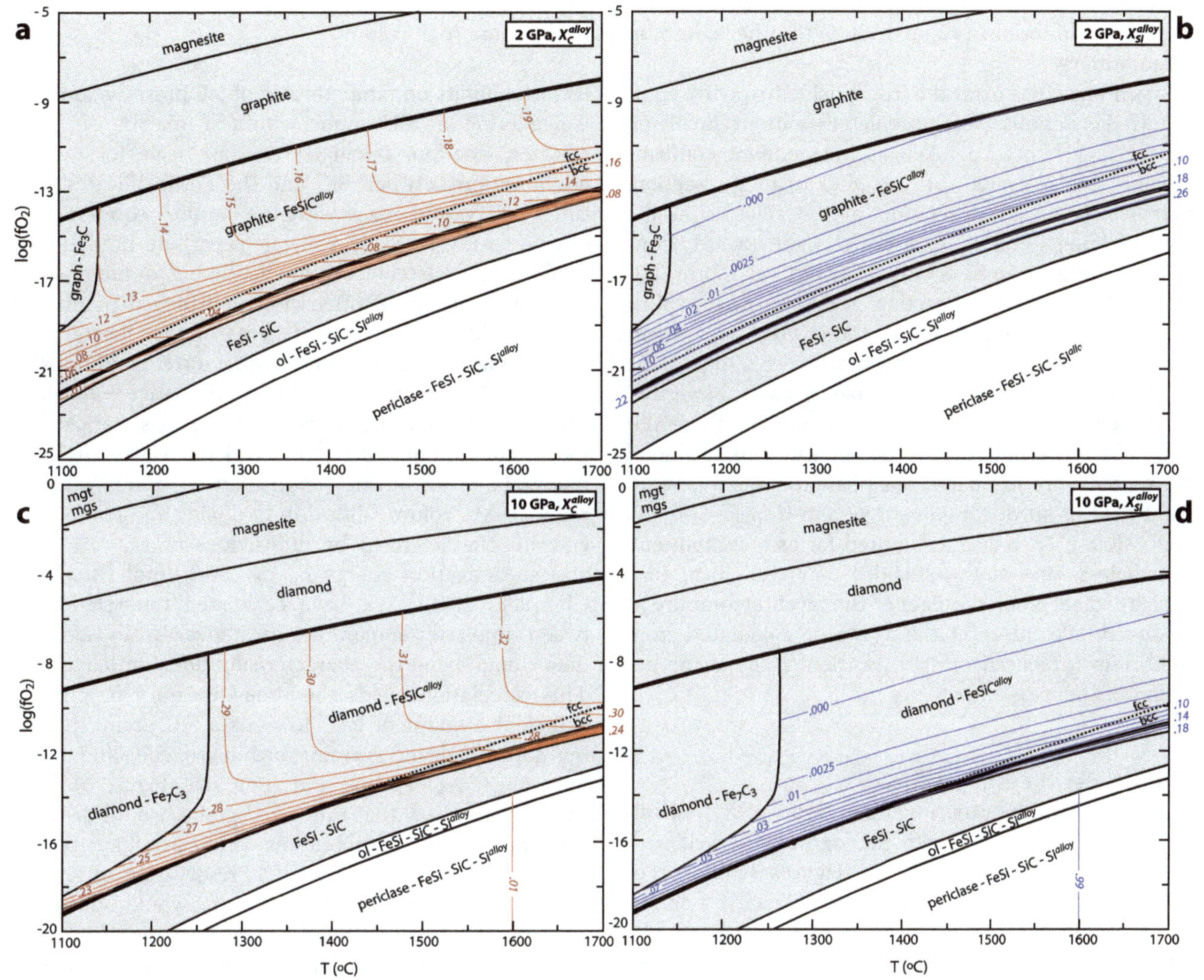

Figure 5 Calculated isopleths of metal composition for the ternary FeSiC alloys at 2 and 10 GPa. (a) and **(c)** C-content in mol-fraction; **(b)** and **(d)** Si-content in mol-fraction. Note the phase transition from the fcc to bcc alloy structure with increasing Si-content when decreasing oxygen fugacity. At the metal-forming reaction, C-contents increase with temperature and decrease with pressure, Si contents are below the lowest increment of the calculations. At 2 GPa, calculated C- and Fe-contents of the Si-alloy stable at ultra-reducing conditions are <1 mol%. Grey fields correspond to olivine+opx saturation.

the silicates is >0.999, almost invariant with pressure and temperature (at least for 1100-1700°C, 2-10 GPa). This confirms the experimental findings. Concomitant with the increase in X_{Mg} of the silicates, an increase of the Si-content of the FeSiC-alloy is predicted (Figures 4 and 5), mirrored by a decrease of X_{Fe}^{metal} from ~0.90 to 0.70 at 2 GPa and from 0.76 to 0.66 at 10 GPa.

Silica carbide stability with temperature and pressure

A major question concerning the stability of SiC in the mantle regards the evolution of the difference in oxygen fugacity between the metal or Fe-carbide forming (10a, 10b, 11) and the SiC-forming reactions (12). This difference, $\Delta \log f_{O2}{}^{(12)\text{-}(10a,10b,11)}$, decreases with increasing temperature but at near-adiabatic temperatures only by -0.1 log-units per 100°C. With increasing pressure there is a slight increase of $\Delta \log f_{O2}$ between the moissanite forming reaction and the equivalent of IW in the mantle such that along a mantle adiabat $\Delta \log f_{O2}{}^{(12)\text{-}(10a,10b,11)}$ remains almost constant at ~ -4.5.

Discussion

Oxygen fugacities calculated from the experiments

In Figure 2a and b we plot the oxygen fugacities calculated for the three experiments with olivine and/or orthopyroxene equilibrated with SiC. The oxygen fugacities calculated from the experiments through the moissanite-forming equilibrium (7) are in perfect agreement with the calculated phase diagram. They are also in good agreement with the values derived from Ulmer et al. (1998) from experiments on a Mg-Si-C-O system that cross-compared equilibrium (12) at 1500°C to other common oxygen buffers.

Oxygen fugacities from the iron silicide involving equilibria (red symbols) were calculated without involving SiC and may hence be taken as independent confirmation of the f_{O2}-value of the SiC-forming reaction. Nevertheless, the interpretation of the silicides in the two experiments at 1500°C is not straightforward. Congruent melting of Fe_2Si occurs at 1212°C at 1 atm, but the dP/dT slope of this reaction is unknown. It is thus well possible that the Fe_2Si in the 2 GPa experiment G#2 crystallized upon quench from a FeSiC-liquid. At 10 GPa, the silicides should be stable as solids. Nevertheless, the experiment at 1500°C (G#3) contains three silicides, FeSi, $FeSi_2$ and $FeSi_4$. Together with opx, diamond and SiC, these are six phases, one phase too much for a divariant field of a five component system (Mg-Fe-Si-C-O), but possible if Ni is also accounted for as a component. Nevertheless, one of the silicides may well form in a transient stage, when the FeO of the fayalite component is reduced. We interpret the f_{O2}-value calculated from equilibrium (1) as reflecting a transient stage at the beginning of the experiment.

X_{Mg} of silicates coexisting with SiC

The results from the experiments in terms of X_{Mg} of silicate minerals coexisting with SiC agree very well with the calculated phase compositions (Figures 3 and 4). Together with the constraints on oxygen fugacity, a coherent picture forms: moissanite can only be expected in a mantle or any rock with silicates essentially devoid of Fe^{2+}. This unsurprising finding has strong implications for the origin of SiC in mantle rocks. It demonstrates that any significantly large domain of mantle at oxygen fugacities of SiC-stability (i.e. void of Fe^{2+} in silicates and oxides) has not yet been identified.

Both the experiments and the calculations demonstrate that for plausible mantle compositions with SiC stable, the X_{Mg} of coexisting olivine or orthopyroxene is >0.993. Hence SiC cannot be in equilibrium with typical mantle that has olivine and pyroxenes with X_{Mg} of 0.88-0.92. The experiments only constrain the X_{Mg} of olivine and orthopyroxene in equilibrium with SiC. Nevertheless, in nature, X_{Mg} of serpentine is generally higher than that of coexisting orthopyroxene, which is similar to that of clinopyroxene and X_{Mg} of olivine is lowest among these phases. One can thus safely assume that for all silicates characteristic for peridotites (except of garnet) similarly high X_{Mg} of >0.993 are to be expected.

As SiC is found in mantle rocks and kimberlites from which anomalous X_{Mg} values for the silicate phases are not reported, the formation mechanism of SiC should explain how average X_{Mg} mantle silicates and SiC persist in the same rock volume.

Diffusion limits on temperatures of SiC preservation

Considering a SiC grain within a matrix of FeMg-silicates, one can calculate the time scale for diffusive equilibration between SiC and the FeMg-silicates. With time, SiC would react with the fayalite component of olivine to form opx and metal or carbide until the SiC grain is completely consumed. Taking a putative SiC grain in a matrix of otherwise undisturbed mantle silicates, Fe-Mg diffusion coefficients of the silicates allow for calculation of the time-temperature conditions for which a given grain size of SiC would react with olivine until its complete destruction. Assuming spherical symmetry, the amount of Fe^{2+} needed to oxidize all Si of a SiC grain in an olivine (Fo_{90}) matrix occurs within 2.3 radii. Fe-Mg volume diffusion in olivine and orthopyroxene are characterized by diffusivities of $D_0 = 10^{-7}$ m^2/s and an activation energy E_A of 250 kJ/mol (Brady and Cherniak, 2010). We have calculated the relation between time and temperature for a given grain size of radius r employing the characteristic diffusion length $2.3r$. This calculation yields the time necessary to diffuse a sufficient amount of Fe^{2+} towards a SiC grain, then being consumed and forming orthopyroxene (and metal), for a given temperature and grain size (Figure 6). These calculations show that mm-sized grains would disappear through diffusive equilibration within a million years at temperatures above 800-900°C. Preservation of SiC in a mantle mineral matrix with $X_{Mg} = 0.9$ would not be possible at adiabatic mantle temperatures, even SiC grains of 3-10 cm would only last for 10'000 years. This conclusion remains robust when employing the experimentally determined range of $D_0 - E_A$ pairs (for summary of diffusion data see Brady and Cherniak, 2010).

These time spans could be prolonged if SiC forms in veins of essentially Fe-free silicates, but such mineral compositions are not reported for peridotites or kimberlites. Regarding the putative high-temperature origin of SiC, at temperatures of the asthenospheric mantle, SiC grains of initially >50 cm would be necessary to persist for >1 million years. It is thus unlikely that SiC would survive any period of sustained high temperatures in the mantle.

A model for SiC formation in nature

Terrestrial environments are typically too oxidizing to permit SiC formation, thus exceptional circumstances need to be explored. One plausible scenario starts with a slightly reduced graphite/diamond-saturated COH-fluid that evolves towards an ultra-reduced composition through the formation of first carbonates or spinel solid solution and then hydroxide minerals, all from anhydrous silicates. In this model, the fluids react with its immediate surroundings while percolating upwards, leaving newly formed carbonates, spinel or hydroxides behind. The fluid-mediated formation of magnetite component in spinel from mantle silicates drives any fluid to more reducing compositions (away from the O_2-corner in Figure 7). The fluid-mediated formation of carbonates (from mantle silicates) leads to more reduced residual fluids when the starting COH-fluid has a composition to the reduced side of the CO_2-H_2O tie-line (Figure 7). Fluids from sediments rich in organic material are expected to already have H:O-ratios slightly to the H-rich side of the H_2O-C tie-line (Connolly JAD 1995; Figure 7c). Nevertheless, as long as they are saturated in graphite, also fluids slightly to the O-rich side of H_2O could be suitable precursors (Figure 7b). The following steps would then lead to a fluid that may mediate SiC formation:

1a) Starting from fluids with some CO_2, the precipitation of carbonates through reaction of the fluid with silicates (e.g. olivine + CO_2 = opx + magnesite; or cpx + olivine + CO_2 = calcite + opx) removes oxygen from the fluid, causing the fluid composition to evolve away from CO_2, the star in Figure 7b representing a hypothetical starting composition.

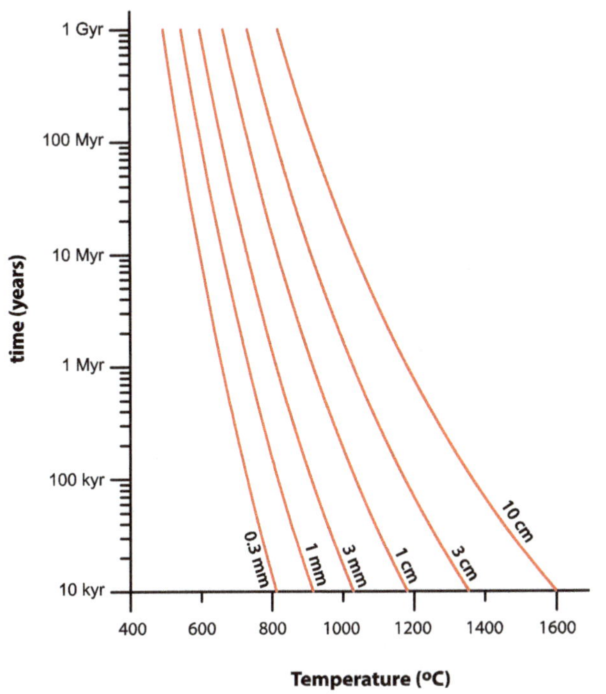

Figure 6 Illustration of the diffusion time – temperature – grain size relation for SiC in an olivine matrix with X_{Mg} = 0.9. For a given grain size, this graph illustrates the time of SiC decomposition through diffusive equilibration with the Fe-component in olivine, effectively delimiting the temperature at which SiC could reside in a mantle with X_{Mg} = 0.9. For details see text.

1b) Alternatively, fluid percolation in mantle rocks may lead to the formation of magnetite component in spinel and hence consumption of oxygen from the fluid. This would lead to co-precipitation of graphite/diamond and the rising fluid would evolve towards the left along the graphite/diamond saturation surface of Figure 7. In both cases, the fluid could only evolve to more reducing compositions if it leaves the zones of carbonate or spinel formation behind.

2) If the initial fluid composition is at H:O <2, steps 1a or b would need to fractionate the fluid composition to H:O >2. Any fluid initially already to the H-rich side of the H_2O-graphite/diamond tie-line or H_2O-maximum (fluids are mostly H_2O at the inflection of the graphite saturation surface, Figure 7) would evolve towards more methane-rich compositions by removal of H_2O from the fluid. This can easily be achieved by the precipitation of hydrous phases when the fluid percolates through mantle rocks. In peridotitic compositions, such hydrous phases could be hydrous alphabet phases at higher pressure, otherwise humites, serpentine, brucite or talc. Removal of H_2O from the fluid forces the fluid to evolve along the graphite/diamond saturation surface continuously precipitating graphite/diamond concomitantly with the formation of hydrous magnesium silicates. Note that step 2 requires temperatures within the stability fields of these phases, which, in equilibrium with olivine, for the most part do not exceed 700-800°C (Fumagalli and Poli, 2005).

3) Finally, still fractionating e.g. serpentine or brucite, such fluids could reach oxygen fugacities of IW-5 to IW-8 where SiC and Fe-silicides become stable. Note that the compositional change in the fluid from moderately reducing conditions near the IW-buffer to ultra-reducing conditions that allow for SiC formation is small in terms of X_O, and is graphically not resolvable in the C-O-H ternary. This final step of fluid fractionation occurs very close to the intersection of the graphite-saturation surface with the C-H sideline (Figure 7).

The simplest case enabling SiC formation is to start with an aqueous fluid that contains more methane than CO_2 (molar), i.e. with a bulk composition more reducing then the H_2O-maximum. Sequestration of oxygen from the fluid by hydroxide minerals would then directly lead to extremely reducing conditions. In mantle compositions, serpentine or brucite are only stable at relatively low temperatures (<700°C) as could be expected for fluid-percolation in supra-subduction mantle or during orogenic metamorphism. In fact, SiC has been reported from serpentinites (Xu et al. 2008). Moissanite accounting for 8% of a rock with a bulk X_{Mg} = 0.998 has been reported from a beach pebble which matrix is brucite-dominated (Di Pierro et al. 2003). This latter rock contains minor (remnant) chromite and orthopyroxene of

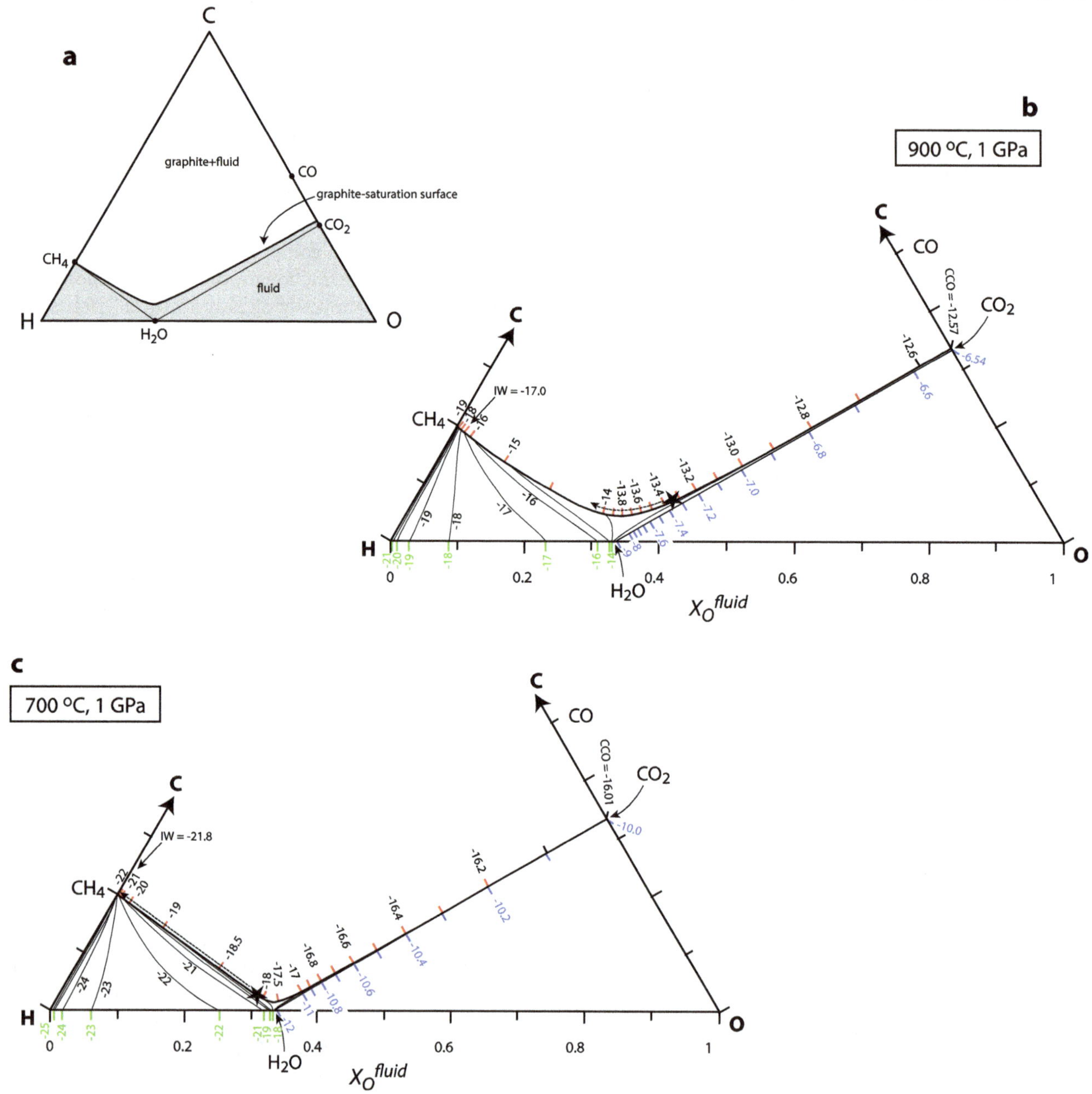

Figure 7 COH-diagrams for fluids illustrating the graphite saturation surface (a) and with calculated graphite-saturation surface and oxygen fugacitiy values and isopleths for 2 GPa, 900°C (b) and 2 GPa, 700°C (c). The solid line traversing the diagram represents the graphite-saturation surface, possible fluid compositions plot below. The CCO-buffer locates per definition where the graphite-saturation surface intersects the C-O side of the triangle. Note that in nature, we do not expect fluid compositions with significantly more oxygen than the H_2O-CO_2 tie-line, rendering the field of oxidized geological fluid compositions extremely narrow, in fact graphically almost not resolvable at 700°C. The red numbers give oxygen fugacities (in log-units) on the graphite saturation surface ($a_C = 1$), the blue numbers on the H_2O-CO_2 tie-line approximated by a C-activity of 10^{-6}, the green numbers those between H and H_2O. Almost the entire width of fluid compositions on the graphite-saturation surface spans a relatively small range in oxygen fugacities equivalent to CCO to IW. More reducing fluids are all on the H-rich side of possible fluid compositions, essentially within a X_O^{fluid} of 0.01. The star in **(b)** indicates a putative fluid from an organic rich sediment that also has some CO_2; to drive this composition to the reducing side, CO_2 has to be removed from the fluid in carbonates. Magnesite and dolomite are only stable to oxygen fugacities of the H_2O-maximum (the inflection in the graphite-saturation surface), but calcite until an approximate $\log f_{O2}$ of -15 (900°C) or -19 (700°C). Fluids from carbonate-free organic-rich sediments are expected to lie on the graphite/diamond saturation surface on the H-side of the H_2O-C tie-line, as indicated by the star in **(c)**. These fluids move away from H_2O, on the graphite-saturation surface co-precipitating graphite or diamond, if reaction with peridotite produces hydrous phases (phase A at high pressure, serpentine or brucite at low pressure). Eventually such fluids would become more reduced than the IW buffer and would possible reach a H:O ratio of >1000 necessary to precipitate SiC. Note that there is no direct proportionality between oxygen fugacity and X_O^{fluid}.

typical mantle composition and may represent a completely metasomatized peridotite. Otherwise, moissanite is rare in most rocks and mostly described from heavy mineral concentrates. Phases grown with or onto moissanite are Fe-silicides, Si- and other metals. Unfortunately, the textural context and relations to silicates remain mostly unknown providing little evidence for the type of reactions leading to moissanite.

Step 3 is likely to occur through grain scale equilibria where new hydrous minerals form in equilibrium with the fluid but where the bulk rock is in disequilibrium with this fluid. Alternatively, discrete metasomatic zones could host SiC and hydrous phases as in the rock from Turkey described by Di Pierro et al. (2003). As SiC is a rare mineral, it appears likely that in most cases either reduced fluids are consumed before extreme fractionation or fluid:rock ratios in percolation zones are too high to change fluid-compositions much. However, when such prolonged fractionation occurs, C-O-H fluids may fractionate to H:O ratios sufficiently high for SiC-precipitation.

Evidence for reduced zones in the mantle wedge

Although the mantle wedge is generally thought to be rather oxidized, evidence for a fairly reduced mantle wedge has been reported for xenoliths from the Avacha volcano, Kamtchatka (Ishimaru et al. 2009). The bulk of minerals in these peridotite xenoliths equilibrated at 870-1040°C and oxygen fugacities of $\Delta QFM = -0.2$ to $+1.5$, and most silicates have X_{Mg} values characteristic for the mantle. Nevertheless, some xenoliths contain small amounts of Ni-, Fe- and possibly Ti-metal and Ti-rich Fe silicides mostly along grain boundaries. Serpentine is absent in these xenoliths. The sporadic occurrence of metal and silicides are ascribed to a reducing fluid phase with oxygen fugacities $\leq$ IW that causes precipitation of the reduced phases under disequilibrium with the host rock. Exact temperatures could not be determined but are suggested to lie below the bulk equilibration temperature and serpentine stability (Ishimaru et al. 2009), probably between 700 and 900°C.

Carbon isotope composition

Natural moissanites have a light carbon isotopic composition of $\delta^{13}C = -20$ to $-37‰$ (Mathez et al. 1995, Trumbull et al. 2009). Generally, the main source material for such light compositions is organic matter typically reaching down to -30‰. Upon burial and heating through subduction or orogenesis such organics will generate fluids with low $\delta^{13}C$. However, the relation between the original fluid composition and the $\delta^{13}C$ of precipitated SiC is not straightforward: The fluid evolves to ultra-reduced compositions through removal of H_2O by hydrous phases, this process is necessarily accompanied by graphite or diamond precipitation (see Figure 7). Poulsen (1996) calculated a fractionation factor $\Delta^{13}C_{graphite\text{-}CH4}$ of +1 to +5‰ (750 to 500°C) and hence graphite fractionation renders an already reduced fluid lighter. On the other hand mixing of the fluid with mantle carbon of $\delta^{13}C \sim -5‰$ (Deines 2002) would lead to heavier compositions. Nevertheless, Deines (2002) points out that the light carbon mode of mantle samples of $\delta^{13}C \sim -20$ to $-25‰$ is not necessarily solely related to surface carbon, also the fractionation factor $\Delta^{13}C_{SiC\text{-}CH4}$ remains unknown. Hence a quantitative model is difficult, but the light isotopic carbon composition of moissanite supports an involvement of initially reduced fluids probably originating from organic material.

Kimberlites

Moissanite is reported from kimberlites (e.g. Shiryaev et al. 2011) which contain light $^{13}\delta$ C diamonds (Trumbull et al. 2009), but their direct paragenetic association remains uncertain (Gurney 1989). In one case, SiC was described as part of a complex inclusion in diamond (Leung 1990), the latter effectively halting diffusive equilibration with the surroundings. Moissanite could survive millions of years in the sub-lithospheric cratonic mantle at low temperatures (<800°C). In principle, the same fluid or melt that leads to diamond formation could also precipitate SiC in a later more fractionated stage, which should occur at shallower pressures and lower temperatures. Regardless of these observations, as kimberlites are CO_2-rich magmas whose eruptions are propelled by CO_2, it is not possible that the kimberlite magma would be equilibrated with SiC. Hence, moissanites in kimberlites represent xenocrysts, the kimberlites sampling cold and reduced parts of lithospheric mantle keels.

Oxygen-content and mass balance

Ultra-reduced mantle To appreciate the amount of oxygen to be removed from a given mantle volume to attain oxygen fugacities of the SiC-buffer, the following mass balance is at order: To move a graphite-bearing mantle (with 8.0 wt% FeO) from just below the CCO-buffer ($\log f_{O2} = -8.5$ or $\Delta IW = +3.3$; all values for 2 GPa, 1200°C) to the metal-forming reaction ($\log f_{O2} = -12.9$ or $\Delta IW = -1.1$) where a few per mill of metal form, about 0.05 wt% O_2 have to be removed from the rock. To further move a mantle volume to just above the SiC-buffer ($\log f_{O2} = -20.1$ or $\Delta IW = -8.3$), where essentially all Fe^{2+} is reduced to Fe^0 and where the metal contains about 26 mol% Si, a total of 2.82 wt% O_2 have to be removed. This is equivalent to 6.4 wt% relative of the total oxygen contained in the mantle volume considered, making it hard to imagine how such a process could possibly operate on a large scale. To remove such an amount of oxygen, an extremely high fluid/rock ratio would be required (as ultra-reduced fluids have little oxygen); this is most easily realized if only very small rock volumes in a zone of high fluid throughput are reduced to oxygen fugacities of the SiC-buffer.

Highly fractionated fluid In the mantle, there is little change in oxygen content between oxygen fugacities just below the olivine-opx-magnetite equilibrium to the olivine-opx-metal equilibrium (i.e. the equivalents to the magnetite-quartz-fayalite (QFM) and iron-wustite (IW) buffers in peridotites), followed by a large change in oxygen content with reduction to the olivine-opx-C-SiC buffer. Exactly the contrary is true for COH-fluid. A fluid from graphite-saturated sediments could have e.g. 20 mol% CO_2 and a X_O^{fluid} (=molar O/(H + O)) of roughly 0.75 (Figure 7). Its oxygen fugacity would be 0.6 to 1.3 log-units below QFM or $\Delta IW = +5$ to +3.5, (all values for 700-900°C, 1 GPa). In order to evolve to an oxygen fugacity just above the IW-buffer which locates at X_O^{fluid} of 0.004 to 0.011 on the graphite-saturation surface, almost 99% of the fluid's oxygen needs to be removed. To further move to a fluid composition that would be in equilibrium with SiC, X_O^{fluid} becomes 0.001, which composition-wise is only a small step.

In summary, oxygen fugacity has no simple proportionality to the oxygen content of a rock. In particular, oxygen fugacity is not a parameter that can be freely varied, one has to carefully consider which changes in oxidation state (mostly of Fe and C, but in this case also Si and Cr) and bulk composition of the rock volumes under consideration are involved.

Conclusions

Mantle silicates in equilibrium with natural moissanite have X_{Mg} values of >0.993, SiC forming only 4.5-6 log units below the IW reference buffer. Large volumes of peridotite with such mineral compositions are unknown, hence, SiC forms locally in disequilibrium with the bulk rock and does not reflect large scale modifications of oxygen contents in the mantle. SiC cannot be in equilibrium with minerals that have typical average mantle X_{Mg} of 0.88-0.92. As Fe-Mg diffusion in silicates is generally fast, SiC could not survive at adiabatic temperatures in the mantle for any reasonable time span (i.e. >100'000 yrs) hence precluding any high temperature (>1000°C) origin of natural moissanite. Consequently, any primordial, core-mantle boundary or lower mantle origin for SiC can be ruled out on the basis of the temperatures involved. Nevertheless, natural moissanite does yet not directly indicate any particular pressure of formation.

We conclude that SiC forms through a relatively low temperature mechanism (<700-800°C) and identify this mechanism as percolation of ultra-reduced fluids. Such temperatures in mantle rocks are only realistic in the subduction realm or during orogenesis. Suitable fluids may originate from sediments rich in organic matter. These fluids evolve from moderately reduced compositions through progressive removal of the oxidized species H_2O through hydrous magnesium silicates, which form through reaction with the host rock during fluid percolation. After >99% fractionation, such fluids may reach H:O ratios of ~1000 that stabilize SiC. Such extreme fractionation is obviously rare but may occur occasionally. The initial fluids are expected to form upon heating of sediments, this could occur in the subduction realm, but also during orogenic burial, possibly precipitating SiC in ophiolitic peridotites upon orogenesis.

Abbreviations

CCO: Graphite/diamond-CO-CO_2, Oxygen fugacity buffers; IW: Iron-wustite; MOOC: Moissanite-olivine-orthopyroxene-graphite/diamond; QFM: Quartz-fayalite-magnetite.; Phases: bcc: Body-centered cubic structure of Fe-alloy; graph: Graphite; fcc: Face-centered cubic structure of Fe-alloy; $FeSiC^{alloy}$: Alloy of ternary composition with dominant iron; mgt: Magnetite; ol: Olivine; opx: Orthopyroxene; Si^{alloy}: Alloy of ternary Si-Fe-C composition, almost pure Si.

Competing interests

The authors declare that they have no competing interest.

Authors' contributions

All authors contributed equally to the study, have read the manuscript and approved the final version.

Acknowledgements

The experiments of this study were performed while C.Gao visited ETH with the State Sponsored Study Abroad Program by the Chinese Scholarship Council. A. Golubkova was supported through SNF grants 200020-140541/1 and 200020-153112/1 to MWS.

Author details

[1]Department of Earth Sciences, ETH, Zurich 8092, Switzerland. [2]State Key Laboratory of Geological Processes and Mineral Resources, School of Earth Science, China University of Geosciences, Wuhan 430074, China. [3]Institut für Mineralogie, Westfälische Wilhelms-Universität Münster, Münster 48149, Germany. [4]School of Ocean and Earth Science, Tongji University, Shanghai 200092, China.

References

Aleksandrov IV, Goncharov AF, Stishov SM, Yakovenko EV (1989) Equation of state and Raman scattering in cubic BN and SiC at high pressures. Jetp Letters 50:127–131

Bai W, Robinson PT, Fang Q, Yang J, Yan B, Zhang Z, Hu XF, Zhou MF, Malpas J (2000) The PGE and base-metal alloys in the podiform chromitites of the Luobasa ophiolite, southern Tibet. Can Mineral 38:585–598

Ballhaus C (1995) Is the upper mantle metal-saturated? Earth Planet Sci Lett 132:75–86

Brady JB, Cherniak DJ (2010) Diffusion in minerals: an overview of published experimental diffusion data. Rev Mineral Geochem 77:899–920

Brosh E, Makov G, Shneck RZ (2009) Thermodynamic analysis of high-pressure phase equilibria in Fe-Si alloys, implications for the inner core. Phys Earth Planet Int 172:289–298

Campbell A, Danielson L, Righter K, Seagle CT, Wang Y, Prakapenka VB (2009) High pressure effects on the iron-iron oxide and nickel-nickel oxide oxygen fugacity buffers. Earth Planet Sci Lett 286:556–564

Chipman J (1972) Thermodynamics and phase diagram of the Fe-C system. Metal Trans 3:55–64

Connolly JAD (1995) Phase-diagram methods for graphitic rocks and application to the system C-O-H-FeO-TiO_2-SiO_2. Contrib Mineral Petrol 119:94–116

Connolly JAD (2009) The geodynamic equation of state: what and how. Geochem Geophys Geosys 10:, doi:10.1029/2009GC002540

Deines P (2002) The carbon isotope geochemistry of mantle xenolioths. Earth Sci Rev 58:247–278

Di Pierro S, Gnos E, Grobety BH, Armbruster T, Bernasconi SM, Ulmer P (2003) Rock-forming moissanite (natural α-silicon carbide). Am Mineral 88:1817–1821

Djurovic D, Hallstedt B, von Appen J, Dronskowski R (2011) Thermodynamic assessment of the Fe-Mn-C system. CALPHAD: Comput Coupling Phase Diagrams Thermochemistry 35:479–491

Frost DJ, McCammon CA (2008) The Redox State of Earth's Mantle. Annual Rev Earth Planet Sci 36:389–420

Fumagalli P, Poli S (2005) Experimentally determined phase relations in hydrous peridotites to 6.5 GPa and their consequences on the dynamics of subduction zones. J Petrol 46:555–578

Gao C, Liu Y (2008) Moissanite-bearing carbonatite xenoliths from Cenozoic basalt, North China: products of ancient oceanic crust subduction ? Geochim Cosmochim Acta 72:A292

Golubkova A, Schmidt MW, Connolly JAD (2014) Simulating reducing and ultra-reducing conditions in the mantle: application to the occurrence of SiC in podiform chromitites. Submitted Contrib Mineral Petrol

Gurney JJ (1989) Diamonds. In: Ross J (ed) Kimberlites and Related Rocks, Vol. 2., pp 935–965

Gustafson P (1985) A thermodynamic evaluation of the Fe-C system. Scand J Metallurgy 14:259–267

Hirayama Y, Fujii T, Kurita K (1993) The melting relation of the system iron and carbon at high pressure and its bearing on the early stage of the Earth. Geophys Res Lett 20:2095–2098

Holland TJB, Powell R (2011) An improved and extended internally consistent thermodynamic dataset for phases of petrological interest, involving a new equation of state for solids. J Metam Geol 29:333–383

Ishimaru S, Arai S, Shukuno H (2009) Metal-saturated peridotite in the mantle wedge inferred from metal-bearing peridotite xenoliths from Avacha volcano, Kamchatka. Earth Planet Sci Lett 284:352–360

Lacaze J, Sundman B (1991) An assessment of the Fe-C-Si system. Metall Transact A 22A:2211–2223

Leung I (1990) Silicon carbide cluster entrapped in a diamond from Fuxian, China. Am Mineral 75:1110–1119

Leung I, Guo W, Friedman I, Gleason J (1990) Natural occurrence of silicon carbide in a diamondiferous kimberlite from Fuxian. Nature 346:352–354

Li Z, Bradt RC (1987) Thermal expansion of the hexagonal (4H) polytype of SiC. J Am Ceramic Soc 70:445–448

Li J, Mao HK, Fei Y, Gregoryanz E, Eremets M, Zha CS (2002) Compression of Fe_3C to 30 GPa at room temperature. Phys Chem Minerals 29:166–169

Lord OT, Walter MJ, Dasgupta R, Walker D, Clark SM (2009) Melting in Fe-C system to 70 GPa. Earth Planet Scie Lett 284:157–167

Lyakhovich W (1980) Origin of accessory moissanite. Geol Rev 22:63–74

Mathez EA, Fogel RA, Hutcheon ID, Marshintsev VK (1995) Carbon isotopic composition and origin of SiC from kimberlites of Yakutia, Russland. Geochim Cosmochim Acta 59:781–791

Miettinen J (1998) Reassessed thermodynamic solution phase data for ternary Fe-Si-C system. CALPHAD: Comput Coupling Phase Diagrams Thermochemistry 22:231–256

Mookherjee M, Nakajima Y, Steinle-Neumann G, Glazyrin K, Wu XA, Dubrovinsky L, McCammon C, Chumakov A (2011) High-pressure behavior of iron carbide (Fe_7C_3) at inner core conditions. J Geophysical Res-Solid Earth 116, B04201

Nakajima Y, Takahashi E, Sata N, Nishihara Y, Hirose K, Funakoshi K, Ohishi Y (2011) Thermoelastic property and high-pressure stability of Fe_7C_3: implication for iron-carbide in the Earth's core. Am Mineral 96:1158–1165

O'Neill H, Wall VJ (1987) The olivine-orthopyroxene-spinel oxygen geobarometer, the nickel precipitation curve, and the oxygen fugacity of the Earth's upper mantle. J Petrol 28:1169–1191

O'Neill HS (1988) The system Fe-O and Cu-O: thermodynamic data for the equilibria Fe-"FeO", Fe-Fe_3O_4, "FeO"-Fe_3O_4, Cu-Cu_2O and Cu_2O-CuO from emf measurements. Am Mineral 73:470–486

Poulsen SR (1996) Equilibrium mineral-fluid stable isotope fractionation factors in graphitic metapelites. Chem Geol 131:207–217

Rohrbach A, Ghosh S, Schmidt MW, Wijbrans CH, Klemme S (2014) The stability of Fe-Ni carbides in the Earth's mantle: evidence for a low Fe-Ni-C melt fraction in the deep mantle. Earth Planet Science Lett 388:211–221

Shiryaev AA, Griffin WL, Stoyanov E (2011) Moissanite (SiC) from kimberlites: polytypes, trace elements, inclusions, and speculations on origin. Lithos 122:152–164

Stagno V, Frost DJ (2010) Carbon speciation in the asthenosphere: experimental measurements of the redox conditions at which carbonate-bearing melts coexist with graphite or diamond in peridotite assemblages. Earth Planet Sci Lett 300:72–84

Trumbull RB, Yang JS, Robinson PT, Di Pierro S, Vennemann T, Wiedenbeck M (2009) The carbon isotopic composition of natural SiC (moissanite) from the Earth's mantle: new discoveries from ophiolites. Lithos 113:612–620

Ulmer GC, Grandstaff DE, Woermann E, Gobbels M, Schonitz M, Woodland A (1998) The redox stability of moissanite (SiC) compared with metal-metal oxide buffers at 1773K and at pressures up to 90 kbar. Neuses Jahrb Min Abhand 172:279–307

Wood IG, Vočadlo L, Knight KS, Dobson DP, Marshall WG, Price GD, Brodholt J (2004) Thermal expansion and crystal structure of cementite, Fe3C, between 4 and 600K determined by time of flight neutron powder diffraction. J Appl Crystallogr 37:82–90

Woodland AB, Koch M (2003) Variation in oxygen fugacity with depth in the upper mantle beneath the Kaapvaal craton, Southern Africa. Earth Planet Sci Lett 214:295–310

Woodland AB, O'Neill HS (1997) Thermodynamic data for Fe-bearing phases obtained using noble metal alloys as redox sensors. Geochim Cosmochim Acta 61:4359–4366

Xu A, Wu W, Xiang W, Yang J, Chen J, Ji S, Liu Y (2008) Moissanite in serpentinites from the Dabie Shan Mountains in China. Mineral Mag 72:899–908

7

A laboratory experiment on salt weathering by humidity change: salt damage induced by deliquescence and hydration

Masato Sato[1*] and Tsuyoshi Hattanji[2]

Abstract

A laboratory experiment on salt weathering was carried out under changing humidity conditions. Three types of rock (tuff and dense and porous sandstone) with sodium chloride, sodium sulfate, or magnesium sulfate were subjected to humidity oscillations from 20%RH to 98%RH during 6 h periods at 20 °C. Prior to the weathering experiment, the rock specimens were oven-dried, and the water supply was restricted to only that from the air during the experiment. Salt on the rock specimens was deliquesced/hydrated at almost 100%RH during the high-humidity period and crystallized/dehydrated during the low-humidity period. Sodium chloride, which has high deliquescence ability, caused the most intensive weathering. Porous sandstone with sodium chloride was completely broken down after 100 cycles of humidity changes. The other two types of rock with sodium chloride showed flaking or swelling on their surfaces. For dense sandstone with sodium chloride, the Equotip rebound value decreased as the humidity cycles increased. Sodium chloride has a high susceptibility of deliquescence at humidities of more than 80%RH. Weathering by sodium chloride would occur extensively on the surfaces of stone heritages and geoheritages in humid coastal spray zones. Magnesium sulfate induced a weight loss in porous sandstone, flaking from the surface of tuff, and cracks on the surface of dense sandstone. Sodium sulfate had almost no effect in the humidity-change experiment, although salt efflorescence was obvious. The results for magnesium sulfate, salt deliquescence, hydration, and crystallization in the humidity-change experiment were easily repeated. Magnesium sulfate might have more influence than sodium sulfate when air humidity fluctuates in a short period.

Keywords: Rock breakdown, Weight loss, Equotip hardness value, Longitudinal wave velocity, Sodium chloride, Sodium sulfate, Magnesium sulfate

Introduction

Salt weathering is one of the major processes of stone decay that causes serious damage to brick walls (Goudie 1977; Kuchitsu et al. 2000), structural foundations (Mottershead 1994; Takahashi et al. 1994), and historical buildings (Gauri 1990; Kamh 2005; Thomachot-Schneider et al. 2011). In natural environments, salt weathering occurs in coastal areas (Matsukura and Matsuoka 1991, 1996; Mottershead 1982), deserts (Cook 1981; Goudie and Cook 1984; Viles and Goudie 2013), and the Antarctic inland (Fahey 1985; Marchant and Head 2007; Miotke 1982; Selby and Wilson 1971). Field observations undertaken globally in a range of environments have suggested the importance of microclimatic conditions in stone decay, including temperature (Hœrlé 2006), insolation (Gunzburger and Merrien-Soukatchoff 2011; McFadden et al. 2005), moisture (Mol 2014), and humidity (Viles 2005).

Many laboratory experiments have simulated salt weathering under the conditions of "cyclic total immersion" or "continuous partial immersion". In cyclic total immersion experiments, rock specimens are repeatedly subjected to total immersion in a saturated salt solution followed by drying (Goudie et al. 1970; Robinson and Williams 1982; Williams and Robinson 1998). In continuous partial immersion experiments, specimens partly immersed in saturated solutions are subjected to environmental conditions (Goudie 1986; Oguchi and Yuasa 2010; Robinson and Williams 2000). However, these conditions, in which

* Correspondence: msato@geoenv.tsukuba.ac.jp
[1]Graduate School of Life and Environmental Sciences, University of Tsukuba, 1-1-1 Tennodai, Tsukuba, Ibaraki 305-8572, Japan
Full list of author information is available at the end of the article

abundant salt water is supplied to the rock specimens, are not likely to occur in the field, with the exception of intertidal zones and places with nearby groundwater. In addition, the drying temperature affects the formation of salt crystals (Aly et al. 2015). The disjuncture between field environments and laboratory experiments has often been pointed out (McGreevy and Smith 1982; Smith et al. 2005).

Air humidity is a water source affecting stone heritages. Arnold and Zehnder (1989) pointed out the importance of monitoring air humidity and temperature for their preservation, as these environmental conditions control salt crystallization and hydration. Goudie and Viles (1997) assumed that air humidity is an important factor in salt crystallization from solution, change of phase (hydration/dehydration), and crystallization pressure. Seasonal salt deliquescence and precipitation have been shown to occur on stone heritages in the presence of air humidity changes (Goudie 1977; Matsukura et al. 2004). In dry areas such as Antarctica and Mars, air humidity is also a major source of water, and salt weathering may be induced by cyclic changes in it (Campbell and Claridge 1987; Gough et al. 2016, 2017; Heinz et al. 2016; Wellman and Wilson 1965). The influence of humidity on hydration/dehydration has been thoroughly examined (Benavente et al. 2004; Rodriguez-Navarro and Doehne 1999; Steiger and Asmussen 2008).

Salt deliquescence occurs when the vapor pressure of a saturated salt solution on salt crystals is less than the vapor pressure of water in the air. Fahey (1985) and Goudie (1993) subjected rock specimens to six different climatic conditions. Both studies suggested that rock breakdown is most intensive under high humidity conditions close to 100%RH. These experimental studies suggested that only salt deliquescence and subsequent salt crystallization could cause the breakdown of rock. However, they partly immersed the rock specimens in water or a salt solution. They also subjected the specimens to humidity changes along with temperature fluctuations.

This study entailed a laboratory experiment on salt damage to rock specimens occurring when humidity fluctuates at a constant temperature. The water source was restricted to only that from air humidity.

Methods/Experimental

Rock properties

Three types of rock (Oya tuff, Aoshima sandstone, and Ubara sandstone) were used. Oya tuff is dacitic or rhyolitic marine pumice tuff of the Miocene age; it was quarried in Utsunomiya, central Japan. In Japan, a variety of buildings and stone statues are made from Oya tuff, which is often damaged by salt weathering (Oguchi and Yuasa 2010). Aoshima sandstone is a very fine-grained, well-cemented sandstone of the Pliocene age; it was collected in the Nichinan Coast, Miyazaki Prefecture, southwest Japan. Ubara sandstone is a fine-grained, weakly cemented, porous sandstone of the Miocene age. The sample was collected on the southeast coast of Chiba Prefecture, central Japan. Tafoni and honeycomb weathering features are well developed on type B shore platforms (Takahashi et al. 1994; Tsujimoto 1987) in these coastal areas and are often associated with salt weathering (Bradley, Hutton, and Twidale 1978; Mustoe 1982; Mottershead 1994; Mottershead and Pye 1994; Strini et al. 2008; Young 1987).

Table 1 shows the physical properties of these rocks. Rock surface hardness was evaluated with the rebound value (L-value) of a type D Equotip hardness tester (Proceq). We used the single impact method, which is one of two methodologies for hardness tests with Equotip (Aoki and Matsukura 2007); this method measures the mean L-values of single impacts at ten different points. The longitudinal wave velocity (P-wave velocity) was measured with an ultrasonic tester (Pundit Plus, CNS Farnell).

Humidity-change weathering experiment

Eight cubic specimens with a side of 5 cm were prepared from the Oya tuff and Aoshima sandstone. Only four cylindrical specimens with a diameter and a height of 3.5 cm were prepared from the Ubara sandstone because of the difficulties in shaping the material into cubes. The initial values of the weight, L-value, and P-wave velocity were measured after oven drying these specimens at 110 °C for 48 h.

Saturated solutions of sodium chloride (NaCl), sodium sulfate (Na_2SO_4), or magnesium sulfate ($MgSO_4$) were prepared at 20 °C. The rock specimens were immersed with one of the three salt solutions or distilled water for 72 h. After immersion, the specimens were oven dried again at 110 °C for 48 h to provide them with salt prior to the humidity-change experiment.

Each specimen was placed on a sieve with 2 mm openings and was subjected to a total of 200 cycles of humidity changes in an ambient chamber, in which the humidity was oscillated from 20 to 98%RH every 6 h (Figs. 1 and 2). The temperature of the air in the chamber was kept at 20 ± 2 °C during the experiment. Air temperature and relative humidity in the chamber were recorded every 5 min with a data logger (TR-72wf, T&D) with accuracies of ± 5% for RH and ± 0.5% for temperature. The sensor required 7 min to reach a 90% response. Weight, L-value, and P-wave velocity of the specimens were measured every 20 cycles. For Oya tuff and Aoshima sandstone, four specimens were used for measuring the weight; the other four specimens were used for measuring the L-value and P-wave velocity. The L-value and P-wave velocity were not measured in the Ubara sandstone because the specimens were small. All fragments remaining on the sieve were weighed. In addition, the weights of the specimens

Table 1 Properties of the rocks used in experiments

	Bulk density, ρ_{bulk} (g/cm^3)	True density, ρ_{true} (g/cm^3)	Effective porosity n_e (%)	Tensile strength, S_t (MPa)*	Equotip rebound value, L	Longitudinal wave velocity, V_p (km/s)
Ubara sandstone	1.56	2.67	41.8	1.67	392	0.84
Oya tuff	1.39	2.39	42.3	1.37	452	2.30
Aoshima sandstone	2.40	2.66	5.3	5.75	593	3.32

*The tensile strengths were taken from Tsujimoto (1987) for Ubara sandstone and Yamada et al. (2005) for Oya tuff and Aoshima sandstone, respectively

were measured every hour only for the first humidity-change cycle in order to estimate the amount of water that absorbed from the air on the specimens in the high-humidity period.

Continuous wetting/drying weathering experiment

Another weathering experiment involving repeated wetting/drying with a salt solution was carried out in order to compare its results with those of the humidity-change experiment. Eight cubic specimens each of Oya tuff and Aoshima sandstone were prepared in the same way as in the humidity-change experiment. Four cubes were used for measuring the weight; the other four cubes were used for measuring the L-value and P-wave velocity. After measuring the initial weight or initial L-value and P-wave velocity, the specimens were subjected to a total of 20 cycles of wetting and drying, each of which was composed of the following treatment steps: the specimens were (1) immersed in one of the three salt solutions or distilled water for 24 h at 20 °C, (2) oven dried at 110 °C for 48 h, and (3) immersed in distilled water for 24 h to remove the salt that had crystallized in the specimens. After cooling, they were (4) oven dried again at 110 °C for 48 h and (5) weighed or their L-value and P-wave velocity measured. As in the humidity-change experiment, all fragments remaining on the sieve were weighed.

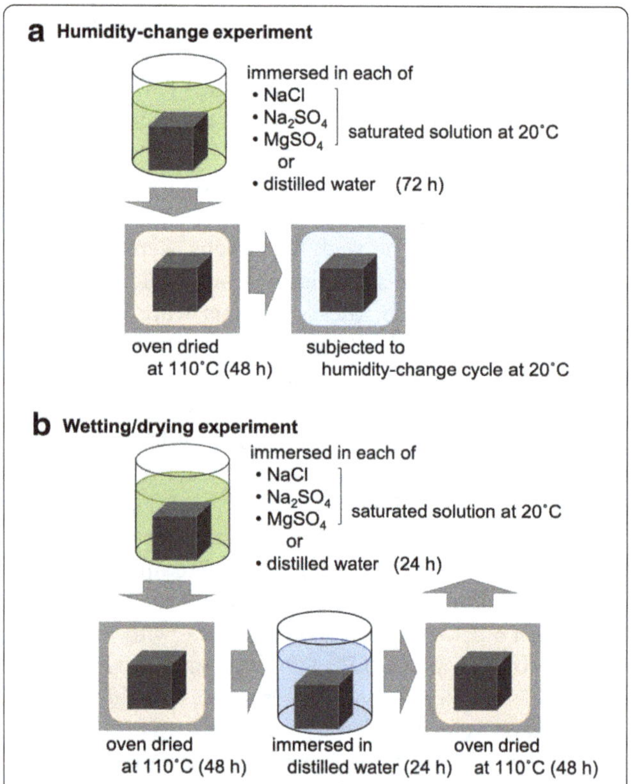

Fig. 1 Experimental design. **a** Humidity-change experiment. The weathering cycle was repeated 200 times. Weight, Equotip rebound value, and longitudinal wave velocity were measured every 20 cycles. **b** Wetting/drying experiment. The weathering cycle was repeated 20 times. Weight and Equotip rebound value were measured at the end of each weathering cycle

Results and discussion

Amount of salt in the humidity-change experiment

Table 2 shows the amount of salt contained in the rock specimens for each combination of rock and salt. The porous Oya tuff and Ubara sandstone contained more salt than the Aoshima sandstone did. The amounts of Na_2SO_4 were smaller than those of the other two salts, which reflected the lower solubility of Na_2SO_4 (19.0 wt%) than that of NaCl (26.4 wt%) and of $MgSO_4$ (25.2 wt%) at 20 °C.

Salt pathway in the humidity-change cycle

The weights of the rock specimens increased in the high-humidity period and decreased in the low-humidity period. Figure 3 shows that there were significant increases

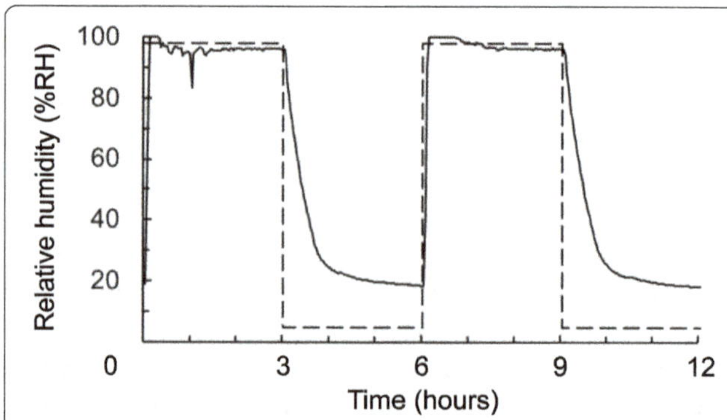

Fig. 2 Details of the humidity-change cycle. A cycle consisted of high- and low-humidity periods lasting 6 h in total. The solid line represents the relative humidity measured at 5 min intervals. The dashed line represents the setting of the ambient chamber

Table 2 Salt content of rock specimens used in the humidity-change experiment

	Sodium chloride	Sodium sulfate	Magnesium sulfate
Ubara sandstone	5.3	2.9	6.1
Oya tuff	7.7	4.2	8.6
Aoshima sandstone	0.5	0.2	1.0

The values represent gram weights of salt in a 100 g salt-free rock specimen

in the weights of the specimens with NaCl during the high-humidity period. The specimens with $MgSO_4$ also showed larger increases than the specimens with Na_2SO_4 or no salt. The weight increases of the specimens with Na_2SO_4 were comparable to the values of the specimens with no salt. The weight oscillation means that the rock specimens absorbed water from the air in the high-humidity period, and they dried out in the low-humidity period. At the beginning of the high-humidity period, all three salts appeared to form a solution by dissolving into the water absorbed from the air, indicating that salt deliquescence occurred. The practical deliquescence relative humidity (DRH) at 20 °C is 75.6%RH for NaCl (halite), 83.3–95.1%RH for Na_2SO_4 (mirabilite), and 92.7%RH for $MgSO_4$ (epsomite) (Apelblat and Korin 1998, 2002; Apelblat and Manzurola 2003). The relative humidity in the ambient chamber rapidly increased and reached almost 100%RH within a few minutes. The high salt content of the rock specimen and low DRH of the NaCl were the causes of the largest weight gain in the high-humidity period.

Hydration also occurred in the high-humidity period. Flatt (2002) and Steiger and Asmussen (2008) described the phase change of Na_2SO_4. Under the experimental conditions (20 °C), Na_2SO_4 exists in two stable phases:

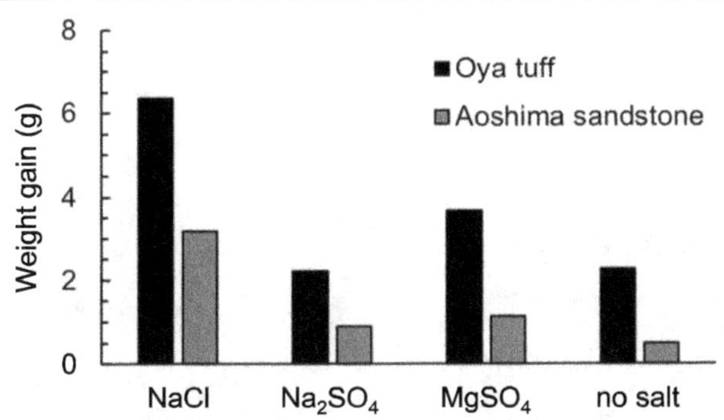

Fig. 3 Weight gains in the first high-humidity period. The value represents the amount of water absorbed into specimens during the high-humidity conditions. The specimens with NaCl showed the largest absorption. The specimens with $MgSO_4$ showed greater gains in weight than the specimens with Na_2SO_4 or without salt (distilled water)

anhydrate (thenardite) and decahydrate (mirabilite). At the beginning of the experiment, thenardite should crystallize on the rock surface because it directly crystallizes from a saturated solution above 32.4 °C. At the beginning of the high-humidity period, salt deliquescence should cause a saturated solution to form. Because the saturated solution is supersaturated to mirabilite, mirabilite crystallizes from a salt solution (Flatt 2002). In fact, mirabilite crystallized in our experiment 30 min after the first high-humidity period started. In the low-humidity period, mirabilite should crystallize from the solution at 95.6%RH and dehydrate to thenardite below 76.4%RH (Steiger and Asmussen 2008). The crystallization of thenardite is dominant below the equilibrium humidity of 86.6%RH, although both mirabilite and thenardite crystallize directly from solution (Steiger and Asmussen 2008).

The sequence of deliquescence and hydration of thenardite should occur at the beginning of every high-humidity period. However, the deliquescence and phase changes of Na_2SO_4 were most distinct in the first high-humidity period. These changes became unclear as the number of humidity cycles increased, and the rock surface became covered with the efflorescence of Na_2SO_4. Kwaad (1970) suggested that the hydration rate of Na_2SO_4 is much slower than that of $MgSO_4$. Dehydration of mirabilite might hardly have occurred under our experimental conditions, in which the air humidity oscillated over a short interval. Deliquescence of Na_2SO_4 might have been hindered by the high DRH of mirabilite (95.6% at 20 °C), and efflorescence of Na_2SO_4 remained on the rock surface.

$MgSO_4$ has three stable hydrated forms: monohydrate (kieserite), hexahydrate, and heptahydrate (epsomite) (Chou and Seal 2003; Vaniman et al. 2004). The hexahydrate is metastable. Though a large number of metastable forms of $MgSO_4$ have been identified besides hexahydrate, the behavior of these metastable forms is still unclear (Chipera and Vaniman 2007). At the beginning of the experiment, kieserite should have crystallized on the rock surface. Steiger et al. (2008) suggested that hydration of kieserite and formation of hexahydrate occur rapidly when the relative humidity exceeds the DRH of kieserite (60%RH at 25 °C), although the reaction proceeds above the equilibrium humidity of kieserite and hexahydrate (42%RH at 25 °C). The rapid reaction occurs as a two-step process of deliquescence and subsequent crystallization. They also reported that epsomite did not form above the equilibrium humidity of hexahydrate and epsomite (51%RH at 25 °C). However, they did not confirm which hydrated crystals form at relative humidities above the DRH of hexahydrate (81%RH at 25 °C). Epsomite would form when the relative humidity exceeds the DRH of hexahydrate. Although we could not distinguish the hydration pathways in our experiment, deliquescence of kieserite and direct crystallization of epsomite would have occurred within a few minutes of exposure to humidity reaching almost 100%RH in the ambient chamber. The epsomite and hexahydrate would have crystallized in the low-humidity period. The kieserite might have occasionally crystallized under the humidity conditions of our experiment (Balboni et al. 2011). Dehydration of epsomite and formation of hexahydrate would also have occurred, whereas the kieserite formation might have been limited by the low equilibrium humidity of hexahydrate and kieserite.

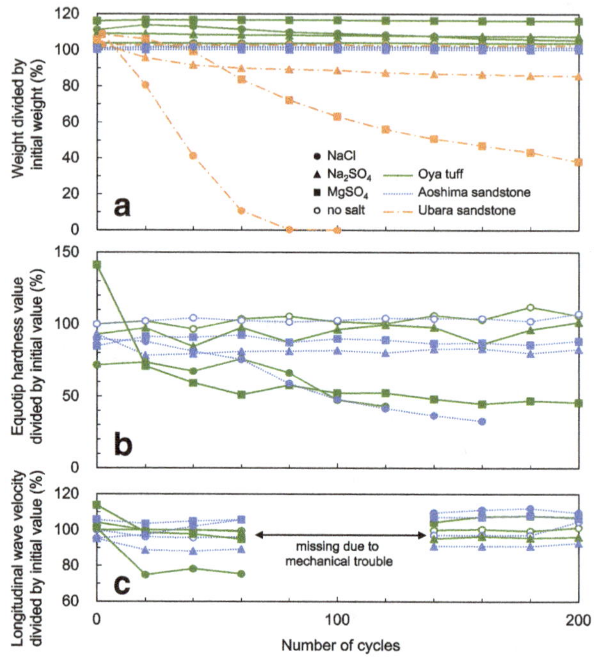

Fig. 5 Weight (**a**), Equotip rebound value (**b**), and longitudinal wave velocity (**c**) measured during the humidity-change experiment. All the measured values are divided by the initial value measured before the specimens were immersed in the salt solutions. Records of longitudinal wave velocity from 80 to 120 cycles went missing because of mechanical trouble. The solid line, dotted line, and dashed line represent the results for Oya tuff, Aoshima sandstone, and Ubara sandstone, respectively. The symbols correspond to the following combinations: filled circles, NaCl; triangles, Na_2SO_4; squares, $MgSO_4$; open circles, no salt (distilled water)

Weight and L-value changes in the humidity-change experiment

Among the three kinds of salt, NaCl caused the most damage to all types of rock in the humidity-change experiment. Ubara sandstone with NaCl was completely broken down after 100 cycles (Figs. 4a and 5a). The cycle 0 data in Fig. 5 refers to the ratio of the weight immediately before the first cycle to the initial salt-free weight; the specimens became heavier from salt precipitation. Oya tuff and Aoshima sandstone did not show any obvious weight losses (Fig. 5a), although a few granular fragments were deposited under the sieve in the cases of the specimens with salt. The samples of Oya tuff with NaCl showed flaking on their surfaces (Figs. 6a), while Aoshima sandstone with NaCl showed swelling on the surface (Fig. 6c). The L-values of the Oya tuff and Aoshima sandstone specimens with NaCl decreased as the number of cycles increased (Fig. 5b). The L-value of these specimens could not be measured after their surfaces became rough. The final L-value was 43.3% of the initial hardness after 120 cycles for Oya tuff and 32.7% after 160 cycles for Aoshima sandstone.

$MgSO_4$ caused the second most amount of damage to the specimens. The weight of Ubara sandstone with $MgSO_4$ decreased to 38.1% of the initial weight after 200 cycles. The samples of Oya tuff with $MgSO_4$ showed flaking on their surfaces (Fig. 6b) and a reduction in their L-value, although the decrease over the first 20 cycles (from 140.3 to 70.9%) was greater than the change in the subsequent 180 cycles (from 70.9% to 45.8%). The decrease in the first 20 cycles was likely induced by salt deliquescence and hydration. The hydration of kieserite and crystallization of the hydrated crystals led to minor rock breakdown and minor changes to the surface micromorphology. These

Fig. 4 Ubara sandstone samples subjected to 100 humidity-change cycles. The initial shape of each sample was cylindrical, 3.5 cm in diameter and height. The internal diameter of the sieve was 7.5 cm, and the opening was 2 mm. **a** NaCl caused the most intensive breakdown. The sandstone completely disintegrated into sand particles. **b** The specimen with Na_2SO_4 kept its initial shape, although the sides of the cylinder showed irregular morphology due to flaking. **c** The specimen with $MgSO_4$ showed moderate weathering, although it obviously disintegrated

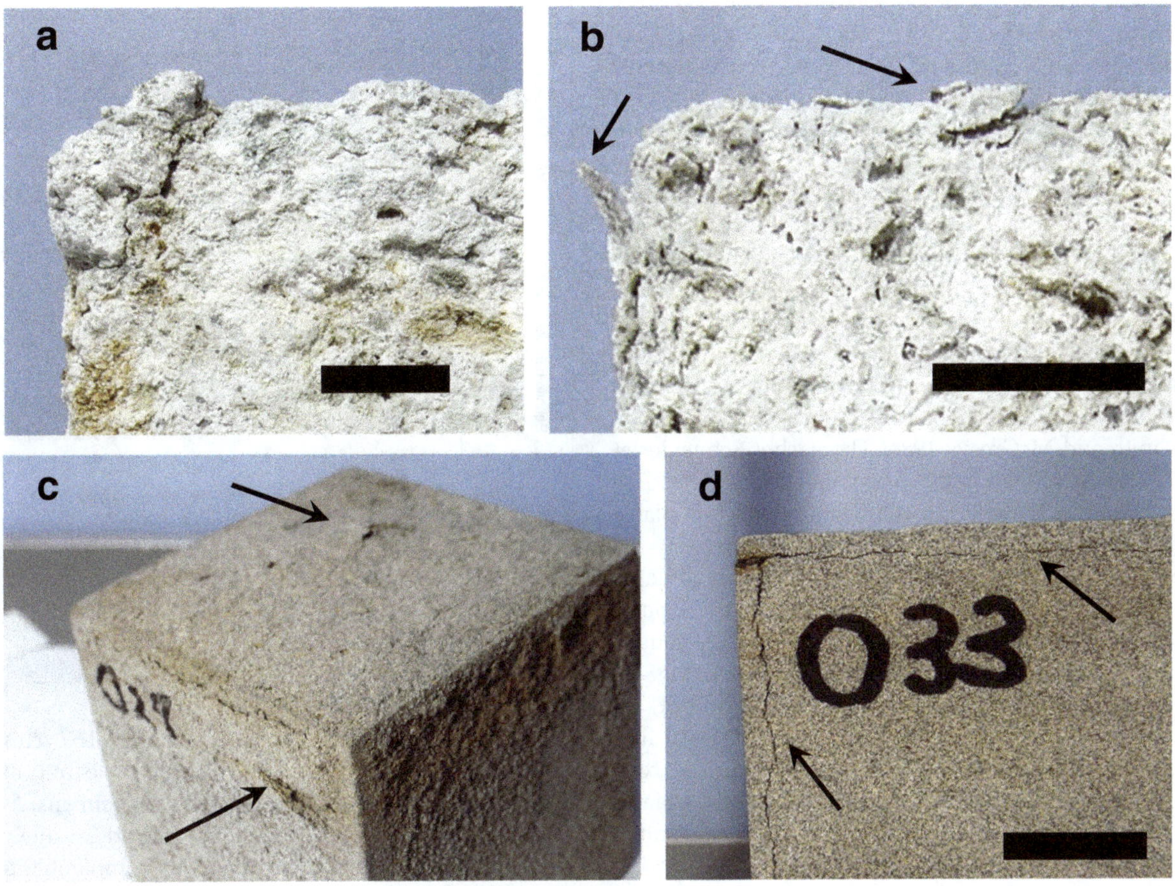

Fig. 6 Oya tuff and Aoshima sandstone with NaCl or $MgSO_4$ subjected to 200 humidity-change cycles. The scale bars are 1 cm long. Oya tuff with NaCl (**a**) and $MgSO_4$ (**b**) showed irregular morphology on their surface. Millimeter-scale fragments flaked from the surface of the Oya tuff with $MgSO_4$. Aoshima sandstone with NaCl (**c**) showed swelling on the surface. Cracks parallel to the rock surface were observed on the edge of the Aoshima sandstone with $MgSO_4$ (**d**)

changes caused the L-value to decrease. Aoshima sandstone with $MgSO_4$ had cracks parallel to the rock surface (Fig. 6d). Cracks occurred on the six faces of the cube.

Na_2SO_4 was relatively inactive as a weathering agent in the humidity-change experiment, with the exception of the weight loss of Ubara sandstone (14.2%) after 200 cycles. The L-value of the Oya tuff with Na_2SO_4 fluctuated within a range from 84.5 to 101.5%. The L-value of Aoshima sandstone showed indistinct changes, except in the first 20 cycles. The L-value for Aoshima sandstone with Na_2SO_4 had a similar trend to that with $MgSO_4$. Tabular fragments flaked from the surface of the Oya tuff with Na_2SO_4 when the specimen was immersed in distilled water after the weathering experiment.

The P-wave velocities of all the specimens were constant or increased slightly during the experiment (Fig. 5c). Only Oya tuff with NaCl recorded a significant decrease from the start to the first 20 cycles.

None of the rock specimens showed any obvious cracks, flaking, or swelling at the beginning of the humidity-change experiment. Therefore, we concluded that the damage caused by the initial immersion in a salt solution and oven drying was relatively weak in comparison with the damage induced by the subsequent humidity changes. The results of the wetting/drying experiment described later supported this conclusion.

The degree of damage induced by each salt was in proportion to the weight gain during the high-humidity period. The lower DRH of NaCl resulted in greater water absorption from the air, and the greater absorption, in turn, induced greater weight losses and L-value reductions. The degree of damage would also have been affected by the fact that NaCl does not hydrate substantially. The hydration of two sulfates involves volume expansion of salt crystals. The hydrated crystals might have filled pores in the rock and hindered infiltration of water (Steiger and Asmussen 2008). The specimens with NaCl could absorb greater amounts of water because NaCl only dissolved and made a solution with deliquescence.

The specimens with Na_2SO_4 were less damaged than the specimens with $MgSO_4$, despite that the DRH of Na_2SO_4 was lower than or comparable to the DRH of

$MgSO_4$. Ruiz-Agudo et al. (2007) carried out a laboratory experiment on the solutions of Na_2SO_4 and $MgSO_4$. They reported that Na_2SO_4 and $MgSO_4$ were both extremely damaging. Balboni et al. (2011) explained that the different behaviors of the two sulfates are due to the different viscosities of their solutions. They assumed that the high viscosity of $MgSO_4$ retards the solution transport and crystal nucleation. High viscosity also causes pores to clog, which retards hydration and deliquescence. In our experiment, as previously mentioned, the slow hydration/dehydration rates of Na_2SO_4 and the high DRH of mirabilite would have restricted their activity. Na_2SO_4 is generally regarded as one of the most aggressive salts in both the field and the laboratory (Goudie and Viles 1995, 1997; Smith and McGreevy 1983; Thomachot-Schneider et al. 2011). However, the relative importance of $MgSO_4$ to Na_2SO_4 might increase in environments in which available water is limited and the humidity is extremely high.

The difference in the water supply method affected how the sulfates induced breakdown of the rocks. Ruiz-Agudo et al. (2007) reported that Na_2SO_4 caused localized damage and detachment, whereas $MgSO_4$ induced cracks and fractures. Their experiment, in which the lower part of the specimen was immersed in a salt solution, induced salt crystallization inside the rock interior. In contrast, in our experiment, both sulfates concentrated in the surface layer because of the limited water supply from the surface of the rock. Cracking parallel to the rock surface, which occurred in the Aoshima sandstone with $MgSO_4$ and in the Oya tuff with Na_2SO_4, suggests that salt weathering occurred only within several millimeters of the surface. The interiors were not weathered much because the P-wave velocities of the specimens did not decrease. López-Arce et al. (2011) suggested that porous dolostone is sensitive to diurnal cycles of air humidity at depths up to 3 cm. In our experiment conducted in a short time interval, the depth at which air humidity could have had an effect would have been less than a centimeter from the rock surface.

The specimens with NaCl did not show any cracks. Because NaCl tends to crystallize on the surface, this crystallization pattern is referred to as efflorescence. In contrast, Na_2SO_4 and $MgSO_4$ tended to crystallize within rock pores, referred to as subflorescence. Salt subflorescence induces greater crystallization pressure than salt efflorescence does (Doehne 2002; Moteershead 2013; Rodriguez-Navarro and Doehne 1999). The difference in the crystallization patterns may be reflected in the absence or presence of cracks on the surface.

Comparison with continuous wetting/drying experiment

The results of the wetting/drying experiment contrasted with those of the humidity-change experiment (Fig. 7). Oya tuff immersed in $MgSO_4$ disintegrated rapidly, and

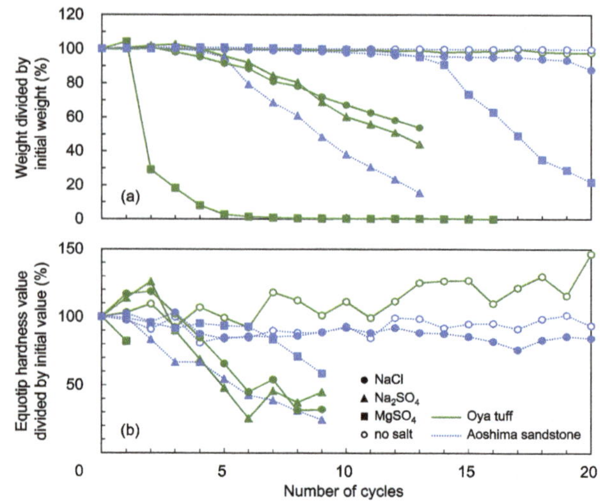

Fig. 7 Weight (**a**) and Equotip rebound value (**b**) measured during the continuous wetting/drying experiment. The solid line represents the results for Oya tuff, and the dotted line represents those for Aoshima sandstone. The symbols correspond to the following combinations: filled circles, NaCl; triangles, Na_2SO_4; squares, $MgSO_4$; open circles, no salt (distilled water)

no fragments remained on the sieve after 16 cycles. The L-value of this specimen could not be measured after 2 cycles. Among the Aoshima sandstone specimens, $MgSO_4$ showed the second greatest weight loss and L-value reduction. Although the effect of NaCl was comparable to that of Na_2SO_4 for Oya tuff, NaCl was obviously inactive for Aoshima sandstone. Na_2SO_4 substantially damaged the Oya tuff and caused the most damage to the Aoshima sandstone.

The weights of some of the specimens could not be measured after the sieve openings became clogged with rock fragments. In addition, the L-values could not be measured after the surfaces of the specimens became rough.

For Aoshima sandstone specimens with Na_2SO_4, the final weight was 15.4% of the initial weight after 13 cycles, and the final L-value was 24.1% of the initial hardness after 9 cycles. Oya tuff specimen with Na_2SO_4 disintegrated rapidly; the final weight was 43.8% of the initial weight after 13 cycles. The L-value of another Oya tuff specimen rapidly decreased; the final value was 44.8% of the initial hardness after 9 cycles. Aoshima sandstone immersed in NaCl did not show any distinct changes in either weight or L-value.

The less hydrated forms of each salt, i.e., halite, thenardite, and kieserite, would crystallize by oven drying at 110 °C. Breakdown and disintegration mainly occurred when the specimens were immersed in distilled water. Hydration of thenardite and kieserite would have occurred during the immersion. We used distilled water in the wetting sequence, whereas the previous experiments often used salt solutions in various concentrations. Thenardite and kieserite in rock specimens dissolve in distilled water, and their solution would have reached saturation as the water penetrated the rock. There was no sign of rock breakdown during the subsequent oven drying and immersion in salt solution, although crystallization and hydration would have been possible if salt remained in the specimens.

As we described above, the dominant weathering processes in the wetting/drying experiment would be salt crystallization of halite and hydration of thenardite or kieserite. These processes are almost the same as those expected in the humidity-change experiment, i.e., deliquescence/recrystallization of halite and hydration of thenardite and hexahydrate of $MgSO_4$. The difference in damage between the two experiments thus did not result from the salt pathways, despite that the crystallization/hydration pressures vary with temperature and humidity (Flatt 2002). The difference between these two experiments was mainly due to the difference in the water supply methods.

Comparison of rock properties

Pore size distribution, micropore volume, and specific surface area generally affect susceptibility to physical weathering (Benavente et al. 2004; Matsukura and Matsuoka 1996; Matsuoka 1990; Ordóñez et al. 1997; Yu and Oguchi 2010; Arnold and Zehnder 1989). The specific surface area probably affects the salt weathering susceptibility by humidity changes because a larger specific surface area causes greater absorption.

Ubara sandstone with higher porosity and lower tensile strength (Table 1) showed the greatest weight losses (Fig. 4). Although Oya tuff has comparable properties to Ubara sandstone, Oya tuff has large pores resulting from the alternation of pumice, while Ubara sandstone is composed of uniform fine sand. These facts imply that the micropore volume and the specific surface area of Ubara sandstone are greater than those of Oya tuff. The differences in micropore volume might have affected the weathering rates in the humidity-change experiment.

Conclusions

This study highlights that salt weathering obviously occurs under conditions in which the water supply is limited to air humidity. The humidity oscillation induced a cycle of repeated salt deliquescence, hydration, and crystallization. The degrees of damage were strongly correlated with the volume of water absorbed in the rock specimens.

NaCl caused the most damage to the specimens. Porous and weak sandstones with NaCl were completely broken down after 100 cycles of humidity change. The other two types of rock also showed concentrated damage on their surfaces. The lower DRH of NaCl resulted in more water being absorbed from the air, which in turn induced greater weight losses and L-value reductions. The absence of hydrated forms would also have facilitated water absorption from the air.

The salt pathways of Na_2SO_4 would be deliquescence of thenardite and subsequent recrystallization of mirabilite. For $MgSO_4$, deliquescence of hexahydrate and recrystallization of epsomite would also occur. $MgSO_4$ induced damage in all of the rock specimens, whereas Na_2SO_4 had almost no effect in our experiment. Na_2SO_4 may be ineffective as an agent of salt weathering in high-humidity conditions. The volume expansion with hydration of Na_2SO_4 might hinder infiltration of water into the rock. The high DRH of mirabilite would also prevent salt deliquescence.

Aoshima sandstone with $MgSO_4$ developed cracks parallel to its surface that reflected the different crystallization patterns (efflorescence or subflorescence). $MgSO_4$ tends to form subflorescence, unlike NaCl and Na_2SO_4. The tendency was observed under the condition in which a limited volume of water from air humidity was supplied to the rock surface.

Tafoni, which is often associated with salt weathering, occasionally appears hundreds of meters away from coastlines. In supratidal zones, salts are provided by sea spray on their surface. In such conditions, NaCl is supplied to and precipitates on the rock surface in stormy weather; subsequent humidity fluctuations would induce salt weathering.

Sulfates are often found on building materials, and sodium sulfate is thought to be one of the most aggressive salts. However, the restricted activity of Na_2SO_4 in our experiment suggests that deliquesce and hydration of Na_2SO_4 are hindered by the clogging of rock pores with the hydrated salts. $MgSO_4$ did not show such behavior, although deliquesce and hydration of $MgSO_4$ would have occurred. $MgSO_4$ would be more effective than Na_2SO_4 under conditions in which the water supply is limited to air humidity.

Abbreviations

DRH: Deliquescence relative humidity; L-value: Equotip rebound value; $MgSO_4$: Magnesium sulfate; Na_2SO_4: Sodium sulfate; NaCl: Sodium chloride; P-wave velocity: Longitudinal wave velocity

Acknowledgements

We are grateful to the anonymous reviewers for their helpful and constructive comments that contributed to improving the manuscript.

Funding

This study was financially supported by a JSPS Grant-in-Aid for Scientific Research (25560143; principal investigator, N. Matsuoka).

Authors' contributions

Both authors designed the experiments, wrote the manuscript, and read and approved the final manuscript.

Competing interests

Both authors declare that they have no competing interests.

Author details
[1]Graduate School of Life and Environmental Sciences, University of Tsukuba, 1-1-1 Tennodai, Tsukuba, Ibaraki 305-8572, Japan. [2]Faculty of Life and Environmental Sciences, University of Tsukuba, 1-1-1 Tennodai, Tsukuba, Ibaraki 305-8572, Japan.

References

Aly N, Gomez-Heras M, Hamed A, Álvarez De Buergo M, Soliman F (2015) The influence of temperature in a capillary imbibition salt weathering simulation test on Mokattam limestone. Mater Constr 65:e044

Aoki H, Matsukura Y (2007) A new technique for non-destructive field measurement of rock-surface strength: an application of an Equotip hardness tester to weathering studies. Earth Surf Proc Land 32:1759–1769

Apelblat A, Korin E (1998) The vapour pressures of saturated aqueous solutions of sodium chloride, sodium bromide, sodium nitrate, sodium nitrite, potassium iodate, and rubidium chloride at temperatures from 227 K to 323 K. J Chem Thermodyn 30:59–71

Apelblat A, Korin E (2002) The vapour pressure of water over saturated solutions of sodium sulfate, calcium bromide, ferric chloride, zinc nitrate, calcium nitrate, and lithium nitrate at temperatures from 278.15 K to 323.15 K. J Chem Thermodyn 34:1621–1637

Apelblat A, Manzurola E (2003) Solubilities and vapour pressures of saturated aqueous solutions of sodium tetraborate, sodium carbonate, and magnesium sulfate and freezing-temperature lowerings of sodium tetraborate and sodium carbonate solutions. J Chem Thermodyn 35:221–238

Arnold A, Zehnder K (1989) Salt weathering on monuments. In: 1st International Symposium on the Conservation of Monuments in the Mediterranean Basin, Bari, 7–10 June 1989

Balboni E, Espinosa-Marzal RM, Doehne E, Scherer GW (2011) Can drying and re-wetting of magnesium sulfate salts lead to damage of stone? Environ Earth Sci 63:1463–1473

Benavente D, García del Cura MA, García-Guinea J, Sánchez-Moral S, Ordóñez S (2004) Role of pore structure in salt crystallisation in unsaturated porous stone. J Cryst Growth 260:532–544

Bradley WC, Hutton JT, Twidale CR (1978) Role of salts in development of granitic tafoni, South Australia. J Geol 86:647–656

Campbell IB, Claridge GGC (1987) Antarctica: soils, weathering processes and environment. Elsevier, Amsterdam

Chipera SJ, Vaniman DT (2007) Experimental stability of magnesium sulfate hydrates that may be present on Mars. Geochim Cosmochim Ac 71:241–250

Chou I-M, Seal RR (2003) Determination of epsomite–hexahydrite equilibria by the humidity–buffer technique at 0.1 MPa with implications for phase equilibria in the system $MgSO_4$–H_2O. Astrobiology 3:619–630

Cooke RU (1981) Salt weathering in deserts. P Geologist Assoc 92:1–16

Doehne E (2002) Salt weathering: a selective review. In: Siegfried S, Thomas W, Axel V (eds) Natural stone, weathering phenomena, conservation strategies and case studies: introduction, Geological Society of London special publication vol 205. Geological Society of London, London, pp 51–64

Fahey BD (1985) Salt weathering as a mechanism of rock breakup in cold climates: an experimental approach. Z Geomorphol 29:99–111

Flatt RJ (2002) Salt damage in porous materials: how high supersaturations are generated. J Cryst Growth 242:435–454

Gauri KL (1990) Decay and preservation of stone in modern environments. Environ Geol Water S 15:45–54

Goudie AS (1977) Sodium sulphate weathering and the disintegration of Mohenjo-Daro, Pakistan. Earth Surf Proc Land 2:75–86

Goudie AS (1986) Laboratory simulation of "the wick effect" in salt weathering of rock. Earth Surf Proc Land 11:275–285

Goudie AS (1993) Salt weathering simulation using a single-immersion technique. Earth Surf Proc Land 18:369–376

Goudie AS, Cooke R, Evans I (1970) Experimental investigation of rock weathering by salts. Area 2:42–47

Goudie AS, Cooke RU (1984) Salt efflorescences and saline lakes; a distributional analysis. Geoforum 15:563–582

Goudie AS, Viles HA (1995) The nature and pattern of debris liberation by salt weathering: a laboratory study. Earth Surf Proc Land 20:437–449

Goudie AS, Viles HA (1997) Salt weathering hazard. John Wiley & Sons, Chichester

Gough RV, Chevrier VF, Tolbert MA (2016) Formation of liquid water at low temperatures via the deliquescence of calcium chloride: implications for Antarctica and Mars. Planet Space Sci 131:79–87

Gough RV, Wong J, Dickson JL, Levy JS, Head JW, Marchant DR, Tolbert MA (2017) Brine formation via deliquescence by salts found near Don Juan Pond, Antarctica: laboratory experiments and field observational results. Earth Planet Sc Lett 476:189–198

Gunzburger Y, Merrien-Soukatchoff V (2011) Near-surface temperatures and heat balance of bare outcrops exposed to solar radiation. Earth Surf Proc Land 36: 1577–1589

Heinz J, Schulze-Makuch D, Kounaves SP (2016) Deliquescence-induced wetting and RSL-like darkening of a Mars analogue soil containing various perchlorate and chloride salts. Geophys Res Lett 43:4880–4884

Hœrlé S (2006) Rock temperatures as an indicator of weathering processes affecting rock art. Earth Surf Proc Land 31:383–389

Kamh GME (2005) The impact of landslides and salt weathering on Roman structures at high latitudes—Conway Castle, Great Britain: a case study. Environ Geol 48:238–254

Kushitsu N, Ishizaki T, Nishiura T (2000) Salt weathering of the brick monuments in Ayutthaya, Thailand. Dev Geotech Eng 84:319–327

Kwaad FJPM (1970) Experiments on the granular disintegration of granite by salt action. University Amsterdam Fysisch Geografi sch en Bodemkundig Laboratorium Publicatie 16:67–80

López-Arce P, Fort R, Gómez-Heras M, Pérez-Monserrat E, Varas-Muriel MJ (2011) Preservation strategies for avoidance of salt crystallisation in El Paular Monastery cloister, Madrid, Spain. Environ Earth Sci 63:1487–1509

Marchant DR, Head JW (2007) Antarctic dry valleys: microclimate zonation, variable geomorphic processes, and implications for assessing climate change on Mars. Icarus 192:187–222

Matsukura Y, Matsuoka N (1991) Rates of tafoni weathering on uplifted shore platforms in Nojima-Zaki, Boso Peninsula, Japan. Earth Surf Proc Land 16:51–56

Matsukura Y, Matsuoka N (1996) The effect of rock properties on rates of tafoni growth in coastal environments. Z Geomorphol Supp 106:57–72

Matsukura Y, Oguchi CT, Kuchitsu N (2004) Salt damage to brick kiln walls in Japan: spatial and seasonal distribution of efflorescence and moisture content. B Eng Geol Environ 63:167–176

Matsuoka N (1990) Mechanisms of rock breakdown by frost action: an experimental approach. Cold Reg Sci Technol 17:253–270

McFadden LD, Eppes MC, Gillespie AR, Hallet B (2005) Physical weathering in arid landscapes due to diurnal variation in the direction of solar heating. Geol Soc Am Bull 117:161–173

McGreevy JP, Smith BJ (1982) Salt weathering in hot desert: observation on the design of simulation experiments. Geogr Ann 64A:161–170

Miotke F (1982) Physical weathering in Taylor Valley, Victoria Land, Antartica. Polar Geogr Geol 6:71–98

Mol L (2014) Investigations into the relationship between changes in internal moisture regimes and rock surface deterioration in cavernous sandstone features. Earth Surf Proc Land 39:914–927

Mottershead DN (1982) Coastal spray weathering of bedrock in the supratidal zone at East Prawle. South Devon Field Studies 5:663–684

Mottershead DN (1994) Spatial variations in intensity of alveolar weathering of a dated sandstone structure in a coastal environment, Weston-super-Mare, UK. In: Robinson DA, Williams RBG (eds) Rock weathering and landform evolution. Wiley, New York, pp 151–174

Mottershead DN (2013) Coastal weathering. Treatise on Geomorphology 4:228–244

Mottershead DN, Pye K (1994) Tafoni on coastal slopes, South Devon, U.K. Earth Surf Proc Land 19:543–563

Mustoe G (1982) The origin of honeycomb weathering. Geol Soc Am Bull 93:108–115

Oguchi CT, Yuasa H (2010) Simultaneous wetting/drying, freeze/thaw and salt crystallization experiments of three types of Oya tuff. In: Přikryl R, Török Á (eds) Natural stone resources for historical monuments, Geological Society of London special publication vol 333. Geological Society of London, London, pp 59–62

Ordóñez S, Fort R, García del Cura MA (1997) Pore size distribution and the durability of a porous limestone. Q J Eng Geol 30:221–230

Robinson DA, Williams RBG (1982) Salt weathering of rock specimens of varying shape. Area 14:293–299

Robinson DA, Williams RBG (2000) Experimental weathering of sandstone by combinations of salts. Earth Surf Proc Land 25:1309–1315

Rodriguez-Navarro C, Doehne E (1999) Salt weathering: influence of evaporation rate, supersaturation and crystallization pattern. Earth Surf Proc Land 24:191–209

Ruiz-Agudo E, Mees F, Jacobs P, Rodriguez-Navarro C (2007) The role of saline solution properties on porous limestone salt weathering by magnesium and sodium sulfates. Environ Geol 52:269–281

Selby M, Wilson A (1971) The origin of the labyrinth, Wright Valley, Antarctica. Geol Soc Am Bull 82:471–476

Smith BJ, McGreevy JP (1983) A simulation study of salt weathering in hot deserts. Geogr Ann 65A:127–133
Smith BJ, Warke PA, McGreevy JP, Kane HL (2005) Salt-weathering simulations under hot desert conditions: agents of enlightenment or perpetuators of preconceptions? Geomorphology 67:211–227
Steiger M, Asmussen S (2008) Crystallization of sodium sulfate phases in porous materials: the phase diagram Na_2SO_4–H_2O and the generation of stress. Geochim Cosmochim Ac 72:4291–4306
Steiger M, Linnow K, Juling H, Gülker G, AEl J, Brüggerhoff S, Kirchner D (2008) Hydration of $MgSO_4{\cdot}H_2O$ and generation of stress in porous materials. Cryst Growth Des 8:336–343
Strini A, Guglielmin M, Hall K (2008) Tafoni development in a cryotic environment: an example from Northern Victoria Land, Antarctica. Earth Surf Proc Land 33:1502–1519
Takahashi K, Suzuki T, Matsukura Y (1994) Erosion rates of sandstone used for a masonry bridge pier in the coastal spray zone. In: Robinson DA, Williams RBG (eds) Rock weathering and landform evolution. John Wiley & Sons, Chichester, pp 175–192
Thomachot-Schneider C, Gommeaux M, Fronteau G, Oguchi CT, Eyssautier S, Kartheuser B (2011) A comparison of the properties and salt weathering susceptibility of natural and reconstituted stones of the Orval Abbey (Belgium). Environ Earth Sci 63:1447–1461
Tsujimoto H (1987) Dynamic conditions for shore platform initiation. Science Report of the Institute of Geoscience, University of Tsukuba, Sec A 8:45–93
Vaniman DT, Bish DL, Chipera SJ, Fialips CI, Carey JW, Feldman WG (2004) Magnesium sulphate salts and the history of water on Mars. Nature 431:663–665
Viles HA (2005) Microclimate and weathering in the central Namib Desert, Namibia. Geomorphology 67:189–209
Viles HA, Goudie AS (2013) Weathering in the central Namib Desert, Namibia: controls, processes and implications. J Arid Environ 93:20–29
Wellman HW, Wilson AT (1965) Salt weathering, a neglected geological erosive agent in coastal and arid environments. Nature 205:1097–1098
Williams RBG, Robinson DA (1998) Weathering of sandstone by alunogen and alum salts. Q J Eng Geol Hydroge 31:369–373
Yamada T, Aoki H, Takahashi M, Matsukura Y (2005) Effect of rock properties on rates of salts weathering: a laboratory experiment. J Japan Soc Eng Geol 46: 72–78 (in Japanese with English abstract)
Young ARM (1987) Salt as an agent in the development of cavernous weathering. Geology 15:962–966
Yu S, Oguchi CT (2010) Role of pore size distribution in salt uptake, damage, and predicting salt susceptibility of eight types of Japanese building stones. Eng Geol 115:226–236

8

Using molluscan assemblages from paleotsunami deposits to evaluate the influence of topography on the magnitude of late Holocene mega-tsunamis on Ishigaki Island, Japan

Akihisa Kitamura[1,2*], Mami Ito[3], Ryoya Ikuta[1,2] and Masayuki Ikeda[1,2]

Abstract

Four ancient tsunami deposits were identified in a trench excavated on Ishigaki Island, Okinawa, Japan. Three of the tsunami deposits (T-I, T-II, and T-IV) consist of calcareous sand beds, whereas the other (T-III, located stratigraphically between T-II and T-IV) consists of boulders. Deposit T-I was caused by a tsunami in 1771. ^{14}C dating, together with the elevations of the landward margins of these sandy tsunami deposits, suggests that tsunamis II and IV were similar in size to the 1771 tsunami, although the influence of local topographic features on the magnitudes of tsunamis has not yet been examined. This study reconstructs the local topographic features by comparing the molluscan assemblages incorporated within the tsunami deposits with those in recent beach deposits. The presence of species that inhabit the intertidal zone in lagoonal settings in all the assemblages indicates that the present-day shallow lagoon has been present off the study area since the occurrence of tsunami T-IV, which supports the previous hypothesis that the magnitudes of the 1771 tsunami and tsunamis II and IV were similar. These molluscan assemblages also suggest that a high relative abundance of large, heavy mollusc shells is a feature of the paleotsunami deposits in the coastal lowlands found along the shallow coral lagoons.

Keywords: 1771 tsunami, Ancient tsunami deposit, Reef system, Southern Ryukyu Islands, Mollusc

Introduction

Paleotsunami research based on coastal geological records is a useful approach to the detection of high-magnitude, low-frequency tsunami events along subduction zones (e.g., Atwater 1987; Minoura et al. 2001; Nanayama et al. 2003; Cisternas et al. 2005; Monecke et al. 2008; Shishikura et al. 2010; Ishimura and Miyauchi 2015; Nelson et al. 2015; Kitamura 2016; Inoue et al. 2017). Although previous paleotsunami research along the Ryukyu Trench has examined coralline boulders (herein, "tsunami boulders") (Kawana and Nakata 1994; Goto et al. 2010; Araoka et al. 2013), Ando et al. (2018) recently identified three sandy tsunami deposits (T-I, T-II, and T-IV) and a layer of buried tsunami boulders (T-III) in a trench on Ishigaki Island (Figs. 1, 2, and 3). Based on their ^{14}C ages and the elevations of the landward margins of these sandy tsunami deposits, the authors estimated that tsunamis II and IV were similar in size to the 1771 tsunami (tsunami deposit T-I), assuming that these paleotsunamis were generated at the same location and by the same mechanism.

It is widely accepted that on coral reefs, the magnitude of tsunamis can be strongly influenced by the morphology of the reef crest and beach ridge, both of which reduce the tsunami's energy. For example, on Taketomi Island, located just 5 km south of Ishigaki Island, damage from the 1771 tsunami was reportedly slight (Makino 1968) (Fig. 1). Taketomi Island is protected by

* Correspondence: kitamura.akihisa@shizuoka.ac.jp
[1]Faculty of Science, Shizuoka University, 836 Ohya Suruga-ku, Shizuoka 422-8529, Japan
[2]Center for Integrated Research and Education of Natural hazards, Shizuoka University, 836 Ohya, Suruga-ku, Shizuoka 422-8529, Japan
Full list of author information is available at the end of the article

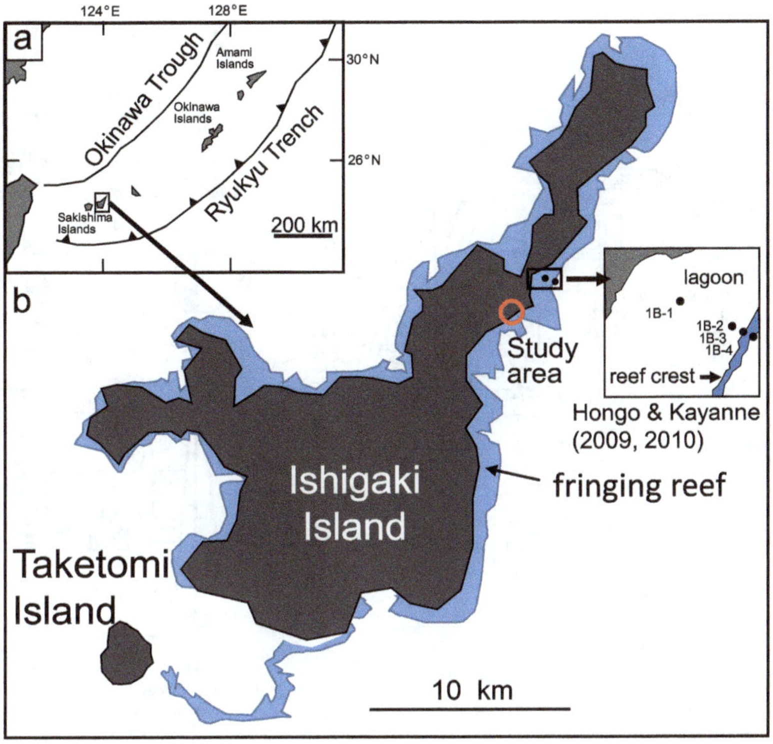

Fig. 1 Study location. **a** Plate boundaries around the southern Ryukyu Islands (after Ando et al. 2009). **b** Ishigaki Island, showing the location of the study area (after Goto et al. 2010)

Fig. 2 **a** Photograph of the trench, showing sampling sites B1–B3. **b** Photograph of the trench, showing sampling sites B2–B6. **c** Photograph of the coastal area at site B2. The circle is around two persons. **d** Two *Nerita squamulata* individuals (arrows) at site B2

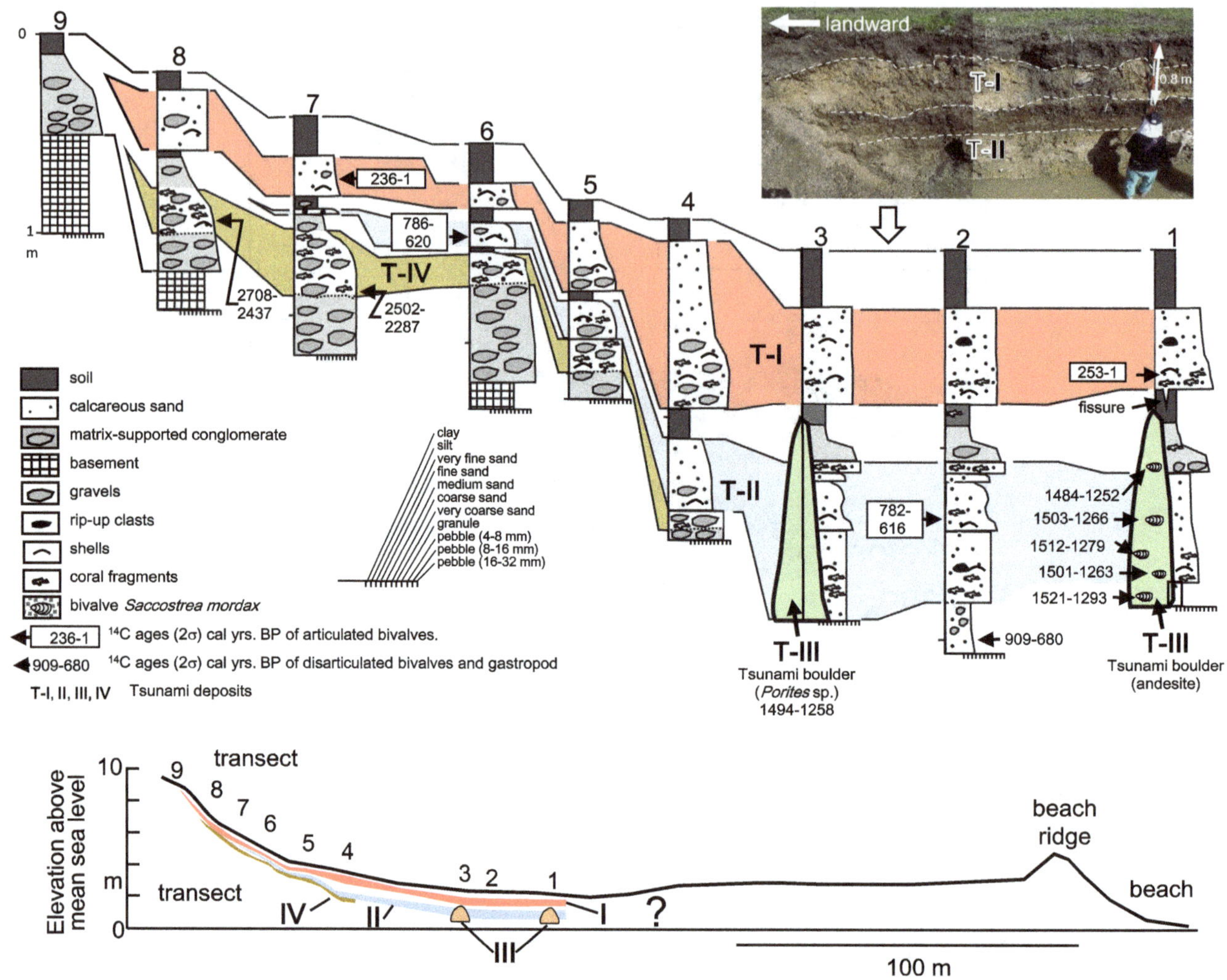

Fig. 3 Columnar sections of the trench. Modified from Kitamura et al. (2018). The inset photograph shows tsunami deposits T-I and T-II

Sekisei Lagoon, which is 10–20 m deep and has formed within the barrier reefs (Kan and Kawana 2006) between the southwestern coast of Ishigaki Island and the eastern coast of Iriomote Island, and includes Taketomi and Kuro Islands (Machida et al. 2001). It is probable that this large lagoon greatly reduced the tsunami's energy before it reached Taketomi Island (Goto et al. 2010). Therefore, the examination of local geomorphological features is essential when evaluating Ando et al.'s (2018) estimate, although, to the best of our knowledge, there have been no previous studies of the local geomorphology.

We collected molluscan shells from tsunami deposits T-I, T-II, and T-IV during the survey reported by Ando et al. (2018), although these specimens have not been analyzed until now. Here, we examine the molluscan assemblages preserved within these tsunami deposits and also those present in recent beach deposits, with the aim of reconstructing the local geomorphological features. Because molluscan species are strongly controlled by the substrate type, which is related in turn to the hydrodynamic conditions (e.g., Kondo et al. 1998), the species compositions provide information on the shallow marine environments in the study area. Our data provide a unique opportunity to evaluate the magnitudes of tsunamis T-I, T-II, and T-IV and the utility of biological proxies in paleotsunami research.

Previous studies of molluscan assemblages within tsunami deposits

Several studies have examined molluscan assemblages within tsunami deposits in coastal lowlands (Morales et al. 2008; Reinhardt et al. 2012; Vött et al. 2011; Goff et al. 2012; Engel et al. 2016; Mannen et al. 2018) and shallow-water inshore areas (Fujiwara et al. 2000; Reinhardt et al. 2006; Donato et al. 2008). Many studies have found that these molluscan assemblages are characterized by a mixture of mollusc shells from a wide range of habitats (e.g., Fujiwara et al. 2000; Donato et al. 2008;

Morales et al. 2008; Richmond et al. 2011). This is also commonly true of microfossils, such as foraminifera (e.g., Nanayama and Shigeno 2006; Sugawara et al. 2009; Rubin et al. 2017) and diatoms (e.g., Sawai et al. 2009; Goff et al. 2012). Kortekaas and Dawson (2007) and Donato et al. (2008) found that molluscan assemblages are characterized by a high percentage of articulated bivalve shells if the sediment source includes a bivalve habitat. However, Fujiwara et al. (2014) examined the species composition and occurrence of molluscan shells in an onshore tsunami deposit formed by the 2011 Tohoku-oki tsunami, and reported a low percentage of articulated bivalve shells.

Recently, Kitamura et al. (2018) analyzed the oxygen isotope profiles of two articulated marine bivalves from tsunami deposit T-II in the present study area. These shells were tightly closed and empty and yielded the same calibrated ages of 920–620 cal. years BP, indicating that the shells were transported and buried alive by the tsunami (Ando et al. 2018). The isotopic analysis showed that tsunami deposit II was caused by a tsunami during the autumn (Kitamura et al. 2018). A combination of radiometric dating and analysis of the $\delta^{18}O$ profiles of articulated bivalves can provide important chronological constraints when examining paleotsunami events.

Tsunami deposits from the 2011 Tohoku-oki tsunami have been examined from many parts of the 900-km-long coastline between southwestern Hokkaido and the northern Boso Peninsula. No marine mollusc shells were detected in most areas (Szczuciński et al. 2012; Takashimizu et al. 2012; Kitamura 2016), except in one area on the Sendai Plain (Fujiwara et al. 2014). Sugawara et al. (2014) suggested that the absence of marine fossils can be explained if the tsunami caused a significant amount of erosion on the beach and in the coastal forest areas, but the erosion was minor on the offshore seafloor.

Study area

The Ryukyu Islands are located along the Ryukyu Trench, which marks the convergent boundary between the Eurasian and Philippine Sea plates. They are divided into three island groups, from north to south: the Amami, Okinawa, and Sakishima Islands (Fig. 1). According to historical descriptions, the area has experienced nine tsunamis since 1644 AD (Watanabe 1985). The 1771 tsunami was an extremely large event, but historical documents show that it damaged only a restricted area, the Sakishima Islands, including Ishigaki Island (Goto et al. 2010). Based on the clast sizes and spatial distributions of boulders on the reefs of the Ryukyu Islands, Goto et al. (2013) suggested that large tsunamis have struck locally but repeatedly around the Sakishima Islands, but not around the Amami and Okinawa Islands, at least during the past 2300 years.

Historical documents show that the coastal areas of Ishigaki Island, located in the southern Ryukyu Islands, Japan (Fig. 1), were inundated by a mega-tsunami (the 1771 tsunami) at ~ 08:00 (Japan time zone) on 24 April 1771 after a large earthquake. The tsunami had a maximum wave height of ~ 30 m (Iwasaki 1927; Goto et al. 2010) and transported tsunami boulders of up to 10 m in diameter (Makino 1968; Shimajiri 1988; Goto et al. 2010). The ^{14}C dates of other tsunami boulders from the southern Ryukyu Islands were examined by Kawana and Nakata (1994) and Goto et al. (2010). Araoka et al. (2013) recently estimated that large tsunamis have occurred at recurrence intervals of 150–400 years since at least 2400 years ago.

More recently, Ando et al. (2018) identified three sandy tsunami deposits (T-I, T-II, and T-IV) and buried tsunami boulders (T-III) in a trench extending inland from the seaward-facing slopes behind a beach ridge on Ishigaki Island. The slopes face a barrier reef lagoon located in the northeastern coastal area of Ishigaki Island. Based on their ^{14}C ages, deposit T-I was caused by the 1771 tsunami and the three older tsunami deposits (T-II, T-III, and T-IV) formed at 920–620, 1670–1250, and between 2700–2280 and 1670–1250 cal. years BP, respectively (Ando et al. 2018).

Ando et al. (2018) have inferred that the boulders of deposit T-III rest on a stratigraphic horizon lower than, but projecting above, that of deposit T-II. Because the bases of the two boulders could not be observed (they were submerged in groundwater), the stratigraphic relationship between these boulders and deposit T-IV is unknown. However, based on the ^{14}C ages, deposit T-III is younger and stratigraphically higher than deposit T-IV (Fig. 3).

The elevations of the landward margins of the sandy tsunami deposits T-I, T-II, and T-IV are up to 9, 6, and 8 m above mean sea level (MSL), respectively (Fig. 3). Based on the ages of these deposits, the mean uplift rate (Yokoyama et al. 2016), and the tidal range, Ando et al. (2018) calculated that the elevations of the land margins associated with deposits I, II, and IV during their deposition were 7.7–9.9 m, 3.9–6.7 m, and 3.9–8.5 m above MSL, respectively. The authors concluded that tsunamis II and IV were similar in size to the 1771 tsunami, assuming that these paleotsunamis were generated at the same location and by the same mechanism.

The study trench was located on the seaward-facing slopes behind a beach ridge (Figs. 1, 2, and 3). The beach ridge faces a reef lagoon that is protected by a reef crest. Hongo and Kayanne (2009) analyzed four drill cores (IB-1 to IB-4; Fig. 1) collected from the same lagoon and a reef crest located 4 km northeast of the study area and concluded that the reef crest had reached sea level and that sea level had stabilized at or around its present level by ca. 7000–6000 years BP, with the burial of the lagoon beginning after that time. On the basis of the stratigraphic distributions of the corals in core IB-3, Hongo and Kayanne (2010) reported that the relative sea-level peaked at ~ 3.0 ± 2.5 m above the present-day MSL at ca. 5000 cal. years BP and gradually decreased thereafter to the level of the present-day MSL.

The shoreface deposits in the study area consist of coarse calcareous sand containing fragments of corals and molluscs. Many coralline and andesitic boulders are present on the shoreface and in the lagoon (Fig. 2c). A 4-m-high beach ridge has developed 30 m landward of the shoreline at MSL and consists of fine calcareous sand. Behind the beach ridge, the 150-m-wide lowland connects to a 7.5% slope (Figs. 2a and 3). The lagoon is ~ 1320 m wide and is less than 4.0 m deep (Hongo and Kayanne 2009). The mean tidal range in the area around Ishigaki Island is ~ 2 m (Goto et al. 2010). Typhoons commonly affect the study area during the summer.

Methods/Experimental

We collected shell specimens by hand from a surface area of 5 m^2 on beach berms (~ 2 m above MSL) at six sites (B1–B6) during the period 18–20 April, 2014 (Fig. 2a, b). We also collected molluscs from tsunami deposits T-I, T-II, and T-IV from a trench excavated in 2014, using two methods. Method 1 involved the sampling of surface exposures in the walls of the trench. The total sampling area was 43.9 m^2 for the T-I deposits, 37.4 m^2 for the T-II deposits, and 10.3 m^2 for the T-IV deposits. Method 2 involved sampling the molluscs from the > 2 mm fractions in blocks of sediment (1.6–6.5 kg) collected from four sites. Methods 1 and 2 are suitable for recording rare and large specimens, and small specimens, respectively. The assemblages in tsunami deposits T-I, T-II, and T-IV are hereafter referred to as assemblages T-I, T-II, and T-IV and were further subdivided into T-I-1 and T-I-2, T-II-1 and T-II-2, and T-IV-1 and T-IV-2, indicating the use of method 1 or method 2 for sample collection, respectively. The taxonomic identification of molluscs was based on Okutani (2000). For the disarticulated or fragmented shells that retained the hinge feature (the beak or umbo portion) on the shell, a separate valve was counted as one individual. Each specimen from assemblages T-I, T-II, and T-IV was weighed, and we also measured the length, height, and width of the shells to the nearest 0.1 mm using digital calipers.

Results

A total of 627 individuals and 53 taxa were identified in the recent assemblages (Additional file 1: Table S1 and Additional file 2: Table S2). The dominant species in the assemblages at sites B1 to B5 were the gastropods *Nerita* (*Theliostyla*) *squamulata* and *Nerita* (*Theliostyla*) *albicilla* and the bivalves *Barbatia* (*Abarbatia*) *lima*, *Fragum fragum*, and *Atactodea striata*, whereas the assemblage at site B6 was exclusively dominated by *A. striata* (66.1%) (Fig. 4). The percentage of articulated bivalve shells ranged from 0% (sites B1, B3, B4, and B5) to 4% (B2; Fig. 4). The assemblages consisted mainly of species that inhabit rocky, sandy, or muddy bottoms in the intertidal zone and weighed less than 5 g (Fig. 4). The heaviest specimen was the bivalve *Tridacna maxima*, with a shell weight of 90 g. Many living individuals of *N. squamulata* and *N. albicilla* were observed on rocky substrates in the intertidal zone at site B2 (Fig. 2c, d).

There were differences in the species compositions of assemblages T-I, T-II, and T-IV collected with the two sampling methods (Fig. 4; Additional file 3: Table S3, Additional file 4: Table S4, Additional file 5: Table S5, Additional file 6: Table S6). The dominant species in assemblage T-I-1 were *F. fragum*, *F. unedo*, and *A. striata*, whereas the dominant species in assemblage T-I-2 were *F. fragum* and *F. loochooanum*. The dominant species in assemblage T-II-1 were *F. fragum*, *F. unedo*, and *Mactra maculate*, whereas the dominant species in assemblage T-II-2 were *Epicodakia bella*, *F. loochooanum*, and *F. unedo*. The dominant species in assemblage T-IV-1 were *F. fragum*, *F. unedo*, *T. maxima*, *A. striata*, and *Glycydonta marica*, whereas the dominant species in assemblage T-IV-2 were *F. loochooanum* and *F. unedo*. The percentages of articulated bivalve shells were less than 1%, except in assemblage T-II-1 (7%) (Fig. 4).

There were no differences in the relative abundances of organisms from specific habitat types among assemblages T-I, T-II, and T-IV (Fig. 4). These habitats were mainly the rocky intertidal zone and the sandy or muddy bottom of the intertidal zone (Fig. 4). The relative abundances of subtidal species were slightly higher in assemblages T-I-1, T-I-2, and T-IV-2 than in the other assemblages. The shell weight of the dominant specimens collected with method 1 (mainly > 1.0 g) was significantly greater than that of specimens collected with method 2 (mainly < 0.5 g) (Fig. 4). We believe that the relatively high abundance of *F. loochooanum* in the assemblages obtained with method 2 can be explained by their smaller shells, and confirm that the combination of these two methods is essential to ensure that reliable fossil data are obtained.

The shell-weight distributions of the specimens collected from the tsunami deposits with methods 1 and 2 are shown in Fig. 5. We observed no systematic landward trends (such as landward reductions) in shell weight.

The molluscan abundances obtained with method 1 (number/m^2) are shown for each 10-m-long section of the trench (Fig. 6). The abundances in deposit T-II were much lower than those in deposits T-I and T-IV. We also observed no systematic landward trends (such as landward reductions) in abundance. However, the molluscan abundances obtained with method 2 (number/kg)

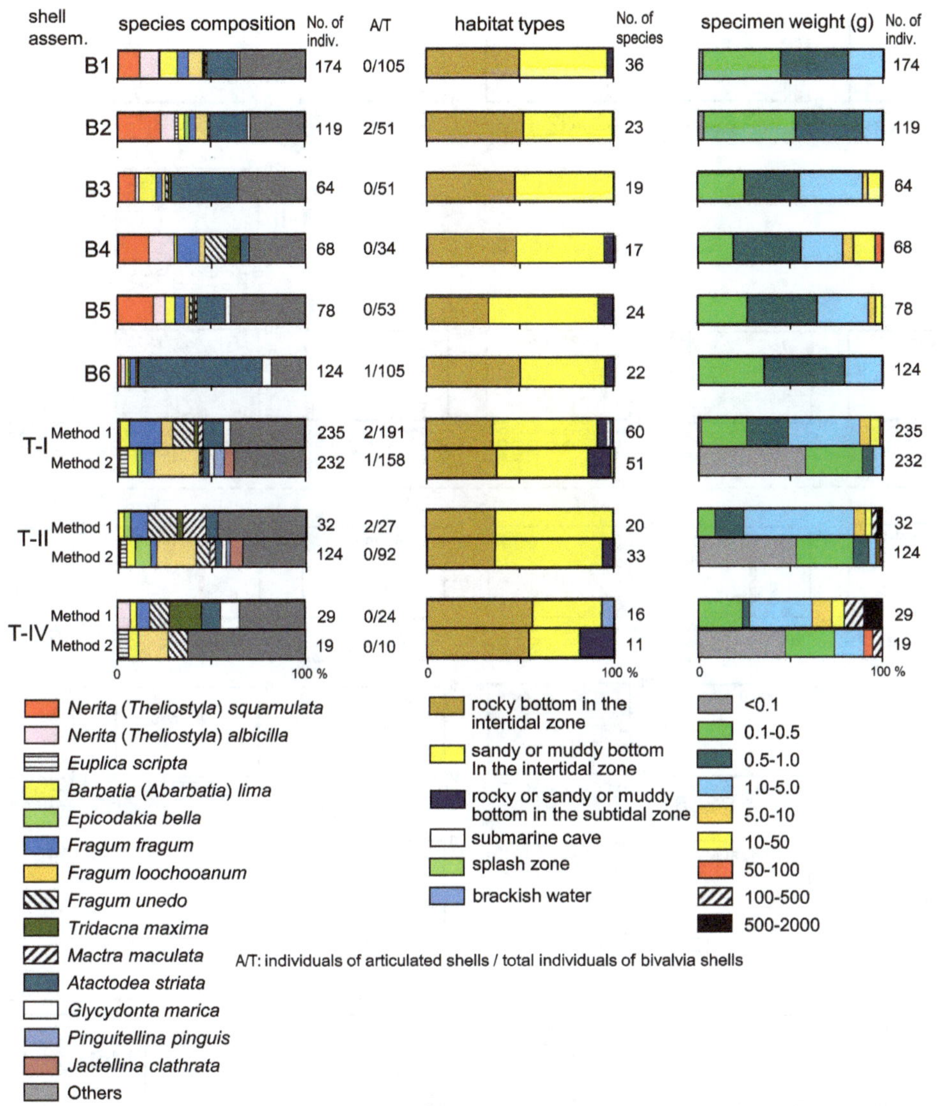

Fig. 4 Species compositions, percentage of articulated bivalve shells, habitat types, and shell weights of molluscan assemblages from recent and tsunami deposits

showed that the abundances in deposit T-II were nearly the same as those in deposit T-I and were significantly greater than the abundances in deposit T-IV (Fig. 6).

Discussion

Influence of local topographic features on tsunami magnitude

As noted above, the species compositions of mollusc assemblages provide information on the shallow marine environments offshore from the study area. The bivalves *F. fragum*, *A. striata*, and *F. unedo* were the dominant species found in the assemblages from tsunami deposits T-I, T-II, and T-IV and were also abundant in the recent assemblage (Fig. 4). Both *F. fragum* and *F. unedo* are burrowing photosymbiotic bivalves (Ohno et al. 1995). The bathymetric range of the former is the lower intertidal zone to a depth of 20 m, whereas that of the latter is the lower intertidal zone (Okutani 2000). *Atactodea striata* inhabits sandy substrata in the intertidal zone of lagoons protected by reef crests (Kondo and Yokogawa 2001). The species occurs over a wide depth range within the intertidal zone, with peak abundances near the mean tide level (Takada et al. 2002). On the basis of the ecology of these species, we conclude that the lagoonal conditions seaward of the study area have not changed significantly since the occurrence of tsunami T-IV, which is consistent with the interpretation of cored sediments near the study area by Hongo and Kayanne (2009). Therefore, differences in the elevations of the land extent of tsunami deposits I (7.7–9.9 m above MSL), II (3.9–6.7 m above MSL), and IV (3.9–8.5 m above MSL) during their deposition, which

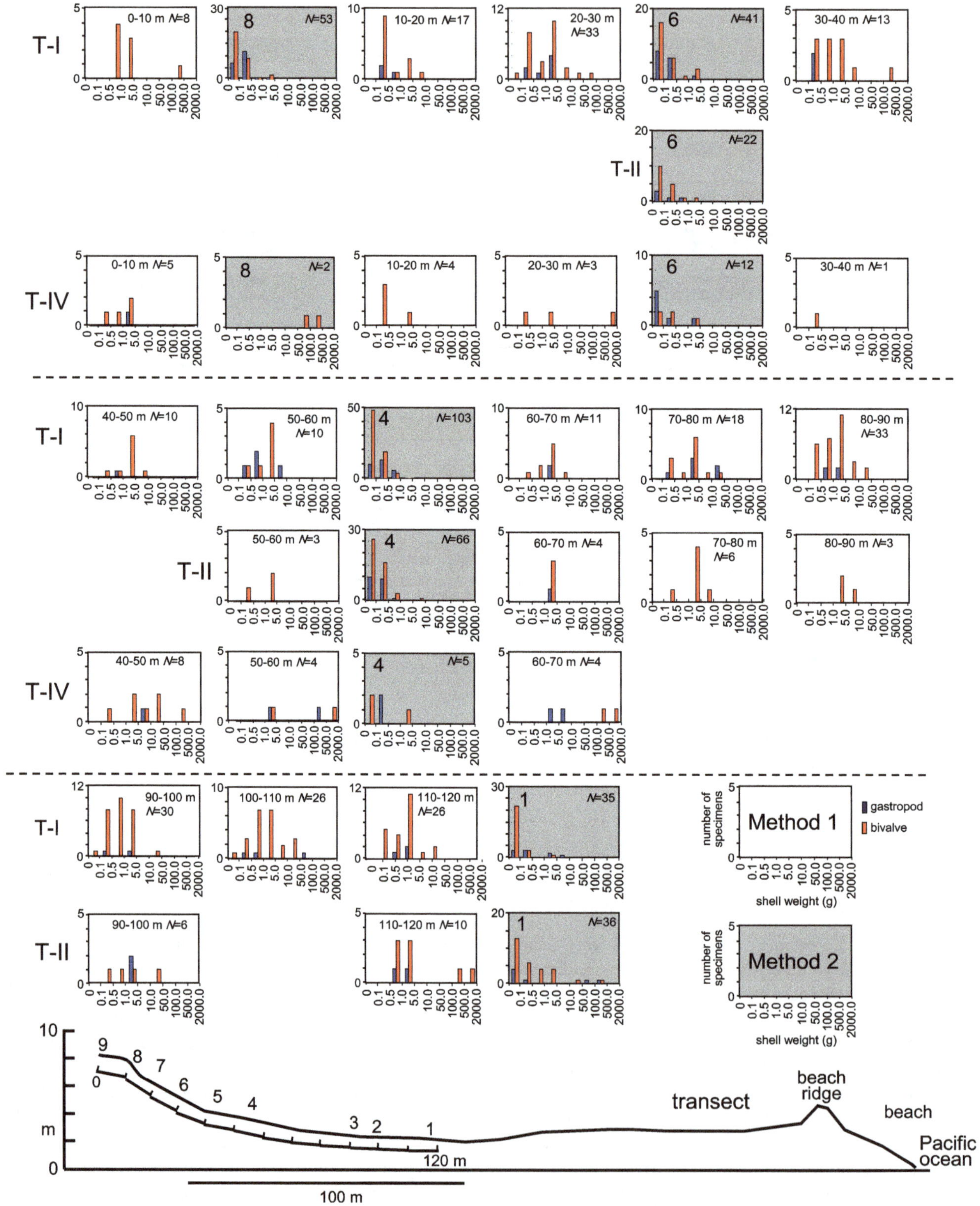

Fig. 5 Histograms of shell weights of specimens collected from the tsunami deposits in a trench using methods 1 and 2

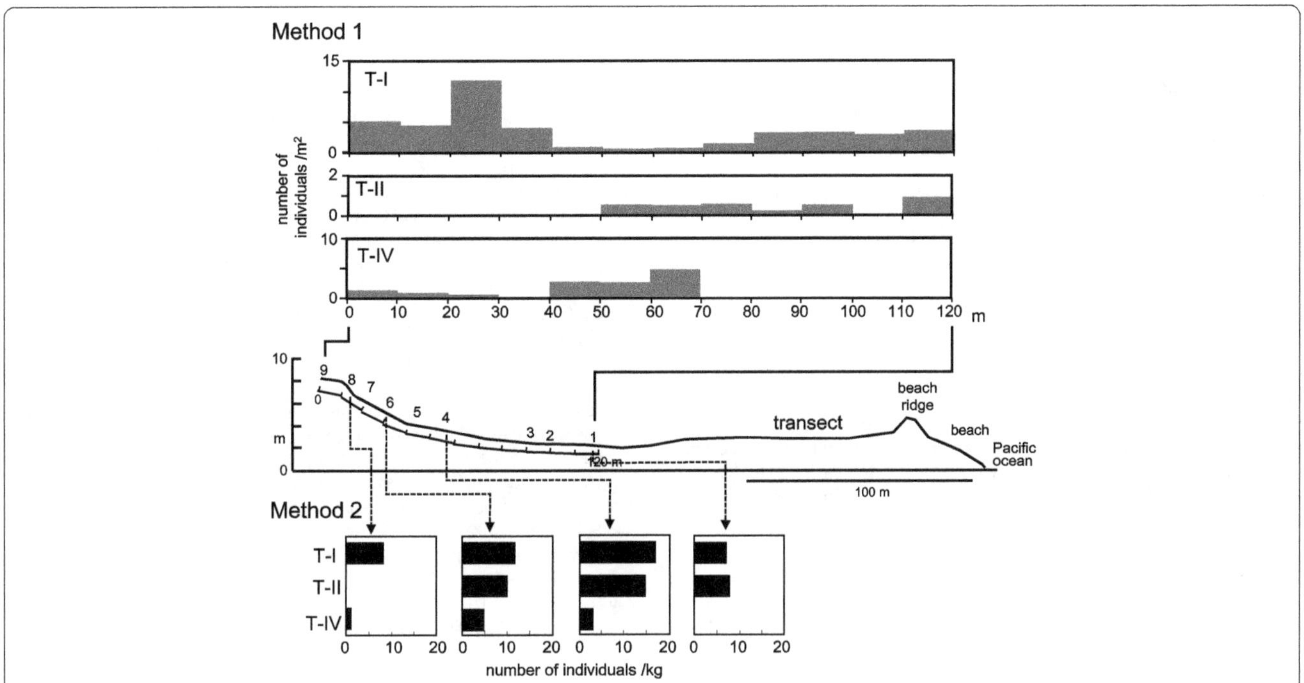

Fig. 6 Molluscan abundances in the tsunami deposits obtained in a trench using methods 1 and 2

were estimated by Ando et al. (2018), were not influenced by the presence or absence of the reef crest. This confirms Ando et al.'s (2018) interpretation that tsunamis II and IV were similar in size to the 1771 tsunami. The present-day lagoonal conditions have persisted since the occurrence of tsunami T-IV, allowing tsunami deposits T-I, T-II, and T-IV to be used as analogues for modeling future tsunamis in the study area.

Utility of biological proxies in the study of paleotsunami deposits

We examined the features of the molluscan assemblages associated with three sandy tsunami deposits by comparing them with a recent assemblage. As noted above, previous studies have identified two key features of the molluscan assemblages found in tsunami deposits: (1) the mixture of mollusc shells is from a wide range of habitats; and (2) the percentage of articulated bivalve shells is high if the sediment source area incorporates a bivalve shell habitat. The former feature is inconsistent with our data, insofar as there were no distinct differences in the patterns of habitat types among assemblages T-II-1, T-II-2, and T-VI-1 and the recent assemblages. However, the lagoon off the study area is less than 4 m deep, so that the environmental heterogeneity in the source area of the molluscan shells may be low. In this situation, it is likely that a wide range of habitats of molluscan shells is not a valid index for the molluscan assemblages in paleotsunami deposits. The very low percentage of articulated bivalve shells in assemblages T-I, T-II, and T-IV (Fig. 4) indicates that the latter feature is not universal.

However, the relative abundance of heavy shells (mainly > 1.0 g) was significantly greater in assemblages T-I-1, T-II-1, and T-IV-1 than in the recent assemblages (Fig. 4). Because strong hydrodynamic flows are required to entrain large, heavy shells, we propose that the relative abundance of large, heavy shells in the molluscan assemblages of paleotsunami deposits is a valid indicator of tsunami strength in coastal lowlands inside shallow coral lagoons.

Conclusions

A comparison of the molluscan assemblages from three sandy tsunami deposits and from a recent beach deposit on Ishigaki Island, southwest Japan, showed that the conditions in the reef systems seaward of the study area have remained unchanged since the occurrence of tsunami T-IV. This confirms that the magnitudes of the 1771 tsunami and tsunamis II and IV were similar. Our data indicate that the proportion of large, heavy shells in the molluscan assemblages in paleotsunami deposits is a valid index of tsunami strength on coastal lowlands inside shallow coral reef lagoons.

Additional files

Additional file 1: Table S1. List of molluscan specimens collected at sites B1–B6. R: right valve; L: left valve.

Additional file 2: Table S2. List of shell sizes and shell weights of specimens collected at sites B1–B6.

Additional file 3: Table S3. List of molluscan specimens collected from the tsunami deposits with method 1.

Additional file 4: Table S4. List of shell sizes and shell weights of specimens collected from the tsunami deposits with method 1. (XLSX 27 kb)

Additional file 5 Table S5. List of molluscan specimens collected from the tsunami deposits with method 2.

Additional file 6 Table S6. List of shell sizes and shell weights of specimens collected from the tsunami deposits with method 2.

Abbreviations

MSL: Mean sea level

Acknowledgements

We thank Dr. Masataka Ando, Dr. Toshiaki Masuda, Imai Takafumi, and Toshiyuki Chinen for our survey on Ishigaki Island. We acknowledge constructive reviews from anonymous reviewers for improvements. This study was supported by the Integrated Center for Education and Research of Natural Hazards, Shizuoka University, and Grants-in-Aid (26287126) awarded by the Japan Society for the Promotion of Science.

Funding

This study was funded by the Integrated Center for Education and Research of Natural Hazards, Shizuoka University, and Grants-in-Aid (26287126) awarded by the Japan Society for the Promotion of Science.

Authors' contributions

AK conceived designed, collected samples and identified specimens of molluscs samples. MIt analyzed molluscs samples. RI and MIk collected samples and examined evaluation of tsunami wave height. All authors read and approved the final manuscript.

Competing interests

The authors declare that they have no competing interests.

Author details

[1]Faculty of Science, Shizuoka University, 836 Ohya Suruga-ku, Shizuoka 422-8529, Japan. [2]Center for Integrated Research and Education of Natural hazards, Shizuoka University, 836 Ohya, Suruga-ku, Shizuoka 422-8529, Japan. [3]Graduate School of Science, Shizuoka University, 836 Ohya Suruga-ku, Shizuoka 422-8529, Japan.

References

Ando M, Kitamura A, Tu Y, Ohashi Y, Imai T, Nakamura M, Ikuta R, Miyairi Y, Yokoyama Y, Shishikura M (2018) Source of high tsunamis along the southernmost Ryukyu trench inferred from tsunami stratigraphy. Tectonophysics 722:265–276.

Ando M, Nakamura M, Matsumoto T, Furukawa M, Tadokoro K, Furumoto M (2009) Is the Ryukyu Subduction Zone in Japan coupled or decoupled? – the necessity of seafloor crustal deformation observation. Earth Planets Space 61:1–9.

Araoka D, Yokoyama Y, Suzuki A, Goto K, Miyagi K, Miyazawa K, Matsuzaki H, Kawahata H (2013) Tsunami recurrence revealed by Porites coral boulders in the southern Ryukyu Islands, Japan. Geology 41:919–922.

Atwater BF (1987) Evidence for great Holocene earthquakes along the outer coast of Washington State. Science 236:942–944.

Cisternas M et al (2005) Predecessors of the giant 1960 Chile earthquake. Nature 437:404–407. https://doi.org/10.1038/nature03943.

Donato S, Reinhardt EG, Boyce JI, Rothaus R, Vosmer T (2008) Identifying paleotsunamis using bivalve shell taphonomy. Geology 36:199–202.

Engel M, Oetjen J, May SM, Brückner H (2016) Tsunami deposits of the Caribbean – towards an improved coastal hazard assessment. Earth Science Rev 163:260–296.

Fujiwara O, Masuda F, Sakai T, Irizuki T, Fuse K (2000) Tsunami deposits in Holocene bay mud in southern Kanto region, Pacific coast of central Japan. Sedi Geol 135:219–230.

Fujiwara O, Tanigawa K, Sato S (2014) Molluscan shell concentration by great tsunami : with special reference to the 2011 Tohoku-oki event. Gekkan Earth 36:36–41 (in Japanese).

Goff J, Chagué-Goff C, Nichol S, Jaffe B, Dominey-Howes D (2012) Progress in palaeotsunami research. Sedi Geol 243–244:70–88.

Goto K, Kawana T, Imamura F (2010) Historical and geological evidence of boulders deposited by tsunamis, southern Ryukyu Islands, Japan. Earth Sci Rev 102:77–99.

Goto K, Miyagi K, Imamura F (2013) Localized tsunamigenic earthquakes inferred from preferential distribution of coastal boulders on the Ryukyu Islands, Japan. Geology 41:1139–1142.

Hongo C, Kayanne H (2009) Holocene coral reef development under windward and leeward locations at Ishigaki Island, Ryukyu Islands, Japan. Sed Geo 214:62–73.

Hongo C, Kayanne H (2010) Holocene sea-level record from corals: reliability of paleodepth indicators at Ishigaki Island, Ryukyu Islands, Japan. Palaeo Palaeo Palaeo 287:143–151.

Inoue T, Goto K, Nishimura Y, Watanabe M, Iijima Y, Sugawara D (2017) Paleo-tsunami history along the northern Japan Trench: evidence from Noda Village, northern Sanriku coast, Japan. Prog Earth Planet Sci 4:42. https://doi.org/10.1186/s40645-017-0158-1.

Ishimura D, Miyauchi T (2015) Historical and paleo-tsunami deposits during the last 4000 years and their correlations with historical tsunami events in Koyadori on the Sanriku Coast, northeastern Japan. Prog Earth Planet Sci 2: 16. https://doi.org/10.1186/s40645-015-0047-4.

Iwasaki T (1927) Tsunami damage at each village. J Climatol Jpn 2:1–57 (in Japanese).

Kan H, Kawana T (2006) "Catch-up" of a high latitude barrier reef by back-reef growth during post-glacial sea-level rise, Southern Ryukyus, Japan. Proc. 10th Int. Coral Reef Symp., Naha, Japan, pp 494–503.

Kawana T, Nakata T (1994) Timing of Late Holocene tsunamis originated around the southern Ryukyu Islands, Japan, deduced from coralline tsunami deposits. Jour Geograph Japan 103:352–376 (in Japanese with English abstract).

Kitamura A (2016) Examination of the largest-possible tsunamis (level 2 tsunami) generated along the Nankai and Suruga troughs during the past 4000 years based on studies of tsunami deposits from the 2011 Tohoku-oki tsunami. Prog Earth Planet Sci 3:12. https://doi.org/10.1186/s40645-016-0092-7.

Kitamura A, Ito M, Sakai S, Yokoyama Y, Miyairi Y (2018) Identification of tsunami deposits using a combination of radiometric dating and oxygen-isotope profiles of articulated bivalves. Mar Geol 403:57–61.

Kondo Y, Abbott ST, Kitamura A, Naish T, Kamp PJJ, Kamataki T, Saul GS (1998) The relationship between shellbed type and sequence architecture: examples from Japan and New Zealand. Sedi Geol 122:109–127.

Kondo Y, Yokogawa K (2001) Substrate preferences and burrowing ability of Donacidae and Mesodesmatidae (Bivalvia) on beaches of southwest Japan. Venus 60(1/2):57–69.

Kortekaas S, Dawson A (2007) Distinguishing tsunami and storm deposits: an example from Martinhal, SW Portugal. Sedi Geol 200:208–221.

Machida H, Ota Y, Kawana T, Moriwaki H, Nagaoka S (2001) Geomorphology of Japan 7: Kyusyu and Southwestern Islands. The Univ. Tokyo Press, Tokyo, p 355.

Makino K (1968) The Meiwa tsunami at Yaeyama, p 462 Published by author, Ishigaki (in Japanese).

Mannen K, Yoong KH, Suzuki S, Matsushima Y, Ota Y, Kain CL, Goff J (2018) History of ancient megathrust earthquakes beneath metropolitan Tokyo inferred from coastal lowland deposits. Sedi Geol 364:258–275.

Minoura K, Imamura F, Sugawara D, Kono Y, Iwashita T (2001) The 869 Jogan tsunami deposit and recurrence interval of large-scale tsunami on the Pacific coast of northeast Japan. J Nat Dis Sci 23:83–88.

Monecke K, Finger W, Klarer D, Kongko W, McAdoo BG, Moore AL, Sudrajat SU (2008) A 1,000-year sediment record of tsunami recurrence in northern Sumatra. Nature 455:1232–1234.

Morales JA, Borrego J, San Miguel EG, Lopez-Gonzalez N, Carro BM (2008) Sedimentary record of recent tsunamis in the Huelva Estuary (Southwest Spain). Quat Sci Rev 27:734–746.

Nanayama F, Satake K, Furukawa R, Shimokawa K, Atwater BF, Shigeno K, Yamaki S (2003) Unusually large earthquakes inferred from tsunami deposits along the Kuril trench. Nature 424:660–663.

Nanayama F, Shigeno K (2006) Inflow and outflow facies from the 1993 tsunami in southwest Hokkaido. Sedi Geol 187:139–158.

Nelson AR, Briggs RW, Dura T, Engelhart SE, Gelfenbaum G, Bradley LA, Forman S, Vane CH, Kelley KA (2015) Tsunami recurrence in the eastern Alaska-Aleutian arc: a Holocene stratigraphic record from Chirikof Island, Alaska. Geosphere 11:1172–1203.

Ohno T, Katoh T, Yamasu T (1995) The origin of algal-bivalve photo-symbiosis. Palaeontology 38:1–21.

Okutani T (2000) Marine mollusks in Japan. University of Tokyo Press, Tokyo, 1173 pp.

Reinhardt EG, Goodman BE, Boyce JI, Lopez G, van Hengstum P, Rink W, Mart Y,

Raban A (2006) The tsunami of 13 December A.D. 115 and the destruction of Herod the Great's harbor at Caesarea Maritima, Israel. Geology 34:1061–1064.
Reinhardt EG, Pilarczyk JE, Brown A (2012) Probable tsunami origin for a shell and sand sheet from marine ponds on Anegada, British Virgin Islands. Nat Hazards 63:101–117.
Richmond BM, Buckley M, Etienne S, Chagué-Goff C, Clark K, Goff J, Dominey-Howes D, Strotz L (2011) Deposits, flow characteristics, and landscape change resulting from the September 2009 South Pacific tsunami in the Samoan islands. Earth Sci Rev 107:38–51.
Rubin CM, Horton BP, Sieh K, Pilarczyk JE, Daly P, Ismail N, Parnell AC (2017) Highly variable recurrence of tsunamis in the 7,400 years before the 2004 Indian Ocean tsunami. Nat Commun 8:16019. https://doi.org/10.1038/ncomms16019.
Sawai Y, Jankaew K, Martin ME, Prendergast A, Choowong M, Charoentitirat T (2009) Diatom assemblages in tsunami deposits associated with the 2004 Indian Ocean tsunami at Phra Thong Island, Thailand. Mar Micro 73:70–79.
Shimajiri K (1988) Historical papers of the large tsunami at Miyako Island. Bull Cult Aff Div Okinawa Prefecture Educ Board 5:1–10 (in Japanese, original title translated).
Shishikura M, Sawai Y, Namegaya Y (2010) Reproduction of mega-tsunami that occurred in Heian period AD 869 Jogan tsunami. AFREC News 16:1–10 (in Japanese).
Sugawara D, Minoura K, Nemoto N, Tsukawari S, Goto K, Imamura F (2009) Foraminiferal evidence of submarine sediment transport and deposition by backwash during 2004 Indian Ocean tsunami. Island Arc 18:513–525.
Sugawara D, Takahashi T, Imamura F (2014) Sediment transport due to the 2011 Tohoku-oki tsunami at Sendai: results from numerical modeling. Mar Geol 358:18–37.
Szczuciński W, Kokociński M, Rzeszewski M, Chagué-Goff C, Cachão M, Goto K, Sugawara D (2012) Sediment sources and sedimentation processes of 2011 Tohoku-oki tsunami deposits on the Sendai Plain, Japan—insights from diatoms, nannoliths and grain size distribution. Sedi Geol 282:40–56.
Takada Y, Kosuge T, Nishihama S, Katoh M (2002) Vertical distributions of Atactodea striata and Latona faba (Bivalvia) on a subtropical sandy beach in Ishigaki Island, Japan. Venus 61(3–4):203–213.
Takashimizu Y, Urabe A, Suzuki K, Sato Y (2012) Deposition by the 2011 Tohoku-oki tsunami on coastal lowland controlled by beach ridges near Sendai, Japan. Sed Geo 282:124–141.
Vött A, Lang F, Brückner H, Gaki-Papanastassiou K, Maroukian H, Papanastassiou D, Giannikos A, Hadler H, Handl M, Ntageretzis K, Willershäuser T, Zander A (2011) Sedimentological and geoarchaeological evidence of multiple tsunamigenic imprint on the Bay of Palairos-Pogonia (Akarnania, NW Greece). Quat Int 242:213–239.
Watanabe H (1985) A comprehensive list of tsunami damage in Japan. University of Tokyo Press, Tokyo, p 206.
Yokoyama Y, Maeda Y, Okuno J, Miyairi Y, Kosuge T (2016) Holocene Antarctic melting and lithospheric uplift history of the southern Okinawa trough inferred from mid- to late-Holocene sea level in Iriomote Island, Ryukyu, Japan. Quat Int 397:342–348.

9

FTIR microspectroscopy of carbonaceous matter in ~ 3.5 Ga seafloor hydrothermal deposits in the North Pole area, Western Australia

Motoko Igisu[1*], Yuichiro Ueno[1,2,3] and Ken Takai[1,3]

Abstract

Carbonaceous matter in ~ 3.5 Ga hydrothermal vein deposits from the Dresser Formation, Western Australia, was analyzed using Fourier transform infrared (FTIR) microspectroscopy. Based on the spectroscopy, the carbonaceous matter was mainly composed of disordered aromatic structures, with minor aliphatic C–H functional groups. Spatially resolved analysis supports that the aliphatic C–H signatures are derived from kerogenous macromolecule and not from free bitumen or other artificial sources. The intensity ratios of the asymmetric aliphatic CH_3 to the asymmetric aliphatic CH_2 ($R_{3/2}$ value) in the carbonaceous clots range from 0.22 to 0.51. Thermal alteration may increase or not change the $R_{3/2}$ value of organic matter, as suggested by previous experiments, but it is unlikely to be the cause of the substantially lower $R_{3/2}$ values when compared with those of primary organic matter. In particular, the low $R_{3/2}$ values (< ~ 0.4) suggest that the carbonaceous matter mainly contains aliphatic C–H bonds derived from bacterial cells. The carbonaceous clots may have been possibly produced by abiotic reaction such as Fischer-Tropsch-type (FTT) synthesis. However, the organic matter source only produced by the FTT synthesis is inconsistent with the $R_{3/2}$ values for the analyzed carbonaceous clots. The results obtained by combining these spectroscopic features of the carbonaceous clots together with the previously reported isotopic features may possibly suggest that both bacteria and archaea were colonized in the ~ 3.5 Ga Dresser hydrothermal system.

Keywords: FTIR microspectroscopy, Carbonaceous matter, Dresser formation, Archean

Introduction

Hydrothermal systems have been considered as the origin of life (Holm and Andersson 1998) and the locations of the earliest evolution of life and metabolism (Nisbet 1995). The geological study of ancient hydrothermal systems is crucial to understanding the potential role of hydrothermal systems as the earliest ecosystems. One of Earth's oldest seafloor hydrothermal deposits occurs in the ~ 3.5 Ga Dresser Formation in Western Australia (Fig. 1). In this formation, prominent silica-barite veins containing carbonaceous matter ubiquitously penetrate pillowed basaltic greenstones (Ueno et al. 2004; Van Kranendonk 2006; Van Kranendonk et al. 2008). The carbonaceous matter in the veins afford an important means of understanding how organic matter was synthesized in the Earth's early hydrothermal systems, although the biological and abiological processes cannot be fully distinguished. The occurrence of biological activity in the Dresser Formation has been inferred from the observation of putative stromatolites (Walter et al. 1980; Van Kranendonk 2006; Van Kranendonk et al. 2008), morphologically preserved putative microfossils (Ueno et al. 2001), microbially induced sedimentary structures (Noffke et al. 2013), and chemical fossils such as ^{13}C-depleted organic matter and methane (Ueno et al. 2001, 2004, 2006) and ^{34}S-depleted pyrites with characteristic $\Delta^{33}S$ (Ueno et al. 2008;

* Correspondence: igisu@jamstec.go.jp
[1]Department of Subsurface Geobiological Analysis and Research, Japan Agency for Marine-Earth Science and Technology (JAMSTEC), 2-15 Natsushima-cho, Yokosuka 237-0061, Japan
Full list of author information is available at the end of the article

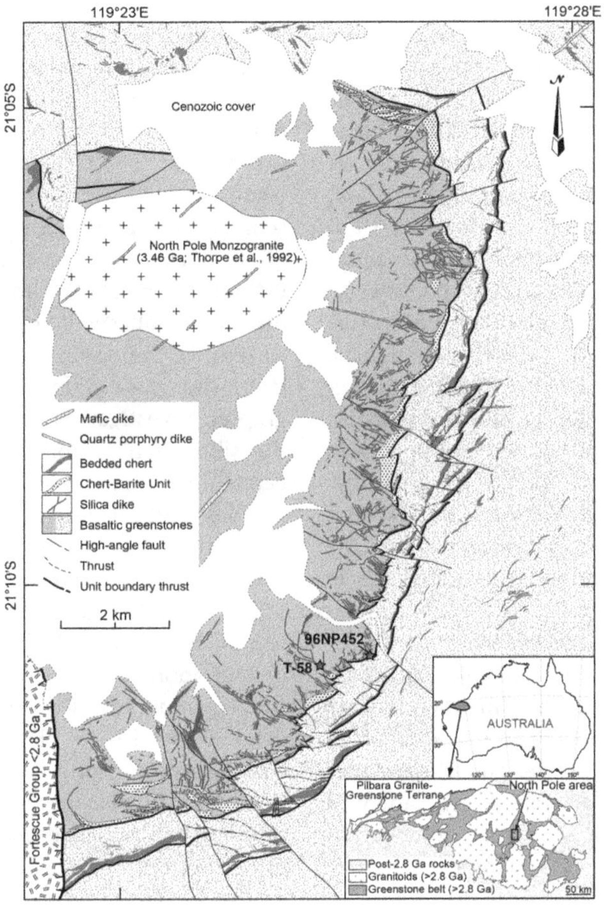

Fig. 1 Geologic map of the Dresser Formation in the North Pole area. The localities of the silica vein samples 96NP452 and T-58 are shown (modified from Kitajima et al. 2001)

Shen et al. 2009). Sulfur isotopic signatures also suggest that microbial sulfate reduction was involved in pyrite formation (Ueno et al. 2008; Shen et al. 2009). In addition, ancient hydrothermal fluid inclusions contain ^{13}C-depleted methane, suggesting the presence of methanogenic archaea (Ueno et al. 2006), although it is arguable if those methane were produced by metal-catalyzed abiotic reactions (Sherwood- Lollar and McCollom 2006). The ^{13}C-depletion of organic matter indicates the predominance of autotrophic carbon fixation (Ueno et al. 2001, 2004). The organic matter has been characterized by Raman microspectroscopy, FTIR spectroscopy, solid-state ^{13}C nuclear magnetic resonance, and pyrolysis analysis (e.g., Ueno et al. 2001; Derenne et al. 2008; Duda et al. 2018). The results of these organic geochemical analyses of isolated insoluble organic matter reveal that the organic matter is dominated by aromatic structures with aliphatic hydrocarbon moieties. In addition, the pyrolysis fragments of the organic matter comprise long-chain aliphatic hydrocarbons and thiophenes, which are likely related to the existence of microbes including sulfate-reducing bacteria (Derenne et al. 2008; Duda et al. 2018).

Chemical analyses of Archean organic matter have provided limited but important information about the physiological and phylogenetic characteristics of the source organic compounds (e.g., Hayes 1994; Brocks et al. 1999). Although the observation of morphologically preserved microfossils is a direct evidence of the existence of life, the identification of microfossils based on only morphological analysis is often insufficient (e.g., Brasier et al. 2002; Schopf et al. 2002). The chemical analysis of carbonaceous matter is necessary for understanding its origin and characteristics. However, the bulk analysis of isolated organic matter generally involves the potential risk of post-depositional and experimental contamination of the organic matter (e.g., Brocks et al. 1999; Derenne et al. 2008; Rasmussen et al. 2008; Bourbin et al. 2012). In situ analysis is a more reliable technique that avoids such contamination problems and extracts chemical signatures specific to individual microstructures in petrographic thin sections (e.g., House et al. 2000, 2013; Kudryavtsev et al. 2001; Ueno et al. 2001; Schopf et al. 2002, 2005; Igisu et al. 2006, 2009; van Zuilen et al. 2007; Wacey et al. 2011; Lepot et al. 2013;

Williford et al. 2013). FTIR microspectroscopy has been used to analyze Proterozoic organic-walled acritarchs and bacterial fossils, as well as Phanerozoic miospores (e.g., Arouri et al. 1999, 2000; Marshall et al. 2005; Igisu et al. 2006, 2009; Steemans et al. 2010; Qu et al. 2015). These studies demonstrated the usefulness of IR signatures for characterizing organic-walled microfossils, including those with unknown affinity such as acritarchs. However, FTIR microspectroscopy has not yet been applied to Archean carbonaceous matter in petrographic thin sections. FTIR spectroscopy has been performed on isolated Archean carbonaceous matter, and reported the presence of aliphatic C-H stretching absorptions (Derenne et al. 2008). This previous attempt did not employ the in situ microscopic method. It has been suggested that isolated and powdered samples may be affected by contamination from volatile hydrocarbons in laboratory air (e.g., Salisbury and Walter 1989) and/or from other aliphatic C-H bearing components during sample preparation and storage (e.g., Kebukawa et al. 2009). Therefore, the data from powdered samples would need to be interpreted with caution.

In this paper, we first report the FTIR microspectroscopic features of early Archean carbonaceous matter obtained from two ~ 3.5 Ga silica veins in the Dresser Formation in the North Pole area of Western Australia. The IR signatures are then compared with the characteristics of a variety of biologically and abiotically produced organic compounds, and the possible origins of the carbonaceous clots are discussed.

Methods/Experimental

Samples

We analyzed two samples taken from the silica veins (96NP452 and T-58) that intrude on the basaltic greenstones of the Dresser Formation in the North Pole area of Western Australia. The sampling localities of the silica veins are shown in Fig. 1. Detailed descriptions of the localities and samples can be found in the works of Ueno et al. (2001, 2004, 2006). The Dresser Formation consists of pillowed basaltic greenstones and chert beds. In the North Pole area, these rocks have only undergone low-grade metamorphism below their greenschist facies (Buick and Dunlop 1990; Kitajima et al. 2001; Ueno et al. 2004). The lowest chert unit of the formation is intercalated with several barite beds (Buick and Dunlop 1990), and numerous silica veins intrude the greenstones below the chert/barite beds. The silica veins are 0.3–20 m in width and generally > 100 m in length (Ueno et al. 2001). Zircon U-Pb dating yielded two distinct age groups of 3458 ± 2 Ma (Thorpe et al. 1992a) and 3660 ± 52 Ma (Kitajima et al. 2008) of the felsic unit that overlies the Dresser Formation. A model lead age of 3490 Ma was obtained for galena taken from the Dresser Formation (Thorpe et al. 1992b), and this may represent the actual depositional age of the formation.

The first doubly polished petrographic thin section sample was made from sample 96NP452 with a thickness of approximately 60 μm. The preparation process was as described by Igisu et al. (2006). We used alumina paste to polish only one side of the thin section to remove interference fringes. To avoid organic contamination during sample preparation, the sample was attached to a glass slide using only a carbon-based dissoluble adhesive (Aron Alpha). The adhesive was used to remove the doubly polished petrographic thin section from the glass slide by immersing in acetone. Petrographic investigation of the thin section indicated that the silica vein sample was mainly composed of microcrystalline quartz (< ~ 10 μm), carbonaceous matter, sulfides, and carbonates (see for example Fig. 2g, h). The filamentous structures described by Ueno et al. (2001) were not observed in the doubly polished thin section.

The second set of doubly polished thin section samples was made from sample T-58 (Ueno et al. 2006) with a thickness of approximately 100 μm. A petrographic examination of the sample revealed three distinctive fabrics, which were denoted by F1–F3 (Fig. 2i). F1 was characterized by coarse-grained quartz (> 1 mm) with abundant fluid inclusions. No carbonaceous matter was observed in F1. F2 comprised microcrystalline quartz chalcedony. Black aggregates in F2 were carbonaceous matter (see Fig. 2i, j) and ion oxides such as hematite. The center of the carbonaceous clot (white part in Fig. 2j) was quartz. F3 consisted mainly of microcrystalline quartz (< ~ 10 μm) and contained carbonaceous matter and ion oxides such as hematite (Fig. 2i). The analyzed carbonaceous clots in the sample were preserved in a microcrystalline quartz matrix or microcrystalline quartz chalcedony, but not in the secondary quartz veins (Fig. 2).

FTIR microspectroscopy

FTIR microspectroscopic analysis was conducted on the two thin section samples using an automated XYZ stage set in an FTIR microspectrometer (JASCO FTIR6200 + IRT7000) with a single-point mercury cadmium telluride detector. The thin section was placed over a hole in the sample holder (Nakashima et al. 1989) and measured as described below.

A reference background spectrum was first measured away from the mounted sample (i.e., in air). A transmission IR spectrum of the sample was then measured. The IR spectrum was described in terms of the IR absorbance, which is a function of the wavenumber (cm^{-1}):

$$\text{Absorbance} = -\log_{10}(T/T_0) \quad (1)$$

where T and T_0 denote the transmitted intensity of the infrared light at each wavenumber for the sample and air

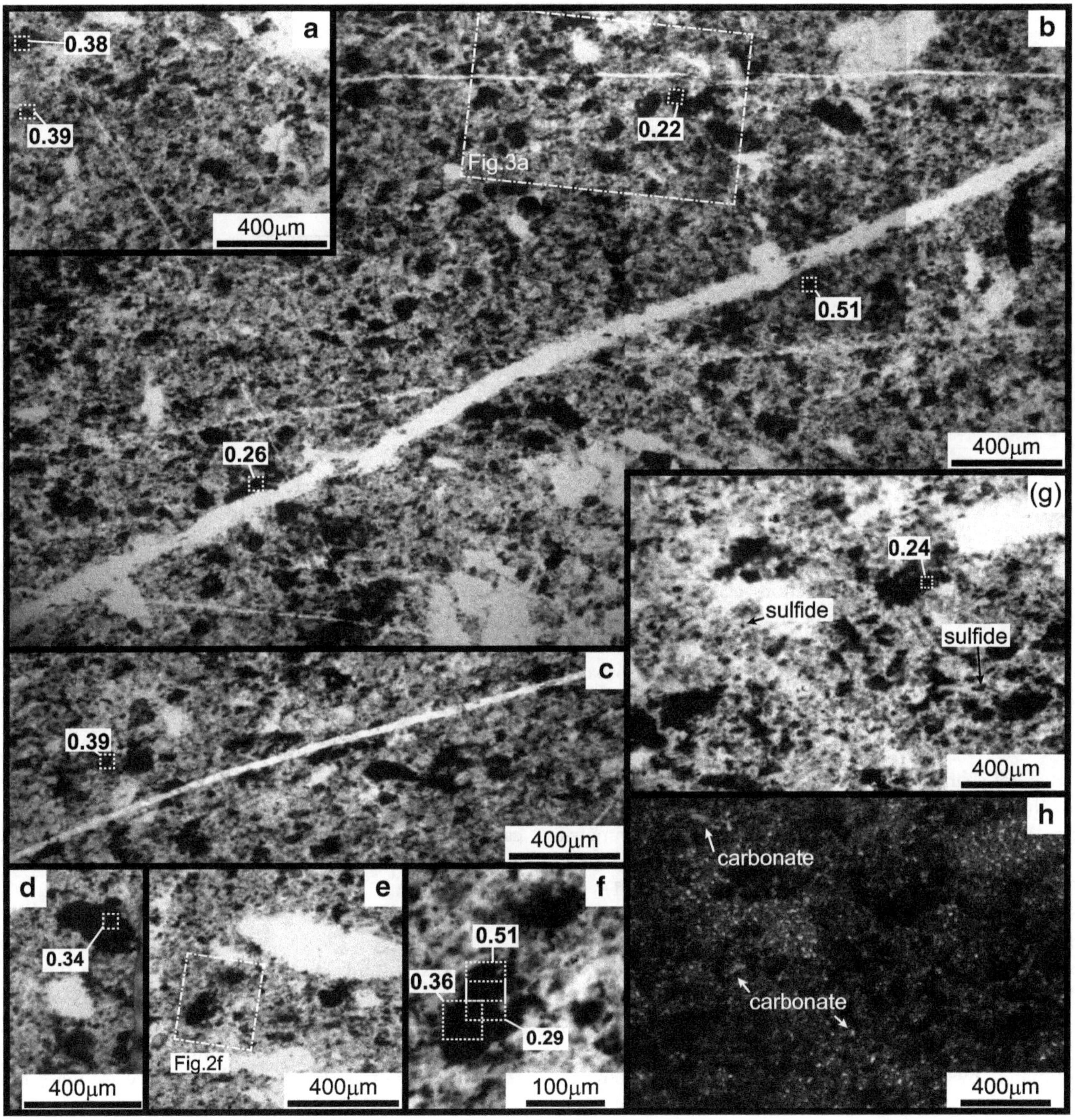

Fig. 2 Optical photomicrographs of the analyzed thin sections. **a–g** Plane-polarized light photomicrographs of the thin section 96NP452. **b** Composite image of the photomicrographs; the dashed rectangle in **b** corresponds to the area shown in Fig. 3a. **f** Enlarged view of the dashed rectangle in **e**. **h** Cross-polarized light photomicrograph of the same area in **g**. **i**, **j** Plane-polarized light photomicrographs of the thin section T-58. **i** Composite image of the photomicrographs. **j** Enlarged view of the dashed rectangle in **i**. The locations ($50 \times 50\ \mu m^2$ dotted squares) of the 12 clots with satisfactory signal/noise ratios and their calculated $R_{3/2}$ ratios (1024 scans) are shown in **a–d**, **f**, **g**, **j**

background, respectively. To search for organic signatures, the IR spectra of about 3500 individual carbonaceous clots were first obtained by collecting 64 scans per spot within a spectral range of 4000–1000 cm^{-1} with a 4 cm^{-1} spectral resolution. Among the ~ 3500 examined spots, 47 had an IR absorbance of ~ 2925 cm^{-1}, which indicated excess analytical error for the adopted 64-scan acquisition mode. These spots were therefore measured again by collecting 1024 scans per spot within a spectral range of 4000–1000 cm^{-1} with a 4 cm^{-1} spectral resolution. This was done to obtain the IR spectra with a better signal-to-noise ratio. Each spectrum was obtained from a spot of $50 \times 50\ \mu m^2$

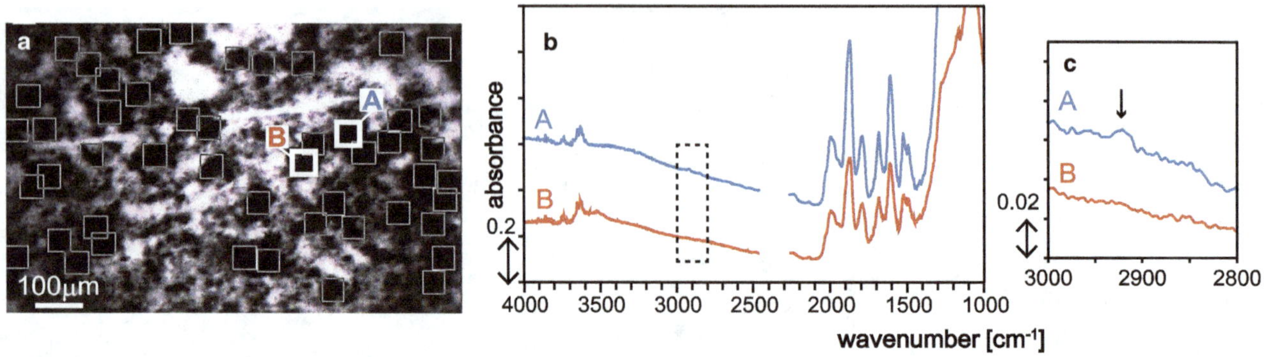

Fig. 3 An example of the optical photomicrographs and transmission IR spectra of the analyzed carbonaceous clots. **a** Optical photomicrograph. The squares indicate the analyzed areas (50 × 50 μm^2). The IR spectra of the two analyzed areas (thick-bordered squares A and B) are shown in **b** and **c**. **b** Transmission IR spectra (64 scans) of the carbonaceous clots with (**a**) and without (**b**) aliphatic CH moieties. One scale unit on the vertical axis corresponds to an absorbance of 0.2. The IR spectrum in **b** is representative of those obtained from the white squares in **a** (without aliphatic CH bonds). **c** An enlarged view of the IR spectra in **b** within 3000–2800 cm^{-1}. The range is indicated by the dotted rectangle in **b**. The arrow represents the IR band at ~ 2925 cm^{-1}. One scale unit on the vertical axis corresponds to an absorbance of 0.02

(Figs. 2 and 3a). Duplicate measurements were performed on the 47 spots in the 1024-scan acquisition mode. The analytical errors were determined from the duplicate measurements at the same location as the background spectrum and were found to be smaller than 0.003 absorbance units in the 3000–2800 cm^{-1} range for the 64-scan acquisition mode and smaller than 0.0007 absorbance units for the 1024-scan acquisition mode. All the spectral data were processed using Spectra Manager (JASCO).

Structural characterization of carbonaceous clots

In the aliphatic C–H stretching region (3000–2800 cm^{-1}), the ~ 2925 cm^{-1} band is due to two asymmetric stretching groups, namely, aliphatic CH_2 and aromatic CH_3 (Jones and Sandorfy 1956; Painter et al. 1981). The band is therefore considered as being the resultant of these two groups. Because the aliphatic CH_2 and aromatic CH_3 groups have symmetric stretching bands with different peak positions at ~ 2850 and ~ 2870 cm^{-1}, respectively (Bellamy 1954; Jones and Sandorfy 1956), the contribution of the aliphatic CH_2 groups can be estimated by elucidating the relationship between the intensities at ~ 2925 cm^{-1} and ~ 2850 cm^{-1}. According to Lambert-Beer's Law:

$$\text{Absorbance} = \varepsilon \cdot c \cdot d, \tag{2}$$

where ε is the molar absorption coefficient, c is the concentration, and d is the sample thickness. Based on Eq. (2), the absorbance ratio between the asymmetric CH_2 and symmetric aliphatic CH_2 is converted into the ratio of their molar absorption coefficients as follows:

$$\text{Abs}_s/\text{Abs}_a = \varepsilon_s/\varepsilon_a, \tag{3}$$

where Abs_s and Abs_a are respectively the absorbances (peak heights) of the symmetric aliphatic CH_2 and asymmetric aliphatic CH_2, and ε_s and ε_a are the corresponding molar absorption coefficients. Thus, if the linear correlation between the peak heights at ~ 2925 and ~ 2850 cm^{-1} is obtained, the ~ 2925 cm^{-1} band would be observed to be mainly due to the aliphatic CH_2, and less attributable to the aromatic CH_3 group.

The IR signatures of the aliphatic CH_3 and CH_2 ratios generally reflect the degree of branching and the chain length of the aliphatic hydrocarbon moiety (e.g., Lin and Ritz 1993; Coates 2000). To evaluate the relative length and degree of branching of the aliphatic chain in the carbonaceous clots, we used the aliphatic CH_3/CH_2 absorbance ratio ($R_{3/2}$):

$$R_{3/2} = \text{Abs}_{\text{CH3}}/\text{Abs}_{\text{CH2}}, \tag{4}$$

where Abs_{CH3} and Abs_{CH2} respectively denote the peak heights of the asymmetric stretching bands of the aliphatic CH_3 (end-methyl, ~ 2960 cm^{-1}) and CH_2 (chain-methylene, ~ 2925 cm^{-1}) after linear baseline correction (Igisu et al. 2009). For a more precise determination of $R_{3/2}$, we used only the data for the absorbances at ~ 2960 and 2925 cm^{-1}, which were more than twice the analytical error.

Raman microspectroscopy

A laser Raman micro-spectrometer (Nanophoton RAMANtouch) was used to characterize the carbonaceous clots. Each sample was exposed to a green laser (532 nm wavelength) for a period of 5 s at a laser power density of 1 × 10^5 W/cm^2. Raman spectra for wavenumbers within ~ 2670–110 cm^{-1} were acquired with a 600 g/mm grating (spectral resolution of ~ 4 cm^{-1}). A × 100 objective lens (NA = 0.9) was used to achieve Raman analysis spatial resolution of < 1 μm, and the incident laser power was estimated to be ~ 0.4 mW at the sample surface. We only analyzed the carbonaceous matter embedded within the rocks below the surface of the thin section to avoid the effect of polishing, which can induce deformation of carbonaceous matter during sample preparation and thus possibly induce artificial modification of the Raman spectroscopic feature (Pasteris 1989). Prior to sample analyses, wavenumber calibration was determined by comparison with the standard silicon spectra (520 cm^{-1}). All the spectral data were processed using RAMAN viewer (Nanophoton).

Raman spectroscopy has been used as a tool for providing a metamorphic grade indicator of carbonaceous matter in geological samples (e.g., Yui et al. 1996; Beyssac et al. 2002; Rahl et al. 2005; Kouketsu et al., 2014). Several indicators have been proposed to estimate peak metamorphic temperatures by using Raman spectral parameters: peak position, intensity ratio, and full width at half maximum (FWHM) of D band and G band. We chose the D/G band intensity ratio (I_{1350}/I_{1600}) for estimation of the degree of thermal alteration of the carbonaceous matter in order to compare with the previous spectral data of the Proterozoic samples (Qu et al. 2015). The analytical errors for the Raman spectral features were determined by taking duplicate measurements within one area analyzed by FTIR microspectroscopy.

Results and discussion

IR spectra

Figures 3 and 4 show examples of the infrared transmission spectra of the carbonaceous clots in the silica veins. We measured more than 3500 locations in the carbonaceous clots in the thin sections, and only 12 clots had absorbances of ~ 2960, ~ 2925, or ~ 2850 cm^{-1}, which are more than twice the analytical error. The obtained absorbances of ~ 2960, ~ 2925, and ~ 2850 cm^{-1} are summarized in Table 1. The spectrum shown by line B in Fig. 3 was the most common feature in the analyzed carbonaceous clots. This is quite similar to that of the quartz matrix (black line in Fig. 4a, b1, c1), and such data were not used. Instead, some carbonaceous clots show spectral patterns such as A in Fig. 3 (with a peak at ~ 2925 cm^{-1}). The detailed peak assignments of the spectra are explained below.

The IR spectrum of the adhesive used to prepare the samples is shown in Fig. 4a, b1, c1. The bands at 2988, 2944, 2908, and 2876 cm^{-1} are due to C–H stretching vibrations. The 1757 cm^{-1} band is due to ester C=O. The 1450 and 1375 cm^{-1} bands are due to C–H bending of $CH_2 + CH_3$ and CH_3 groups, respectively. The bands at 1260 and 1180 cm^{-1} are due to C–N and C–O, respectively. In the 3000–2800 cm^{-1} region, the most intense band for the adhesive is observed at 2988 cm^{-1} (shown by a dotted arrow in Fig. 4b1). Based on the 4 cm^{-1} wavenumber resolution of the IR spectrometer, the bands of the carbonaceous clots (~ 2960, ~ 2925, and/or ~ 2850 cm^{-1}) are considered to be different from the IR bands of the adhesive. The adhesive used to attach the thin section to the glass has been almost completely washed by acetone. The difference between the absorbances at ~ 2960, ~ 2925, and ~ 2850 cm^{-1} of individual analyzed spots (Table 1) would reflect the amounts of carbonaceous matter and/or its chemical structure. The saturated bands (1400–1000 cm^{-1}) and seven bands (1995, 1870, 1793, 1684, 1610, 1525, and 1492 cm^{-1}) in the figures are due to the Si–O bonds of quartz (Nakashima et al. 1995; Ito and Nakashima 2002; Igisu et al. 2006, 2009; Qu et al. 2015). The IR spectrum around 2300 cm^{-1} is not shown because this range is affected by absorption by CO_2 in the air. In the region of lower wavenumbers of less than 2500 cm^{-1}, the many intense bands of the quartz matrix mask the small signal from the organic matter (Figs. 3 and 4). Hence, we focused on the IR bands with wavenumbers of > 2500 cm^{-1}.

The contribution of the aliphatic CH_2 group to the intensity in the ~ 2925 cm^{-1} band was examined by comparison with those in the ~ 2925 and ~ 2850 cm^{-1} band. Figure 5 shows the relationship between the intensities in the ~ 2925 and ~ 2850 cm^{-1} bands. The strong correlation between these two bands suggests that they are due to the same component(s) and that the ~ 2925 cm^{-1} band of the carbonaceous clots is mostly due to the aliphatic CH_2 and is less attributable to the aromatic CH_3 groups. These results indicate that the Dresser carbonaceous clots contain small amounts of aliphatics.

Raman spectral data

The carbonaceous clots in sample 96NP452 have disordered carbonaceous structures, which were observed as ~ 1600 cm^{-1} (graphite, G) and ~ 1350 cm^{-1} (disordered, D) bands in the Raman spectra (Fig. 6). Their Raman spectral features are shown in Table 1 and are similar to each other for this sample. They are also consistent with the previously reported data (Ueno et al. 2001, 2004) and those of carbonaceous matter from lower greenschist facies metasediments (Yui et al. 1996). In addition, it is inconsistent with the carbonaceous matter with different thermal alteration phases observed in ~ 3.46 Ga hydrothermal Apex Chert (Olcott-Marshall et al. 2012; Sforna et al. 2014). These Raman and IR spectral features of the carbonaceous clots suggest that the clots mainly consist of randomly arranged layered polyaromatic domains with minor aliphatic C–H bonds. On the other hand, the carbonaceous clot in sample T-58 does not show clear G or D bands. Instead, it shows a sloping baseline (Fig. 6), possibly suggesting fluorescence produced by hydrogen-rich saturated aliphatic hydrocarbons (Marshall et al. 2005). However, this appears inconsistent

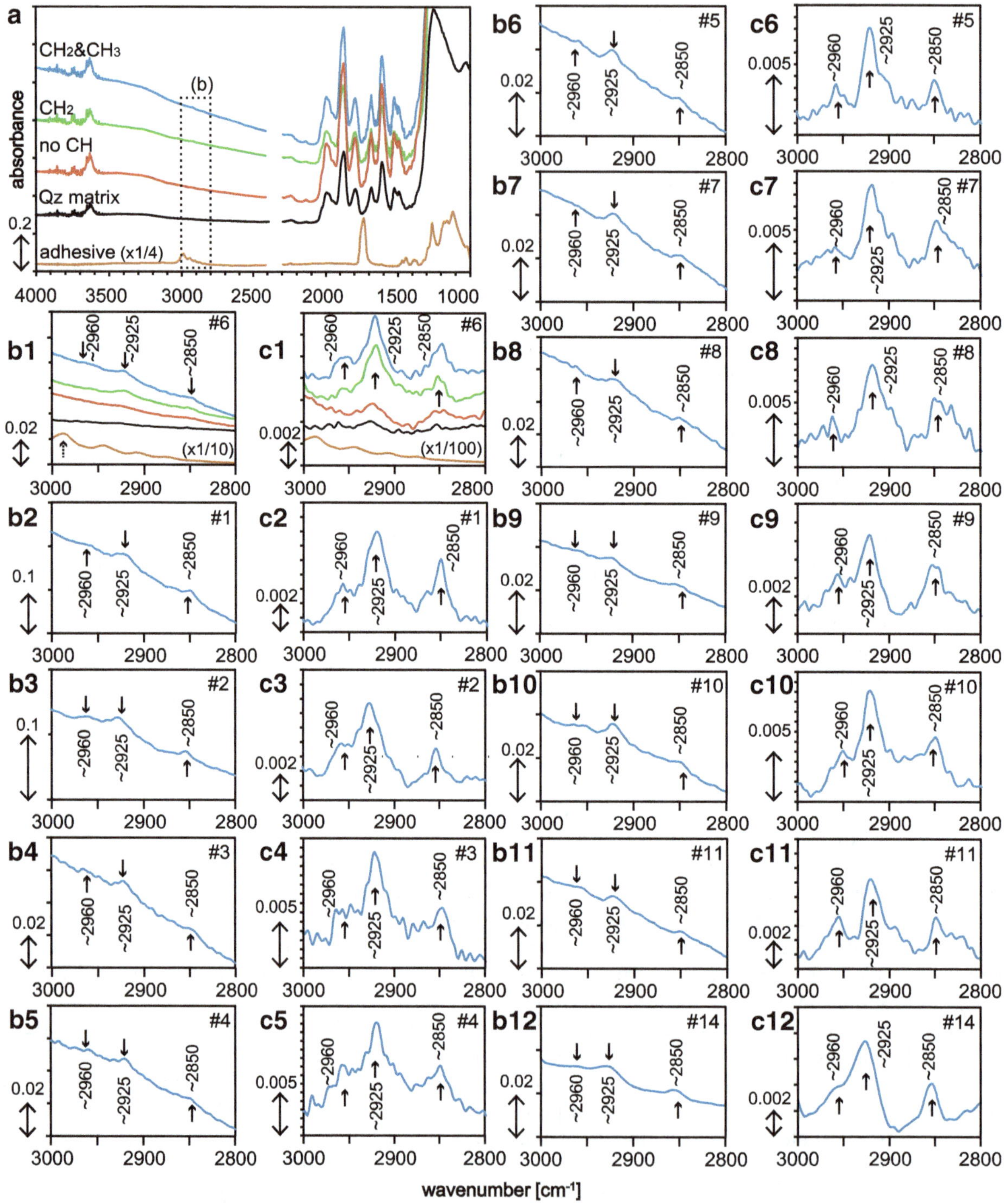

Fig. 4 The transmission IR absorbance spectra (1024 scans) of the carbonaceous clots. **a** Examples of the IR spectra of the carbonaceous clots with aliphatic CH_2 and CH_3 bonds (blue line: spot #6 in Table 1), only aliphatic CH_2 bond (green line: spot #12 in Table 1), and no CH_2 bond (red line: spot #13 in Table 1) together with quartz matrix (black line). IR transmission-reflection spectrum of the carbon-based dissolvable adhesive used in this study on an aluminum foil is shown by the brown line (modified from Igisu et al. 2006). One scale unit on the vertical axis corresponds to an absorbance of 0.2. (b) An enlarged view of the IR spectra within 3000–2800 cm^{-1}. The range for **b1** is indicated by the dotted rectangle in **a**. Spot numbers (#1, #2, etc.) are shown in the upper right in each figure. One scale unit on the vertical axis corresponds to an absorbance of 0.02 in **b1** and **b4**–**b12**, and 0.1 in **b2** and **b3**. **c** The IR spectra after linear baseline correction of the spectra in **b** within 3000–2800 cm^{-1}. Spot numbers (#1, #2, etc.) are shown in the upper right in each figure. One scale unit on the vertical axis corresponds to an absorbance of 0.002 in **c1** and **c12**, and 0.001 in **c2**–**c11**. The spectra of the adhesive are re-scaled to a quarter in **a**, one-tenth in **b1**, and one-hundredth in **c1** for clarity

Table 1 IR and Raman parameters for analyzed carbonaceous clots

Sample #	Spot #	Microphotograph	IR				Raman				
			Abs at ~ 2960 cm^{-1}	Abs at ~ 2925 cm^{-1}	Abs at ~ 2850 cm^{-1}	$R_{3/2}$	I_{1350}/I_{1600}	A_{1350}/A_{1600}	$FWHM_{1350}$	$FWHM_{1600}$	*N*
96NP452	#1	Figure 2a	0.0024 ± 0.0003	0.0062 ± 0.0005	0.0043 ± 0.0000	0.38 ± 0.02	1.54 ± 0.07	2.21 ± 0.06	65.0 ± 4.9	45.1 ± 3.0	27
	#2	Figure 2a	0.0019 ± 0.0001	0.0050 ± 0.0002	0.0018 ± 0.0001	0.39 ± 0.00	1.46 ± 0.07	2.24 ± 0.07	67.6 ± 4.3	44.2 ± 2.9	28
	#3	Figure 2b	0.0025 ± 0.0000	0.0088 ± 0.0010	0.0046 ± 0.0007	0.29 ± 0.03	1.45 ± 0.08	2.24 ± 0.05	68.3 ± 5.1	44.2 ± 2.2	27
	#4	Figure 2b	0.0051 ± 0.0003	0.0100 ± 0.0005	0.0055 ± 0.0000	0.51 ± 0.01	1.46 ± 0.06	2.23 ± 0.03	67.1 ± 2.9	43.8 ± 1.3	27
	#5	Figure 2b	0.0025 ± 0.0001	0.0070 ± 0.0006	0.0039 ± 0.0008	0.36 ± 0.05	1.52 ± 0.08	2.22 ± 0.08	63.2 ± 4.3	43.2 ± 3.1	27
	#6	Figure 2c	0.0017 ± 0.0000	0.0052 ± 0.0001	0.0026 ± 0.0001	0.34 ± 0.01	1.39 ± 0.07	2.20 ± 0.08	70.8 ± 5.5	44.8 ± 3.0	27
	#7	Figure 2d	0.0016 ± 0.0004	0.0066 ± 0.0005	0.0034 ± 0.0004	0.24 ± 0.04	1.47 ± 0.08	2.24 ± 0.04	65.1 ± 3.1	42.6 ± 1.2	27
	#8	Figure 2f	0.0018 ± 0.0005	0.0073 ± 0.0009	0.0040 ± 0.0005	0.26 ± 0.10	1.48 ± 0.06	2.25 ± 0.06	66.3 ± 2.7	43.6 ± 2.2	30*
	#9	Figure 2f	0.0014 ± 0.0001	0.0038 ± 0.0001	0.0024 ± 0.0000	0.39 ± 0.05	1.48 ± 0.06	2.25 ± 0.06	66.3 ± 2.7	43.6 ± 2.2	30*
	#10	Figure 2f	0.0021 ± 0.0003	0.0094 ± 0.0001	0.0048 ± 0.0002	0.22 ± 0.03	1.46 ± 0.04	2.26 ± 0.04	66.7 ± 1.6	42.9 ± 0.9	30
	#11	Figure 2g	0.0026 ± 0.0002	0.0050 ± 0.0003	0.0024 ± 0.0002	0.51 ± 0.00	1.49 ± 0.08	2.25 ± 0.03	65.8 ± 3.9	43.4 ± 1.4	28
	#12		n.d.	0.0042 ± 0.0001	0.0015 ± 0.0002		1.40 ± 0.06	2.24 ± 0.03	69.6 ± 2.5	43.6 ± 0.9	27
	#13		n.d.	n.d.	n.d.		1.36 ± 0.13	2.00 ± 0.37	77.9 ± 8.8	52.4 ± 12.7	27
T-58	#14	Figure 2j	0.0021 ± 0.0000	0.0056 ± 0.0004	0.0025 ± 0.0001	0.38 ± 0.02					

Absorbances at ~ 2960 cm^{-1}, ~ 2925 cm^{-1}, and 2850 cm^{-1} and $R_{3/2}$ values of the analyzed carbonaceous matter. Errors show standard deviations. *n.d.* not detected
*The same data are shown because the analyzed area for FTIR microspectroscopy of the clots were overlapped

with the IR data that the absorbance of aliphatic CH_2 in the carbonaceous clots in sample T-58 falls within the range for those in sample 96NP452 (Fig. 5). The inconsistency between the Raman and IR data of the carbonaceous clots in sample T-58 and sample 96NP452 may be due to difference in the amounts of organic matter. Raman spectral data suggest that the analyzed carbonaceous clots at least in sample 96NP452 are moderately graphitized by metamorphism, supporting indigenous origin in ca. 3.5 Ga rocks rather than contaminations.

Spatial distributions of aliphatic C–H groups

The $R_{3/2}$ values of the carbonaceous clots range from 0.22 ± 0.03 to 0.51 ± 0.01 (Fig. 7 and Table 1). In sample 96NP452, some of the carbonaceous clots containing aliphatic C–H signatures were located relatively close (< 100 μm) to the secondary quartz veins (Fig. 2b, c), while others were far (> 100 μm) from the veins (Fig. 2e, g). Both low (~ 0.3) and high (~ 0.5) $R_{3/2}$ values were observed within one carbonaceous clot (Fig. 2f) and also along one secondary quartz vein (Fig. 2b). In sample T-58, one carbonaceous clot with aliphatic C–H signatures was observed in the quartz chalcedony (F2; Fig. 2i), while the other carbonaceous clots in the quartz chalcedony and microcrystalline quartz did not have aliphatic C–H signatures. These results indicate the absence of an apparent distribution pattern among the carbonaceous clots with and without aliphatic C–H signatures. In addition, there was significant variation among the $R_{3/2}$ values within a given sample, and even within a given carbonaceous clot. The aliphatic C–H signatures of the carbonaceous clots are spatially heterogeneous on a micrometer to millimeter scale.

Origin of the carbonaceous matter

Post-depositional thermal alteration of the primary organic matter may have altered the $R_{3/2}$ values. The varying $R_{3/2}$ values of the carbonaceous clots may also be explained by a mixture of different sources of biological organic compounds and/or abiotically

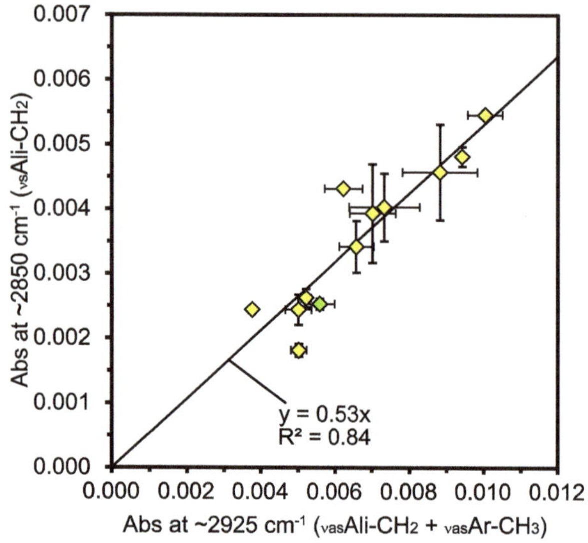

Fig. 5 Relationship between the intensities of the carbonaceous clots at ~ 2925 cm^{-1} and ~ 2850 cm^{-1}. The ~ 2925 cm^{-1} band is due to asymmetric stretching vibrations of the aliphatic CH_2 and aromatic CH_3, and the ~ 2850 cm^{-1} band is due to symmetric stretching vibration of the aliphatic CH_2. The yellow diamonds indicate the data for sample 96NP452, while the green diamond indicates that for sample T-58. The error bars indicate the standard deviations of the duplicate measurements

produced compounds. The possible occurrence of anaerobic chemoautotrophic microorganisms was suggested by the carbon isotopic signatures of the carbonaceous matter in the silica veins in the Dresser Formation (Ueno et al. 2004). The carbon isotopic signatures provide the potential chemical evidence of microbial activity in the environment at the time, but do not

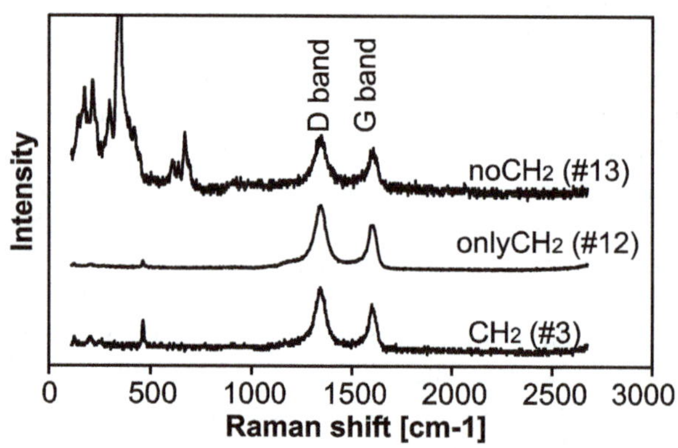

Fig. 6 Examples of Raman spectra of the carbonaceous clots. The spectra marked as CH_2, CH_2 only, and no CH_2 were obtained from sample 96NP452 (CH_2: both aliphatic CH_2 and CH_3 bonds were observed in micro-FTIR measurements; CH_2 only: CH_2 bond was observed but CH_3 was not; no CH_2: neither CH_2 nor CH_3 were observed). Spot numbers (#3, #12, and #13) corresponding to Table 1 are also shown. The band around 464 cm^{-1} is due to the quartz matrix of the silica veins

necessarily offer clues for distinguishing whether the products are derived from the microbial functions of the bacterial and/or the archaeal populations. The existence of sulfate-reducing bacteria has been proposed by the pyrolysis fragments of the organic matter (Derenne et al. 2008). However, the potential risk of post-depositional and experimental contamination has been found in the pyrolysis analysis of the isolated Archean organic matter (e.g., Bourbin et al. 2012). The presence of ancient methanogenic archaea has been inferred from the fluid inclusions in the silica veins containing ^{13}C-depleted methane (Ueno et al. 2006), although the abiotic origin of the methane has also been proposed (Sherwood-Lollar and McCollom 2006). We will discuss the origin of the varying $R_{3/2}$ values for the carbonaceous clots in the following order: (1) abiological origin, (2) effect of thermal alteration, and (3) biological origin.

Abiological origin

It is known that an FTT synthesis under specific hydrothermal conditions (e.g., in ultramafic rock-associated hydrothermal systems) may produce organic compounds (e.g., Yanagawa and Kobayashi 1992; Holm and Andersson 1998; McCollom et al. 1999; McCollom and Seewald 2006). The organic products of the FTT synthesis are mainly methane and normal alkanes with only a small amount of branched hydrocarbons because the relative abundance of these hydrocarbons decreases with increasing carbon number in the alkyl chains (McCollom and Seewald 2006, 2007). McCollom and Seewald (2006) reported that isobutane was the only branched product of FTT synthesis, but the yield of isobutane (~ 0.02 mmol/kg) was much lower than those of the normal alkanes (e.g., 0.64 mmol/kg for n-butane). Therefore, it is supposed that mixing of the organic matter produced by the synthesis (mainly the short-chain products) increases the $R_{3/2}$ value. Although we did not directly determine the carbon number of the aliphatic components in the carbonaceous clots, the $R_{3/2}$ values for the normal C_{10} alkane are estimated to be ~ 0.6 (Fig. 7a; data from Igisu et al. 2009). The relative abundance of the saturated $> C_{10}$ hydrocarbons produced by the FTT synthesis has been shown to be several orders of magnitude lower than those of the saturated $< C_6$ hydrocarbons (McCollom and Seewald 2006). Thus, it appears to be difficult to explain the observed $R_{3/2}$ values of the carbonaceous clots (0.22–0.51) at this stage based solely on abiotically produced organic matter. Moreover, it is unlikely that the FTT synthesis of organic compounds occurs during the formation of the silica veins. This is because native Fe–Ni minerals and other preferred catalysts for FTT reactions were not observed in the thin sections of the present study.

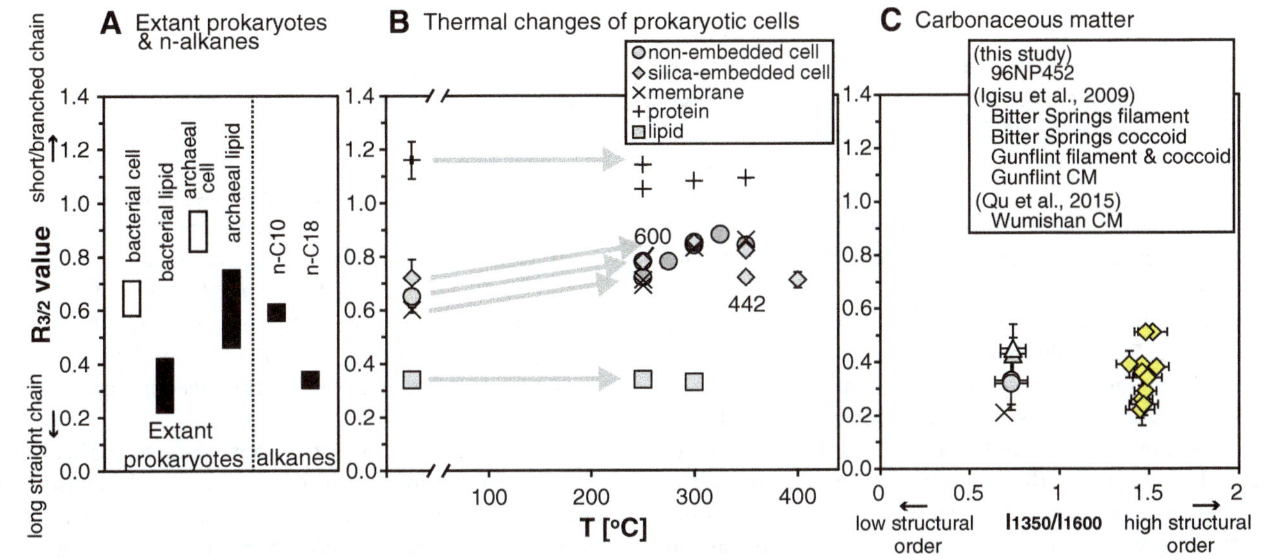

Fig. 7 $R_{3/2}$ values of the extant prokaryotes, experimentally maturated cyanobacteria, and the carbonaceous clots. **a** $R_{3/2}$ values of the extant bacterial and archaeal cells and lipids (data from Igisu et al. 2012) together with the standard n-alkane samples (n-C_{10} and n-C_{18}) (data from Igisu et al. 2009). The $R_{3/2}$ values for the representative extant untreated bacterial and archaeal cells range from 0.59 ± 0.01 to 0.70 ± 0.01 and from 0.86 ± 0.04 to 0.95 ± 0.02, and those for bacterial and archaeal lipids range from 0.31 ± 0.09 to 0.41 ± 0.01 and from 0.51 ± 0.05 to 0.67 ± 0.08, respectively. **b** Results of the heating experiments for the non-embedded and silica-embedded cyanobacterial cells (modified from Igisu et al. 2009, 2018). The $R_{3/2}$ values for the heated samples are plotted against the heating temperature. All the experimental run products of the silica-embedded samples were obtained after heating for 600 min except the one indicated by a number (the sample marked was heated for 442 min). Others were obtained after heating for 300 min except the one indicated by a number (the sample marked was heated for 600 min). **c** $R_{3/2}$ values of the carbonaceous clots obtained from the Dresser Formation together with the Proterozoic prokaryotic fossils and carbonaceous matter (CM) in stromatolite. The error bars for the carbonaceous clots indicate the standard deviations of the duplicate measurements. The data for Bitter Springs and Gunflint samples were obtained from Igisu et al. (2009), and that for Wumishan sample was obtained from Qu et al. (2015)

Effect of thermal alteration

Thermal alteration may increase or not change the $R_{3/2}$ value of organic matter. This indicates the preferential loss of the methylene group and/or formation of branched CH_3 in the aliphatic structures during the degradation of the organic matter, as revealed by FTIR measurements of coal (Wang and Griffiths 1985; Ibarra et al. 1996), and heating experiments performed on kerogens (Robin et al. 1977; Huang and Otten 1998) and extant cyanobacteria (Igisu et al. 2009, 2018). Recently, a kinetic approach applied to experimental maturation of cyanobacteria revealed that decreasing rate constants for aliphatic CH_3 were generally similar to or lower than aliphatic CH_2 at 250–450 °C under both non-embedded and silica-embedded conditions. This suggests that aliphatic CH_3 groups degraded more slowly than aliphatic CH_2 groups (Igisu et al. 2018). The metamorphic grade of basaltic greenstones surrounding the silica veins is generally below the greenschist facies (< 350 °C; Kitajima et al. 2001). Thermal alteration may therefore explain the varying $R_{3/2}$ values, but is unlikely to be the cause of the substantially lower $R_{3/2}$ values when compared with those of primary organic matter.

In addition, the coexistence of carbonaceous matter with different degrees of thermal alteration may cause variation in the $R_{3/2}$ value and the presence/absence of aliphatic C–H signatures in the carbonaceous clots. There is the possibility of contribution of some molecules and/or some functional groups, which are very resistant to heat even if they suffered from the same heating processes, to the carbonaceous clots. Assuming that some parts of the molecules and groups were graphitized but some parts were not, there may be differences in the $R_{3/2}$ values and Raman spectral features observed in the carbonaceous clots in samples 96NP452 and T-58. Such heterogeneity was reported in Phanerozoic eukaryotic fossils (Qu et al. 2015).

Biological origin

The $R_{3/2}$ values for the carbonaceous clots in the present study (0.22–0.51) generally fall within the range for extant bacterial and archaeal cellular compounds (Fig. 7a, c). The observed low $R_{3/2}$ values below ~ 0.4 cannot be explained by the degraded cellular compounds other than degraded bacterial lipids, possibly representing the relatively bacteria-dominated fractions of the microbial communities. On the other hand, there are carbonaceous clots with higher $R_{3/2}$ values (> ~ 0.4). Such carbonaceous clots may be caused by thermal alteration of organic matter with lower $R_{3/2}$ values or be derived from the archaea-dominated assemblages as well as compounds with short-chain aliphatic components. Such a possible contribution of aliphatic components to geomacromolecules by living organisms without resistant aliphatic precursors (e.g., algaenan) has been suggested by Gupta et al. (2007a, 2007b, 2014).

A comparison of the $R_{3/2}$ values of the carbonaceous clots with those of other Proterozoic prokaryotic fossils and carbonaceous matter in the stromatolites (~ 830 Ma Bitter Springs Group, ~ 1500 Ma Wumishan Group, and ~ 1900 Ma Gunflint Formation) reveals that the carbonaceous clots have similar $R_{3/2}$ values as other Proterozoic samples (Fig. 7c), while the Raman spectra of the carbonaceous clots show a more graphitized nature than those of the Proterozoic microfossils (i.e., narrower bandwidth of especially the D band and higher D/G band intensity ratio). This suggests the possible interpretation of the prokaryotic origin of the carbonaceous clots, assuming that the $R_{3/2}$ value either increases or remains unchanged during the thermal degradation processes.

Given that the aliphatic moieties in the carbonaceous clots might indicate the presence of cellular organic aggregates of the prokaryotic population in the ancient microbial communities, the IR spectroscopic features of the carbonaceous clots likely suggest a contribution from the bacterial lipids in the ancient microbial communities associated with the hydrothermal activities. Our results thus afford additional important insights into the indigenous microbial communities in the ~ 3.5 Ga Dresser seafloor hydrothermal system that consists of not only archaeal components, but also significant bacterial populations. This supports the previous pyrolysis analysis of the isolated organic matter, suggesting microbial origin of the organic matter (Derenne et al. 2008; Duda et al. 2018).

Conclusions

FTIR microspectroscopic analysis was conducted on carbonaceous clots taken from two silica veins in the ~ 3.5 Ga Dresser Formation. We found the following:

1. The chemical structure of the carbonaceous clots comprises randomly arranged layered polyaromatic domains with very minor aliphatic straight chain moieties.
2. It appears unlikely at this stage that the observed $R_{3/2}$ values of the carbonaceous clots and the mineral assemblages in the analyzed samples do not support the only abiogenic origin of the carbonaceous clots.
3. Thermal alteration may increase or not change the $R_{3/2}$ value of organic matter. Therefore, it may explain the varying $R_{3/2}$ values, but is unlikely to be the cause of the substantially lower $R_{3/2}$ values when compared with those of primary organic matter.
4. The $R_{3/2}$ values of the carbonaceous clots (0.22–0.51) generally fall within the range of the values for extant prokaryotic cellular components. In particular, the low $R_{3/2}$ values ($< \sim 0.4$) could have originated from bacterial lipid, assuming that the $R_{3/2}$ value would either increase or remain unchanged during the thermal degradation processes. This suggests the preservation of remnants of the bacterial population, mainly as aliphatic C–H groups, in the macromolecular structure of the analyzed carbonaceous matter.

The new FTIR microspectroscopic data of the carbonaceous clots combined with the previous studies based on their isotopic features (e.g., Ueno et al. 2001, 2004, 2006) may possibly suggest that both bacteria and archaea were colonized in the ~ 3.5 Ga Dresser hydrothermal system.

Abbreviations

FTIR: Fourier transform infrared (spectroscopy); FTT: Fischer-Tropsch-type; Ga: Giga-annum; IR: Infrared

Acknowledgements

We would like to thank S. Maruyama, Y. Isozaki, M. Terabayashi, Y. Kato, K. Okamoto, T. Ota, T. Kabashima, K. Kitajima, and K. Shimizu for their assistance in the field work; A. Thorne, K. J. McNamara, and A. H. Hickman for the field collaboration; and Editage (www.editage.jp) for English language editing. This work was supported by JSPS KAKENHI (24·7955, JP15J40157).

Funding

This work was supported by JSPS KAKENHI (24·7955, JP15J40157).

Authors' contributions

MI and YU proposed and designed the study. YU conducted the fieldwork and sampling. MI carried out the measurements, analyzed the data, and drafted the manuscript. YU and KT collaborated with the corresponding author in the interpretation of the data and the construction of the manuscript. All authors read and approved the final manuscript.

Competing interests

The authors declare that they have no competing interests.

Author details

[1]Department of Subsurface Geobiological Analysis and Research, Japan Agency for Marine-Earth Science and Technology (JAMSTEC), 2-15 Natsushima-cho, Yokosuka 237-0061, Japan. [2]Department of Earth and Planetary Sciences, Tokyo Institute of Technology, 2-12-1 O-okayama, Meguro-ku, Tokyo 152-8551, Japan. [3]Earth-Life Science Institute, Tokyo Institute of Technology, 2-12-1 O-okayama, Meguro-ku, Tokyo 152-8551, Japan.

References

Arouri K, Greenwood PF, Walter MR (1999) A possible chlorophycean affinity of some Neoproterozoic acritarchs. Org Geochem 30:1323–1337

Arouri K, Greenwood PF, Walter MR (2000) Biological affinities of Neoproterozoic acritarchs from Australia: microscopic and chemical characterisation. Org Geochem 31:75–89

Bellamy LJ (1954) The infra-red spectra of complex molecules. John Wiley & Sons, New York

Beyssac O, Goffé B, Chopin C, Rouzaud JN (2002) Raman spectra of carbonaceous material in metasediments: a new geothermometer. J Metamorph Geol 20:859–871

Bourbin M, Derenne S, Robert F (2012) Limits in pyrolysis-GC-MS analysis of kerogen isolated from Archean cherts. Org Geochem 52:32–34

Brasier MD, Green OR, Jephocoat AP, Kleppe AK, Van Kranendonk MJ, Lindsay JF, Steele A, Grassineau NV (2002) Questioning the evidence for Earth's oldest fossils. Nature 416:76–81

Brocks JJ, Logan GA, Buick R, Summons RE (1999) Archean molecular fossils and the early rise of eukaryotes. Science 285:1033–1036

Buick R, Dunlop JSR (1990) Evaporitic sediments of early Archean age from the Warrawoona Group, North Pole, Western Australia. Sedimentology 37:247–277

Coates J (2000) Interpretation of infrared spectra, a practical approach. In: Meyers RA (ed) Encyclopedia of analytical chemistry. John Wiley & Sons Ltd, UK

Derenne S, Robert F, Skrzypczak-Bonduelle A (2008) Molecular evidence for life in the 3.5 billion year old Warrawoona chert. Earth Planet Sci Lett 272:476–480

Duda JP, Thiel V, Bauersachs T, Mißbach H, Reinhardt M, Schäfer N, Van Kranendonk MJ, Reitner J (2018) Ideas and perspectives: hydrothermally driven redistribution and sequestration of early Archaean biomass—the "hydrothermal pump hypothesis". Biogeosciences 15:1535–1548

Gupta NS, Briggs DEG, Collison ME, Evershed RP, Michels R, Jack KS, Pancost RD (2007b) Evidence for the in situ polymerization of labile aliphatic organic compounds during the preservation of fossil leaves: implications for organic matter preservation. Org Geochem 38:499–522

Gupta NS, Michels R, Briggs DEG, Collison ME, Evershed RP, Pancost RD (2007a) Experimental evidence for the formation of geomacromolecules from plant leaf lipids. Org Geochem 38:28–36

Gupta NS, Steele A, Fogel M, Griffin P, Adams M, Summons RE, Yang H, Cody GD (2014) Experimental formation of geomacromolecules from microbial lipids. Org Geochem 67:35–40

Hayes JM (1994) Global methanotrophy at the Archaean-Proterozoic transition. In: Bengtson S (ed) Early life on earth. Columbia University Press, New York

Holm NG, Andersson EM (1998) Hydrothermal systems. In: Brack A (ed) The molecular origins of life. Cambridge University Press, Cambridge

House CH, Oehler DZ, Sugitani K, Mimura K (2013) Carbon isotopic analyses of ca. 3.0 Ga microstructures imply planktonic autotrophs inhabited Earth's early oceans. Geology 41:651–654

House CH, Schopf JW, McKeegan KD, Coath CD, Harrison TM, Stetter KO (2000) Carbon isotopic composition of individual Precambrian microfossils. Geology 28:707–710

Huang WL, Otten GA (1998) Oil generation kinetics determined by DAC-FS/IR pyrolysis: technique development and preliminary results. Org Geochem 29:1119–1137

Ibarra JV, Muñoz E, Moliner R (1996) FTIR study of the evolution of coal structure during the coalification process. Org Geochem 24:725–735

Igisu M, Nakashima S, Ueno Y, Awramik SM, Maruyama S (2006) In situ infrared microspectroscopy of ~850 million-year-old prokaryotic fossils. Appl Spectrosc 60:1111–1120

Igisu M, Takai K, Ueno Y, Nishizawa M, Nunoura T, Hirai M, Kaneko M, Naraoka H, Shimojima M, Hori K, Nakashima S, Ohta H, Maruyama S, Isozaki Y (2012) Domain-level identification and quantification of relative prokaryotic cell abundance in microbial communities by micro-FTIR spectroscopy. Environ Microbiol Rep 4:42–49

Igisu M, Ueno Y, Shimojima M, Nakashima S, Awramik SM, Ohta H, Maruyama S (2009) Micro-FTIR signature of bacterial lipids in Proterozoic microfossils. Precambrian Res 173:19–26

Igisu M, Yokoyama T, Ueno Y, Nakashima S, Shimojima M, Ohta H, Maruyama S (2018) Changes of aliphatic C–H bonds in cyanobacteria during experimental thermal maturation in the presence or absence of silica as evaluated by FTIR microspectroscopy. Geobiology 16:412–428

Ito Y, Nakashima S (2002) Water distribution in low-grade siliceous metamorphic rocks by micro-FTIR and its relation to grain size: a case from the Kanto Mountain region, Japan. Chem Geol 189:1–18

Jones RN, Sandorfy C (1956) The application of infrared and Raman spectrometry to the elucidation of molecular structure. In: Weissberger A (ed) Technique of organic chemistry Vol. IX. Interscience Publishers, New York

Kebukawa Y, Nakashima S, Otsuka T, Nakamura-Messenger K, Zolensky ME (2009) Rapid contamination during storage of carbonaceous chondrites prepared for micro FTIR measurements. Meteorit Planet Sci 44:545–557

Kitajima K, Hirata T, Maruyama S, Yamanashi T, Sano Y, Liou JG (2008) U-Pb zircon geochronology using LA-ICP-MS in the North Pole Dome, Pilbara craton, Western Australia: a new tectonic growth model for the Archean chert/greenstone succession. Int Geol Rev 50:1–14

Kitajima K, Maruyama S, Utsunomiya S, Liou JG (2001) Seafloor hydrothermal alteration at an Archaean mid-ocean ridge. J Metamorph Geol 19:583–599

Kouketsu Y, Mizukami T, Mori H, Endo S, Aoya M, Hara H, Nakamura D, Wallis S (2014) A new approach to develop the Raman carbonaceous material geothermometer for low-grade metamorphism using peak width. Island Arc 23:33–50

Kudryavtsev AB, Schopf JW, Agresti DG, Wdowiak TJ (2001) In situ laser-Raman imagery of Precambrian microscopic fossils. Proc Natl Acad Sci 98:823–826

Lepot K, Williford KH, Ushikubo T, Sugitani K, Mimura K, Spicuzza MJ, Valley JW (2013) Texture-specific isotopic compositions in 3.4 Gyr old organic matter support selective preservation in cell-like structures. Geochim Cosmochim Acta 112:66–86

Lin R, Ritz GP (1993) Reflectance FT-IR microspectroscopy of fossil algae contained in organic-rich shales. Appl Spectrosc 47:265–271

Marshall CP, Javaux EJ, Knoll AH, Walter MR (2005) Combined micro-Fourier transform infrared (FTIR) spectroscopy and micro-Raman spectroscopy of Proterozoic acritarchs: a new approach to palaeobiology. Precambrian Res 138:208–224

McCollom TM, Ritter G, Simoneit BRT (1999) Lipid synthesis under hydrothermal conditions by Fischer-Tropsch-type reactions. Origins of Life and Evolution of Biospheres 29:153–166

McCollom TM, Seewald JS (2006) Carbon isotope composition of organic compounds produced by abiotic synthesis under hydrothermal conditions. Earth Planet Sci Lett 243:74–84

McCollom TM, Seewald JS (2007) Abiotic synthesis of organic compounds in deep-sea hydrothermal environments. Chem Rev 107:382–401

Nakashima S, Matayoshi H, Yuko T, Michibayashi K, Masuda T, Kuroki N, Yamagishi H, Ito Y, Nakamura A (1995) Infrared microspectroscopy analysis of water distribution in deformed and metamorphosed rocks. Tectonophysics 245:263–276

Nakashima S, Ohki S, Ochiai S (1989) Infrared microspectroscopy analysis of the chemical state and spatial distribution of hydrous species in minerals. Geochem J 23:57–64

Nisbet EG (1995) Archaean ecology: a review of evidence for the early development of bacterial biomes and speculations on the development of a global-scale biosphere. In: Coward MP, Ries AC (eds) Early Precambrian processes. Geological Society Special Publication, London

Noffke N, Christian D, Wacey D, Hazen RP (2013) Microbially induced sedimentary structures recording an ancient ecosystem in the ca. 3.48 billion-year-old Dresser Formation, Pilbara, Western Australia. Astrobiology 13:1103–1124

Olcott-Marshall A, Emry JR, Marshall CP (2012) Multiple generations of carbon in the Apec chert and implications for preservation of microfossils. Astrobiology 12:160–166

Painter PC, Snyder RW, Starsinic M, Coleman MM, Kuehn DW, Davis A (1981) Concerning the application of FT-IR to the study of coal: a critical assessment of band assignments and the application of spectral analysis programs. Appl Spectrosc 35:475–485

Pasteris JD (1989) In situ analysis in geological thin-sections by laser Raman microprobe spectroscopy: a cautionary note. Appl Spectrosc 43:567–570

Qu Y, Engdhl A, Zhu S, Vajda V, McLoughlin N (2015) Ultrastructural heterogeneity of carbonaceous material in ancient cherts: investigating biosignature origin and preservation. Astrobiology 15:825–842

Rahl JM, Anderson KM, Brandon MT, Fassoulas C (2005) Raman spectroscopic carbonaceous material thermometry of low-grade metamorphic rocks: calibration and application to tectonic exhumation in Crete, Greece. Earth Planet Science Letters 240:339–354

Rasmussen B, Fletcher IR, Brocks JJ, Kilburn MR (2008) Reassessing the first appearance of eukaryotes and cyanobacteria. Nature 455:1101–1104

Robin PL, Rouxhet PG, Durand B (1977) Caracterisation des kérogènes et de leur evolution par spectroscopie infrarouge. In: Campos R, Goni J (eds) Advances in organic geochemistry 1977. Enadimsa, Madrid

Salisbury JW, Walter LS (1989) Thermal infrared (2.5-13.5 μm) spectroscopic remote sensing of igneous rock types on particulate planetary surfaces. J Geophys Res 94:9192–9202

Schopf JW, Kudryavtsev AB, Agresti DG, Czaja AD, Wdowiak TJ (2005) Raman imagery: a new approach to assess the geochemical maturity and biogenicity of permineralized Precambrian fossils. Astrobiology 5:333–371

Schopf JW, Kudryavtsev AB, Agresti DG, Wdowiak TJ, Czaja AD (2002) Laser-Raman imagery of Earth's earliest fossils. Nature 416:73–76

Sforna MC, van Zuilen MA, Philippot P (2014) Structural characterization by Raman hyperspectral mapping of organic carbon in the 3.46 billion-year-old

Apex chert, Western Australia. Geochim Cosmochim Acta 124:18–33
Shen Y, Farquhar J, Masterson A, Kaufman AJ, Buick R (2009) Evaluating the role of microbial sulfate reduction in the early Archean using quadruple isotope systematics. Earth Planet Sci Lett 279:383–391
Sherwood-Lollar B, McCollom TM (2006) Biosignatures and abiotic constraints on early life. Nature 444:E18
Steemans P, Lepot K, Marshall CP, Le Hérissé A, Javaux EJ (2010) FTIR characterisation of the chemical composition of Silurian miospores (cryptospores and trilete spores) from Gotland, Sweden. Rev Palaeobot Palynol 162:577–590
Thorpe RI, Hickman AH, Davis DW, Mortensen JK, Trendall AF (1992a) U-Pb zircon geochronology of Archaean felsic units in the marble Bar region, Pilbara craton, Western Australia. Precambrian Res 56:169–189
Thorpe RI, Hickman AH, Davis DW, Mortensen JK, Trendall AF (1992b) Constraints to models for lead evolution from precise zircon U-Pb geochronology for the Marble Bar region, Pilbara Craton, Western Australia. In: Glover JE, Ho SE (eds) The Archaean: Terrains, Processes and Metallogeny. The University of Western Australia, Publication 9, Perth
Ueno Y, Isozaki Y, Yurimoto H, Maruyama S (2001) Carbon isotopic signatures of individual Archean microfossils (?) from Western Australia. Int Geol Rev 43:196–212
Ueno Y, Ono S, Rumble D, Maruyama S (2008) Quadruple Sulphur isotope analysis of ca. 3.5 Ga Dresser Formation: new evidence for microbial sulphate reduction in the early Archean. Geochim Cosmochim Acta 72:5675–5691
Ueno Y, Yamada K, Yoshida N, Maruyama S, Isozaki Y (2006) Evidence from fluid inclusions for microbial methanogenesis in the early Archaean era. Nature 440:516–519
Ueno Y, Yoshioka H, Maruyama S, Isozaki Y (2004) Carbon isotopes and petrography of kerogens in ~3.5-Ga hydrothermal silica dikes in the North Pole area, Western Australia. Geochim Cosmochim Acta 68:573–589
Van Kranendonk MJ (2006) Volcanic degassing, hydrothermal circulation and the flourishing of early life on Earth: a review of the evidence from c. 3490–3240 Ma rocks of the Pilbara Supergroup, Pilbara Craton, Western Australia. Earth-Sci Rev 74:197–240
Van Kranendonk MJ, Philippot P, Lepot K, Bodorkos S, Pirajno F (2008) Geological setting of Earth's oldest fossils in the ca. 3.5 Ga Dresser Formation, Pilbara Craton, Western Australia. Precambrian Res 167:93–124
van Zuilen MA, Chaussidon M, Rollion-Bard C, Marty B (2007) Carbonaceous cherts of the Barberton Greenstone Belt, South Africa: isotopic, chemical and structural characteristics of individual microstructures. Geochim Cosmochim Acta 71:655–669
Wacey D, Kilburn MR, Saunders M, Cliff J, Brasier MD (2011) Microfossils of sulphur-metabolizing cells in 3.4-billion-year-old rocks of Western Australia. Nat Geosci 4:698–702
Walter MR, Buick R, Dunlop JSR (1980) Stromatolites 3,400–3,500 Myr old from the North Pole area, Western Australia. Nature 284:443–445
Wang SH, Griffiths PR (1985) Resolution enhancement of diffuse reflectance i.r. spectra of coals by Fourier self-deconvolution. 1. C-H stretching and bending modes. Fuel 64:229–236
Williford KH, Ushikubo T, Schopf JW, Lepot K, Kitajima K, Valley JW (2013) Preservation and detection of microstructural and taxonomic correlations in the carbon isotopic compositions of individual Precambrian microfossils. Geochim Cosmochim Acta 104:165–182
Yanagawa H, Kobayashi K (1992) An experimental approach to chemical evolution in submarine hydrothermal systems. In: Holm NG (ed) Marine hydrothermal systems and the origin of life. Kluwer, Dordrecht
Yui TF, Huang E, Xu J (1996) Raman spectrum of carbonaceous material: a possible metamorphic grade indicator for low-grade metamorphic rocks. J Metamorph Geol 14:115–124

10

Global evaluation of erosion rates in relation to tectonics

Hagar Hecht[1*] and Takashi Oguchi[1,2]

Abstract

Understanding the mechanisms and controlling factors of erosion rates is essential in order to sufficiently comprehend bigger processes such as landscape evolution. For decades, scientists have been researching erosion rates where one of the main objectives was to find the controlling factors. A variety of parameters have been suggested ranging from climate-related, basin morphometry and the tectonic setting of an area. This study focuses on the latter. We use previously published erosion rate data obtained mainly using ^{10}Be and sediment yield and sediment yield data published by the United States Geological Survey. We correlate these data to tectonic-related factors, i.e., distance to tectonic plate boundary, peak ground acceleration (*PGA*), and fault distribution. We also examine the relationship between erosion rate and mean basin slope and find significant correlations of erosion rates with distance to tectonic plate boundary, *PGA*, and slope. The data are binned into high, medium, and low values of each of these parameters and grouped in all combinations. We find that groups with a combination of high *PGA* (> 0.2.86 g) and long distance (> 1118.69 km) or low *PGA* (< 0.68 g) and short distance (< 94.34 km) are almost inexistent suggesting a strong coupling between *PGA* and distance to tectonic plate boundary. Groups with low erosion rates include long distance and/or low *PGA*, and groups with high erosion rates include neither of these. These observations indicate that tectonics plays a major role in determining erosion rates, which is partly ascribable to steeper slopes produced by active crustal movements. However, our results show relatively unclear correlation of slope with erosion rates, pointing to problems with using mean basin-wide slope as a slope indicator because it does not represent the complex slope distribution within a basin.

Keywords: Erosion rate, Sediment yield, Basin morphometry, Tectonic plates, Peak ground acceleration, Slope, Geographic information system

Introduction

Understanding the mechanisms and controlling factors of erosion rates is of great importance as they are related to bigger processes such as tectonic plate movements and climate change. Primarily, erosion is part of the sediment cycling processes. The material eroded on earth surfaces is transported over time into the oceans, subducted into the mantle, and then returns to the surface through volcanism and tectonic uplift. In this way, estimates of erosion rates are a vital component of both sediment and geochemical mass balance studies. The geochemical mass balance includes the carbon cycle as the erosion of silicates and carbonate minerals serves as a transformer of carbon dioxide between the atmosphere and lithosphere. For this reason, understanding the controlling factors of erosion rates will not only help understand landscape evolution but also climate change (Kump et al. 2000).

Much work has been done and a lot of energy has been invested in understanding erosion rates. Most studies have concentrated on investigating controlling factors on a local scale and have come to a variety of conclusions regarding what most affects erosion rates ranging from precipitation (e.g., Galy and France-Lanord 2001; Griffiths 1979; Griffiths 1981) to mass wasting events (Hovius et al. 1997) and lithology (Matsushi et al. 2006). In addition, global data have been compiled with the intention of identifying the controlling factors of erosion rates. Ahnert (1970) found that erosion rates are dependent on basin relief, and Milliman and Syvitski (1992) found a link between erosion rates and basin area. Portenga and Bierman (2011) found that while basin-wide erosion rates are correlated best to basin slope, erosion rates for

* Correspondence: hhecht@csis.u-tokyo.ac.jp
[1]Department of Natural Environmental Studies, The University of Tokyo, 5-1-5 Kashiwanoha, Kashiwa City, Chiba 277-8563, Japan
Full list of author information is available at the end of the article

outcrops are most strongly correlated to mean annual precipitation.

Another dominant controlling factor that shows up in many studies, both global and local, is the rate of tectonic uplift. A study in South Africa concluded that the reason for the very slow erosion rates despite alpine topography is tectonic stability (Scharf et al. 2013). A similar example is Sri Lanka where even though there are high temperatures, precipitation, and relief, erosion rates are relatively low because the country is situated in a cratonic area that lacks tectonic activity (Von Blanckenburg 2005; Von Blanckenburg et al. 2004). Cyr et al. (2010) found that ^{10}Be-derived erosion rates increase linearly with increasing rock uplift rates. Some studies even derived uplift rates from observed denudation rates assuming that they are primarily controlled by rock uplift (Morell et al. 2015; Roux-Mallouf et al. 2015).

Despite the strong relationship between erosion rate and tectonic activity, a simpler way to make this connection is still missing since tectonic activity and uplift rates are difficult to measure. This study comprehensively compiles previously published erosion rate data derived by various methods, namely, ^{10}Be and sediment yield, and erosion rates calculated from suspended sediment load published by the United States Geological Survey (USGS) and correlates them to data related to tectonics, i.e., distance to tectonic plate boundary, peak ground acceleration, fault distribution, and slope.

Methods/experimental

This work uses several datasets. The largest dataset is the ^{10}Be-derived erosion rate data compiled by Portenga and Bierman (2011) ($n = 1555$). We also added newer published ^{10}Be-derived data ($n = 546$), erosion rates derived from sediment yield ($n = 447$), and erosion rates derived from other methods such as channel incision rates based on ^{36}Cl and ^{14}C ($n = 88$). In addition, we downloaded suspended sediment loads from the USGS sediment data portal (U.S. Geological Survey) ($n = 47$). Only sites with at least 30 years of records were chosen. Of these, sites with no information about upstream drainage area or records for only several months of the year were discarded. Following previous studies (e.g., Cyr and Granger 2008; Olen et al. 2016), a rock density of 2.6 g cm^{-3} was used in order to estimate basin-wide erosion rates from the sediment yield data. Figure 1 shows the distribution of all data points used in this compilation.

It is important to note that the two main types of data used in this compilation, i.e., ^{10}Be- and sediment yield-derived erosion rates, are inherently different in that they represent different time scales. While sediment yields are measured over the past few decades, ^{10}Be-derived data represent erosion rates from 10^3 to 10^6 years ago, both of which refer to shorter time scales than large-scale tectonic processes; therefore, they deal with generally unchanged tectonic settings.

Another important point is the skewed distribution of the data. Most of the data originate from the USA, the west of South America, central Europe, the Himalayas, and the Yangtze basin. These areas are studied more frequently than others due to many reasons, some being convenient and easily accessible and some are points of special interest, partly because they are located in tectonically active regions. Therefore, an effort was made to find data from less represented areas such as central Asia and Africa. Altogether, the area of the basins analyzed in this study represents about 10% of the Earth's land surface.

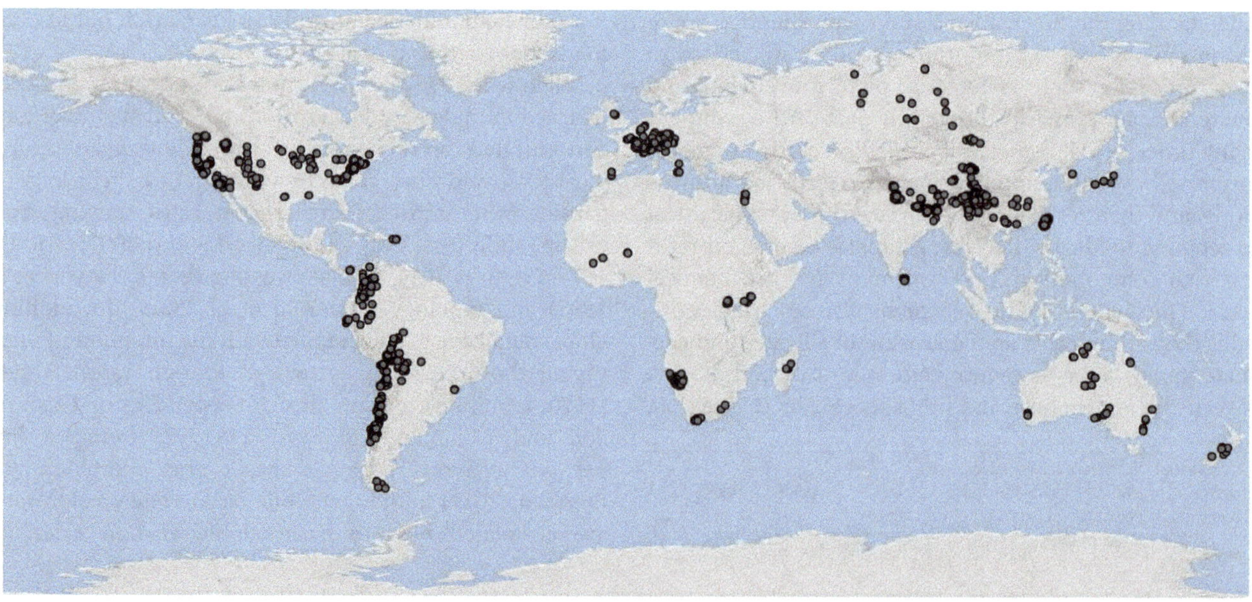

Fig. 1 Geographical distribution of data used for this study

Basin boundaries were extracted using the basin tool of the hydrology toolset in ArcGIS. In most cases, the high-resolution filled 1 arcsecond Advanced Spaceborne Thermal Emission and Reflection Radiometer (ASTER) Global Digital Elevation Model (GDEM) (U.S./Japan ASTER Science Team 2011) was used for the extraction, but the filled 3 arcsecond STRM DEM (Jarvis et al. 2008) was used for the larger basins. Since there are cases of multiple studies in the same area or data representing sub-basins, 211 selected basins were delineated with an area ranging from about 2.5×10^0 to 6.5×10^6 km^2. For each basin, a basin-wide mean slope was calculated using the ArcGIS slope tool. Fault data derived from a map published by ESRI and attributed to the USGS (ESRI 2014) were used to calculate fault density for each basin. Additional data used in this study include tectonic plate boundary data published by Nordpil (2014) based on Bird (2003) and peak ground acceleration (*PGA*) data published by the Global Seismic Hazard Assessment Program (Giardini et al. 1999). This is a map produced by compiling regional maps representing the peak ground acceleration with a 10% chance of being exceeded within 50 years at a certain location due to earthquakes.

Using JMP pro 13.0, we examined the bilinear relationship of erosion rates with five tectonic-related factors. These include distance to fault, *PGA*, distance to tectonic plate boundaries, fault density, and mean basin-wide slope. From these, we chose those that correlated well to erosion rates and defined high and low value ranges for each as the upper and lower quartiles, respectively. Then, we grouped them in all possible combinations and calculated the mean erosion rate for each group.

Results

Erosion rates in relation to tectonics

Figure 2 and Table 1 show bivariate correlations of erosion rates with tectonic-related factors. The correlation between erosion rates and distance to tectonic plate boundary shows a statistically significant relationship.

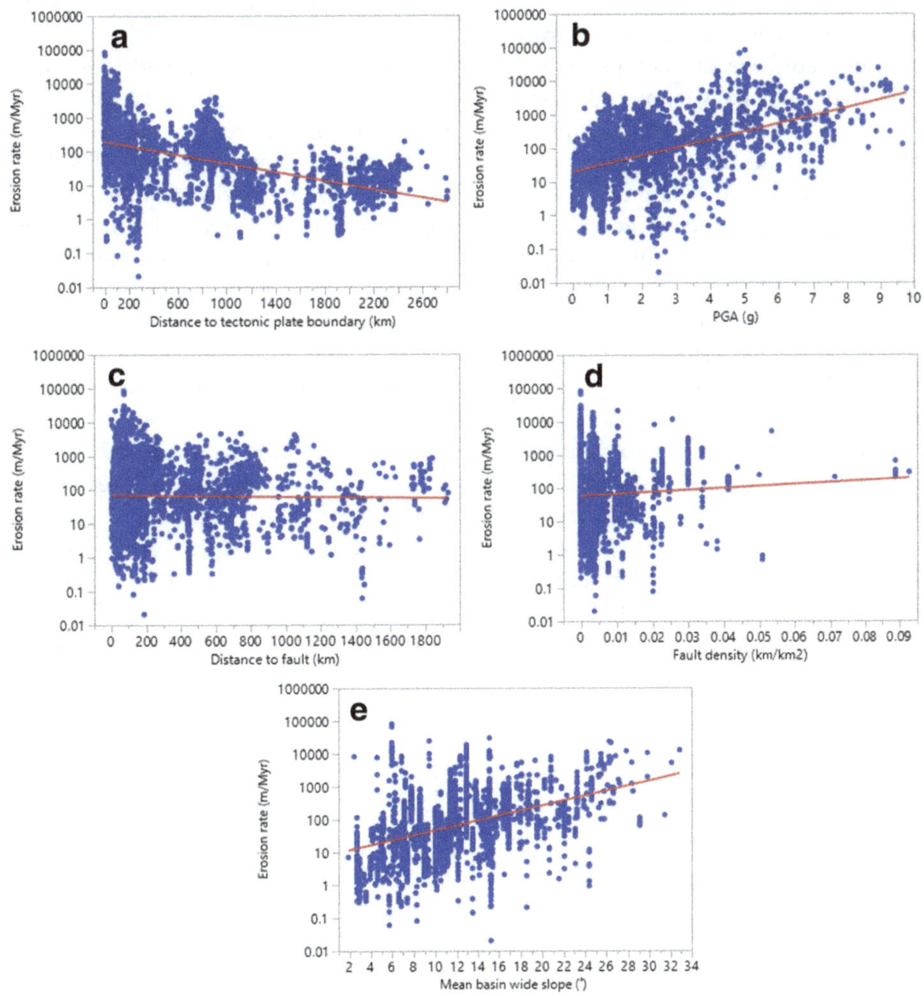

Fig. 2 Bivariate correlations of erosion rates with the different tectonic-related factors: **a** distance to tectonic plate boundary, **b** peak ground acceleration (PGA), **c** distance to fault, **d** fault density, and **e** mean basin-wide slope

Table 1 Bivariate correlations of tectonic-related factors to erosion rate

	Distance to the tectonic plate boundary (km)	*PGA* (g)	Distance to the fault (km)	Fault density (km/km^2)	Mean basin-wide slope (°)
*P**	< 0.0001	< 0.0001	0.3341	0.0127	< 0.0001
*R***	− 0.4861	0.4816	− 0.0173	0.0480	0.4093

*P value
**Square root of R squared, coefficient of determination

The correlation is negative, which means that the further away from tectonic plate boundaries, the slower the erosion rates. There is a statistically significant positive correlation between erosion rate and *PGA*. Higher *PGA* values correspond to higher erosion rates, and a similar *R* value is obtained for the correlation between erosion rate and distance to boundary. The correlation between erosion rate and distance to fault is statistically insignificant. Although a slightly positive correlation with fault density is more significant, the *R* value is very low. There is a statistically meaningful positive correlation between erosion rate and mean basin-wide slope.

Grouping analysis

Taking into account the poor correlations of fault distribution data with erosion rates, we did not use the distance to fault and fault density data in our discussion. The high, middle, and low values of the three remaining factors, i.e., distance to tectonic plate boundary, *PGA*, and mean basin-wide slope, were defined using the upper quartile, interquartile range, and lower quartile of data distribution: long, middle, and short distances from tectonic plate boundary (LD, MD, and SD respectively); high, middle, and low *PGA* (HP, MP, and LP, respectively); and steep, middle, and gentle slopes (SS, MS, and GS, respectively) (Table 2). The complete dataset consists of 2683 data points with a mean erosion rate of 559.16 m/Myr. Out of the 27 possible combinations, four had no data points that met all the criteria and are empty groups. They are steep slope, high *PGA*, and long distance (SS HP LD); middle slope, high *PGA*, and long distance (MS HP LD); middle slope, low *PGA*, and short distance (MS LP SD); and gentle slope, high *PGA*, and long distance (GS HP LD). In addition, six other groups had 20 data points or less, which are not included for further analysis.

Table 3 shows the mean erosion rates of the remaining 17 groups, arranged from low to high mean erosion rates. The groups with the highest mean erosion rates ranging from 1085.27 to 11,642.11 m/Myr are middle slope, high *PGA*, and short distance (MS HP SD); steep slope, high *PGA*, and short distance (SS HP SD); and gentle slope, high *PGA*, and short distance (GS HP SD). The groups with the lowest erosion rates ranging from 10.13 to 18 m/Myr are gentle slope, middle *PGA*, and long distance (GS MP LD); middle slope, low *PGA*, and long distance (MS LP LD); and gentle slope, low *PGA*, and long distance (GS LP LD. The group with the largest number of data points has a mean erosion rate of 267.22 m/Myr and is middle slope, middle *PGA*, and middle distance (MS MP MD).

Discussion

The mean erosion rate for this study (about 559 m/Myr, Table 3) is significantly higher than the previously reported rate by Portenga and Bierman (2011): a mean of 218 m/Myr for basins and 12 m/Myr for outcrops. There is one major difference between this study and that of Portenga and Bierman (2011). The latter used a compilation of ^{10}Be-derived data only, but this study compiled different types of data including sediment yield. Erosion rates from sediment yield are those for recent years when human activity is considered to accelerate erosion rates (e.g., Saunders and Young 1983; Von Blanckenburg 2005; Walling and Webb 1996) and could account for the overall higher mean erosion rate.

Interpretation of bivariate correlation

The correlation of erosion rates to distance to tectonic plate boundary is negative (Fig. 2a), suggesting that enhanced tectonic activity along the contact lines of the plates is favorable for erosion. *PGA* is an expression of probable ground acceleration due to earthquakes; thus, it is a direct indicator of tectonic activity. Figure 2b shows that higher erosion rates tend to occur at locations with higher *PGA* values. The strongest correlation of erosion rates is that with distance to tectonic plate boundary, although its *R* value is similar to that of *PGA* (Table 1). This emphasizes the strength of the former as a simple geographical predictor of erosion rates.

Table 2 High, middle, and low range values for each parameter defined using quartiles

	Distance to tectonic the plate boundary (km)		*PGA* (g)		Mean basin-wide slope (°)	
Upper quartile	LD	> 1118.69	HP	> 2.86	SS	> 15.03
Interquartile range	MD	94.34–1118.69	MP	0.68–2.86	MS	8.1–15.03
Lower quartile	SD	< 94.34	LP	< 0.68	GS	< 8.1

Table 3 Mean erosion rate (m/Myr) and its standard deviation (SD), median, minimum, and maximum for each of the groups that contain more than 20 data points as well as the complete data set

	n	Mean	SD	Median	Minimum	Maximum
All data	2683	559.16	2656.37	65.7	0	80,700
GS MP LD	130	10.13	15.27	3.04	0.29	113.25
MS LP LD	292	16.27	15.14	11.6	1	115.68
GS LP LD	154	18	25.06	9.93	0.88	186.48
MS MP LD	72	26.34	11.3	25.01	4.85	57.2
GS LP MD	97	65.2	173.36	11.73	0.28	1513.6
MS LP MD	61	70.53	58.76	47.81	0.32	223
SS LP MD	36	71.39	50	55.1	14.6	233.6
GS MP MD	165	183.86	295.73	74.61	0.06	2190
SS HP MD	100	193.96	286.49	93.38	0.2	1773.83
MS MP SD	133	215.18	433.57	43	0.24	4131
MS MP MD	468	267.22	516.68	98.33	0.08	4151.28
GS MP SD	66	333.88	403.15	180.97	0.22	1987.16
SS MP MD	306	357.01	456.15	174.345	0.02	3236.7
MS HP MD	95	829.49	2740.38	118.49	0.56	18,400
MS HP SD	248	1085.27	2360.73	357.7	15.43	24,300
SS HP SD	178	2553.63	4131.26	800	0	29,500
GS HP SD	28	11,642.11	18,242.28	7000	80	80,700

As noted, the erosion rate data we used show a slightly positive correlation to distance to fault and a slightly negative correlation to fault density (Fig. 2c, d), but they are statistically insignificant. This may be because faults have quite different levels of activity even locally, and hence, each fault does not represent regional tectonics well, in contrast to each plate boundary. On a local scale, a connection between erosion rates and faults has been shown in the past, for example, erosion has been shown to influence seismicity along faults (Steer et al. 2014). If enough data are collected from a single basin, a correlation could be found between erosion rate and fault distribution. In addition, more information regarding the activity level of the faults might also improve the correlation. However, the global fault data used here are not comprehensive and tectonic activity on a larger scale related to plate boundaries correlates better to erosion rates.

Implication of group analysis

Groups with the combinations of high *PGA* and long distance and low *PGA* and short distance are either empty or include very few data points. Groups with a relatively large number of data points are generally those with the combination of low *PGA* and long distance, high *PGA* and short distance, and middle *PGA* and middle distance (Table 3). This confirms that *PGA*, and therefore tectonic activity, is directly affected by distance to tectonic plate boundary and is also apparent in the distribution of *PGA* as a function of distance to plate (Fig. 3).

Table 3 has implications on the effect of each parameter on erosion rates. The seven groups with the lowest erosion rates all include long distance (LD) and/or low *PGA* (LP), whereas the other 10 groups with highest erosion rates did not include any of these. In addition, the top part of Table 3 with low erosion rates is dominated by groups including long distance (LD), while the bottom part of the table with high erosion rates is dominated by groups including short distance (SD). A similar trend is observed for groups with the correlation of low *PGA* with low erosion rates and high *PGA* with high erosion rates but to a smaller effect. No such trend exists for groups with gentle and steep slopes. This points out that distance to tectonic plate boundary is the most important factor and emphasizes its strong effect on erosion rates. Although there is no apparent correlation between slope and erosion rates, numerous studies have shown that a steeper terrain is more erodible than a gentler terrain, which is naturally acceptable in relation to change in slope stability. Indeed, the data we used show a significant relationship between erosion rate and slope when examining a bivariate correlation of all the data, even though they are scattered. However, when the data are separated into groups with different characteristics and sections of the data are separately examined, this relationship becomes less apparent. It is important to note that the slope parameter used in this study is the mean basin-wide slope, and its lumped characteristics do not reflect the actual complex distribution of slope within a basin. To address this problem, we take a closer look at the group with the highest mean erosion rate, i.e., GS HP SD, which has a mean erosion rate of 11,642.11 m/Myr that is an order of magnitude higher than the next group (SS HP SD, 2553.63 m/Myr). In this

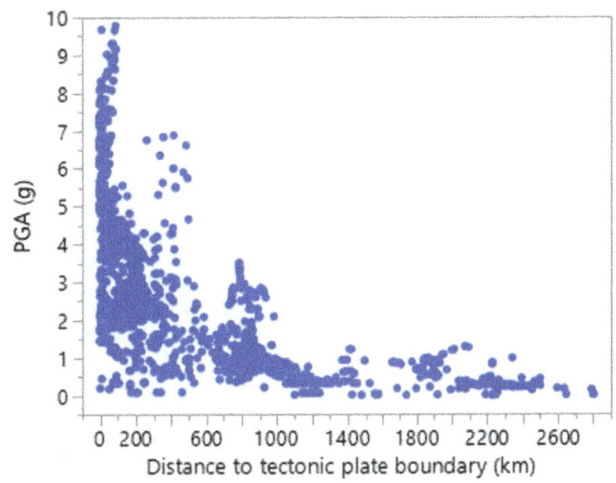

Fig. 3 PGA as a function of distance to tectonic plate boundary

group, all but two data points are located on the southwestern side of Taiwan. Taiwan is known for its exceptionally high erosion rates due to a combination of extremely wet climate and a tectonically active environment (Chen-Feng et al. 1988; Dadson et al. 2003). Taiwan's topography consists of roughly north–south orienting mountain belts with peaks of almost 4000 m in elevation. This means that even though the mean basin-wide slope of the southwest basins is lower due to the gentle slopes of the lower extensive depositional plains, the rivers originate higher up in the mountain range where slopes are steeper and are able to produce abundant sediment. A distribution map of all data points from Taiwan shows that the majority of the points belong to the SS HP SD group, which is theoretically the group with the highest expected erosion rates as it represents steep slopes and active tectonics. The other points mainly belong to the gentle slope and active tectonic group (GS HP SD) (Fig. 4). The two data points of the GS HP SD group from areas other than Taiwan are located on a tributary of the Orinoco basin in Colombia basin and in the upstream part of a tributary of the Indus River. These cases also have steeper slopes in the upstream areas, but the overall basin slope is low due to wide depositional areas. The same applies to the case of the next group in line, i.e., middle slope and active tectonics (MS HP SD). Here, most data points are from the Himalayan origins of the Ganges River, the Andean origins of the Magdalena River in Colombia, the San Gabriel mountain and the San Bernardino mountains

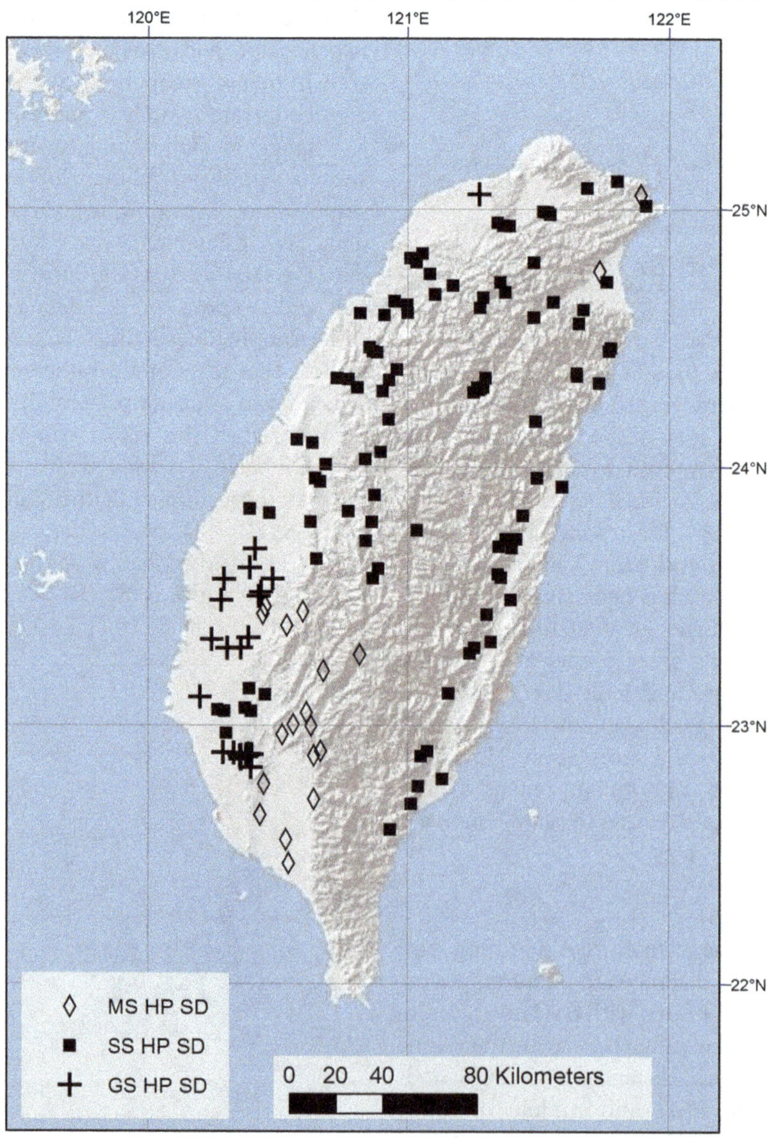

Fig. 4 Distribution of data points in Taiwan: SS HP SD (squares), GS HP SD (crosses), and MS HP SD (diamonds)

along the San Andreas Fault, and the southwest part of Taiwan. As indicated, steep slopes in these areas are not properly represented when referring to mean basin-wide slope because broad depositional areas are included. It is quite common that steep mountainous river basins with high erosion rates produce broad depositional landforms such as alluvial fans in piedmont areas. Therefore, while a dissonance between existing steep slopes in a basin and its significantly lower mean basin-wide slope can be smoothed when examining large data, this is more problematic when looking at smaller data sets and superficial correlations of lower mean basin slope with high erosion rates can occur. This should be taken into account in future erosion rate studies.

It is expected that the group with the lowest erosion rates would be that of gentle slope, low *PGA*, and long distance (GS LP LD). Results show that this group indeed has a low mean erosion rate, but it is not the lowest. The MS LP LD group has a very similar mean erosion rate and is also dominated by a tectonically inactive setting (low *PGA* and long distance), and the GS MP LD group has the lowest mean erosion rate even though it is from middle *PGA* areas. This group is dominated by data originating from the middle and southern parts of Australia, where basins have a low mean annual precipitation ranging from about 200 to 425 mm/year. Here, the dry climate might contribute to the slow erosion rates.

Effect of tectonic activity on erosion rates

The results of this study indicate that erosion rates are strongly related to tectonic activity. A simple direct explanation would be that tectonic activity with ground motion weakens and fractures bedrock, creates debris, and hence incites physical erosion (Molnar et al. 2007). In addition, influences of erosion on tectonics may cause their correlation. Whipple (2009) has shown that erosion in active mountain belts tends to influence tectonic processes by thinning the crust, which accelerates uplift through isostasy. This process would be further accelerated with the help of increased precipitation on mountain slopes due to the orographic effect (Barros and Lettenmaier 1994). Furthermore, tectonic activity produces high relief topography prone to erosion, resulting in higher erosion rates. The effects of slope and relief on erosion rates have been discussed extensively (e.g., Ahnert 1970; Portenga and Bierman 2011; Summerfield and Hulton 1994). This study also shows that higher erosion rates generally occur in basins with steeper slopes (Fig. 2e), and basins closer to tectonic plate boundaries and/or with higher *PGA* values have higher mean slopes (Fig. 5). This means that stronger tectonic activities increase slope steepness and thus increase slope instability and erosion rates.

This study utilized data from more than 2600 points and concluded that erosion rates are strongly dependent on tectonic activity. However, many of the data are from tectonically active areas, so lower erosion rates in areas of weak tectonics need to be confirmed using more data from such areas.

Conclusions

Understanding erosion rate distribution and its controlling factors is necessary for discussing and predicting landscape formations. Global-scale studies provide general ideas about these processes. As part of these efforts, this study has utilized global DEM data and an extensive compilation of erosion rates derived from various methods. The results of this research show that erosion rates are strongly related to tectonic activity. A strong coupling between *PGA*, which is a direct indicator of tectonic activity, and distance to tectonic plate boundary suggests that the latter can also be an indicator of tectonic activity. This means that the easy-to-measure parameter of distance to plate boundary can be used to make general correlations between erosion rates and tectonic activity on a global scale, despite the compilations in accurately measuring tectonic activity. The power of this parameter is further emphasized through the group analysis where the strongest correlation is between high

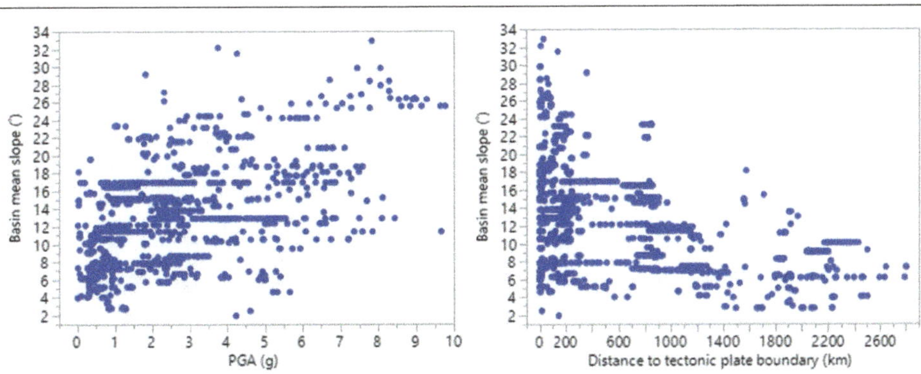

Fig. 5 Mean basin slope as a function of PGA and distance to tectonic plate boundary

and low erosion rates and short and long distances to tectonic boundaries, respectively, and not between erosion rates and *PGA* or mean basin-wide slope, even though the latter is well known as a parameter that affects erosion rates. The precise way in which tectonic activity is connected to erosion rates is yet unknown; however, some possibilities include influence through seismic shattering and fracturing of rocks, a relation between erosion and isostatic uplift, which is increased by orographic rainfall on uplifted mountains, and increased slope steepness and instability in such mountains. These are supported by the strong connections among high *PGA*, proximity to tectonic boundaries, and steep slopes. Similar to some previous studies, our study has found that on a global scale, a significant bivariate correlation between erosion rates and mean basin-wide slope exists. However, this relationship is not reflected when smaller data sets in the group analysis are taken into account; therefore, this parameter should be used with caution and careful consideration since it cannot account for the complex distribution of slopes within a basin. Although this study also discusses some regional issues, such as some unique drainage basins in Taiwan, analyses are essentially global and deal with only general trends. Further research is needed to understand more detailed mechanisms that determine regional and local erosion rates.

Abbreviations

ASTER: Advanced Spaceborne Thermal Emission and Reflection Radiometer; GS: Gentle slope; HP: High *PGA*; LD: Long distance; LP: Low *PGA*; MD: Middle distance; MP: Middle *PGA*; MS: Middle slope; *PGA*: Peak ground acceleration; SD: Short distance; SS: Steep slope; USGS: United States Geological Survey

Acknowledgements

We thank the editor and two anonymous reviewers for their helpful comments that improved the manuscript.

Funding

This work was supported by JSPS KAKENHI Grant Numbers JP15K12452 and JP15H01782.

Authors' contributions

Both authors read and approved the final manuscript.

Competing interests

The authors declare that they have no competing interests.

Author details

[1]Department of Natural Environmental Studies, The University of Tokyo, 5-1-5 Kashiwanoha, Kashiwa City, Chiba 277-8563, Japan. [2]Center for Spatial Information Science, The University of Tokyo, 5-1-5 Kashiwanoha, Kashiwa City, Chiba 277-8568, Japan.

References

Ahnert F (1970) Functional relationships between denudation, relief, and uplift in large, mid-latitude drainage basins. Am J Sci 268:243–263

Barros AP, Lettenmaier DP (1994) Dynamic modeling of orographically induced precipitation. Rev Geophys 32:265–284

Bird P (2003) An updated digital model of plate boundaries. Geochem Geophy Geosy:4. https://doi.org/10.1029/2001GC000252

Chen-Feng Y, Typhoon L, Brown L, Shen JJ, Ju-Chin C (1988) ^{10}Be study of rapid erosion in Taiwan. Geochim Cosmochim Acta 52:2687–2691

Cyr AJ, Granger DE (2008) Dynamic equilibrium among erosion, river incision, and coastal uplift in the northern and central Apennines, Italy. Geology 36:103–106

Cyr AJ, Granger DE, Olivetti V, Molin P (2010) Quantifying rock uplift rates using channel steepness and cosmogenic nuclide-determined erosion rates: examples from northern and southern Italy. Lithosphere 2:188–198

Dadson SJ, Hovius N, Chen H, Dade WB (2003) Links between erosion, runoff variability and seismicity in the Taiwan orogen. Nature 426(6967):648–651

ESRI. (2014) Global Faults layer from ArcAtlas. http://www.arcgis.com/home/item.html?id=a549601 1fa494b99810e4deb5c618ae2#overview. Accessed August 2016

Galy A, France-Lanord C (2001) Higher erosion rates in the Himalaya: geochemical constraints on riverine fluxes. Geology 29:23–26

Giardini D, Grünthal G, Shedlock KM, Zhang P (1999) The GSHAP global seismic hazard map. Ann Geophys 42:1225–1230

Griffiths GA (1979) High sediment yields from major rivers of the Western Southern Alps, New Zealand. Nature 282:61–63

Griffiths GA (1981) Some suspended sediment yields from South Island catchments, New Zealand. J Am Water Resour As 17:662–671

Hovius N, Stark CP, Allen PA (1997) Sediment flux from a mountain belt derived by landslide mapping. Geology 25:231–234

Jarvis A, Reuter HI, Nelson A, Guevara E (2008) Hole-filled SRTM for the globe version 4. Available from the CGIAR-CSI SRTM 90m database (http://srtm.csi.cgiar.org)

Kump LR, Brantley SL, Arthur MA (2000) Chemical weathering, atmospheric CO_2, and climate. Annu Rev Earth Planet Sci 28:611–667

Matsushi Y, Wakasa S, Matsuzaki H, Matsukura Y (2006) Long-term denudation rates of actively uplifting hillcrests in the Boso Peninsula, Japan, estimated from depth profiling of in situ-produced cosmogenic ^{10}Be and ^{26}Al. Geomorphology 82:283–294

Milliman JD, Syvitski JP (1992) Geomorphic/tectonic control of sediment discharge to the ocean: the importance of small mountainous rivers. J Geol 100:525–544

Molnar P, Anderson RS, Anderson SP (2007) Tectonics, fracturing of rock, and erosion. J Geophys Res Earth 112

Morell KD, Sandiford M, Rajendran C, Rajendran K, Alimanovic A, Fink D, Sanwal J (2015) Geomorphology reveals active décollement geometry in the central Himalayan seismic gap. Lithosphere 7:247–256

Nordpil (2014) Tectonic plate GIS data. https://nordpil.com/resources/tectonic-plates-gis-data/. Accessed Sept 2016

Olen SM, Bodo B, Strecker MR (2016) Role of climate and vegetation density in modulating denudation rates in the Himalaya. Earth Planet Sc Lett 445:57–67

Portenga EW, Bierman PR (2011) Understanding Earth's eroding surface with ^{10}Be. GSA Today 21:4–10

Roux-Mallouf L, Godard V, Cattin R, Ferry M, Gyeltshen J, Ritz JF, Drupka D, Guillou V, Arnold M, Aumaître G (2015) Evidence for a wide and gently dipping Main Himalayan Thrust in western Bhutan. Geophys Res Lett 42: 3257–3265

Saunders I, Young A (1983) Rates of surface processes on slopes, slope retreat and denudation. Earth Surf Proc Land 8:473–501

Scharf TE, Codilean AT, De Wit M, Jansen JD, Kubik PW (2013) Strong rocks sustain ancient postorogenic topography in southern Africa. Geology 41:331–334

Steer P, Simoes M, Cattin R, Shyu JB (2014) Erosion influences the seismicity of active thrust faults. Nat Commun 5:5564

Summerfield M, Hulton N (1994) Natural controls of fluvial denudation rates in major world drainage basins. J Geophys Res Solid Earth 99:13871–13883

USGS Sediment Data Portal. https://cida.usgs.gov/sediment/#. Accessed September 2015

U.S./Japan ASTER Science Team (2011) ASTER Global Digital Elevation Model V002. https://asterweb.jpl.nasa.gov/gdem.asp . Accessed August 2014

Von Blanckenburg F (2005) The control mechanisms of erosion and weathering at basin scale from cosmogenic. nuclides in river sediment. Earth Planet Sc Lett 237:462–479.

Von Blanckenburg F, Hewawasam T, Kubik PW (2004) Cosmogenic nuclide evidence for low weathering and denudation in the wet, tropical highlands of Sri Lanka. J Geophys Res-Earth 109, https://doi.org/10.1029/2003JF000049.

11

Subducted oceanic crust as the origin of seismically slow lower-mantle structures

Timothy D. Jones[1*], Ross R. Maguire[2,3], Peter E. van Keken[1], Jeroen Ritsema[4] and Paula Koelemeijer[5]

Abstract

Mantle tomography reveals the existence of two large low-shear-velocity provinces (LLSVPs) at the base of the mantle. We examine here the hypothesis that they are piles of oceanic crust that have steadily accumulated and warmed over billions of years. We use existing global geodynamic models in which dense oceanic crust forms at divergent plate boundaries and subducts at convergent ones. The model suite covers the predicted density range for oceanic crust over lower mantle conditions. To meaningfully compare our geodynamic models to tomographic structures, we convert them into models of seismic wavespeed and explicitly account for the limited resolving power of tomography. Our results demonstrate that long-term recycling of dense oceanic crust naturally leads to the formation of thermochemical piles with seismic characteristics similar to the LLSVPs. The extent to which oceanic crust contributes to the LLSVPs depends upon its density in the lower mantle for which accurate data is lacking. We find that the LLSVPs are not composed solely of oceanic crust. Rather, they are basalt rich at their base (bottom 100–200 km) and grade into peridotite toward their sides and top with the strength of their seismic signature arising from the dominant role of temperature. We conclude that recycling of oceanic crust, if sufficiently dense, has a strong influence on the thermal and chemical evolution of Earth's mantle.

Keywords: Geodynamical modeling, Mantle structure, Seismic tomography

Introduction

Tomographic models reveal the presence of two large anomalous structures in the lowermost mantle. These large low-shear velocity provinces (LLSVPs) are surrounded by high shear-wave velocity regions that are associated with past subduction (Fig. 1; Tanaka et al. 2009; Ritsema et al. 2011; and and Cottaar and Lekic 2016). The origin of the LLSVPs and their connection to mantle dynamics and the long-term chemical evolution of the planet remain unclear (see McNamara (2019) for a recent overview). Suggested origins include large-scale thermal upwellings (e.g.,Schubert et al. 2004; Davies et al.2015; and Koelemeijer et al. 2017), dense piles of primordial and compositionally distinct mantle (e.g., Garnero and McNamara 2008; Burke et al. 2008), and accumulations of recycled oceanic crust (e.g., Christensen and Hofmann 1994; Nakagawa et al. 2009, 2010). Markedly different interpretations of LLSVPs persist because there is no unique cause of seismic velocity anomalies in the mantle, but there are many plausible ones. For example, the low-shear velocity anomaly ($\delta \ln V_s \approx -2\%$) associated with LLSVPs may be equally well explained by an increase in temperature (Schuberth et al. 2009), a compositional change (McNamara and Zhong 2005), the presence of melt (Lee et al. 2010), or a combination thereof (Li and McNamara 2013; Ballmer et al. 2016). The elevated ratio between shear and compressional wavespeed anomalies within LLSVPs (Robertson and Woodhouse 1995; Masters et al. 2000; Romanowicz 2001; Moulik and Ekström 2016) is often regarded as evidence that they are thermochemical in origin, but it can also indicate the presence of post-perovskite without requiring a contribution from large-scale chemical variations (Koelemeijer et al. 2018). Measurements of solid-Earth tides (Lau et al. 2017) and

*Correspondence: tjones@carnegiescience.edu
[1]Department of Terrestrial Magnetism, Carnegie Institution for Science, 5241 Broad Branch Road NW, Washington, DC, 20015 USA
Full list of author information is available at the end of the article

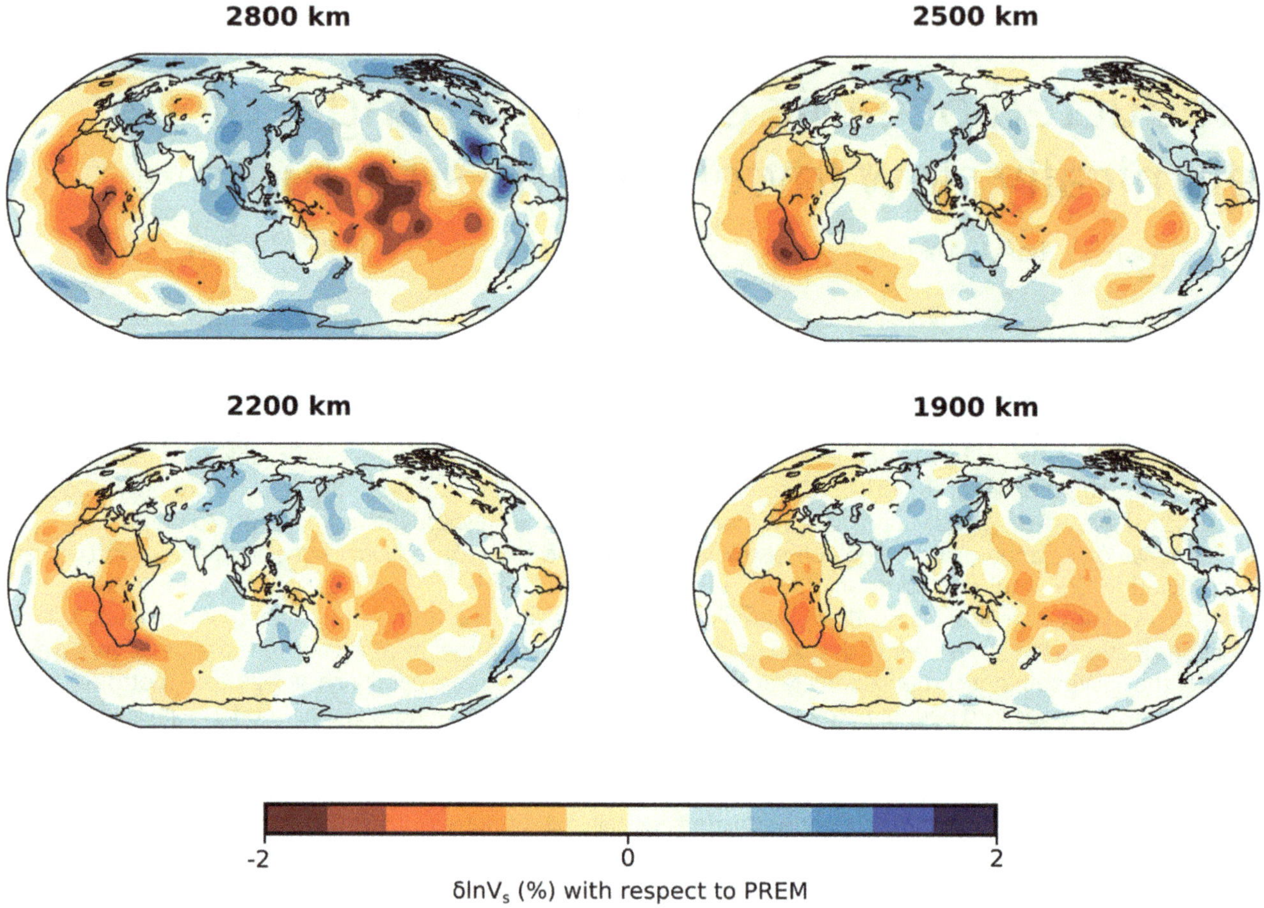

Fig. 1 Depth slices of S40RTS (Ritsema, 2011) showing variations in shear-wave velocity, $\delta \ln V_S$, with respect to the radial average. Both positive and negative velocity anomalies are highest toward the CMB. The color intensity is proportional to the velocity amplitude up to a maximum value of 2%

normal mode splitting functions (e.g., Ishii and Tromp 1999; Trampert et al. 2004; Moulik and Ekström 2016; and Koelemeijer et al. 2016) suggest that the LLSVPs have a higher than average density across the bottom two thirds of their depth range. However, constraining the density from low-frequency data depends on the chosen set of normal modes, on how CMB topography is modeled, and on the theoretical simplifications made to model the data (e.g., Koelemeijer et al. 2017; Akbarashrafi et al. 2018).

In this study, we will test the hypothesis that LLSVPs are accumulations of dense recycled oceanic crust. We are motivated in particular by geodynamical studies that show that the long-term recycling of oceanic crust can significantly contribute to the formation of warm but dense pools of compositionally distinct material at the base of the mantle (Fig. 2; Christensen and Hofmann 1994; Brandenburg and van Keken 2007; Brandenburg et al. 2008; and Nakagawa and Tackley 2008). We will focus on how well thermochemical model simulations predict the seismological characteristics of LLSVPs by projecting the temperature and composition to seismic velocity. We will then test how well such velocity structures are resolved in the shear-wave velocity model, S40RTS (Ritsema et al. 2011), and the combined shear- and compressional-wave velocity model, SP12RTS (Koelemeijer et al. 2016).

As oceanic lithosphere subducts, three distinct components are recycled back into the mantle: a thin basaltic crust and a thicker base of harzburgite on top of the more primitive peridotite from which the crust was extracted. Due to convective stirring of these components, the mantle composition is an evolving mixture of the three end-member compositions. To simplify the description below, we will refer to end-member basaltic, harzburgitic, and peridotitic compositions as "basalt", "harzburgite", and "peridotite", respectively, and, where relevant, will use subscripts B, H, and P to identify them.

Mineral physics experiments indicate that the basaltic component is denser than ambient mantle through most of Earth's mantle with a relative density contrast $\delta \ln \rho$ of up to 2–7%, except near the surface and in the uppermost lower mantle (Ringwood and Irifune 1988;

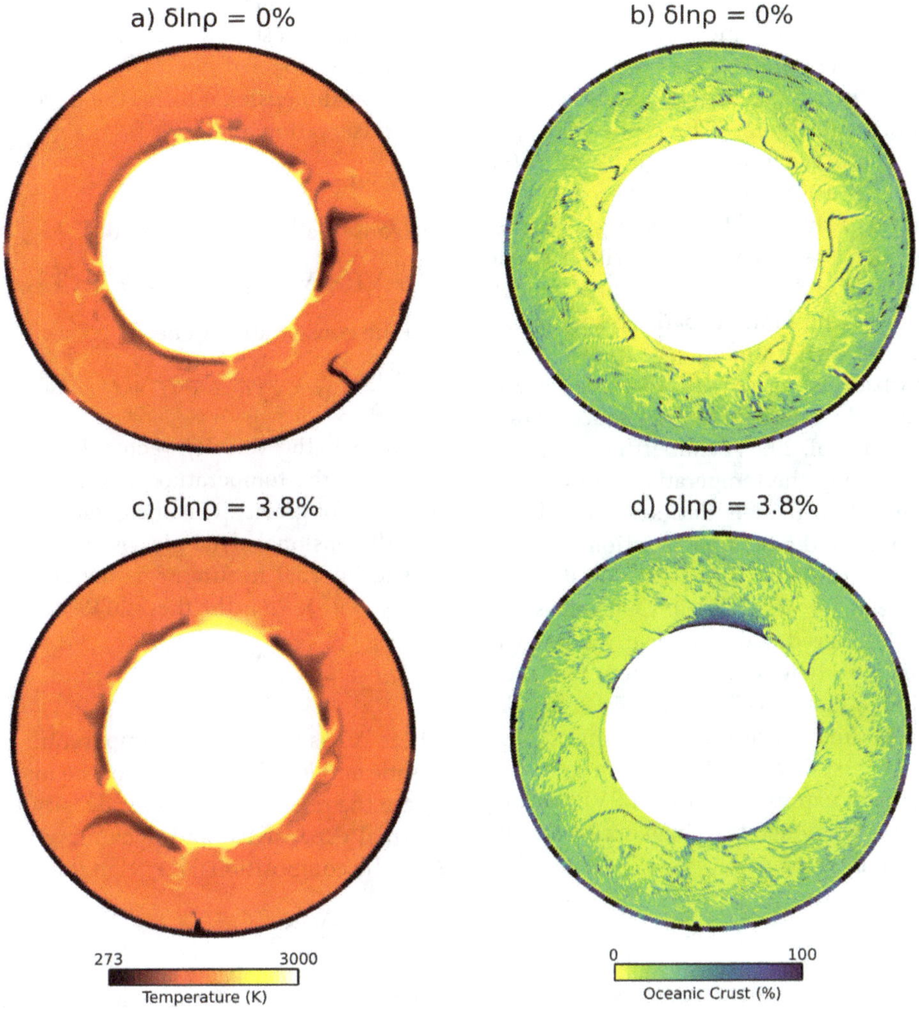

Fig. 2 Plot of temperature and composition fields for Brandenburg et al. (2008) models for $\delta \ln \rho$ of 0% (**a**, **b**) and 3.8% (**c**, **d**), where we use the non-linear radial projection of the models onto Earth coordinates. In both cases, recently subducted oceanic crust meets the CMB and the low viscosity upper mantle has been homogenized by convective stirring relative to the lower mantle. When $\delta \ln \rho = 0\%$, oceanic crust is somewhat evenly distributed throughout the lower mantle. In contrast, when $\delta \ln \rho = 3.8\%$, oceanic crust accumulates and the CMB forms sharp piles

Ringwood 1990; Hirose et al. 1999; Aoki and Takahashi 2004; Hirose et al. 2005; Ricolleau et al. 2010; Tsuchiya 2011). Under these conditions, numerical experiments predict that the oceanic crust will accumulate at the core-mantle-boundary (CMB) (Christensen and Hofmann 1994; Davies 2002; Xie and Tackley 2004; Brandenburg and van Keken 2007). The presence of recycled oceanic crust as a major component of mantle plumes (Hofmann and White 1982) is empirical evidence that at least some portion of subducted material reaches the CMB.

The models presented in this study follow the approach of Christensen and Hofmann (1994) (herein CH94) who assumed that the basalt remains denser throughout the mantle and retains a constant density contrast with the ambient mantle. CH94 found that at sufficiently large density excess, the subducted oceanic crust accumulates at the CMB where it is swept into piles and mixes with the ambient mantle. They concluded that this process could not only account for important features of Earth's chemical evolution but may also explain seismic heterogeneity in the D" layer.

Due to computational limitations at the time, CH94 performed dynamical computations at a convective vigor that is significantly below that of present day Earth and used a scaling relationship to extrapolate their findings to models with present-day plate velocities. Brandenburg and van Keken (2007) explicitly showed that this extrapolation is an accurate approximation. Brandenburg et al. (2008) (herein BB08) further extended the study of CH94 by exploring the effects of crustal recycling in models with force-balanced plates and with Earth-like convective vigor. Their results confirmed the findings of CH94. They also showed that the formation and recycling of oceanic crust could contribute to the HIMU-DMM-EMI diversity of Ocean Island Basalt (OIB) and Mid-Oceanic Ridge Basalt (MORB) in multiple isotope systems.

The BB08 study confirmed that recycled dense oceanic crust forms thermochemical piles. These piles naturally form sharp edges, as observed in seismological studies (Ni et al. 2002; Ritsema et al. 1998), and they are broadly consistent with the size and shape of LLSVPs (Fig. 2 frames c and d). However, any assessment of similarity between thermochemical models and global tomographic models must take into account the limited spatial resolution of seismic tomography. As earlier studies have shown, shear-wave velocity perturbations can be reduced, smeared, and spatially shifted due to the effects of uneven data coverage and damping (Ritsema et al. 2007; Schuberth et al. 2009). In this paper, we will map the temperature and compositional fields of the BB08 models into seismic velocity and will then account for the parameterization and regularization parameters of S40RTS (Ritsema et al. 2011) and SP12RTS (Koelemeijer et al. 2016). Our analysis of these seismologically filtered geodynamical models will include a comparison of their spectral content and radially averaged profiles to that of S40RTS and SP12RTS. We conclude that the recycling of oceanic crust, by itself or in addition to other processes, can explain the imaged seismic heterogeneity that is expressed by the LLSVPs as long as the oceanic crust has a moderate to high excess density with respect to the ambient mantle.

Methods

Geodynamic models

The BB08 models are finite element solutions of convective flow in an incompressible fluid at an infinite Prandtl number driven by thermal and chemical buoyancy. The models are based on a 2D cylindrical geometry with a reduced core radius to mimic the heat loss characteristics of Earth (van Keken 2001).

The models use a force-balanced plate approach (Gable et al. 1991) that yields average surface velocities and surface heatflow consistent with values observed today. This sets up energetically consistent convection with prescribed zones of divergence and convergence where oceanic crust forms and subduct, respectively. The resolution of the models varies from 50 km in the middle-lower mantle to 13 km at both the surface and CMB, such that the thermal boundary layers, of characteristic thickness ~100 km are adequately resolved. At divergent zones, a melting parameterization generates basaltic crust with a harzburgite residue from the peridotitic mantle to form oceanic lithosphere. When this lithosphere subducts at convergent zones, it sinks to the base of the mantle under negative thermal buoyancy before mixing with the surrounding mantle. We represent the peridotitic composition by tracers that are spaced, on average, at 5 km and that concentrate by a factor of 8 when the oceanic crust is formed such that a ~10-km-thick basaltic crust is adequately modeled.

In non-dimensional form, the governing equations are the conservation of mass:

$$\nabla \cdot \boldsymbol{u} = 0 \tag{1}$$

the conservation of momentum:

$$-\nabla P + \nabla \cdot \left(\eta \underline{\dot{\epsilon}}\right) = [\, RaT - RcC\,]\, \hat{\boldsymbol{g}} \tag{2}$$

and the conservation of heat:

$$\frac{\partial T}{\partial t} + (\boldsymbol{u} \cdot \nabla)T = \nabla^2 T + Q \tag{3}$$

where $\boldsymbol{u}$ is the velocity vector, P the dynamic pressure, t time, T the temperature, and $\hat{\boldsymbol{g}}$ the unit vector in the direction of gravity. C is the chemical composition, η the non-dimensional dynamic viscosity, and Q is the volumetric internal heating. $\underline{\dot{\epsilon}}$ is the strain-rate tensor $\underline{\dot{\epsilon}} = \left(\nabla u + \nabla u^T\right)$. Ra is the thermal Rayleigh number:

$$Ra = \frac{\rho_0 g \alpha_0 \Delta T h^3}{\kappa_0 \eta_0} \tag{4}$$

where ΔT is the assumed temperature contrast across the mantle and h is the thickness of the mantle. ρ_0, κ_0, α_0, and η_0 are the reference values for density, thermal diffusivity, thermal expansivity, and dynamic viscosity, respectively. Rc is the compositional Rayleigh number:

$$Rc = \frac{\Delta \rho_0 g h^3}{\kappa \eta_0} \tag{5}$$

where $\Delta\rho_0$ is related to the density contrast between the oceanic crust and the depleted mantle (see below). For reference values, see Table 1. The models assume constant expansivity and diffusivity, and a temperature- and depth-dependent viscosity of the non-dimensional form

$$\eta(T, z) = \eta_z(z) \exp(-bT) \tag{6}$$

where z is depth, $b = \ln(1000)$, and η_z is 1000 in the lithosphere, 1 in the upper mantle, and 30 in the lower mantle.

Table 1 Parameters common to all cases examined and their reference values

Symbol	Parameter	Reference value	Units
h	Thickness	2885	km
α_0	Thermal expansion coefficient	3×10^{-5}	K^{-1}
η_0	Lower mantle viscosity	10^{22}	Pa s
ρ_0	Density	4500	kg m^{-3}
κ_0	Thermal diffusivity	10^{-6}	$m^2\ s^{-1}$
ΔT	Temperature contrast	3000	K

BB08 unfortunately introduced a number of notation errors that include typos in the equations (e.g., their equation (10) should read $f = N/(N + 7M)$, not $f = 1/(N + 7M)$) and inverted the meaning of the term "extraction coefficient" where "retention coefficient" was intended. It also obscured the definition of the eclogite excess density (EED) as it was not fully defined. The EED used in their paper was defined as EED=Rc/Ra=$\Delta\rho/(\rho_0\alpha_0\Delta T)$ and ranged from 0 to 10%. The composition function C (Eq. 2) is 1 for peridotite, 0 for harzburgite (which has density ρ_0), and 8 for basalt. This means that the dimensional density contrast of the oceanic crust to depleted mantle $\Delta\rho_{B-H}= \rho_B - \rho_H$ can be found by multiplying EED by $8\rho_0\alpha_0\Delta T$. The more commonly used expression for relative excess density $\partial \ln \rho = (\rho_B - \rho_P)/\rho_P$ is with respect to ambient mantle found by multiplying EED by $7\rho_0\alpha_0\Delta T$. Using ρ_0=4500 kg/m^3, α_0=3 × 10^{-5}/K, and ΔT=3000 K, we find $\Delta\rho_{B-P}$=EED×2840 kg/m^3. For EED=10% (the maximum value considered), we find a dimensional density of the oceanic crust to be ρ_B=4824 kg/m^3 which translates to a relative density increase of $\delta \ln \rho = 6.2\%$.

We select seven simulations that are similar to those in BB08 except that the models have been evaluated on a higher resolution finite element mesh (with 70 K vs. 46 K nodal points) and a larger number of tracers (3.7 M vs. 2.5 M). Between each model, we vary only the the density difference between basalt and harzburgite. This has some control over the dynamics, which leads to differences in the spatial distribution of chemical heterogeneity throughout the mantle.

Each simulation starts from snapshot of a purely thermal convection simulation at Ra=5 × 10^6 that was evolved for a sufficiently long time to let any of the transients associated with the initial conditions disappear. We then model oceanic crust formation and recycling for a period equivalent to 4.5 Byr and for various values for Rc (between 0 and 5 × 10^5). Piles of recycled crust build up over time for $Rc > 0$. Since we keep Rc/Ra fixed for each simulation, the accumulation of oceanic crust for the higher $\delta \ln \rho$ models tends to make the convection slightly more sluggish towards the end of the model evolution (consistent with BB08).

Model selection and seismic wavespeed conversion

We first select a time slice from the past 2.5 Gyr of each simulation in which a broad thermochemical pile is present (for Rc > 0). Piles tend not to approach the size of LLSVPs earlier than about 2 Ga and are more robust for higher $\delta \ln \rho$. We also use a time-slice for $Rc = 0$ for comparison.

Two steps are required to create an Earth-like geometry from the 2D cylindrical models. We use a non-linear radial projection of the models onto Earth coordinates (i.e., the non-dimensional core radius 0.492 maps onto a CMB depth of 2885 km and the non-dimensional surface radius 1.492 corresponds to depth 0 km; van Keken (2001)). We then project a 180° section of the geodynamic model into an axisymmetric spherical shell representation that is representative of a fully 3D spherical model with the important constraint that the thermochemical pile is centered about the rotation axis. This effectively imposes the dome-like geometry characteristic of LLSVPs. Structures away from the rotation axis thus become ring-like.

The 3D spherical model is then converted from temperature and composition to seismic wavespeed using the following approach. First, we convert potential temperature to absolute temperature by adding an adiabatic gradient for a pyrolite composition. This step is required because the geodynamic simulations assume the incompressible Boussinesq approximation.

Next, we map the pressure P, temperature T, and basalt fraction f to seismic shear-wave velocity V_s and compressional-wave velocity V_p using the method of Xu et al. (2008), where it is assumed that the mantle can be represented as a mechanical mixture of basalt and harzburgite. The bulk composition of the two end members are defined in terms of the proportions of the oxides SiO_2, Na_2O, CaO, FeO, MgO, and Al_2O_3.

We use the composition of basalt from Workman and Hart (2005) and the composition of harzburgite from Baker and Beckett (1999). The stable mineral assemblage and associated anharmonic seismic wavespeeds are computed for both basalt and harzburgite using Perple_X (Connolly 2005) and the mineral elastic parameter database of Stixrude and Lithgow-Bertelloni (2011).

The anharmonic seismic wavespeeds are corrected for frequency-dependent effects of attenuation using Maguire et al. (2016) attenuation model. The seismic wave speed V (V_s or V_p) of intermediate compositions is taken as the linear combination of the end member compositions basalt and harzburgite:

$$\mathrm{V}(P, T, f) = f\mathrm{V}^B(P, T) + (1 - f)\mathrm{V}^H(P, T) \qquad (7)$$

The velocity anomaly $\delta \ln \mathrm{V}$ is computed relative to the radially averaged seismic structure.

Reparameterization and the seismic resolution operator

To meaningfully compare geodynamic models with tomographic models, we must account for the parameterization and limited resolution of the latter. This process is illustrated in Fig. 3. We begin by reparameterizing our geodynamically predicted wavespeed model onto the basis of S40RTS. Lateral variations of $\delta \ln \mathrm{V}_s$ are described using spherical harmonics up to degree and order 40, and depth variations by 21 vertical splines.

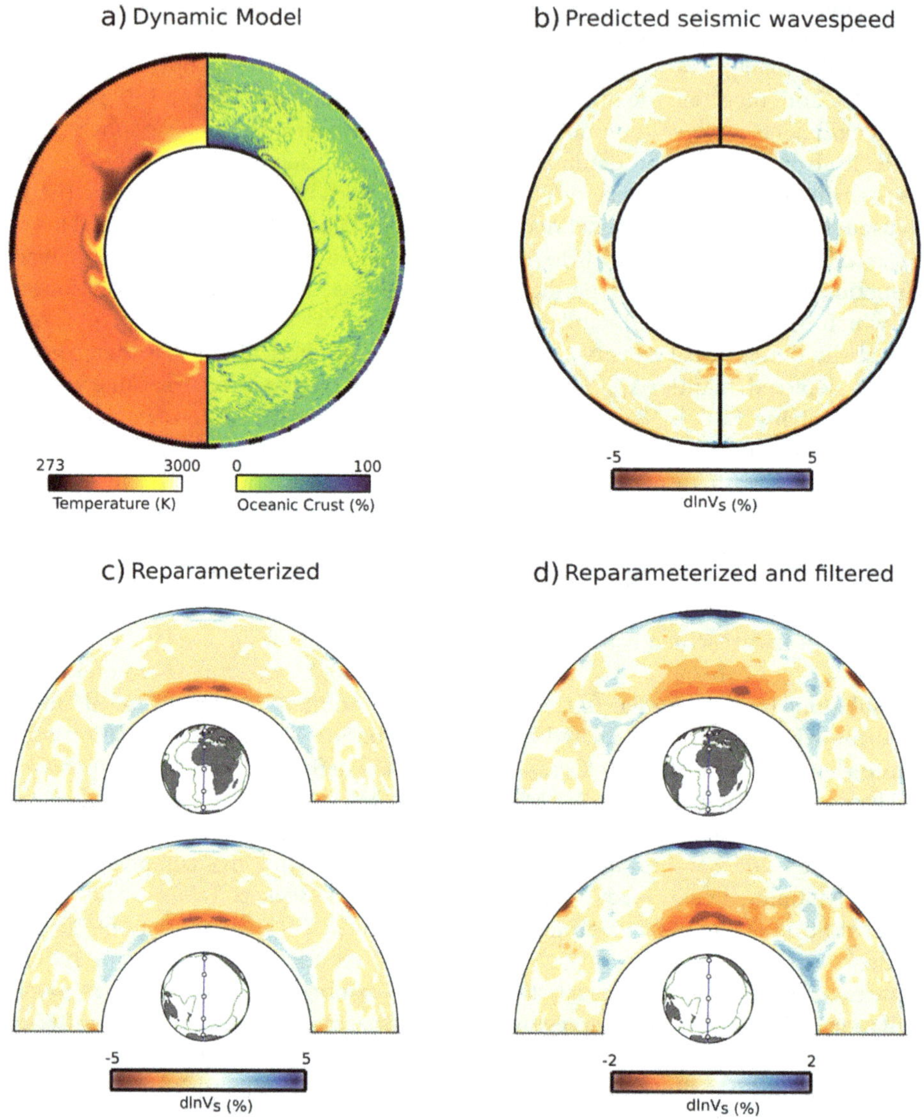

Fig. 3 Illustration of steps in our method to convert dynamic models into seismic velocity models and account for the parameterization and resolution of S40RTS. **a** Dynamic model, where $\delta \ln \rho$=3.8%, showing temperature and composition fields. **b** Predicted seismic wavespeed model showing $\delta \ln V_S$ variations with respect to the radial average. **c** Reparameterized model of $\delta \ln V_S$ on the basis of S40RTS. **d** Recovered $\delta \ln V_S$ after accounting for damping and uneven sampling of S40RTS. The geographic location of depth slice is provided in by the black vertical line through the globe

Following the approach of Ritsema et al. (2007), we define the resolution operator to be $\mathcal{R} = \mathbf{G}^{\dagger}\mathbf{G}$, where $\mathbf{G}$ is the operator of the seismic forward problem and $\mathbf{G}^{\dagger}$ is its generalized inverse. We then modify our reparameterized model of seismic wavespeed (the input model $\mathbf{m}^{\mathrm{IN}}$) by multiplying it with $\mathcal{R}$ to obtain a filtered output model, $\mathbf{m}^{\mathrm{OUT}}$, as if imaged by tomographic inversion:

$$\mathbf{m}^{\mathrm{OUT}} = \mathcal{R}\mathbf{m}^{\mathrm{IN}} \tag{8}$$

where $\mathcal{R}$ is a complete description of the variable spatial resolution of S40RTS. We compute $\mathcal{R}$ for the same damping parameter as that used in S40RTS.

To explore how both $\delta \ln \mathrm{V}_s$ and $\delta \ln \mathrm{V}_p$ are recovered after tomographic filtering, we repeat this processes using $\mathcal{R}$ from the joint V_s and V_p tomography model SP12RTS. SP12RTS uses a similar model parameterization to S40RTS but only up to spherical harmonic degree 12. Since SP12RTS incorporates information from both V_s and V_p, the model vector $\mathbf{m}$ consists of two parts:

$$\mathbf{m} = \begin{pmatrix} \mathbf{S} \\ \mathbf{P} \end{pmatrix} \tag{9}$$

where $\mathbf{S}$ and $\mathbf{P}$ are model vectors describing $\mathrm{dln}V_s$ and $\mathrm{dln}V_p$, respectively. Therefore, Eq. 8 becomes:

$$\begin{pmatrix} \mathbf{S}^{\mathrm{OUT}} \\ \mathbf{P}^{\mathrm{OUT}} \end{pmatrix} = \begin{pmatrix} \mathcal{R}_{\mathrm{SS}} & \mathcal{R}_{\mathrm{SP}} \\ \mathcal{R}_{\mathrm{PS}} & \mathcal{R}_{\mathrm{PP}} \end{pmatrix} \begin{pmatrix} \mathbf{S}^{\mathrm{IN}} \\ \mathbf{P}^{\mathrm{IN}} \end{pmatrix}, \tag{10}$$

where the diagonal blocks, $\mathcal{R}_{SS}$ and $\mathcal{R}_{PP}$, describe how V_s and V_p map into V_s and V_p, respectively, while the off-diagonal blocks, $\mathcal{R}_{SP}$ and $\mathcal{R}_{PS}$, describe how V_s and V_p map into each other.

Results

Effects of reparameterization and tomographic filtering

We first look at the effects of the filtering processes itself on $\delta \ln Vs$. Figure 3 illustrates how reparameterization and $\mathcal{R}$ modify the input model to an output model with the resolution of S40RTS. Only the broadest scale of seismic heterogeneity of the input model is preserved following the filtering process. The highest $\delta \ln V_s$ anomalies are in the uppermost mantle, and the relative position of high and low seismic velocity features are well preserved in the lower mantle.

The most pronounced effects are amplitude reduction and smoothing. These effects are accentuated where data coverage is sparse. This is for example of the case beneath the African plate (Fig. 4, left column), where $\delta \ln V_s$ anomalies in the same filtered geodynamic input model are smaller and smoother than beneath the Pacific plate (Fig. 4, right column). Consequently, cold subducted lithosphere and the thermochemical piles are more dominant features of the lower mantle in our Pacific cross-sections. Where data coverage is particularly poor, low $\delta \ln V_s$ anomalies disappear entirely after filtering. For example, subducted lithosphere is imaged along the CMB in the Pacific cross-section with $\delta \ln V_s \sim 0.5\%$ (Fig. 4, $\delta \ln \rho$=3.1%), but the same structure is absent in the African cross-section.

Effects of $\delta \ln \rho$

The spatial patterns of $\delta \ln Vs$ for all filtered dynamic models and along radial segments of S40RTS are illustrated in Fig. 4. In the isochemical convection case, plume clusters (Fig. 4, bottom row) are imaged as regions of low $\delta \ln V_s$ in the lower mantle, but these are significantly smaller and of weaker amplitude than LLSVPs in S40RTS. Furthermore, regions of high $\delta \ln V_s$ are absent from the lowermost mantle.

By comparison, regions of strongly negative $\delta \ln V_s$ for the thermochemical convection cases are broader and of stronger amplitude. Their size does not reflect that of the basaltic crust piles. Instead, the margins of the low $\delta \ln V_s$ regions map onto the thermal anomaly associated with the basaltic crust piles. This implies that LLSVPs should not be interpreted as pure oceanic crust. Rather, under the parameters explored here, they are basalt rich at their base (bottom 100–200 km) and grade into peridotite toward their sides and top, which can reach mid-mantle depths.

The range of $\delta \ln V_s$ in all thermochemical simulations is comparable to that of S40RTS after tomographic filtering is applied. However, the low velocity structures resemble LLSVPs in their size and strength only when $\delta \ln \rho \geq 3.1\%$ so that enough basaltic crust accumulates at the CMB.

The thermochemical piles are generally flanked by moderate-to-high $\delta \ln V_s$ anomalies, corresponding to cold lithosphere that has recently subducted. In these models, the subducted lithosphere sweeps the hot, dense material into piles. This is consistent with S40RTS, and other tomographic models, that show LLSVPs surrounded by a ring of high seismic velocity (e.g., Dziewonski et al. 1977; Grand et al. 1997; and Kennett et al. 1998).

In contrast to S40RTS, subducted lithosphere in a number of the filtered models is coherently imaged through the entire mantle. This indicates that subducted lithosphere in those cross sections has a higher temperature anomaly or are broader than subducted slabs in Earth's mantle. Our models have important limitations that may explain this difference. We force an axisymmetric spherical symmetry on the models causing the downwellings in the geodynamical models to be projected as ring-like features which likely are recovered with higher amplitude and geographical extent than subduction zones in the Earth. Additionally, on Earth, subduction is one-sided and occurs at a range of angles while in the geodynamic models, subduction is two-sided and the slabs sink almost exclusively vertical.

Given that our models are two-dimensional and have dynamic plates, we can neither draw meaning from geographical correlations nor impose past plate motions. We therefore compare the similarity of cross sections of S40RTS and filtered geodynamic models in two ways. First, we compare the heterogeneity spectra of $\delta \ln V_s$ by computing the power spectral density (PSD) at each depth (Figs. 5 and 6). This gives the power as a function of wavenumber k that represents the number of wavelengths per 360° and thus are analogous to spherical harmonic degree. The heterogeneity spectrum is normalized by the maximum of the PSD at each depth. Second, we compare the root-mean-square (RMS) of $\delta \ln V_s$ as a function of radius (Figs. 8 and 9).

In the lower mantle, the spectral characteristics of our filtered models are similar to the two S40RTS slices (Figs. 5a and 6a) when $\delta \ln \rho \geq 3.1$. The dominance of power at lower k in such cases is a consequence of the long-wavelength structure of the thermochemical piles. For lower values of $\delta \ln \rho$, power is somewhat randomly spread over wavenumbers $k < 20$ beneath the African plate and $k < 40$ beneath the Pacific plate, since basaltic piles do not form in as coherent a fashion as at larger density contrasts. Spectral characteristics diverge between the filtered models and S40RTS in the upper mantle, except when $\delta \ln \rho = 6.2\%$, which is the highest value used in

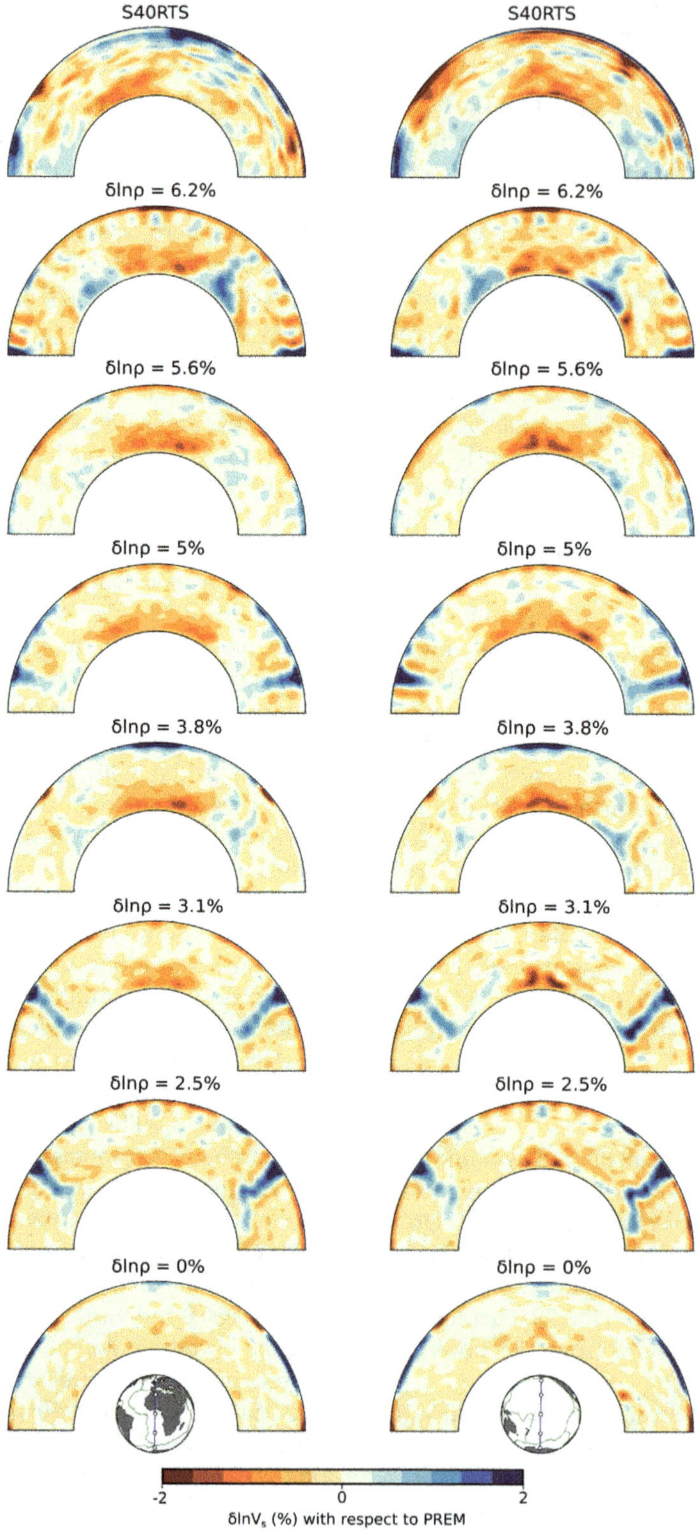

Fig. 4 Top row: cross-sections through S40RTS beneath the African (left column) and Pacific (right column) plates. Each subsequent row is a cross-section through the recovered $\delta \ln V_S$ for dynamic models with different densities for the basaltic component after tomographic filtering. Geographic location of depth slice is provided in by the black vertical line through the globe. Broad low $\delta \ln V_S$ structures form in models where $\delta\rho \geq 3.1\%$. In many cases, high velocity features, which map onto recently subducted lithosphere, flank the low $\delta \ln V_S$ structures

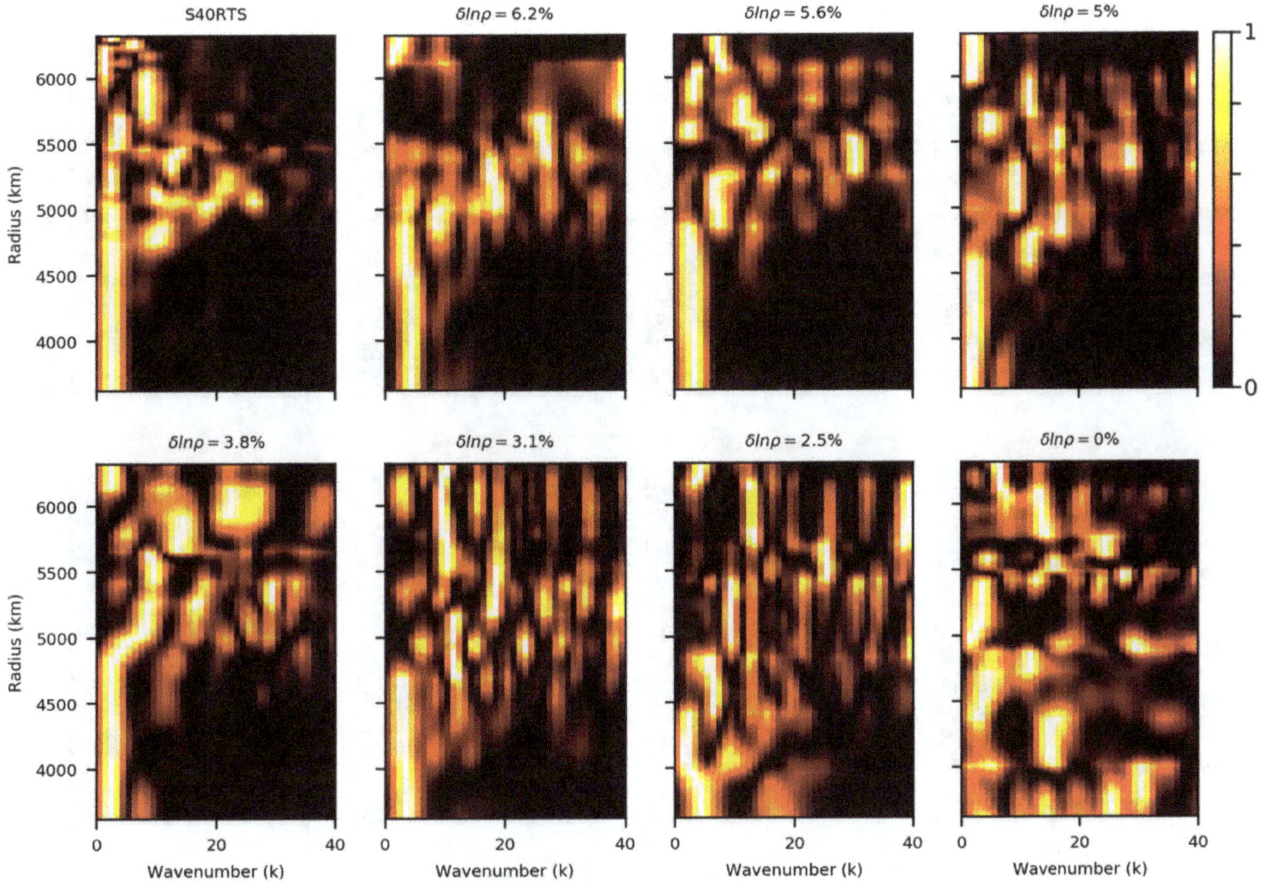

Fig. 5 Power spectral density plots for S40RTS and the tomographically filtered models as imaged beneath the Pacific plate. Spectral power is normalized to the maximum power at each depth and plotted as a function of wavenumber and radius. High power at low wavenumbers characterize the lower mantle in S40RTS and the filtered geodynamic models when $\delta\rho \geq 3.1\%$. This is due to the formation of large thermochemical structures at this density contrast. Below this value, power in the lower mantle is more randomly distributed across high and low wavenumbers due to a lack of broad-scale structure. The upper mantle of all geodynamic models contains more power at higher wavenumbers compared to S40RTS

this study. At the scale of our interpretation, the spectral characteristics of LLSVPs in S40RTS do not vary longitudinally. This is illustrated in plots of the spectral content through the Pacific and African LLSVPs spanning 30° (Fig. 7).

Similar correlations can be observed in the RMS $\delta \ln V_s$ profiles for our filtered models (Figs. 8 and 9). The S40RTS profile is given for comparison. As before, models with $\delta \ln \rho \geq 3.1\%$ show good correlation to S40RTS in the lower mantle. The larger amplitudes close to the thermal boundary layers and small peaks around the mid-to-upper mantle in S40RTS are poorly matched.

Lastly, we compute the expected $\delta \ln V_s/\delta \ln V_p$ ratio (S/P) of our models after filtering with the resolution operator of SP12RTS. S/P is calculated by dividing the RMS of $\delta \ln V_s$ by the RMS of $\delta \ln V_p$ at each depth. Figure 10 illustrates S/P as a function of radius for the purely thermal case, for the thermochemical cases $\delta \ln \rho$=3.1% and $\delta \ln \rho$=6.2% and for SP12RTS along the same radial segment to which filtering is applied (Fig. 10). Model slices are taken along the same longitude as Pacific mantle sections above but are centered at 15° latitude where P-wave coverage is higher. In each case, we include the effect of post-perovskite (pPv), whose transition from bridgmanite (brg) increases the amplitude of Vs and thus S/P.

In all filtered models, the increase in S/P throughout the lower mantle is comparable to SP12RTS. In the profile shown, the magnitude of S/P of SP12RTS is higher than average due to the presence of the Pacific LLSVP. S/P is weak in the purely thermal case and strengthens in the thermochemical cases as $\delta \ln \rho$ is increased. When $\delta \ln \rho$=6.2%, the magnitude of S/P far exceeds the maximum values observed in SP12RTS. Peak S/P consistently occurs at $\sim$ 2500-km depth in our filtered models, which is $\sim$ 100 km deeper than in SP12RTS. The same result was

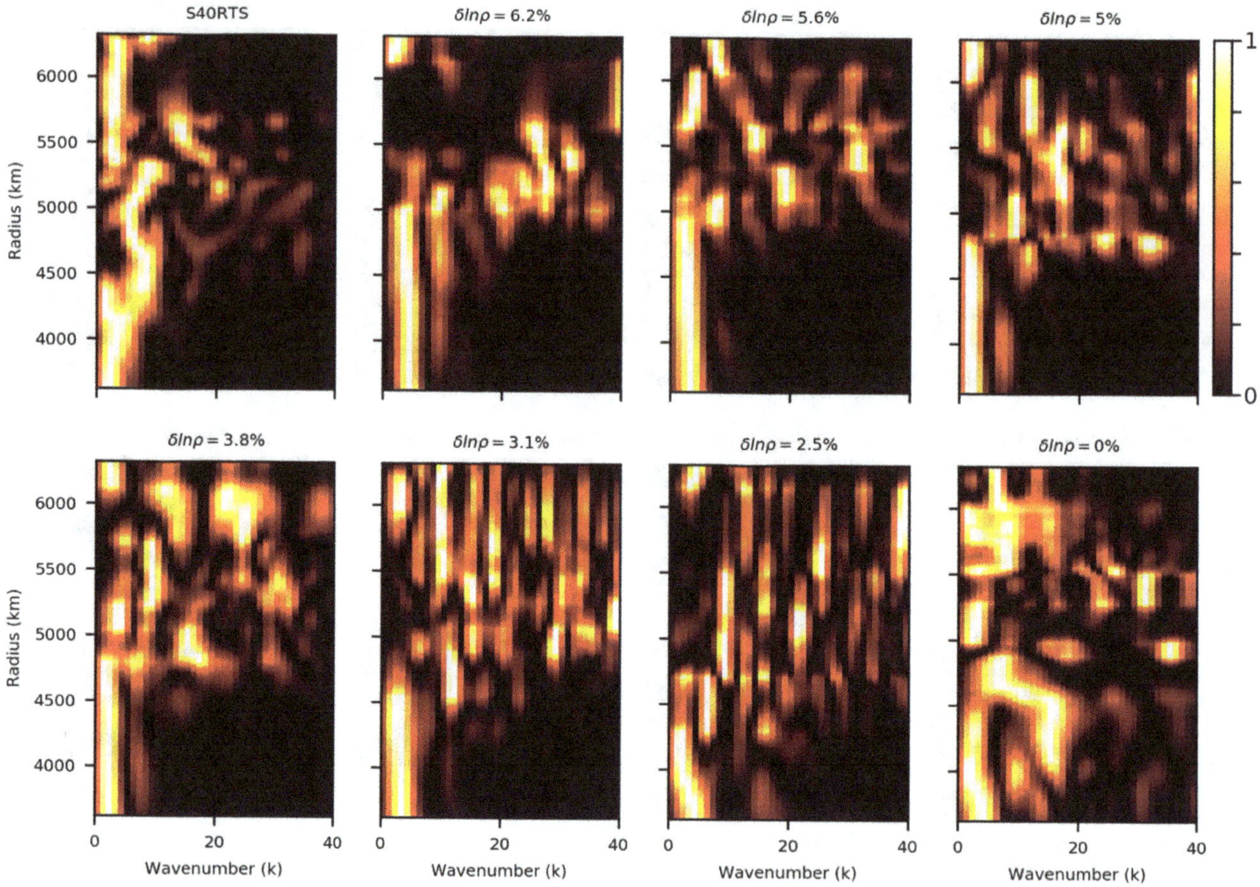

Fig. 6 Power spectral density plots for S40RTS and the tomographically filtered models as imaged beneath the African plate. Spectral power is normalized to the maximum power at each depth and plotted as a function of wavenumber and radius. Same basic features as for mantle beneath the Pacific except that S40RTS has more power at low wavenumbers in the upper mantle as well as in the lower mantle

found by Koelemeijer et al. (2018), who proposed two scenarios to resolve the offset: either the Clapeyron slope of the brg-pPv transition shallows at high temperatures or the entire stability field of pPv occurs at shallower depths.

Discussion

Our results support the hypothesis that long-term oceanic crust formation and recycling leads naturally to the formation of relatively dense thermochemical piles in the lowermost mantle with characteristics similar to the LLSVPs in S40RTS. We find the best match if the density contrast between basaltic crust and ambient mantle is around 3% or larger. Note that the precise threshold will depend upon parameters that were not varied in this study but are nonetheless uncertain, such as the temperature dependence of viscosity and the mantle's radial viscosity structure (e.g., Lau et al. 2016).

This interpretation implies a distinct thermal and chemical evolution for the seismic structure of the lower mantle. Piles of basaltic crust have two main attributes that influence V_S: their long residence time at the CMB and their distinct composition. The latter modifies the recovered V_S directly through its effect on the elastic parameters. The former is more important. Initially cold, subducted lithosphere has a high V_S. As it warms, basaltic crust segregates from harzburgite and accumulates at the CMB. Stabilized by an excess density, these accumulations become hotter than the ambient mantle and develop a broad thermal halo. The resulting thermochemical structure has seismic characteristics similar to LLSVPs after accounting for tomographic filtering.

Although both former basaltic crust and depleted harzburgite are eventually returned to the upper mantle, the latter has a much shorter residence time in the lower mantle. This has an interesting implication for the evolution of mantle structure. While regions of the lower mantle become increasingly enriched with the accumulation of basalt, the upper mantle becomes increasingly depleted with the addition of harzburgite. Such processes will contribute to the development of the modern depleted upper mantle alongside the extraction of continental crust; a hypothesis also proposed by Campbell (2002).

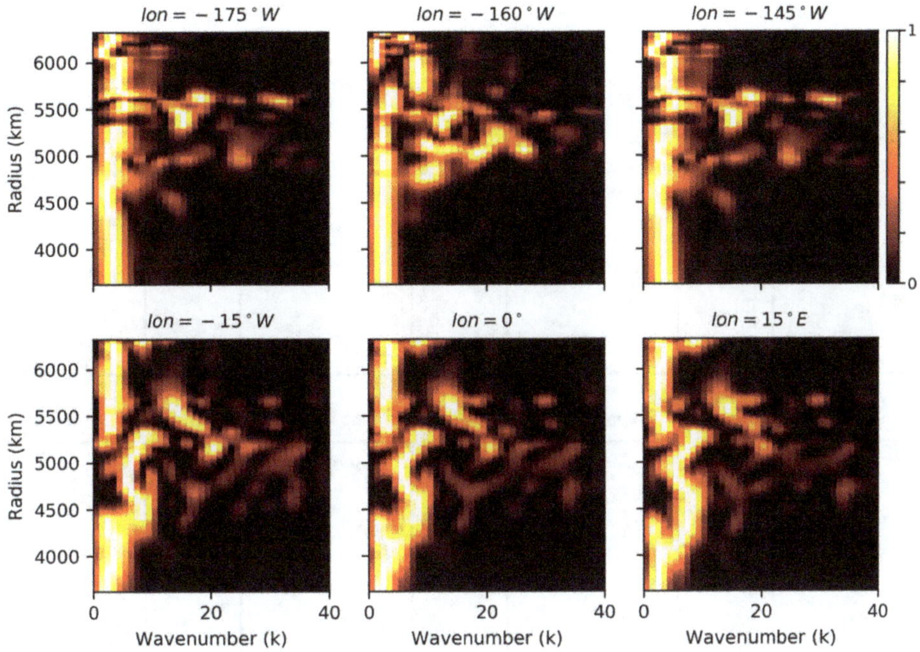

Fig. 7 Power spectral density plots for S40RTS over 30° longitude across the Pacific (top row) and African (bottom row) LLSVPs. Spectral power is normalized to the maximum power at each depth and plotted as a function of wavenumber and radius. Over 30°, the broad scale spectral characteristics of both LLSVPs is relatively consistent

Our model predicts that LLSVPs are not fixed features of the mantle in several respects. For one, they are not merely accumulations of basaltic crust but mixtures of basalt, harzburgite, and peridotite, with the concentration of basalt increasing with depth and changing in time. Their shape and size are robust although they become less so with decreasing $\delta \ln \rho$. With lower $\delta \ln \rho$, it becomes easier for subduction to displace the accumulations of oceanic crust. There is no clear indication of the robustness of LLSVPs one way or the other; therefore, a range of dynamic behavior remains plausible.

The influence of long-wavelength thermochemical piles remains pronounced after filtering to the even lower spatial resolution of SP12RTS. However, in the presence of pPv, purely thermal models and thermochemical models produce an increase in lower mantle S/P comparable to SP12RTS. The increase is stronger in thermochemical models and proportional to the density of oceanic crust. For the highest $\delta \ln \rho$ examined here, the amplitude of S/P is unrealistically large. Thus S/P, while not indicative of large scale compositional variation, as demonstrated here and first by Koelemeijer et al. (2018), restricts the maximum density contrast of thermochemical piles. Future work should examine the effect of basaltic crust on S/P in the absence of pPv and over the full range of experimentally predicted $\delta \ln \rho$.

Oceanic crust recycling also provides an explanation for seismic properties of the lower mantle on a much smaller scale. Haugland et al. (2018) mapped recycled oceanic crust as P velocity anomalies in the mantle. Their analysis predicts wave-front deflections by fragments of recycled crust. The magnitude of such distortions is sufficient to explain PKIKP precursors between epicentral distances of 130–142° in both arrival time and magnitude when $\delta \ln \rho \approx 4\%$ (see Haugland et al. (2018); their Figure 6).

Our study highlights the need for large-scale chemical variation to explain the seismic properties of LLSVPs. This is in disagreement with previous work that has shown thermochemical piles to overpredict the $\delta \ln V_s$ amplitudes of LLSVPs (Bull et al. 2009; Davies et al. 2012, 2015). We suggest that a difference in how thermochemical piles form in each model could explain the discrepancy. In the models of (Bull et al. 2009; Davies et al. 2012, 2015), thermochemical piles form from an initial dense layer above the CMB. The volume of chemically anomalous material is chosen to be $\sim$ 3% of the mantle so that it is consistent with volumetric estimates for regions with a pronounced decrease in V_s relative to V_p (Hernlund and Houser 2008). In our models, thermochemical piles form by the steady accumulation and subsequent warming of dense oceanic crust. After filtering, the resulting V_s anomaly is consistent with the size of LLSVPs. However, less than half the size of the anomalous region is chemically distinct from the ambient mantle, and only the bottom $\sim$ 100 km approaches pure oceanic crust. Thus, the influence of composition on the seismic properties of thermochemical piles is weakened for two reasons: a smaller portion of their size is chemically distinct and the portion

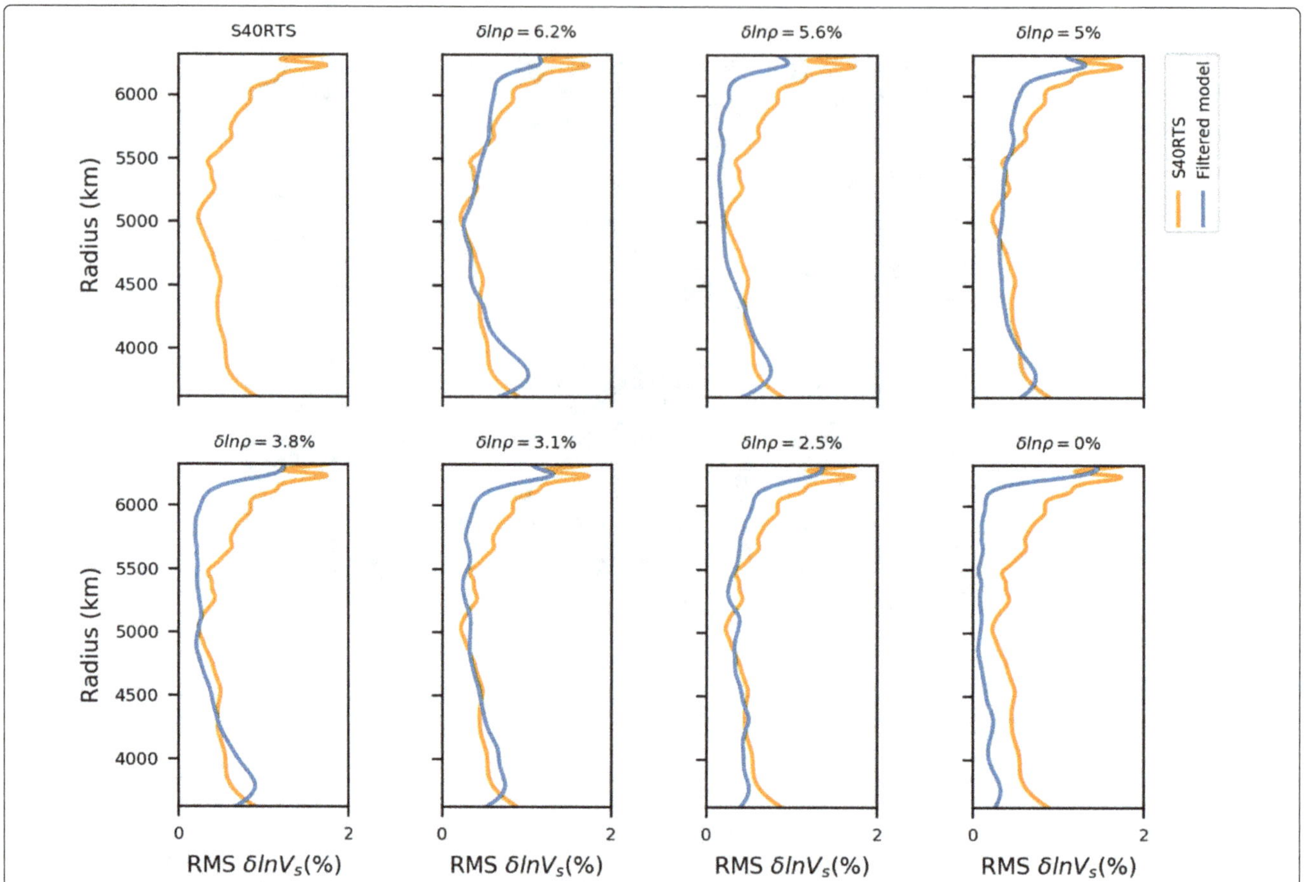

Fig. 8 RMS δ ln V_S as a function of radius for S40RTS slices (orange) and dynamic models (blue) after tomographic filtering beneath the Pacific plate. RMS δ ln V_S is poorly matched for all models in the upper mantle. When $\delta\rho \geq 3.1\%$, the amplitude of RMS δ ln V_S in the lower mantle can be matched except for the small inflection at the CMB. Overall, RMS δ ln V_S is smoother for the filtered models than for S40RTS

that becomes diluted due to mixing with the ambient mantle.

Indeed, without dilution, accumulations of oceanic crust are predicted to yield positive V_S anomalies (Deschamps et al. 2012). Deschamps et al. (2012) found that the sensitivity of V_S to oceanic crust in the lowermost mantle is mostly positive and that in order to induce a δ ln V_S of – 2.0%, typical of LLSVPs, oceanic crust would have to be 1600 K above ambient mantle temperature. They concluded that this is unrealistically high and that LLSVPs must therefore not be piles of pure oceanic crust. In contrast, we find that piles consisting of a mixture of oceanic crust, peridotite, and harzburgite require an excess temperature of only 400–700 K to explain the highest (negative) δ ln V_S amplitudes of LLSVPs.

Although we find that thermal heterogeneity plays a dominant role in LLSVP formation, our study is also in contrast with previous work suggesting that the LLSVPs could be purely thermal (Schubert et al. 2004; Schuberth et al. 2009; Bull et al. 2009; Davies et al. 2012, 2015; Koelemeijer et al. 2018) as we find that some dense compositional component is required. This discrepancy may be explained by the organizing effect of plate-motion history. In previous studies where past plate motions are imposed at the model surface, the formation of long-wavelength and strong amplitude velocity structures is attributed to the organizing force of Earth's particular tectonic history and form whether compositional variations are present or not. The influence of past plate motions cannot be explored in our 2D geometry.

We cannot exclude the importance of primordial heterogeneity that may still be sequestered near the base of the mantle (e.g., Tackley 1998; and Labrosse et al. 2007). In fact, geochemical arguments based on mantle noble gas chemistry (e.g., Allègre et al. 1996; and Albarède 1998) and studies of short-lived isotope systems (e.g., Boyet and Carlson 2005; and Rizo et al. 2016) strongly suggest that primordial heterogeneity has been preserved in the Earth's mantle. From these studies, it is not clear however what the required volume for preservation is. We share the view, suggested in previous studies (e.g., Tackley 2012; Li et al. 2014; and Ballmer et al. 2016), that there likely is a combination of both primordial chemical heterogeneity and recycled oceanic crust that contributes

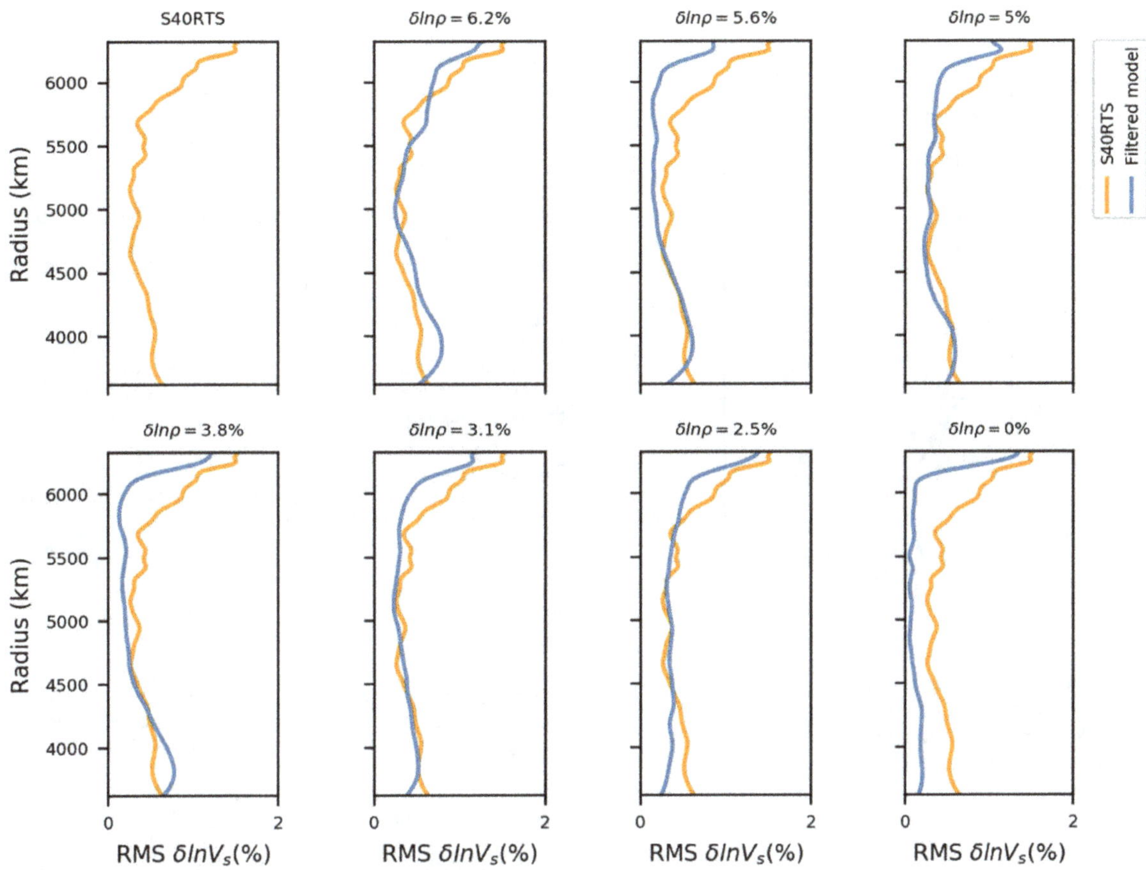

Fig. 9 RMS δ ln V_S as a function of radius for S40RTS slices (orange) and dynamic models (blue) after tomographic filtering beneath the African plate

to the observed structure of the LLSVPs. Teasing out their relative contributions will require further interdisciplinary work combining geodynamics and geochemistry (as in CH94; van Keken and Ballentine 1998, 1999); Xie and Tackley 2004; Jackson et al. 2018) with further seismological analyses as, for example, in this paper or Haugland et al. (2018) .

Clear improvements to the models that can be made include (i) compressible convection, (ii) better representation of the spherical geometry, and (iii) more flexible implementation of plate tectonics. As for (i), the use of full compressible mantle convection (e.g., Nakagawa et al. 2010; and Bossmann and van Keken 2013) will allow for the self-consistent implementation of phase changes and realistic equation of state. This will allow for an improved and more consistent comparison with the seismological observations and models. As for (ii), while we expect eventually to be able to perform full 3D simulations such as in Nakagawa et al. (2010) and Barry et al. (2017), the required high resolution and time evolution over the age of the Earth makes the use of 2D geometries still highly attractive. In BB08, a cylindrical "shrunken core" approach was used to better represent the heat loss properties of the Earth. The "spherical annulus" geometry of Hernlund and Tackley (2008) provides a more natural approach that, while 2D, represents the additional curvature by mapping of the equations onto a spherical cross section that retains the appropriate scaling between CMB radius and Earth's radius. As for (iii), we use a formulation for introducing tectonic plate into the models that is based on defining the plate geometry and computing the velocity of each plates by integrating the stress that the underlying convection exerts on them. This approach, first introduced by Gable et al. (1991), has the advantage over kinematically prescribing plates (e.g., Christensen and Hofmann 1994; and Davies et al. 2012) that the formulation is energy conservative and that one can reach Earth-like convective vigor without significant computational issues. An interesting alternative is the use of the "yield-stress" rheology (e.g., Tackley 2000) which has been demonstrated to lead to plate-like behavior without the requirement of prescribing the geometry of the plates. While it is still a question whether plate-like velocities can be modeled with this approach, recent advances using continental rafts suggest

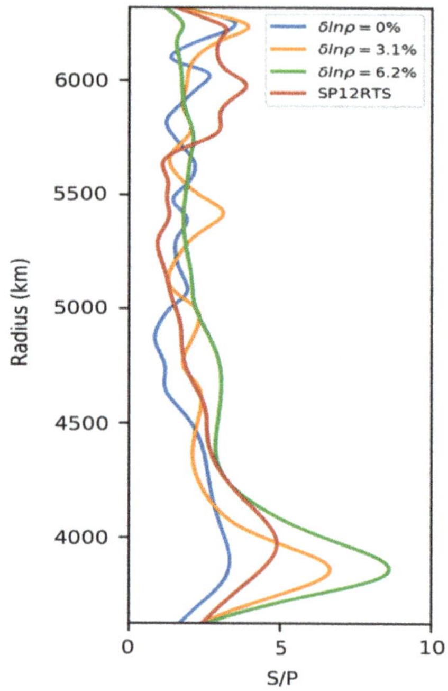

Fig. 10 S/P plotted against radius, where S/P is the ratio RMS $\delta \ln V_S$/RMS $\delta \ln V_P$. Model slices are taken along the same longitude as Pacific mantle segments in previous plots except they are centered at 15° latitude where P-wave coverage is higher

Earth-like convective vigor may be in reach (e.g., Arnould et al. 2018; Coltice et al. 2013; and Rolf et al. 2018).

Conclusions

In summary, we have shown that oceanic crust formation and recycling as modeled by BB08 can satisfactorily explain the LLSVPs in both the size and amplitude of the shear-wave anomalies as long as oceanic crust is denser than ambient mantle by 3% or more. The LLSVPs formed in our models are not composed solely of oceanic crust. Rather, they are basalt rich at their base (bottom 100–200 km) and grade into peridotite toward their sides and top. The strength of their seismic signature arises from warming over billions of years. These models appear to be well suited as base models for further exploration of the interplay between geodynamical and geochemical processes that lead to the long-term chemical and thermal evolution of the Earth. These models have been shown to match the present-day surface velocities and heat flow characteristics of the Earth, match the required outgassing efficiency of the crust and mantle as evidenced from ^{40}Ar, and match to a reasonable extent the HIMU-DMM-EM1 characteristics observed in OIBs and MORBs (BB08). This latter point has been further demonstrated by including the Lu-Hf isotope system (Jones et al. 2019).

Abbreviations

CMB: Core-mantle boundary; DMM: Depleted MORB mantle; EED: Eclogite excess density as defined by Rc/Ra (as in Brandenburg et al. 2008); EMI: Enriched mantle type 1; HIMU: High original μ where μ=U/Pb; MORB: Mid-oceanic ridge basalt; OIB: Oceanic island basalt

Acknowledgements

We thank two anonymous reviewers for their constructive comments and questions, which led to better versions of the ideas presented in this paper. PvK and JR gratefully acknowledge partial travel support from the Japanese Geosciences Union to the 2014 SEDI meeting in Kanagawa, Japan.

Authors' contributions

TJ and RM performed the modeling, analysis, and interpretation of the results. PvK and JR designed the study and contributed to the dynamical models. JR and PK provided the seismic filtering software. The authors read and approved the final manuscript.

Funding

This work was partly funded through the National Science Foundation grants EAR CSEDI 1664642 and EAR CSEDI 0855487.

Competing interests

The authors declare that they have no competing interests.

Author details

[1]Department of Terrestrial Magnetism, Carnegie Institution for Science, 5241 Broad Branch Road NW, Washington, DC, 20015 USA. [2]Department of Geology, University of Maryland, 8000 Regents Drive, College Park, MD, 20742, USA. [3]Department of Earth and Planetary Sciences, University of New Mexico, 221 Yale Blvd NE, Albuquerque, NM, 87131, USA. [4]Department of Earth and Environmental Sciences, University of Michigan, 1100 North University Avenue, Ann Arbor, MI, 40819, USA. [5]Department of Earth and Sciences, Royal Holloway University of London, Egham Hill, Egham, Surrey, TW20 0EX, UK.

References

Akbarashrafi F, Al-Attar D, Deuss A, Trampert J, Valentine A (2018) Exact free oscillation spectra, splitting functions and the resolvability of Earth's density structure. Geophys J Int 213(1):58–76

Albarède F (1998) Time-dependent models of U-Th-He and K-Ar evolution and the layering of mantle convection. Chem Geol 145:413–429

Allègre C. J, Hofmann A, O'Nions K (1996) The argon constraints on mantle structure. Geoph Res Lett 23:3555–3557

Aoki I, Takahashi E (2004) Density of MORB eclogite in the upper mantle. Phys Earth Planet Inter 143:129–143

Arnould M, Coltice N, Flament N, Seigneur V, Müller RD (2018) On the scales of dynamic topography in whole-mantle convection models. Geochem Geophys Geosyst 19:3140–3163

Baker MB, Beckett JR (1999) The origin of abyssal peridotites: a reinterpretation of constraints based on primary bulk compositions. Earth Planet Sci Lett 171(1):49–61

Ballmer M. D, Schumacher L, Lekic V, Thomas C, Ito G (2016) Compositional layering within the large low shear-wave velocity provinces in the lower mantle. Geochem Geophys Geosyst 17(12):5056–5077

Barry TL, Davies JH, Wolstencroft M, Millar IL, Zhao Z, Jian P, Safonova I, Price M (2017) Whole-mantle convection with tectonic plates preserves long-term global patterns of upper mantle geochemistry. Sci Rep 7:1870

Bossmann AB, van Keken PE (2013) Dynamics of plumes in a compressible mantle with phase changes: implications for phase boundary topography. Phys Earth Planet Inter 224:21–31

Boyet M, Carlson RW (2005) ^{142}Nd evidence for early (>4.53 Ga) global differentiation of the silicate Earth. Science 309:576–581

Brandenburg JP, van Keken PE (2007) Deep storage of oceanic crust in a vigorously convecting mantle. J Geophys Res. 112: B06403. https://doi.org/10.1029/2006jb004813

Brandenburg JP, Hauri EH, van Keken PE, Ballentine CJ (2008) A multiple-system study of the geochemical evolution of the mantle with force-balanced plates and thermochemical effects. Earth Planet Sci Lett 276:1–13

Bull AL, McNamara AK, Ritsema J (2009) Synthetic tomography of plume clusters and thermochemical piles. Earth Planet Sci Lett 278(3-4):152–162

Burke K, Steinberger B, Torsvik TH, Smethurst MA (2008) Plume generation zones at the margins of large low shear velocity provinces on the core-mantle boundary. Earth Planet Sci Lett 265:49–60

Campbell IH (2002) Implications of Nb/U, Th/U and Sm/Nd in plume magmas for the relationship between continental and oceanic crust formation and the development of the depleted mantle. Geochim Cosmochim Acta 66(9):1651–1661

Christensen UR, Hofmann AW (1994) Segregation of subducted oceanic crust in the mantle. J Geophys Res 99:19867–19884

Coltice N, Seton M, Rolf T, Müller RD, Tackley PJ (2013) Convergence of tectonic reconstructions and mantle convection models for significant fluctuations in seafloor spreading. Earth Planet Sci Lett 383:92–100

Connolly JAD (2005) Computation of phase equilibria by linear programming: a tool for geodynamic modeling and its application to subduction zone decarbonation. Geophys J Int 236:524–541

Cottaar S, Lekic V (2016) Morphology of seismically slow lower-mantle structures. Geophys J Int 207:1122–1136

Davies GF (2002) Stirring geochemistry in mantle convection models with stiff plates and slabs. Geochim Cosmochim Acta 66:3125–3142

Davies DR, Goes S, Davies JH, Schuberth BA, Bunge H, Ritsema J (2012) Reconciling dynamic and seismic models of Earth's lower mantle: the dominant role of thermal heterogeneity. Earth Planet Sci Lett 353:253–269

Davies DR, Goes S, Lau HCP (2015) Thermally dominated deep mantle LLSVPs: a review. In: Khan A, Deschamps F (eds). The Earth's Heterogeneous Mantle. Springer, Switzerland. pp 441–477

Deschamps F, Cobden L, Tackley PJ (2012) The primitive nature of large low shear-wave velocity provinces. Earth Planet Sci Lett 349:198–208

Dziewonski AM, Hager BH, O'Connell RJ (1977) Large-scale heterogeneities in the lower mantle. J Geophys Res 82:239–255

Gable CW, O'Connell RJ, Travis BJ (1991) Convection in three dimensions with surface plates: generation of toroidal flow. J Geophys Res 96:8391–8405

Garnero E, McNamara AK (2008) Structure and dynamics of Earth's lower mantle. Science 320:626–628

Grand S, van der Hilst RD, Widiyantoro S (1997) Global seismic tomography: a snapshot of mantle convection in the Earth. GSA Today 7:1–7

Haugland SM, Ritsema J, van Keken PE, Nissen-Meyer T (2018) Analysis of PKP scattering using mantle mixing simulations and axisymmetric 3D waveforms. Phys Earth Planet Inter 276:226–233

Hernlund JW, Houser C (2008) On the statistical distribution of seismic velocities in Earth's deep mantle. Earth and Planet Sci Let 265(3-4):423–437

Hernlund JW, Tackley PJ (2008) Modeling mantle convection in the spherical annulus. Phys Earth Planet Inter 171:48–54

Hirose K, Fei Y, Ma Y, Mao H-K (1999) The fate of subducted basaltic crust in the Earth's lower mantle. Nature 397:53–56

Hirose K, Takafuji N, Sata N, Ohishi Y (2005) Phase transition and density of subducted MORB crust in the lower mantle. Earth Planet Sci Lett 237:239–251

Hofmann AW, White WM (1982) Mantle plumes from ancient oceanic crust. Earth Planet Sci Lett 57:421–436

Ishii M, Tromp J (1999) Normal-mode and free-air gravity constraints on lateral variations in velocity and density of Earth's mantle. Science 285(5431):1231–1236

Jackson MG, Becker TW, Konter JG (2018) Geochemistry and distribution of recycled domains in the mantle inferred from Nd and Pb isotopes in oceanic hot spots: implications for storage in the large low shear wave velocity provinces. Geochem Geophys Geosyst 19:3496–3519

Jones RE, van Keken PE, Hauri EH, Vervoort J, Ballentine CJ (2019) Origins of the terrestrial Hf-Nd mantle array: evidence from a combined geodynamical-geochemical approach. Earth Planet Sci Lett 518:26–39

Kennett BLN, Widiyantoro S, van der Hilst RD (1998) Joint seismic tomography for bulk sound and shear wave speed in the Earth's mantle. J Geophys Res 103:12469–12493

Koelemeijer P, Ritsema J, Deuss A, Van Heijst H-J (2016) SP12RTS: a degree-12 model of shear-and compressional-wave velocity for Earth's mantle. Geophys J Int 204(2):1024–1039

Koelemeijer P, Deuss A, Ritsema J (2017) Density structure of Earth's lowermost mantle from Stoneley mode splitting observations. Nature Commun. 8: 15241. https://doi.org/10.1038/ncomms15241

Koelemeijer P, Schuberth BS, Davies DR, Deuss A, Ritsema J (2018) Constraints on the presence of post-perovskite in Earth's lowermost mantle from tomographic-geodynamic model comparisons. Earth Planet Sci Lett 494:226–238

Labrosse S, Hernlund JW, Coltice N (2007) A crystallizing dense magma ocean at the base of the Earth's mantle. Nature 450:866–869

Lau HC, Mitrovica JX, Austermann J, Crawford O, Al-Attar D, Latychev K (2016) Inferences of mantle viscosity based on ice age data sets: radial structure. J Geophys 121(10):6991–7012

Lau HCP, Mitrovica JX, Davis JL, Tromp J, Yang H-Y, Al-Attar D (2017) Tidal tomography constrains Earth's deep-mantle buoyancy. Nature 551:321–326

Lee C-TA, Luffi P, Höink T, Li J, Dasgupta R, Hernlund J (2010) Upside-down differentiation and generation of a 'primordial' lower mantle. Nature 463:930–933

Li M, McNamara AK (2013) The difficulty for subducted oceanic crust to accumulate at the Earth's core-mantle boundary. J Geophys 118:1807–1816

Li M, McNamara AK, Garnero EJ (2014) Chemical complexity of hotspots caused by cycling oceanic crust through mantle reservoirs. Nat Geosci 7(5):366

Maguire R, Ritsema J, van Keken PE, Fichtner A, Goes S (2016) P- and S-wave delays caused by thermal plumes. Geophys J Intern 206:1169–1178

Masters G, Laske G, Bolton H, Dziewonski A (2000) The relative behavior of shear velocity, bulk sound speed, and compressional velocity in the mantle: implications for chemical and thermal structure. In: Karato S-I, Forte A, Liebermann R, Masters G, Stixrude L (eds). Earth's Deep Interior: Mineral Physics and Tomography from the Atomic to the Global Scale, Geophysical Monograph Series, vol 117. American Geophysical Union, Washington DC. pp 63–87

McNamara AK, Zhong S (2005) Thermo-chemical structures beneath Africa and the Pacific Ocean. Nature 437:1136–1139

McNamara AK (2019) A review of large low shear velocity provinces and ultra low velocity zones. Tectonophysics 760:199–220

Moulik P, Ekström G (2016) The relationships between large-scale variations in shear velocity, density, and compressional velocity in the Earth's mantle. J Geophys Res 121(4):2737–2771

Nakagawa T, Tackley PJ (2008) Lateral variations in CMB heat flux and deep mantle seismic velocity caused by a thermal-chemical-phase boundary layer in 3D spherical convection. Earth Planet Sci Lett 271:348–358

Nakagawa T, Tackley PJ, Deschamps F, Connolly JA (2009) Incorporating self-consistently calculated mineral physics into thermochemical mantle convection simulations in a 3-D spherical shell and its influence on seismic anomalies in Earth's mantle. Geochem Geophys Geosys. 10: Q03004. https://doi.org/10.1029/2008gc002280

Nakagawa T, Tackley PJ, Deschamps F, Connolly JAD (2010) The influence of MORB and harzburgite composition on thermo-chemical mantle convection in a 3-D spherical shell with self-consistently calculated mineral physics. Earth Planet Sci Lett 296:403–412

Ni S, Tan E, Gurnis M, Helmberger D (2002) Sharp sides to the African superplume. Science 296:1850–1852

Ricolleau A, Perrillat J-P, Fiquet G, Daniel I, Matas J, Addadd A, Menguy N, Cardon H, Mezouar M, Guignot N (2010) Phase relations and equation of state of a natural MORB: implications for the density profile of subduction oceanic crust in the Earth's lower mantle. 115: B08202. https://doi.org/10.1029/2009jb006709

Ringwood A. E, Irifune T (1988) Nature of the 650-km seismic discontinuity: implications for mantle dynamics and differentiation. Nature 331:131–136

Ringwood A. E (1990) Phase transformations and differentiation in subducted lithosphere: implications for mantle dynamics, basalt petrogenesis, and crustal evolution. J Geol 90:611–643

Ritsema J, Lay T, Garnero EJ, Benz H (1998) Seismic anisotropy in the lowermost mantle beneath the Pacific. Geophys Res Lett 25(8):1229–1232

Ritsema J, McNamara AK, Bull A (2007) Tomographic filtering of geodynamic models: implications for model interpretation and large-scale mantle structure. J Geophys Res. 112: B01303.

Ritsema J, Deuss A, van Heijst HJ, Woodhouse JH (2011) S40RTS: a degree-40 shear-velocity model for the mantle from new Rayleigh wave dispersion, teleseismic traveltime and normal-mode splitting function measurements. Geophys J Int 184:1223–1236

Rizo H, Walker RJ, Carlson RW, Horan MF, Mukhopadhyay S, Manthos V, Francis D, Jackson MG (2016) Preservation of earth-forming events in the tungsten

isotopic composition of modern flood basalts. Science 352:809–812
Robertson GS, Woodhouse JH (1995) Evidence for proportionality of P and S heterogeneity in the lower mantle. Geophys J Int 123:85–116
Rolf T, Capitanio FA, Tackley PJ (2018) Constraints on mantle viscosity structure from continental drift histories in spherical mantle convection models. Tectonophysics 746:339–351
Romanowicz B (2001) Can we resolve 3-D density heterogeneity in the lower mantle? Geophys Res Lett 28:1107–1110
Schubert G, Masters G, Olson P, Tackley P (2004) Superplumes or plume clusters? Phys Earth Planet Inter 146:147–162
Schuberth BSA, Bunge H, Steinle-Neumann G, Moder C, Oeser J (2009) Thermal versus elastic heterogeneity in high-resolution mantle circulation models with pyrolite composition: high plume excess temperatures in the lowermost mantle. Geochem Geophys Geosys. 10: Q01W01. https://doi.org/10.1029/2008gc002235
Stixrude L, Lithgow-Bertelloni C (2011) Thermodynamics of mantle minerals - II, phase equilibria. Geophys J Int 184:1180–1213
Tackley PJ (1998) Three-dimensional simulations of mantle convection with a thermo-chemical basal boundary layer: D. Core-Mantle Bound Reg Geodyn Ser 28:231–253
Tackley PJ (2000) Mantle convection and plate tectonics: toward an integrated physical and chemical theory. Science 288:2002–2007
Tackley PJ (2012) Three-dimensional simulations of mantle convection with a thermo-chemical basal layer: D"?. In: Gurnis M, Wysession ME, Knittle E, Buffett BA (eds). The Core-Mantle Boundary. AGU, Washington. pp 231–53
Tanaka S, Suetsugu D, Shiobara H, Sugioka H, Kanazawa T, Fukao Y, Barruol G, Reymond D (2009) On the vertical extent of the large low shear velocity province beneath the South Pacific Superswell. Geophys Res Lett. 36: L07305. https://doi.org/10.1029/2009gl037568
Trampert J, Deschamps F, Resovsky J, Yuen D (2004) Probabilistic tomography maps chemical heterogeneities throughout the lower mantle. Science 124(5697):942–926
942-926
Tsuchiya T (2011) Elasticity of subducted basaltic crust at the lower mantle pressures: insights on the nature of deep mantle heterogeneity. Phys Earth Planet Inter 188:142–149
van Keken PE (2001) Cylindrical scaling for dynamical cooling models of the Earth. Phys Earth Planet Inter. https://doi.org/10.1016/s0031-9201(01)00195-9
van Keken PE, Ballentine CJ (1998) Whole-mantle versus layered mantle convection and the role of a high-viscosity lower mantle in terrestrial volatile evolution. Earth Planet Sci Lett 156:19–32
van Keken PE, Ballentine CJ (1999) Dynamical models of mantle volatile evolution and the role of phase transitions and temperature-dependent rheology. J Geophys Res 104:7137–7151
Workman RK, Hart SR (2005) Major and trace element composition of the depleted MORB mantle (DMM). Earth Planet Sci Lett 231:53–72
Xie S, Tackley PJ (2004) Evolution of U-Pb and Sm-Nd systems in numerical models of mantle convection. J Geophys Res. 109: B11204.
Xu W, Lithgow-Bertelloni C, Stixrude L, Ritsema J (2008) The effect of bulk composition and temperature on mantle seismic structure. Earth Planet Sci Lett 275:70–79

12

Evidence for natural molecular hydrogen seepage associated with Carolina bays (surficial, ovoid depressions on the Atlantic Coastal Plain, Province of the USA)

Viacheslav Zgonnik[1,4*], Valérie Beaumont[1], Eric Deville[1], Nikolay Larin[2], Daniel Pillot[1] and Kathleen M. Farrell[3]

Abstract

A study of soil gases was made in North Carolina (USA) in and around morphological depressions called "Carolina bays." This type of depression is observed over the Atlantic coastal plains of the USA, but their origin remains debated. Significant concentrations of molecular hydrogen (H_2) were detected, notably around the bays. These measurements suggest that Carolina bays are the surficial expression of fluid flow pathways for hydrogen gas moving from depth to the surface. The potential mechanisms of H_2 production and transport and the geological controls on the fluid migration pathways are discussed, with reference to the hypothesis that Carolina bays are the result of local collapses caused by the alteration of rock along the deep pathways of H_2 migrating towards the surface. The present H_2 seepages are comparable to those in similar structures previously observed in the East European craton.

Keywords: Hydrogen; Seepage; Carolina bays

Background

Carolina bays are surficial, consistently oriented, oval-shaped depressions that occur widely across the southeastern Atlantic Coastal Plain, Province of eastern North America (Brooks et al. 2010). They are well defined on satellite images (Figs. 2, 3, and 4) and densely cover parts of the Coastal Plain in North Carolina (NC) and South Carolina (SC). They vary in size, ranging from ~100 m to 8 km in diameter (Lake Waccamaw, NC, USA). Slightly elevated rims (~1–3 m), commonly consisting of sand, surround these features. Although some bays have continuous elevated rims, the rims do not usually completely encircle the bays but often form a crescent. The long axes of these elliptical features are preferentially oriented NW–SE (Figs. 1, 2, 3, and 4). Bays of various sizes may overlap, and small bays are frequently present inside larger bays (Fig. 2). In areas undisturbed by anthropogenic activities, these bays can include densely vegetated wetlands or open water lakes. These features were originally called "bays" because of the bay trees that inhabit these wet depressions or pocosins. Now, the term "bay" indicates a wet oval-shaped depression. Locally, these features are also called cypress domes, Grady ponds, citronelle ponds, wet prairies, sandhill ponds, etc. (Folkerts 1997). In anthropogenically modified areas, the bays are commonly drained and cleared for agriculture or other purposes. Even when modified, most of the bays are still easily discernable in satellite and Light Detection and Ranging (LiDAR) images because of their characteristic morphology and relief and the soil bleaching on their rims. Hundreds of thousands of bays occur along the Atlantic Coastal Plain from New Jersey to Florida (Prouty 1952), and in NC, the bays cover as much as 65 % of the land surface of the Coastal Plain (Prouty 1952). Eyton and Parkhurst (1975) summarized the physical characteristics of Carolina bays.

The origin of Carolina bays is still unclear. Many theories have been proposed to explain their formation,

* Correspondence: Zgonnik@UAHE.net.ua
[1]IFP Energies Nouvelles, 1 et 4 avenue de Bois Préau, 92852 Rueil-Malmaison Cedex, France
[4]Present address: Natural Hydrogen Energy Ltd, 819-35th Ave., Greeley, CO 80634, USA
Full list of author information is available at the end of the article

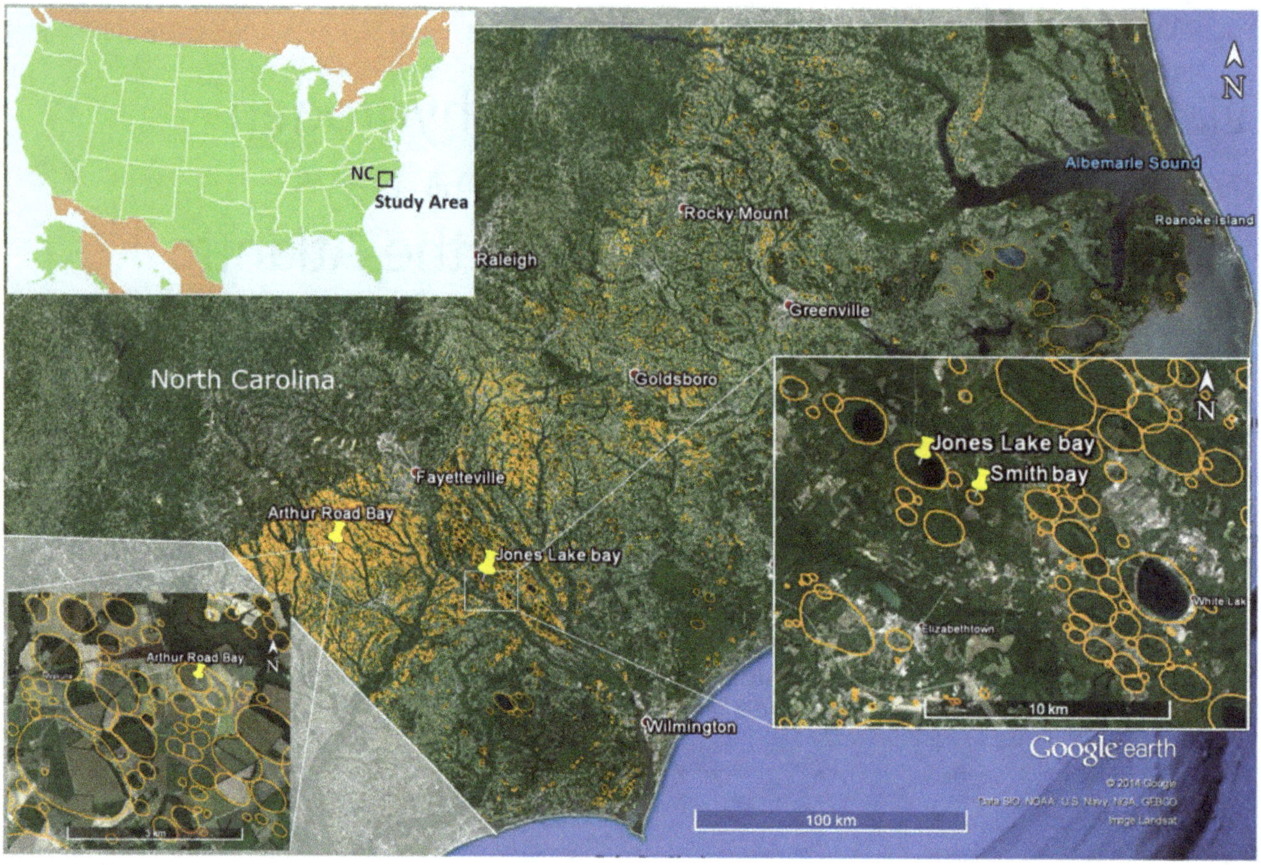

Fig. 1 Location of the study area. *NC* North Carolina. Carolina bays are outlined by *orange polygons*. This and all other satellite images in this article were downloaded from Google Earth and Google Earth Pro

including meteorite impact, comet impact, wind action, water table fluctuations, solution phenomena, and even fish nests (Ross 1987), but no general consensus has yet been reached.

In the present study, we examined the chemical compositions of the gases from the soils in selected Carolina bays and their vicinities to investigate the contents of H_2 and other reduced gases. Our aim was to compare the bays with similar geomorphological features recently described in the East European craton (EEC) in Russia, which were shown to emit H_2 gas in an area where the occurrence of H_2 at depth (several hundred meters) has been confirmed in boreholes (Larin et al. 2014). A variety of mechanisms can produce H_2, and there may be a causal link between these H_2 emissions and the formation of these structures. The possible relationships between the H_2 seepages and the observed geomorphological features were investigated.

Geological framework

The Atlantic Coastal Plain, Province of eastern North America, is the emergent part of a platform on a passive continental margin and is currently characterized by incised valleys that separate flat interfluves (Farrell et al. 2013). The continental shelf is underlain by a seaward-thickening wedge of sediment that thins westward to a feather edge along its boundary with the crystalline bedrock. This clastic wedge attains a maximum thickness of 3009 m (9854 ft) on land at Cape Hatteras (Lawrence and Hoffman 1993) and includes deposits of Cretaceous to Quaternary age (Brown et al. 1972). The sediment composition and recurring sedimentary facies on the NC Coastal Plain have been summarized by Farrell et al. (2012, 2013).

The NC Coastal Plain is predominantly a relict Plio-Pleistocene landscape, characterized by a series of progressively younger scarps, or paleoshorelines, and intervening terraces that step down in elevation and age towards the coast and towards the river incisions (see Fig. 1 in Farrell et al. 2013 and Figs. 1, 2, 3, and 4 in Abbott et al. 2011), forming a stair-step topography. Over the past 5 Ma, glacio-eustatic changes in sea level drove the transgressive–regressive cycles that sculpted this landscape. The surficial deposits that underlie the relict landscape include a complex assemblage of marine, barrier island, estuarine, fluvial, and other coastal plain deposits, which are predominantly Pliocene, Pleistocene, and Holocene in age. Abbott et al. (2011) summarized the surficial geology and

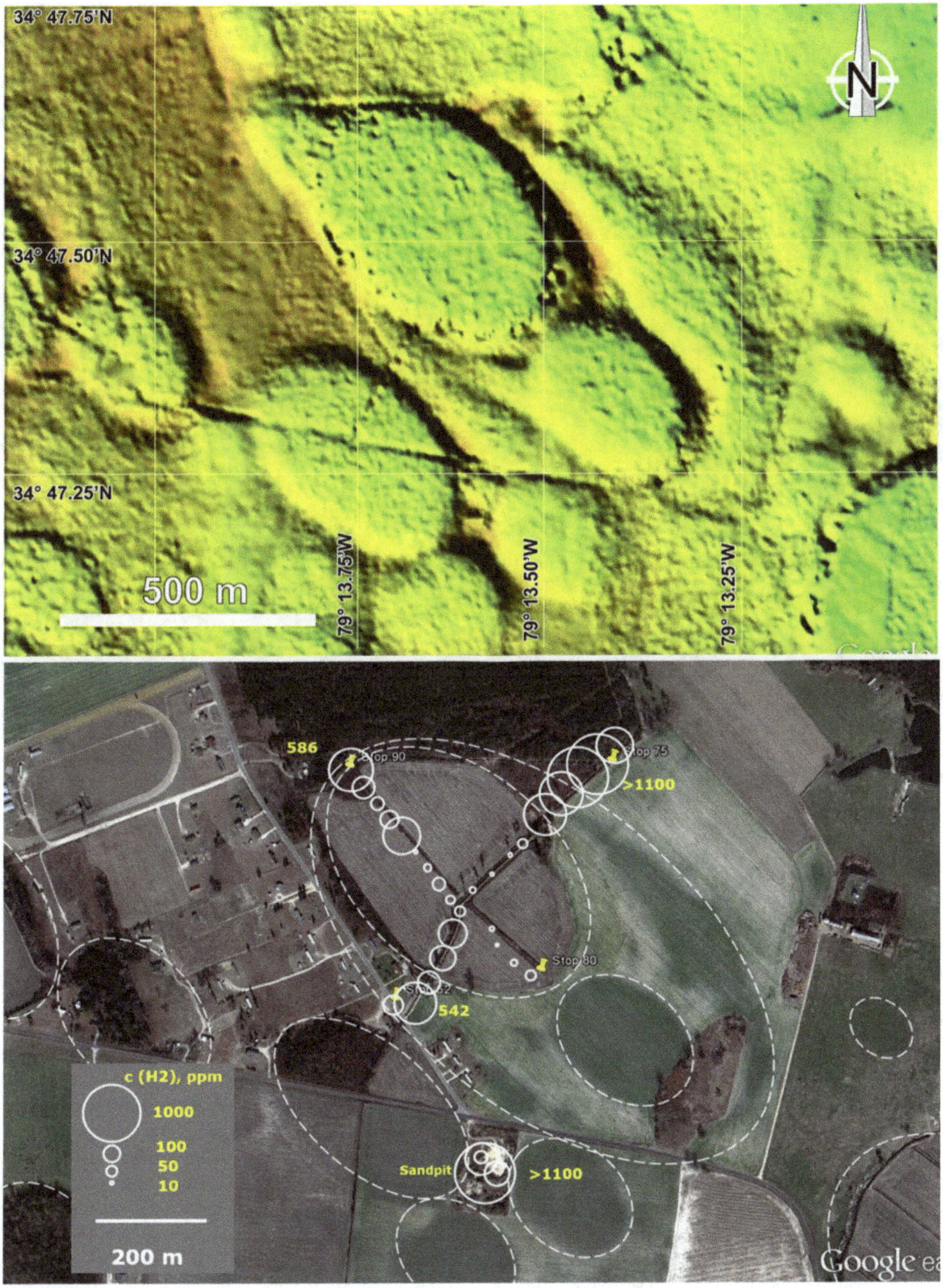

Fig. 2 Subsoil H_2 concentration measurements at Arthur Road Bay and Arthur Road Sandpit (*lower image*). *Dashed lines* outline the bays. *Upper image* is a LiDAR image showing the relief of the lower image. Profile lines follow ditches (*dark lines* in the lower photograph). Date of measurements 14 March 2012. LiDAR image is from the site cintos.org

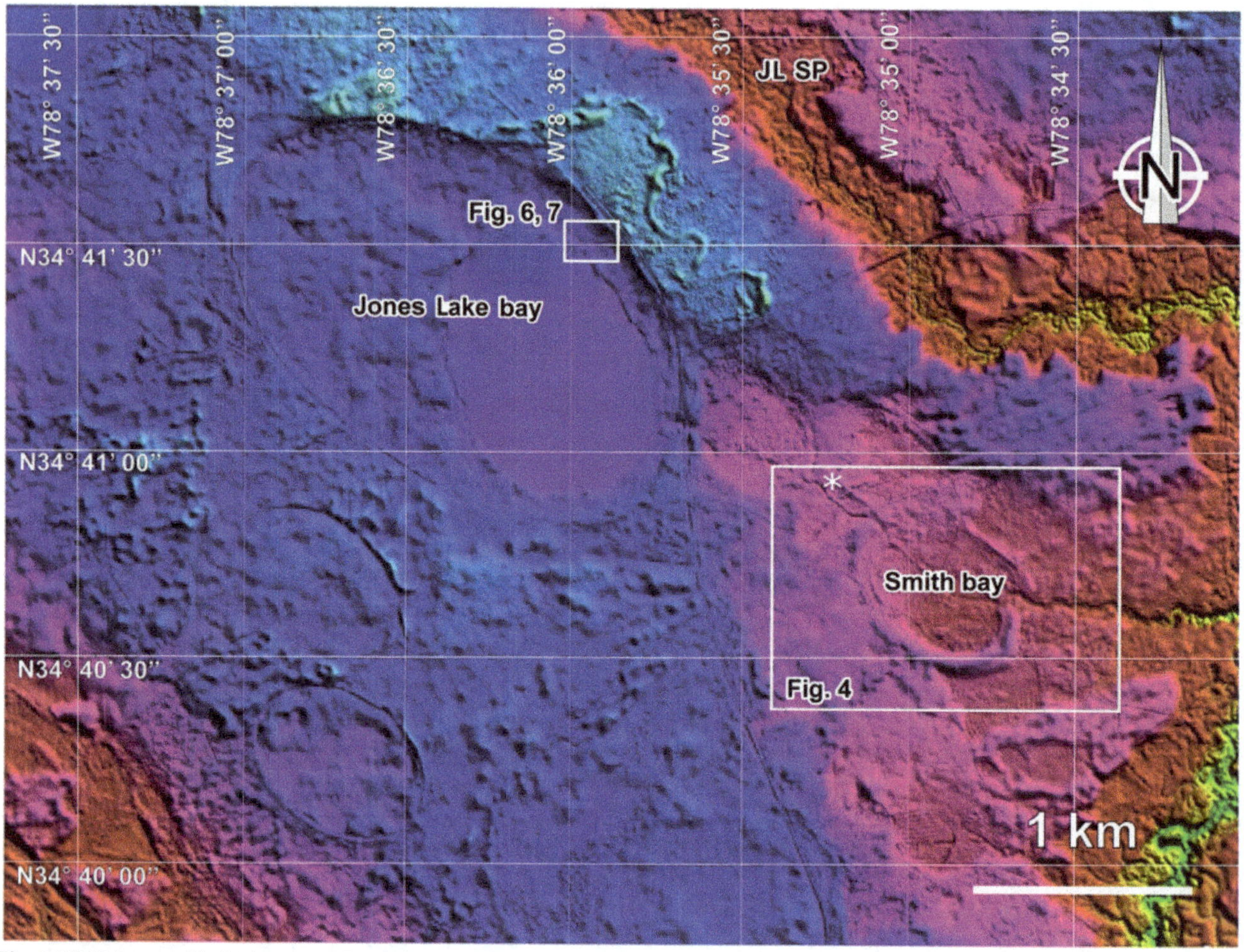

Fig. 3 LiDAR image with the locations of sites shown in Figs. 4, 6, and 7. *JLSP* Jones Lake Sandpit. The *asterisk* is the intersection zone for Smith Bay and unnamed bay. LiDAR image is from the site cintos.org

distribution of the shallow subcrops of the NC Coastal Plain. The features known as Carolina bays postdate the formation of the terraced topography.

Carolina bays are dynamically active features. According to Cohen et al. (1999), they originated during the Pleistocene. A relatively new technique for dating quartz grains, optically stimulated luminescence (OSL), has helped to refine the time frame of bay formation. Ivester et al. (2004) used this dating method to establish the multiple phases of bay evolution during the past 100,000 years. Brooks et al. (2010) concluded that most of the bays formed during the late Pleistocene, and Frey (1954) and Thom (1970) provided evidence for recent enlargement of existing bays. Moore et al. (2014) demonstrated geomorphologically and stratigraphically that these bays migrate through their own lacustrine deposits and calculated the migration rate from a series of OSL dates. In this study, we document an actively forming bay.

Methods

Study sites

The study sites (Figs. 1 and 3 and Table 1) are three prominent Carolina bays (Arthur Road Bay (Fig. 2), Smith Bay (Figs. 3 and 4), and Jones Lake Bay (Fig. 3)), and four smaller-scale detailed study sites associated with them. The detailed study sites include two sandpits (Arthur Road Sandpit and Jones Lake Sandpit), a small, recently formed bay-like feature associated with Jones Lake and a bay intersection area (between Smith Bay and an unnamed bay). All the study sites are on public land in Bladen County, except Arthur Road Bay and the sandpit located in Robeson County, which are privately owned. These public lands include Bladen Lakes State Forest, Turnbull Creek Educational State Forest, and Jones Lake State Park. The bays are naturally well preserved in these areas.

The Arthur Road sites are situated on surficial units that are Pliocene in age. The Smith Bay and Jones Bay

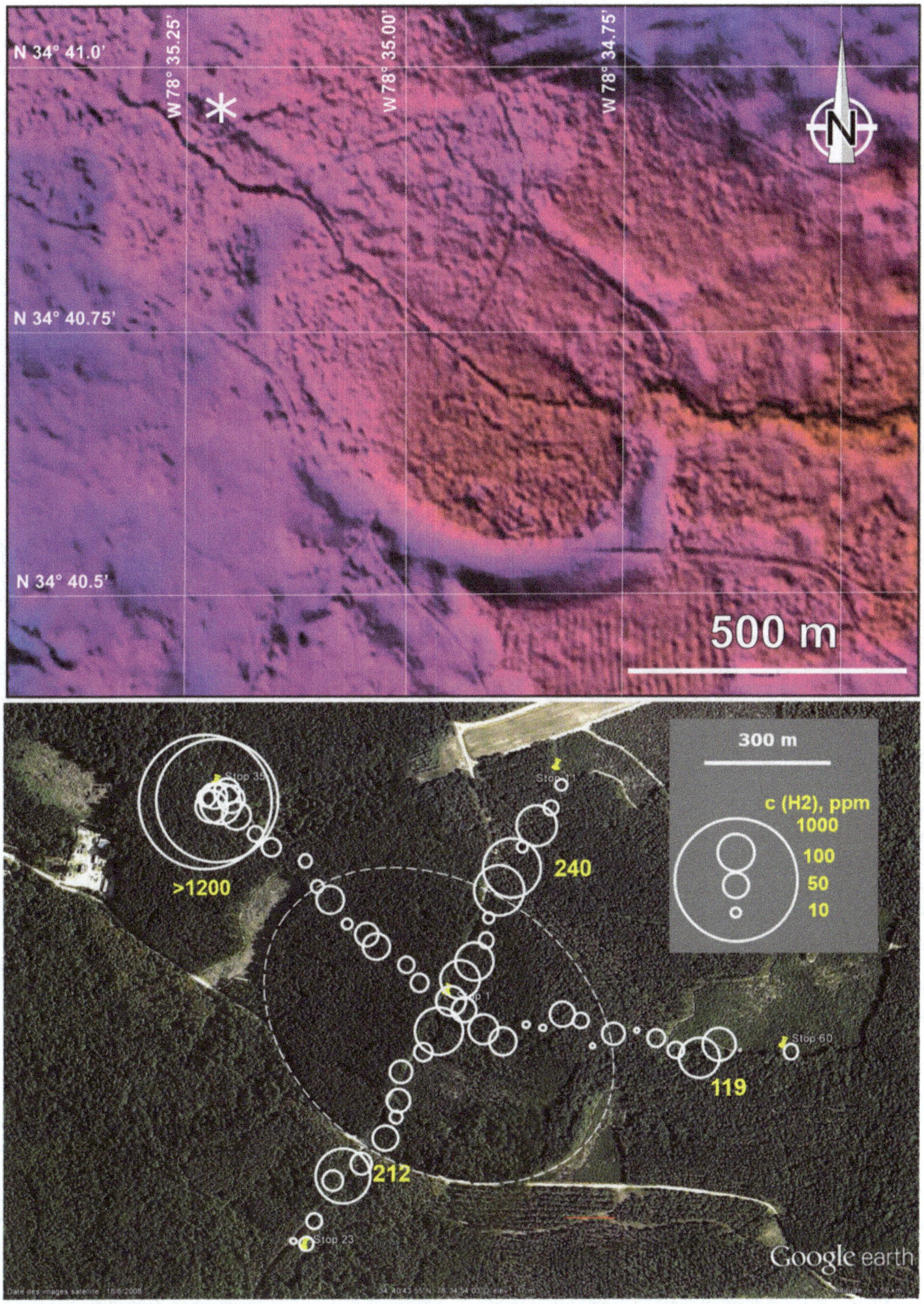

Fig. 4 An example of the subsoil hydrogen concentration measurements made in Smith Bay (*lower image*). *Upper image* is the corresponding LiDAR image. The *asterisk* is the intersection zone for Smith Bay and unnamed bay. *Dashed lines* outline Smith Bay. Date of measurements 09 March 2012. LiDAR image is from the site cintos.org

Table 1 Gas chromatography (GC) measurements of soil gas samples

Work site	Sample index number	°dec N	°dec W	H_2 measured in the field (ppm)	GC measurements						
					H_2 (ppm)	O_2 + Ar (%)	N_2 (%)	CH_4 (ppm)	CO_2 (%)	C_2H_6 (ppm)	C_3H_8 (ppm)
Arthur Road Bay		34.7939167	−79.2296667	586	275	20.05	79.30	11	0.62	1	2
Arthur Road Sandpit		34.7869444	−79.2266667	SAT	605	19.37	79.28	735	1.21	1	1
Arthur Road Sandpit, bubbles		34.7870556	−79.2269167	–	0	2.84	35.04	53.57 %	8.56	0	0
Smith Bay	1	34.6824722	−78.5870694	659	179	20.23	79.26	15	0.50	2	2
	2			715	146	20.32	79.15	11	0.51	0	0
	3			574	296	20.38	79.40	17	0.20	2	2
Small new structure inside Jones Lake Bay	1	34.6930278	−78.6004722	210	107	20.12	79.61	194	0.23	0	0
	2	34.6930556	−78.6008056	391	167	16.51	79.35	27468	1.38	0	0
	3	34.6928131	−78.6003308	477	202	20.00	78.11	5875	1.28	0	0
	4	34.6928232	−78.6003460	815	463	18.67	78.66	13783	1.24	0	0
Jones Lake Sandpit	1	34.7001111	−78.5867222	3700*	698	14.54	84.02	244	1.34	0	0
	2	34.7001111	−78.5872500	719	245	20.34	79.30	15	0.33	0	0
	3	34.7001111	−78.5867222	SAT	1043	14.78	82.12	392	2.96	0	0

Data for all field H_2 measurements were obtained with a GA2000 Plus detector, except in fields marked with *, which were measured with the MDS detector. Total gas measured for each sample is equal to 100 %. Discrepancies between field measurements and laboratory measurements are attributable to the sampling method (see explanation in "Field methods" subsection). Field measurements were usually about twice the laboratory measurements, except where sampling failed. For all field measurements, see Additional file 1: Table SI-1
SAT saturation (overload) of the detector, *°dec* coordinates in decimal degrees

localities occur on Middle to Early Pleistocene deposits (see Abbott et al. 2011 for their Figs. 4 and 5). The surficial units at all the sites overlie Cretaceous formations (see Fig. 6 in Abbott et al. 2011).

Site selection

We chose the study sites using satellite images, by reviewing areas that included a high density of bays with a variety of dimensions, that were easy to access and represented various ages deduced from the geomorphology. We identified the bays using satellite images from Google Earth and Google Earth Pro. LiDAR images were downloaded from the site cintos.org. We measured the H_2 concentrations along profiles that transected the entire bays, starting outside the bay rims at a distance equivalent to half the diameters of the bays. This allowed us to determine the point on the profile at which H_2 started to increase around the bays.

Field methods

Field measurements of the soil gas concentrations were made with a method modified from previously described techniques (Battani et al. 2010; Rogozhin et al. 2010; Garcia et al. 2012; Sukhanova et al. 2013; Larin et al. 2014). Small holes were drilled with a portable perforator using a 120 cm long drilling bit with a 12 mm diameter. A thin steel perforated tube was then inserted into the hole and the upper part was plugged with Plasticine. We adapted the length of the tube (10, 30, 100, or 120 cm) to the depth of the water table. The soil gas samples were collected in Vacutainer® tubes. Before sampling, a vacuum (10^{-4} mbar) was established in each tube using a turbomolecular pump. Experiments with a 2 m long probe were conducted with a Gas Vapor Probe Kit from AMS Inc., American Falls, ID, USA. This kit has a retractable tip, which allows the soil gas to be sampled at a specific depth.

Two hydrogen gas analyzers were used to measure H_2 in the field. The first was a palladium-based metal-insulator semiconductor detector provided by the Moscow Engineering Physics Institute, Russia, which includes three sensors with different ranges of sensitivity to H_2 (0.5–50, 50–1000, and 1000–16,000 ppm). This detector is selective for H_2 and is not affected by H_2S, CO, or any other gas. It has demonstrated excellent efficiency in several studies in which H_2 was measured or monitored (Firstov and Shirokov 2005; Rogozhin et al. 2010; Sukhanova et al. 2013). The second gas analyzer was the landfill GA2000 Plus multi-gas detector by Geotech® Leamington Spa, UK, equipped with a H_2-sensitive electrochemical cell with a sensitivity for H_2 ranging from 1 to 1100 ppm. Water filters (Pall Acro® 50, 0.45 μm) were used to prevent humidity penetrating the detectors. Detector accuracy was first checked in the laboratory using gas mixtures with H_2 concentrations of 100 and 1000 ppm in air-like mixtures (80 % N_2, 20 %

O_2). The relative accuracy of the detectors was "within ±10 %." The results of all field tests are shown in Additional file 1: Table SI-1.

Laboratory analyses

Laboratory analyses were performed to determine the molecular compositions of gas samples in Vacutainer® tubes using gas chromatography (GC), and the results were compared with the measurements made in the field. The chromatograph used was a VARIAN 3800 by Agilent Technologies, Santa Clara, CA, USA, which is equipped with several columns in series and three detectors: two thermal conductivity detectors and one flame ionization detector. The analytical results are given with a precision of ±0.1 %. Each analysis was bracketed with blanks.

The compositions of the collected gas samples showed that air made an important contribution, which is very normal for soil gas. The values determined with the GC analysis were always lower than the field measurements, because of the sampling method used. The samples were measured after the maximum concentration of H_2 was reached on the field detector, when its value started to decrease. Therefore, the samples were always diluted because the soil gas mixes with air present in the H_2 detector system.

Results

Arthur Road Bay

The Arthur Road Bay is an oval-shaped depression (580 m long and 360 m wide) with a sand rim (Fig. 2). It has been modified by anthropogenic activity (draining and clearing). Two ditches drain water from this bay. Data were collected along two profiles that transected the bay perpendicularly (Fig. 2). Gas measurements were made every 50 m along both profiles (Fig. 2). The maximum detected H_2 concentration was >0.11 % (detector saturation), with an average of 233 ppm across 23 measurements (Additional file 1: Table SI-1). Traces of C_2+ hydrocarbons were detected in the laboratory analysis of one soil gas sample (Arthur Road Bay, Table 1).

Arthur Road sandpit

Additional measurements were collected in a sandpit located 500 m south of the Bay's center (Fig. 2). The pit was surrounded to its north, east, and west by at least three other bays (Fig. 2). The maximum H_2 concentration detected at the pit was also >0.11 % (detector saturation), with an average of 313 ppm across six measurements. The laboratory analysis of one soil gas sample provided evidence of C_2+ hydrocarbons (Table 1). Some parts of the sandpit included stagnant water pools that were degassing natural bubbles of gas. A GC analysis revealed a high concentration of methane (54 %) in the bubbles (Table 1), which could suggest high biological activity of methanogens using the H_2.

Smith Bay

Smith Bay is an oval-shaped depression with a sand rim (Figs. 3 and 4). This depression is 1–3 m deep, 720 m long, and 545 m wide. The bay formed on a probable middle Pleistocene landscape. Inside its rim, it is mantled with 1–2 m of Holocene peat. The vegetation patterns, swamp distribution, and the topography sharply delimit the bay's outline and its rim. Water drains from the bay in two ditches that cross near the center of the bay. Data were collected along two profiles that followed the ditches (Fig. 4). A maximum concentration of H_2 >1200 ppm (saturation of the GA2000 detector) was detected outside the northwestern border of the bay (see Fig. 4), near the intersection of Smith Bay and a smaller unnamed bay to the northwest. Three other sites with H_2 peaks >1200 ppm were situated where the transects coincide with the sand rims bordering the bays (Fig. 4). Elevated concentrations of H_2 were also detected in the center of the bay. The H_2 concentration decreased to near-zero values outside the sand rim.

Intersection zone for Smith Bay and unnamed bay

The maximum H_2 concentration (>0.12 %) occurred at a site that coincided with the overlap or intersection of the rims of the two bay (Smith Bay and an unnamed bay situated to the northwest) (Fig. 4). Here, the surface was mantled with peat (70–80 cm thick)-overlying sand. In a small area of the bay overlap where high H_2 concentrations were detected, ten measurements were made within a circumference of 5 m. Only one other point inside this circumference displayed a similarly high H_2 concentration (Additional file 1: Table SI-1). The subsoil H_2 concentrations here were highly variable, probably because the porosity and macropore distribution of the soil were heterogeneous.

We performed an experiment with a 2 m long probe near the recorded maximum H_2 concentration to measure the gradient of the H_2 concentration. Measurements were made at several depths (every 50 cm) using a system of retractable tips (Gas Vapor Probe Kit). An increase in the H_2 concentration with depth (Fig. 5) was observed at the transition from peat to sand (within a 50–100 cm interval), and the maximum concentration was in the coarse sand. As well as H_2 and CH_4 anomalies, C2+ hydrocarbons were also detected in samples from this location (Smith Bay 1–3, Table 1) with laboratory measurements.

Jones Lake Bay

Jones Lake Bay is a very large bay (2400 × 1500 m), with an almost perfect inner elliptical lake (Jones Lake) in the

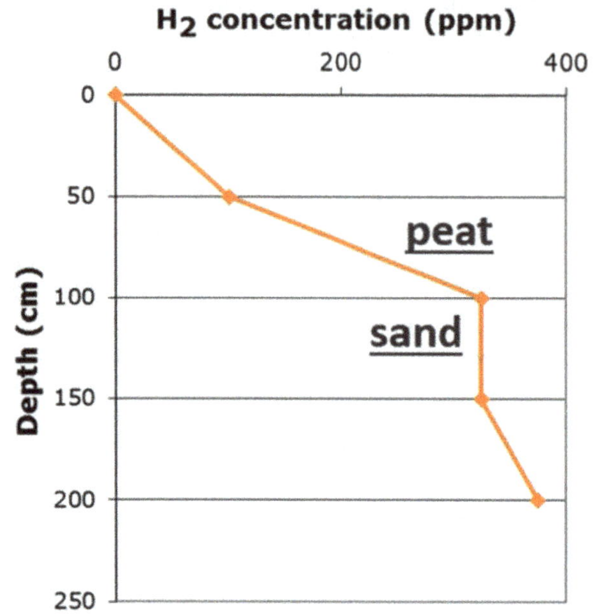

Fig. 5 Increasing hydrogen concentrations with depth in the intersection zone of Smith Bay and unnamed bay. Its location is shown in Fig. 4. Date of measurements 10 March 2012

southeastern part (Fig. 3). Like the other bays studied, the limits of the bay are clearly detectable from the vegetation. The bay is covered with peat and is densely vegetated on the inside, whereas outside it, the soils are sandy and have rare undergrowth. The bay rim consists of soils with coarse sand derived from the underlying deposits of probable Early Pleistocene age. H_2 was only detected within the confines of the bay border. The maximum concentration of H_2 was 815 ppm, detected in coarse sand. We detected no traces of H_2 outside the bay limits (Additional file 1: Figure SI-1 and Additional file 1: Table SI-1).

Jones Lake sandpit

We also measured the H_2 concentrations in a sandpit near the Jones Lake Bay (Fig. 3 and Additional file 1: Figure SI-1, upper right part). We detected very high concentrations of H_2 at the bottom of the peat, 3–4 m below the ground surface, in clayey sand. The maximum concentration was 0.37 % H_2, measured with the Gas Vapor Probe Kit. The average H_2 concentration was 0.1 % across seven measurements (Additional file 1: Table SI-1).

Small new structure inside Jones Lake Bay

Detailed studies were conducted in a small area outside the lake in Jones Lake Bay (Fig. 3, insets in Figs. 6 and 7), where high H_2 levels (up to 815 ppm) were detected in the soil. This site is a relatively small near-circular wet area (diameter ~60 m) where the tall trees suddenly withered naturally in 2008–2009, which was confirmed by the park rangers who manage the reserve. Satellite images taken in 2008 were compared with images taken in 2009 (Fig. 6). The trees probably withered (see panoramic picture, Additional file 1: Figure SI-2) in response to the increase in soil water saturation (water table depth of only 10–15 cm) or from the negative impact of another unknown process affecting the vegetation. Despite the very high water table, an H_2 concentration of 210 ppm was detected in the upper few centimeters (<10 cm) of the soil in the center of the newly forming bay (Fig. 7). The depression of the newly forming bay was less than 1 m, according to field measurements made with an altimeter.

Two detailed cross-sections of the new bay, showing the H_2 concentrations and distances, with measurements made every meter, are shown in Fig. 7. In both sections, the highest H_2 concentrations were detected on the new bay's borders. Laboratory measurements of the chemical composition of the gas revealed that relatively high concentrations of methane were associated with H_2 (Table 1). No heavier hydrocarbons were detected. This wet area with H_2 anomalies is very similar to a recently formed round feature studied by our team in Russia, near the city of Electrostal, close to Moscow (Larin et al. 2014).

Estimation of H_2 fluxes

In this study, the maximum H_2 contents occurred in coarse sand along the rims of the bays and were clearly not associated with wetlands. Because H_2 is a highly diffusive gas, it cannot remain for long in a porous medium-like coarse sand. We used a 2 m long probe (Fig. 5) to demonstrate the gradient in the H_2 concentration, and we interpret this H_2 gradient as the result of the diffusive flow of H_2.

We calculated the H_2 flow using the classical Fick's law of diffusion:

$$F = \Phi \; D \; \text{grad} \; [C] \tag{1}$$

where F is the gas flow, ϕ is the exchange area or the effective porosity of a homogeneous porous medium, D is the diffusion coefficient of the gas in air, $[C]$ is the gas concentration, and grad is the gradient.

The diffusion coefficient (D) for H_2 in air was taken to be 7.7×10^{-5} m^2/s (Cussler 2009).

The gradients were estimated to be twice the average concentration measured in the perforation, depending on the level of the groundwater. Effective porosity was estimated as follows: for wet swamp, 10–20 %; for dried swamp, 10–25 %; for sand, 35–45 %; and for soil, 25–35 % (Hough 1969; Swiss standard 1999; Das 2008).

To establish the zonation in the H_2 concentrations observed in specific bays, the average H_2 concentrations were estimated for each defined zone in each bay. Four concentric oval-shaped rings were defined for each bay,

Fig. 6 Appearance of new structure inside the Jones Lake Bay. *Upper panel*: satellite image of part of Jones Lake Bay in 2008. *Lower panel*: satellite image of the same part of Jones Lake Bay 1 year later. Its location is shown in Fig. 3. Note the appearance of a *circle* about 60 m in diameter where the forest is damaged

corresponding to the center of the bay, the internal shore, the external shore, and the outside zone. For each zone, the average H_2 concentration was estimated specifically from field measurements. The daily H_2 flow was calculated for each zone, and the sum of the flows calculated for each zone provided the total flow within the bay.

In the Smith Bay, the estimated daily H_2 flow was 750–1000 m^3/day over an area of 1.14 km^2 (660–880 m^3/day/km^2). In the Arthur Road Bay, the estimated daily H_2 flow was 1000–1370 m^3/day over an area of 0.48 km^2 (2240–3060 m^3/day/km^2). For Jones Lake Bay, the daily H_2 flow was estimated to be 1120–2740 m^3/day for a surface area of 6.25 km^2 (180–440 m^3/day/km^2). For the small structure inside the large Jones Lake Bay, the daily H_2 flow was estimated to be 21–31 m^3/day over a surface area of 0.007 km^2 (3000–4400 m^3/day/km^2).

Discussion

The occurrence of molecular hydrogen in the soils of Carolina bays was discovered after noting the remarkable geomorphological similarity to depressions that emit H_2 in the EEC in Russia (Sukhanova et al. 2013; Larin et al. 2014). In both cases, relatively high H_2 concentrations were found in the soil gas of the bay-like features. This study suggests that Carolina bays are geomorphological features related to the occurrence of

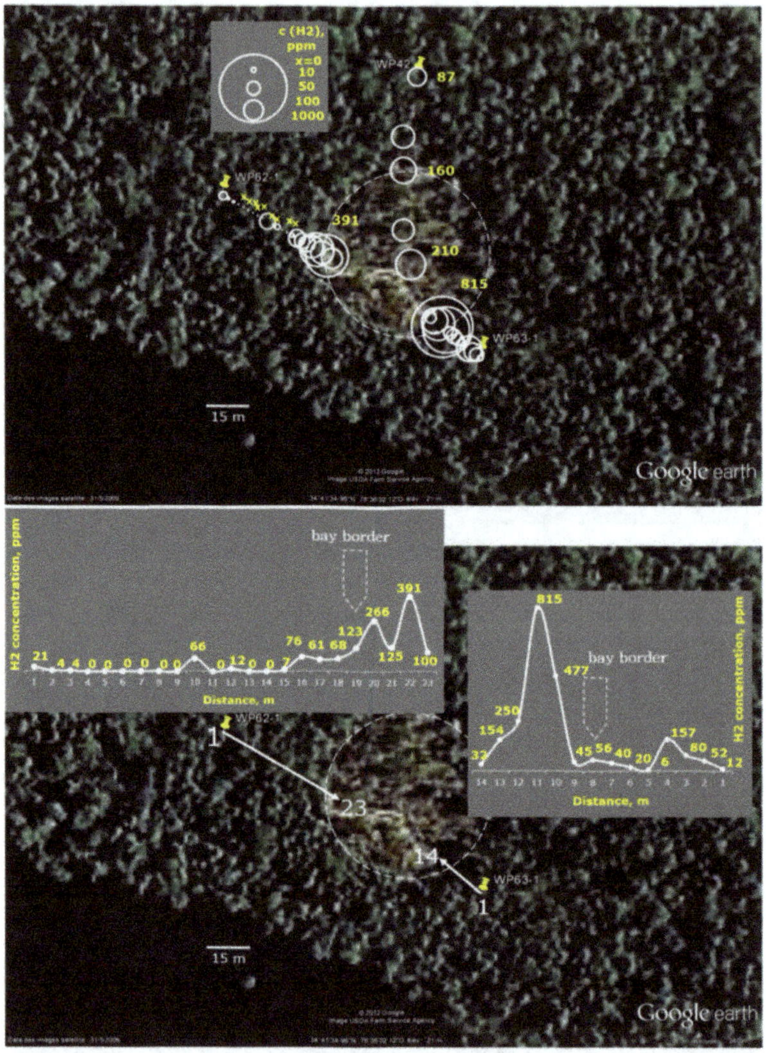

Fig. 7 Subsoil hydrogen concentrations in Jones Lake Bay. Its location is shown in Fig. 3. *Dashed line* shows the limits of the new structure. Geochemistry and geomorphology correlate perfectly, defining the borders of the structure. Date of measurements 13 March 2012

H_2 rather than the results of simple depositional processes along coastal plains (Stolt and Rabenhorst 1987). Therefore, the causal link between the origin of these morphological features and their association with molecular hydrogen must be established. This bay–hydrogen association suggests a connection between fluid seepage from depth towards the atmosphere (deep geological control) and the surficial geomorphic expression of Carolina bays. A possible interpretation might be that Carolina bays are not only focal points for groundwater recharge, as suggested by Grant et al. (1998), but also a morphological expression of fluid seepage from depth toward the atmosphere.

Origin of hydrogen

Carolina bays are natural depressions, commonly filled with water and organic deposits (peat). Consequently, we should consider whether the source of H_2 might be attributable to superficial biological activity under reducing conditions (Sugimoto and Fujita 2006). However, our study clearly shows that the highest concentrations of H_2 systematically occurred in the sand deposits on the rims of the bays, where there was no peat cover (Figs. 2, 4, and 7). We also detected H_2 at a depth of 2 m, 120 cm below the level of the peat (Fig. 5), which reduces the likelihood of a very-near-surface origin for the H_2. Furthermore, the H_2 concentrations measured in the wetlands were low compared with the measurements made in sand. The H_2 concentrations also increased with depth (Fig. 5, Additional file 1: Table SI-1), and the maximum concentration measured, 0.37 vol.%, was detected at a depth of 5 m below ground level in sandpit 2 (Additional file 1: Table SI-1), in a clay and sand substrate. Although slight background H_2 associated with biological activity cannot be excluded, these observations make a purely superficial origin for the H_2 highly improbable. It is also generally believed that the free H_2 produced by bacteria remains at low concentrations because it is rapidly consumed by methanogenic bacteria and soil enzymes (Conrad and Seiler 1981). Indeed, the biological production of H_2 occurs within bacterial consortia, where this valuable chemical energy source is constantly consumed and consequently remains at low concentrations (high H_2 concentrations inhibit H_2-producing bacteria). H_2 is considered to be the most energetic substrate, able to sustain lithoautotrophic ecosystems in subsurface environments, where it is readily consumed (Nealson et al. 2005). Therefore, such high concentrations in soil gases are more probably linked to geochemical H_2-producing processes.

Inventories of the geological controls on the sources of natural hydrogen have been made by Apps and Van De Kamp (1993) and Smith et al. (2005). The natural settings for hydrogen include hydrocarbon-bearing basins, young organic-rich sediments, coal beds, fault zones, extrusive igneous rocks, alkaline igneous complexes, geothermal fields, crystalline basements, potash-bearing strata, salt-bearing strata, and ultramafic rocks. The geologically controlled sources of natural H_2 can be grouped according to four main families of processes: (1) water hydrolysis processes (several processes that include the oxidation of ferrous minerals, radiolysis, cataclasis, and metamorphism); (2) organic matter decay (including thermal maturation); (3) methane and/or ammonia decomposition during metamorphism; and (4) deep Earth degassing. The alteration of Fe(II)-bearing minerals is the most commonly reported source of natural H_2 seepages on Earth, notably at mid-oceanic ridges and in ophiolitic massifs, where mafic and ultramafic are altered. Moreover, a recent study suggested that the H_2 production from the Precambrian continental lithosphere has been underestimated (Sherwood Lollar et al. 2014).

Several ultramafic suites in the eastern Piedmont Province of NC are interpreted as parts of ophiolite sequences (Butler 1989). In particular, the Halifax County complex, described in detail by Kite (1982) and Kite and Stoddard (1984), and interpreted as ophiolitic, is locally overlain by Coastal Plain deposits. Kite and Stoddard (1984) also proposed that an extensive ophiolitic belt might occur beneath the Coastal Plain (also see Lawrence and Hoffman 1993). However, the available deep drilling data are insufficient to define exactly where these rocks are located (and at what depth). The alteration of peridotite, in both oceanic and continental contexts, produces H_2 with the reduction of water by Fe(II), contained especially in fayalite (olivine ferric phase). This process, which is associated with serpentinization, consumes water and is responsible for mineral hydration. Other minerals, as well as olivine, contain Fe(II). In particular, clays (ferrous illites, chlorites, or smectites) can release Fe^{2+} ions in solution under certain conditions. The oxidation of dissolved ferrous ions by water produces H_2, and this type of process could be a potential source of H_2 if the sedimentary pile is clay-rich and provides a substantial reservoir of Fe(II).

It has been suggested that H_2 forms during rock–water reactions (e.g., cataclasis) between fresh rock surfaces containing radicals (Sugisaki et al. 1983) and by the redox conversion of hydroxyls to peroxy groups in silicates (Freund et al. 2002). H_2 can also be formed in uranium-, thorium-, and potassium-rich geological settings, by the radiolysis of water and/or organic matter, and also by the reaction of water with newly formed elements (Savchenko 1958; Lin et al. 2005). However, the quantities of H_2 that can be produced by these mechanisms are limited, and therefore they cannot explain the H_2 flows estimated in this study.

Another possible source of H_2 is the decay of solid or dissolved organic matter, either by bacterial decomposition in sediments during diagenesis and/or later during thermal maturation, which produces H_2 during the ultimate cracking of organic matter. Thermal maturation can be ruled out on the coastal plain of eastern USA because the sedimentary cover of NC and SC is not thick enough to produce the burial depth necessary for H_2 production by thermal maturation.

The decomposition of methane and/or ammonia at high temperatures (above 600 °C) during metamorphism can produce H_2. Such reactions can also produce dinitrogen. In the present case, we observed a diffusive flow of H_2 in the soil, but because it was already mixed with atmospheric components, it was difficult to evaluate the gas components that evolved with it, which may help to clarify the origin of this molecular hydrogen.

The degassing of the Earth's mantle is usually ruled out based on our present understanding of the oxidation state of the upper mantle, and it is therefore not compatible with the high concentrations of dihydrogen detected here. However, because the amount of hydrogen that was originally incorporated into the Earth's interior is unknown, the oxidation state of the upper mantle may actually differ from the conclusions drawn from surface observations, which are necessarily indirect. A recently revised geochemical concept (Toulhoat et al. 2015), initially proposed by Larin (1993), suggests that the Earth's interior is enriched in hydrogen, which is supported by calculations and experiments with iron hydride (Isaev et al. 2007). Such a model is consistent with Ohmoto's suggestion (Ohmoto 1997) that the oxygenation of the atmosphere is linked to the evolution of the continental crust. Under these conditions, deeply stored hydrogen could slowly seep to the surface.

The daily H_2 flows estimated in this study are quite important. For the largest structure, Jones Lake, they varied from 1120 to 2740 m^3/day. If the total number of Carolina bays spread along the Atlantic coast of the USA (about 500,000) is considered, it is clear that a large-scale process is involved, and the mechanism that produces this hydrogen must be efficient enough to sustain the observed H_2 flow.

Geometry and origin of Carolina bay structures

Whatever the source of this hydrogen, the subsurface migration of the ultimate reduced gas must induce reactions with oxidized subsurface rocks and fluids. These reactions and the seepage routes taken might be associated with the initial formation of the elliptical geomorphic structures that are characteristic of Carolina bays. These landscape features develop exclusively in areas where unconsolidated sediments are present at the surface. This is the case in both Carolina and Russia, where the weathered bedrock below the soil consists of unconsolidated granular sediments (Larin et al. 2014). These structures are rarely seen in river valleys, where they are likely to be obscured by fluvial processes. However, a creek that crosses Smith Bay (Fig. 4) does not seem to greatly affect the form of the bay. This may indicate either the recent age of the creek or of Smith Bay.

A variety of compounds can form with the hydrogenation of rocks along the migration pathways of H_2, including water, hydrocarbons, and acids. All these compounds are susceptible to mobilization and to migration out of the reaction zone. In this way, hydrogen might induce an increase in the bulk porosity of the rocks during its vertical migration, and as a consequence, it is likely to create its own channel for vertical migration. All the associated processes (degassing, dewatering, and volume loss at depth) will generate subsidence at the land surface, thus forming rounded (circular or elliptical) depressions.

A study of similar depressions in the Ukraine suggested that water seepage from them is possible (Bixio et al. 2002). Therefore, water with dissolved hydrogen might be discharged onto the surface, creating swamps in some bays, such as the newly formed, small structure in Jones Lake Bay. This small depression, which appeared during the late 2000s, suggests a still active mechanism underlying the formation of some of the bays. Satellite images taken in 2008 show no sign of this structure, whereas 1 year later in 2009, it had become clearly visible (Fig. 6), with the loss of all the trees within it, within less than 1 year, which was probably related to the flooding of the area. The same rapid formation of a new structure, within only several years, has been observed in the Moscow region of Russia (Larin et al. 2014).

The highest concentrations of H_2 occurred on the external slopes of the sandy rims of the Carolina bays, with moderate concentrations inside the bays and no H_2 recorded at some distance from the bay limits (Figs. 2, 4, and 7). This distribution of H_2 is similar to that in the structures studied in the EEC (Larin et al. 2014). The sand rims that border the bays (1–3 m higher than surrounding surface elevation; visible on LiDAR images in Fig. 3 and highlighted in Additional file 1: Figures SI-3 and SI-4) often showed the highest concentrations of H_2. These zones appear to be the preferential drainage sites for H_2-rich fluids moving toward the surface. Although the mechanisms of their formation and their association with higher concentrations of H_2 are not yet understood, they may be linked to fault networks around the features or their peculiar petrological properties (notably the porosity of the rocks surrounding Carolina bays).

Field studies in the EEC have shown that bay-like features emitting H_2 gas sometimes occur along structural trends, very probably corresponding to basement faults (Larin et al. 2014). Many studies have suggested that H_2 anomalies are commonly related to faults (Wakita et al. 1980; Jones and Pirkle 1981; Ware et al. 1985; Sato et al. 1986; Shcherbakov and Kozlova 1986; McCarthy and McGuire 1998; Rogozhin et al. 2010), which act as fluid conduits. This suggests that H_2-emitting features are genetically related to structural features of the crystalline basement. The available information on the deep geology of NC is insufficient to exactly define the locations of faults or their distances from these depressions (Lawrence and Hoffman 1993). This theory, together with the potential alignment of the bays along structural trends, suggests a close relationship between Carolina bays and the observed molecular hydrogen seepages from these potential geological structures.

The elliptical shape of Carolina bays is the feature that most significantly distinguishes them from the hydrogen-seeping structures in the EEC, which usually have rounded shapes. The elliptical shape of Carolina bays could be interpreted as a consequence of the local stress regime. The long and short axes of the bays appear to occur parallel to the minimum and maximum horizontal stresses, as is the case for many calderas. The formation of stress-induced oval structures is well documented in calderas around mud volcano systems (Bonini 2012). Stress discharge will determine the predominant orientation of the vertical (or subvertical) fractures of crystalline basement rocks (Lawrence and Hoffman 1993). Indeed, on geological maps, faults are almost always parallel to the bays' orientation (Brown 1985). Consequently, the initially round (isometric) shape of the hydrogen stream would gradually become elliptical as it ascends through the upper layers of the lithosphere.

When Carolina bays are compared with the structures in the EEC (Larin et al. 2014), the chemical composition of the gases seeping from them and the flow rates of the gases are quite similar. The H_2 concentrations range from tens to hundreds of parts per million. Small quantities of CH_4 and its close homologues (C2+) are also sometimes present locally. We observed similar links between the geochemistry, geomorphology, and the distributions of the H_2 concentrations in the Carolina bays and the Russian structures: the highest concentrations commonly occurred on the external shores of these structures, with high H_2 concentrations inside the structures and almost zero H_2 outside the depressions. According to our estimates, larger bays showed greater absolute H_2 flows, whereas smaller bays showed greater flows per unit of surface.

When soil scientists studied the hydrogen-seeping features in the EEC in Russia (Sukhanova et al. 2013; Polyanskaya et al. 2014), they found that molecular hydrogen seeps from these structures and that this seepage affects the soil layers by disturbing the vegetation and the microbial biomass. In areas of H_2 seepage, the humus content decreases by a factor of 2–3 and the optical density of the humic acids is lower than in the surrounding areas. The fertility of arable lands decreases significantly, and they often become unsuitable for cultivation (Sukhanova et al. 2013). The total quantities and biomasses of bacterial cells and fungal spores and the lengths of fungal and actinomycete mycelia decrease (Polyanskaya et al. 2014). The soil bleaching associated with Carolina bays could also be similar to the bleaching phenomenon associated with the bay-like features in the EEC.

The size distribution of the Carolina bays indicates that the number of bays decreases as the surface area of the bays increases (Semlitsch 2000). In this, their size distribution has the same characteristics as the bay-like features in the EEC (Larin et al. 2014), indicating that they may have a common origin.

In summary, the rounded (elliptical) depressions in the Atlantic Coastal Plain, Province of the USA, and the rounded structures in the EEC show similar geomorphic features and size distributions, and they emit H_2 at similar flow rates. All these similarities suggest a common origin and the same mechanism of formation for these features. We interpret them as the surficial marks of pathways of hydrogen-rich fluid migration. We interpret these bays as the results of local structural collapses associated with the rock alterations induced by H_2-enriched fluid flowing from the crust or deeper. Similar features can be seen in satellite images on all continents (except ice-covered Antarctica), suggesting that other analogous structures exist elsewhere in various settings.

Conclusions

The elliptical depressions known as Carolina bays in the USA are associated with H_2 flows, as observed in similar surface features in the East European craton. This molecular hydrogen probably originates from geochemical processes taking place under the sedimentary pile and migrates towards the surface. If such H_2 migration pathways exist, it is possible that this flow of H_2 induces gas–rock interactions, forming shallow pathways. These pathways might link these H_2 flows with the formation of surficial topographic anomalies that correspond to the Carolina bays, and this process might contribute to the active and ongoing development of these bays today. Our observations provide an alternative explanation to the former controversial theories on the formation of Carolina bays. We estimated the daily hydrogen flow to be quite high, up to 2700 m^3, in some of these features. Evidence of the diffusive flow of H_2 seeping from Carolina bays and its abundance suggests that the role of molecular hydrogen in the processes of the Earth's surface environments must be reconsidered. This reconsideration may influence our understanding of the dynamics and chemical processes of the biosphere and the atmosphere (Syvorotkin 2010).

Abbreviations

EEC: East European craton; GC: gas chromatography; NC: North Carolina; SC: South Carolina.

Competing interests

The authors declare that they have no competing interests.

Authors' contributions

VZ, VB, ED, and NL proposed the topic and conceived and designed the study. All authors performed the experimental study. VZ, VB, ED, and NL analyzed the data and participated in their interpretation. VB, ED, and KF collaborated with the corresponding author VZ in the preparation of the manuscript. All authors have read and approved the final manuscript.

Acknowledgments

We thank North Carolina State geologist, Dr. Kenneth B. Taylor, for facilitating this multinational oil- and gas-related investigation in North Carolina and the staff of Bladen Lakes State Forest, Turnbull Creek Educational State Forest, and Jones Lake State Park for assisting with the logistics of the field work and providing permits for us to access and collect data at the study sites.

Author details

[1]IFP Energies Nouvelles, 1 et 4 avenue de Bois Préau, 92852 Rueil-Malmaison Cedex, France. [2]Russian Academy of Science, Schmidt Institute of Physics of the Earth, B. Gruzinskaya St. 10, 123995 Moscow, Russia. [3]North Carolina Geological Survey, Coastal Plain Office and Core Repository, MSC 1620, Raleigh, NC 27699-1620, USA. [4]Present address: Natural Hydrogen Energy Ltd, 819-35th Ave., Greeley, CO 80634, USA.

References

Abbott LE, Farrell KMJ, Nickerson JG, Gay NK. Lithic resources of the North Carolina Coastal Plain: prehistoric acquisition and utilization patterns. In: Ewen CR, Whyte TR, Davis RPSJ, editors. The archaeology of North Carolina: three archaeological symposia. North Carolina: North Carolina Archaeological Council Publication No. 30; 2011. p. 2-1–2-51.

Apps JA, Van De Kamp PC. Energy gases of abiogenic origin in the Earth's crust. Futur Energy Gases United States Geol Surv Prof Pap. 1993; 81–130.

Battani A, Deville E, Faure JL, Jeandel E, Noirez S, Tocqué E et al (2010) Geochemical study of natural CO_2 emissions in the French Massif Central: how to predict origin, processes and evolution of CO_2 leakage. Oil Gas Sci Technol—Rev l'Institut Français du Pétrole 65:615–633. doi:10.2516/ogst/2009052

Bixio AC, Gambolati G, Paniconi C, Putti M, Shestopalov VM, Bublias VN et al (2002) Modeling groundwater-surface water interactions including effects of morphogenetic depressions in the Chernobyl exclusion zone. Environ Geol 42:162–177. doi:10.1007/s00254-001-0486-7

Bonini M (2012) Mud volcanoes: indicators of stress orientation and tectonic controls. Earth Sci Rev 115(3):121–152. doi:10.1016/j.earscirev.2012.09.002

Brooks MJ, Taylor BE, Ivester AH (2010) Carolina Bays: time capsules of culture and climate change. Southeast Archaeol 29:146–163

Brown PM. Geological map of North Carolina. Department of Natural Resources and Community Development of North Carolina. 1985.

Brown PM, Miller JA, Swain FM. Structural and stratigraphic framework and spatial distribution of the permeability of the Atlantic Coastal Plain, North Carolina to New York: US Geological Survey Professional Paper 796. 1972.

Butler JR. Review and classification of ultramafic bodies in the Piedmont of the Carolinas. In: Mittwede SK, Stoddard EF, editors. Ultramafic rocks of the Appalachian Piedmont. Geological Society of America Special Paper 231; 1989. p. 19–31.

Cohen AD, Gage CP, Moore WS, VanPelt RS (1999) Combining organic petrography and palynology to assess anthropogenic impacts on peatlands: part 2. An example from a Carolina Bay wetland at the Savannah River Site in South Carolina. Int J Coal Geol 39:47–95. doi:10.1016/S0166-5162(98)00040-8

Conrad R, Seiler W (1981) Decomposition of atmospheric hydrogen by soil microorganisms and soil enzymes. Soil Biol Biochem 13:43–49. doi:10.1016/0038-0717(81)90101-2

Cussler EL (2009) Diffusion mass transfer in fluid systems, 3rd edition. University Press, Cambridge

Das B (2008) Advanced soil mechanics. Taylor & Francis, London & New York

Eyton JR, Parkhurst JI. A re-evaluation of the extraterrestrial origin of the Carolina Bays. Geography Graduate Student Association, University of Illinois at Urbana-Champaign. 1975, 92 p.

Farrell KM, Harris WB, Mallinson DJ, Culver SJ, Riggs SR, Pierson J et al (2012) Standardizing texture and facies codes for a process-based classification of clastic sediment and rock.
J Sediment Res 82:364–378. doi:10.2110/jsr.2012.30

Farrell KM, Harris WB, Mallinson DJ, Culver SJ, Riggs SR, Wehmiller JF et al (2013) Graphic logging for interpreting process-generated stratigraphic sequences and aquifer/reservoir potential: with analog shelf to shoreface examples from the Atlantic Coastal Plain Province, USA. J Sediment Res 83:723–745. doi:10.2110/jsr.2013.52

Firstov PP, Shirokov VA (2005) Dynamics of molecular hydrogen and its relation to deformational processes at the Petropavlovsk-Kamchatskii geodynamic test rite: evidence from observations in 1999–2003. Geochemistry Int 43:1056–1064

Folkerts GW (1997) Citronelle ponds: little-known wetlands of the Central Gulf Coastal Plain, USA. Nat Areas J 17:6–16

Freund F, Dickinson JT, Cash M (2002) Hydrogen in rocks: an energy source for deep microbial communities. Astrobiology 2:83–92. doi:10.1089/153110702753621367

Frey DG (1954) Evidence for the recent enlargement of the "bay" lakes of North Carolina. Ecology 35:78. doi:10.2307/1931408

Garcia B, Billiot JH, Rouchon V, Mouronval G, Lescanne M, Lachet V et al (2012) A geochemical approach for monitoring a CO_2 pilot site: Rousse, France. A major gases, CO2 -carbon isotopes and noble gases combined approach. Oil Gas Sci Technol – Rev d'IFP Energies Nouv 67:341–353. doi:10.2516/ogst/2011154

Grant JA, Brooks MJ, Taylor BE (1998) New constraints on the evolution of Carolina Bays from ground-penetrating radar. Geomorphology 22:325–345. doi:10.1016/S0169-555X(97)00074-3

Hough B (1969) Basic soil engineering. Ronald Press Company, New York

Isaev EI, Skorodumova NV, Ahuja R, Vekilov YK, Johansson B (2007) Dynamical stability of Fe-H in the Earth's mantle and core regions. Proc Natl Acad Sci U S A 104:9168–71. doi:10.1073/pnas.0609701104

Ivester AH, Godfrey-Smith DI, Brooks MJ, Taylor BE. The timing of Carolina Bay and inland activity on the Atlantic coastal plain of Georgia and South Carolina. In: Geological Society of America. Denver Annual Meeting. Geological Society of America; 2004. p 69

Jones VT, Pirkle RJ (1981) Helium and hydrogen soil gas anomalies associated with deep or active faults. 181st ACS National Meeting, Atlanta, GA

Kite LE (1982) The Halifax County complex: oceanic lithosphere in the northeastern Piedmont, North Carolina. North Carolina State University, Raleigh, NC

Kite LE, Stoddard EF (1984) The Halifax County complex: oceanic lithosphere in the eastern North Carolina Piedmont. Bull Geol Soc Am 95:422–432. doi:10.1130/0016-7606(1984)95<422:THCCOL>2.0.CO;2

Larin VN, Warren Hunt C. Hydridic Earth: the new geology of our primordially hydrogen-rich planet. Alberta: Polar Publishing; 1993.

Larin NV, Zgonnik V, Rodina S, Deville E, Prinzhofer A, Larin VN. (2014) Natural molecular hydrogen seepage associated with surficial, rounded depressions on the European craton in Russia. Nat Resour Res 24(3):369–383. doi:10.1007/s11053-014-9257-5

Lawrence DP, Hoffman CW, Geology of basement rocks beneath the North Carolina Coastal Plain. North Carolina Geological Survey Bulletin 95. 1993.

Lin L-H, Hall J, Lippmann-Pipke J, Ward JA, Sherwood Lollar B, DeFlaun M, et al (2005) Radiolytic H_2 in continental crust: nuclear power for deep subsurface microbial communities. Geochemistry Geophys Geosystems 6:1–13. doi:10.1029/2004GC000907

McCarthy H, McGuire E. Soil gas studies along the Carlin trend, Eureka and Elko counties, Nevada. In: Tosdal RM, editor. Contributions to the gold metallogeny of Northern Nevada. Open-File Report 98–338 1998. US Dept. of the Interior, US Geological Survey, Menlo Park ; 1998. pp 243–250

Moore CR, Brooks MJ, Mallinson DJ, Parham PR, Mester AH, Feathers JK. (2014) Rapid scour, sand rim construction, and basin migration of a Carolina Bay in Southeastern North Carolina. Geological Society of America, Abstracts with Programs V 46(3):96

Nealson KH, Inagaki F, Takai K (2005) Hydrogen-driven subsurface lithoautotrophic microbial ecosystems (SLiMEs): do they exist and why should we care? Trends Microbiol 13:405–10. doi:10.1016/j.tim.2005.07.010

Ohmoto H (1997) When did the Earth's atmosphere become toxic. Geochemical News 93(12–13):26–27

Polyanskaya LV, Sukhanova NI, Chakmazyan KV, Zvyagintsev DG (2014) Specific features of the structure of microbial biomass in soils of annular mesodepressions in Lipetsk and Volgograd oblasts. Eurasian Soil Sci 47:904–909. doi:10.1134/S1064229314090105

Prouty WF (1952) Carolina bays and their origin. Bull Geol Soc Am 63:167–224. doi:10.1130/0016-7606(1952)63

Rogozhin EA, Gorbatikov AV, Larin NV, Stepanova MY (2010) Deep structure of the Moscow Aulacogene in the western part of Moscow. Izv Atmos Ocean Phys 46:973–981. doi:10.1134/S0001433810080062

Ross TE (1987) A comprehensive bibliography of the Carolina bays literature. J Elisha Mitchell Sci Soc 103:28–42

Sato M, Sutton AJ, McGee KA, Russell-Robinson S (1986) Monitoring of hydrogen along the San Andreas and Calaveras Faults in central California in 1980–1984.
J Geophys Res 91:12315–12326. doi:10.1029/JB091iB12p12315

Savchenko VP (1958) The formation of free hydrogen in the Earth's crust as determined by the reducing action of the products of radioactive transformations of isotopes. Geochemistry Int 1:16–25

Semlitsch RD (2000) Size does matter: the value of small isolated wetlands. Natl Wetl Newsl 22:5–6

Shcherbakov AV, Kozlova ND (1986) Occurrence of hydrogen in subsurface fluids and the relationship of anomalous concentrations to deep faults in the USSR. Geotectonics 20:120–128

Sherwood Lollar B, Onstott TC, Lacrampe-Couloume G, Ballentine CJ (2014) The contribution of the Precambrian continental lithosphere to global H_2 production. Nature 516:379–382. doi:10.1038/nature14017

Smith NJP, Shepherd TJ, Styles MT, Williams GM (2005) Hydrogen exploration: a review of global hydrogen accumulations and implications for prospective areas in NW Europe, Petroleum Geology: North-West Europe and Global Perspectives—Proceedings of the 6th Petroleum Geology Conference. Geological Society, London, pp 349–358

Stolt MH, Rabenhorst MC (1987) Carolina Bays on the eastern shore of Maryland: II. Distribution and origin. Soil Sci Soc Am J 51:399. doi:10.2136/sssaj1987.03615995005100020026x

Sugimoto A, Fujita N (2006) Hydrogen concentration and stable isotopic composition of methane in bubble gas observed in a natural wetland. Biogeochemistry 81:33–44. doi:10.1007/s10533-006-9028-4

Sugisaki R, Ido M, Takeda H, Isobe Y, Hayashi Y, Nakamura N et al (1983) Origin of hydrogen and carbon dioxide in fault gases and its relation to fault activity. J Geol 91:239–258

Sukhanova NI, Trofimov SY, Polyanskaya LM, Larin NV, Larin VN. (2013) Changes in the humus status and the structure of the microbial biomass in hydrogen exhalation places. Eurasian Soil Sci 46:135–144. doi:10.1134/S1064229313020142

Swiss Standard SN 670 010b. Characteristic coefficients of soils. Association of Swiss Road and Traffic Engineers, Zurich 1999.

Syvorotkin VL (2010) Hydrogen degassing of the Earth: natural disasters and the biosphere, Man and the Geosphere. Nova Science Publishers, New York, pp 307–347

Thom BG (1970) Carolina Bays in Horry and Marion Counties, South Carolina. Geol Soc Am Bull 81:783–814. doi:10.1130/0016-7606(1970)81[783:CBIHAM]2.0.CO;2

Toulhoat H, Beaumont V, Zgonnik V, Larin NV, Larin VN. Chemical differentiation of planets: a core issue. Submitted for publication. Available at: http://arxiv.org/abs/1208.2909v2 (2015).

Wakita H, Nakamura Y, Kita I, Fujii N, Notsu K. (1980) Hydrogen release: new indicator of fault activity. Science 210:188–90. doi:10.1126/science.210.4466.188

Ware RH, Roecken C, Wyss M (1985) The detection and interpretation of hydrogen in fault gases. Pure Appl Geophys PAGEOPH 122:392–402. doi:10.1007/BF00874607

13

Paleo-tsunami history along the northern Japan Trench: evidence from Noda Village, northern Sanriku coast, Japan

Taiga Inoue[1*], Kazuhisa Goto[2], Yuichi Nishimura[3], Masashi Watanabe[4], Yasutaka Iijima[1,5] and Daisuke Sugawara[2,6]

Abstract

Throughout history, large tsunamis have frequently affected the Sanriku area of the Pacific coast of the Tohoku region, Japan, which faces the Japan Trench. Although a few studies have examined paleo-tsunami deposits along the Sanriku coast, additional studies of paleo-earthquakes and tsunamis are needed to improve our knowledge of the timing, recurrence interval, and size of historical and pre-historic tsunamis. At Noda Village, in Iwate Prefecture on the northern Sanriku coast, we found at least four distinct gravelly sand layers based on correlation and chronological data. Sedimentary features such as grain size and thickness suggest that extreme waves from the sea formed these layers. Numerical modeling of storm waves further confirmed that even extremely large storm waves cannot account for the distribution of the gravelly sand layers, suggesting that these deposits are highly likely to have formed by tsunami waves. The numerical method of storm waves can be useful to identify sand layers as tsunami deposits if the deposits are observed far inland or at high elevations. The depositional age of the youngest tsunami deposit is consistent with the AD 869 Jogan earthquake tsunami, a possible predecessor of the AD 2011 Tohoku-oki tsunami. If this is the case, then the study site currently defines the possible northern extent of the AD 869 Jogan tsunami deposit, which is an important step in improving the tsunami source model of the AD 869 Jogan tsunami. Our results suggest that four large tsunamis struck the Noda site between 1100 and 2700 cal BP. The local tsunami sizes are comparable to the AD 2011 and AD 1896 Meiji Sanriku tsunamis, considering the landward extent of each tsunami deposit.

Keywords: Paleo-tsunami, Sanriku coast, Japan, Tsunami deposit identification, AD 869 Jogan tsunami, Storm wave, Numerical modeling

Introduction

The Tohoku-oki earthquake (Mw = 9.0) and tsunami that struck on March 11, 2011, generated severe damage along the Pacific coast of eastern Japan. The threat of large tsunamis is of great concern to Japan and other countries (e.g., Goto et al. 2014). Studies of tsunami deposits are important in estimating the size and recurrence interval of historical and pre-historical tsunamis, and in assessing the risk of large low-frequency tsunamis (e.g., Goto et al. 2014). Many studies have described tsunami deposits around the world (e.g., Dawson and Shi 2000; Pinegina and Bourgeois 2001; Peterson et al. 2011).

* Correspondence: taiga.inoue.r7@dc.tohoku.ac.jp
[1]Department of Earth Sciences, Graduate School of Science, Tohoku University, Sendai 980-8578, Japan
Full list of author information is available at the end of the article

Studies of paleo-tsunami deposits along the Pacific coast of the Tohoku region (Fig. 1), which faces the Japan Trench, are the result of global collaboration. The Sendai and Ishinomaki plains, located in the southern part of the Sanriku coast, are representative locations examined in those studies and have yielded evidence for historical and pre-historical tsunamis (Minoura and Nakaya 1991; Minoura et al. 2001; Sugawara et al. 2001, 2012, 2013; Sawai et al. 2007, 2008, 2012, 2015; Shishikura et al. 2007; Satake et al. 2008; Namegaya et al. 2010; Matsumoto et al. 2012). Shishikura et al. (2007), for example, identified five possible tsunami deposits that formed during the past 3000 years and estimated a tsunami recurrence interval of 500–1000 years. Sawai et al. (2015) proposed that the tsunami recurrence period in this region is ~ 600 years. The Sendai and Ishinomaki plains also contain deposits of the AD 869 Jogan tsunami (e.g., Abe et al. 1990; Minoura

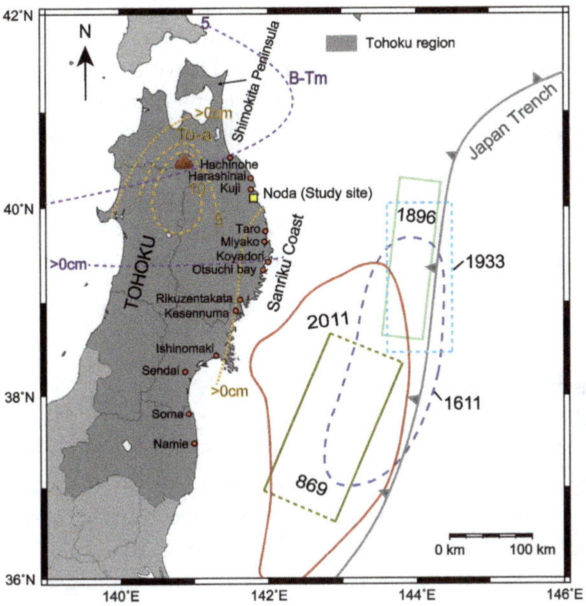

Fig. 1 Location of the study area. *Dark gray* area denotes the Tohoku region. Tsunami source models of the 869 Jogan (Sawai et al. 2012), 1611 Keicho (Hatori 2009), 1896 Meiji Sanriku (Tanioka and Satake 1996; Tanioka and Seno 2001), 1933 Sanriku (Kanamori 1971), and 2011 Tohoku-oki earthquakes (Ozawa et al. 2011) are shown by colored *lines* off the Sanriku coast. The *broken lines* indicate that the extent of the rupture model remains unknown. Thickness distributions (cm) of the Towada-a (To-a) tephra of AD 915 (Machida and Arai 2003) and the Baegdusan-Tomakomai (B-Tm) tephra of AD 946 (Machida and Arai 1992) are shown by *broken orange* and *purple lines*, respectively

and Nakaya 1991; Sawai et al. 2012), which was a possible predecessor of the 2011 Tohoku-oki tsunami (e.g., Goto et al. 2011; Namegaya and Satake 2014).

In contrast to the well-studied Sendai and Ishinomaki regions, few geological studies have described tsunami deposits along the Sanriku coast. Consequently, it is necessary to elucidate the long-term tsunami history along the Sanriku coast from geological evidence. However, this region has the disadvantage of low sedimentary potential and a lack of lowlands necessary to form tsunami deposits, due to the development of a ria coast. It has also been intensely modified by agricultural development. Observation records and other historical documents indicate that large tsunamis have frequently occurred in this region (e.g., Utsu 2004; Fig. 1). In fact, within the past 400 years, the AD 1611, 1896, 1933, and 2011 tsunamis have severely affected the Sanriku coast, with maximum recorded inundation heights of 25.0, 38.2, 29.2, and 33.0 m, respectively (Japan Tsunami Trace database: http://tsunami-db.irides.tohoku.ac.jp).

Along the Sanriku coast, Goto et al. (2015) studied the north area of Taro (Fig. 1) and identified possible tsunami deposits from the last 600 years. Ishimura and Miyauchi (2015) and Ishimura (2017) reported 11 tsunami deposits, including a possible AD 869 Jogan tsunami deposit, from the past 4000 years at Koyadori (Fig. 1). Ishimura (2017) estimated that the average recurrence interval of these deposits is 350–390 years. Farther north, on the Shimokita Peninsula (Fig. 1), Tanigawa et al. (2014) reconstructed the geological record of the past ~ 6000 years. Minoura et al. (1994, 2013) and Tanigawa et al. (2014) reported evidence of tsunami deposits that were possibly formed by the AD 1611 Keicho tsunami. Between Miyako and the Shimokita Peninsula, only two studies, Hirakawa (2012) and Takada et al. (2016), have described possible tsunami deposits. It is necessary to collect more geological evidence along the northern Sanriku coast to elucidate the size and recurrence interval of paleo-tsunamis that are generated along the Japan Trench.

It is difficult to distinguish tsunami deposits from storm and flood deposits (e.g., Morton et al. 2007; Goff et al. 2012). Therefore, a multi-proxy approach is recommended (Goff et al. 2012). Although diatom analysis might be useful to infer the origins of deposits (Henphill-Haley 1996), it cannot distinguish tsunami deposits from storm deposits. Numerical methods can be used to distinguish deposits formed by tsunami and other extreme wave events. For instance, the inundation distance of storm waves is known to be generally shorter than that of tsunami waves because of the difference in wavelength (e.g., Morton et al. 2007; Watanabe et al. 2017). Therefore, in a lowland sedimentary environment, the distance that event deposits can be transported will differ for storms and tsunamis. If the event layers were deposited far beyond the numerically estimated maximum limit of inundation by storm waves, then it can be inferred that the deposits were formed by tsunami waves.

In this study, we conducted two field surveys at Noda Village, Iwate Prefecture, on the northern Sanriku coast (Fig. 1). We coupled a sedimentological approach, to accurately identify the origin of event layers at the study site, with applied numerical modeling of storm waves rather than tsunami modeling, which allowed us to test whether the distribution of event deposits can be explained by large storm waves.

Study area

Noda Village is located on the northern Sanriku coast (Fig. 1), where the geology is dominated by the Upper Cretaceous Kuji Group (Terui and Nagahama 1995). The alluvial lowland developed along rivers and marine terraces, corresponding to marine isotope stage (MIS) 1:14 ka across the Sanriku coast area (Koike and Machida 2001). In coastal areas, cliffs and shore reefs extend seaward. The beach consists of sand and gravel, which is a rare sedimentary environment on the northern Sanriku coast, where a ria coast is developed. Paddy fields and urban areas are distributed in the lowland area.

The study site, in the Maita area, is a narrow valley among forested hills (Fig. 2a). Present-day beach sediments consist of well-rounded gravel and sand (Fig. 2b). Wave breakers are set about 50 m from the shoreline. The Sanriku railway runs north–south at ~ 300 m inland from the shoreline. These artificial objects were built well before the 2011 Tohoku-oki tsunami. The Maita River is ~ 5 m wide, and the banks are covered by revetments. The study site is used as a rice paddy field and contains agricultural terraces (Fig. 3).

The historical tsunamis that have struck the Sanriku coast are summarized by Ishimura and Miyauchi (2015). Among historical tsunamis, it is known that the AD 1611, 1896, 1933, and 2011 tsunamis were particularly large. In our study area specifically, the 1896 Meiji Sanriku tsunami reached up to ~ 880 m inland and 16 m in elevation, according to local people. No written records of large tsunamis are available before the 1896 event. The 2011 Tohoku Earthquake Tsunami Joint Survey (TTJS) Group reported that tsunami height was 22.84 m near the shoreline at this site and reached up to ~ 720 m inland and 12 m in elevation (http://www.coastal.jp/ttjt/) (Fig. 3b). The inundation distances and run-up heights of the 1896 and 2011 tsunamis cannot be compared directly because the wave breakers were not constructed until after the 1896 tsunami.

The gently sloping topography and low-energy wave environment of the study area promotes the preservation of tsunami deposits. In July 2011, we confirmed that the 2011 tsunami deposits were formed in the area, as shown in Fig. 2c. The 2011 tsunami deposit consists of gravel to very fine sand layers of 5–10 cm thickness. The deposits show normal grading and lamination and contain shell fragments. The 2011 tsunami deposit was removed during the recovery process, several months after this field survey.

Methods/experimental

Field survey

In November 2014 and March 2015, we conducted geological surveys and examined sediments in the coastal lowland of the study area. We set a survey transect perpendicular to the shoreline, 400–700 m from the sea. To retrieve cores, we used a Handy Geoslicer sampling tool of 0.6 and 1 m long (Takada et al. 2002) and a peat sampler. Coring sites were selected to evaluate the continuity of event deposits (NM1 to NM20 and NM-A to NM-L; Fig. 3a). Samples for analyses were taken from the cores using rectangular plastic cases (5 × 20 cm). In the field and laboratory, we conducted lithological observations, made sketches, and took photographs of the cores. As for some coring sites, also used a peat sampler that can obtain samples from greater depths than the ones by Handy Geoslicer, we made a composite columnar section from both cores.

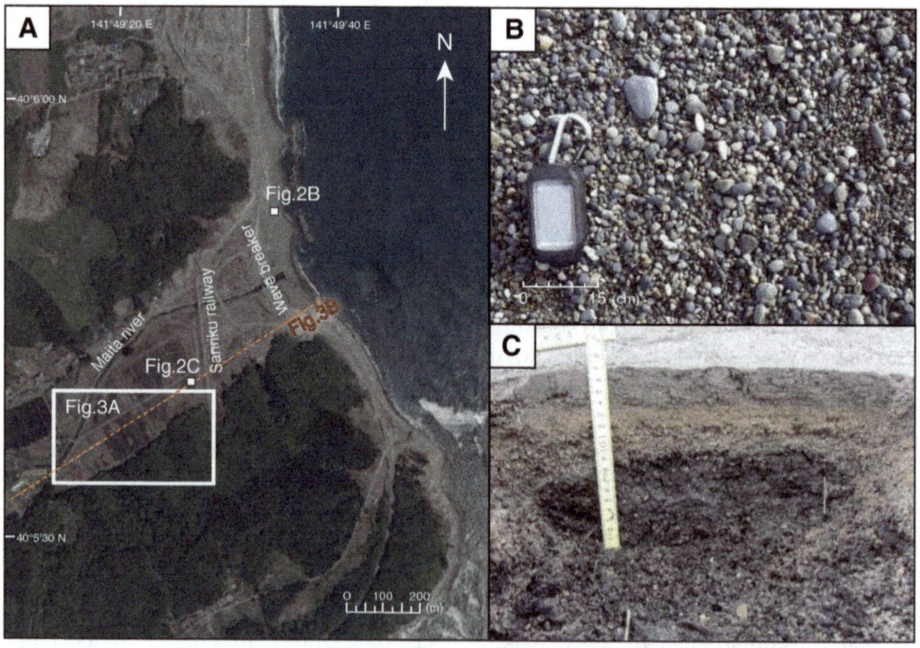

Fig. 2 **a** Aerial photograph of the survey area at Noda Village (CTO20112-C23-22, provided by the Geospatial Information Authority of Japan). **b** Sediments deposited on the nearby beach. **c** Typical 2011 Tohoku-oki tsunami deposits (photograph taken July 29, 2011)

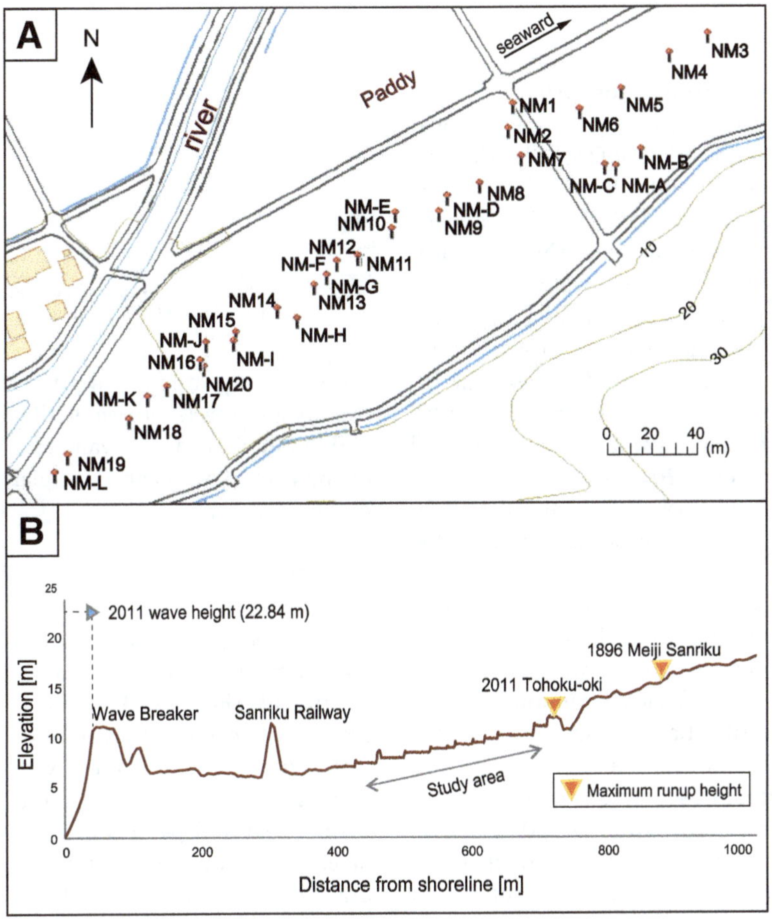

Fig. 3 **a** Locations of survey points NM1 to NM20 and NM-A to NM-L. **b** Cross-sectional topography along the transect shown in Fig. 2a. Maximum run-up heights of the 2011 Tohoku-oki and 1896 Meiji Sanriku tsunamis are shown. Tsunami height at the coast of the 2011 Tohoku-oki tsunami is based on the 2011 Tohoku Earthquake Tsunami Joint Survey (TTJS) Group

Laboratory analyses

We analyzed the grain size of the gravelly sand deposits using a CAMSIZER (Verder Scientific) at Tsukuba University, Japan. This instrument provides particle size data from 30 μm to 30 mm (Yamada and Fujino 2013). Samples for the analysis were collected from the gravelly sand layers at 1 cm intervals vertically, although some samples were collected as bulk samples because they were thin (0.5–1.5 cm in thickness).

We first dried samples and added 10% hydrogen peroxide to remove organic material. Then, ultrasonic cleaning was performed after drying. We removed the mud component using a 63-μm sieve, as the target sediment is composed mostly of coarse sand to gravel. We chose to investigate the sand and gravel grain sizes only. After removing the mud component, we dried the samples and measured the weight. The grain size analysis results show the volume frequency of the grain size distribution.

We used radiocarbon dating to correlate event deposits among cores and to estimate the depositional ages of event layers. We performed age-dating on 20 bulk peat sediments obtained below and above the gravelly sand layers. Radiocarbon dating was conducted using accelerator mass spectroscopy (AMS) at Beta Analytic Inc. in Miami, USA. Radiocarbon ages were calibrated to calendar years using the approach of Vogel et al. (1993) and the calibration curve IntCal13 (Reimer et al. 2013).

We recognized and analyzed tephra layers in some cores, which are useful in determining the absolute age. The tephra layers were sampled from cores at 600–650 m from the shoreline. To ascertain the origin of tephra quantitatively, the major element compositions were determined by energy dispersive X-ray spectrometer (EDS, JSM-5310; JEOL) at Hokkaido University, Japan, following the method of Nakamura (2016). We used TiO_2 and K_2O contents to identify the tephra, employing the database of Quaternary marker tephra provided by Aoki and Machida (2006) and data of Nakamura (2016).

Numerical method to identify tsunami deposits

We performed a numerical simulation of wave propagation and run-up using a combination of Delft-3D (Deltares 2011) and SWAN (Booij et al. 1999) software. The numerical method followed Bricker et al. (2014) and Watanabe et al. (2017). When using Delft-3D and SWAN, the hydrodynamic model repeatedly passes the water level and current fields to the wave model, which then calculates the wave field considering currents and storm surges (Bricker et al. 2014). Then, the wave model passes the radiation stress field back to the hydrodynamic model, which calculates the breaking-wave-induced setup and nearshore currents (Bricker et al. 2014).

To ensure coverage of the survey area, the numerical domain was 3-km long and 6-km wide. Topographic data on a 5-m grid, provided by the Geospatial Information Authority of Japan, was used for the land area, and bathymetric data on a 4-m grid, from the Central Public Disaster Prevention Council of Japan, was used for the marine area. We removed artificial embankments such as wave breakers, the Sanriku Railway, and the artificial topography created by reclamation and banking from the whole numerical domain as they are contemporary features. The computational grid size was 10 m. Time steps of 0.01 s were used to ensure computational stability. The propagation and inundation extent of storm waves were simulated over a period of 4 h.

According to the measurement records of NOWPHAS (1970–2013) (Kawaguchi et al. 2015), the previous maximum significant wave observed at Kuji, located just north of the study area (Fig. 1), had a 9.53-m wave height, with a 12.1 s period (recorded in 2010). Our study site is located in the northern part of Japan, where the impacts of typhoons are generally lower than in the southern subtropical region (Kawaguchi et al. 2015). Other regions in southern Japan have been affected by much larger waves: the maximum recorded significant wave heights and periods are 13.5 m and 15.8 s at Murotsu (33°16′18′′N, 134°08′50′′E) in 2004, 12.49 m and 16.4 s at Kochi (33°2857′′N, 133°35′13′′E) in 2004, 14.44 m and 16.1 s at Omaezaki-oki (34°24′12′′N, 138° 16′30′′E) in 2009, and 15.14 m and 14.4 s at Owase-oki (33°54′08′′N, 136°15′34′′E) in 2004, respectively.

The highest sea-level rise caused by low atmospheric pressure during a typhoon since 1950 was recorded by the Japan Meteorological Agency at 3.89 m at Nagoya port (35°05′00′′N, 136°53′00′′E) in 1959. Using the storm records along the entire coast of Japan, we set a parameter, assuming a large size and an extremely large size of storm waves, and the sea-level rise for boundary conditions (Table 1). We also adopted several unrealistically large assumptions for the storm wave conditions and sea-level rise at the study site (e.g., cases 1C and 2C in Table 1) to evaluate whether the lateral extent of the event layers can be explained by such exaggerated circumstances.

Table 1 Boundary conditions for the numerical simulation

Case	Sea level (m)	Wave height (m)	Period (s)
1A	0	10	10
1B	0	15	15
1C	0	20	20
2A	5	10	10
2B	5	15	15
2C	5	20	20

Sea level, significant wave height, and period are set as the main conditions on the boundary of the numerical domain

Results

Lithology

The stratigraphy in the study region consists mainly of black peat and black peaty silt deposited in a marsh environment (Fig. 4). The 2011 tsunami deposits were already removed at that time of this study, and the surface sediment was artificially disturbed. At each coring site, one or several gravel sand layers were interbedded within peat and peaty silt. At 20–30 cm below the surface, single-layer or patchy tephra were found. The gravelly sand consists of well-rounded gravel and coarse to medium sand (Fig. 5a, b) that locally shows normal grading. At site NM-K, we identified four gravelly sand layers that are named layers I to IV in descending order (Fig. 6).

Differences were found in the lithostratigraphy between the seaward and landward sides of the transect. Here, we describe the details of the lithology of NM2 (seaward side: 467 m inland; Fig. 5a) and NM-K (landward side: 655 m inland; Fig. 6) as representative cores. The characteristics of all the gravelly sand layers, such as thickness, mean grain size, and sedimentary structures, are presented in Additional file 1, Table A1.

NM2 (Fig. 5a): In this 50-cm-long core, agriculturally disturbed sediment and black peat lie at respective depths of 0–22 and 22–50 cm from the surface. At depths of 28–30 and 38–46 cm, layers of gravelly sand are interbedded with the peat. Normal grading was observed in the lower layer of the gravelly sand.

NM-K (Fig. 6): In this 58-cm-long core, agriculturally disturbed sediment and black peat occur at respective depths of 0–6 and 6–58 cm from the surface. At depths of 17–18, 33–34, 43–45, and 53–55 cm, gravelly sand layers are interbedded with peat or peaty silt. A thin (< 1 cm) tephra layer lies 1 cm above the uppermost gravelly sand layer. The peat layer at 22–29 cm depth is

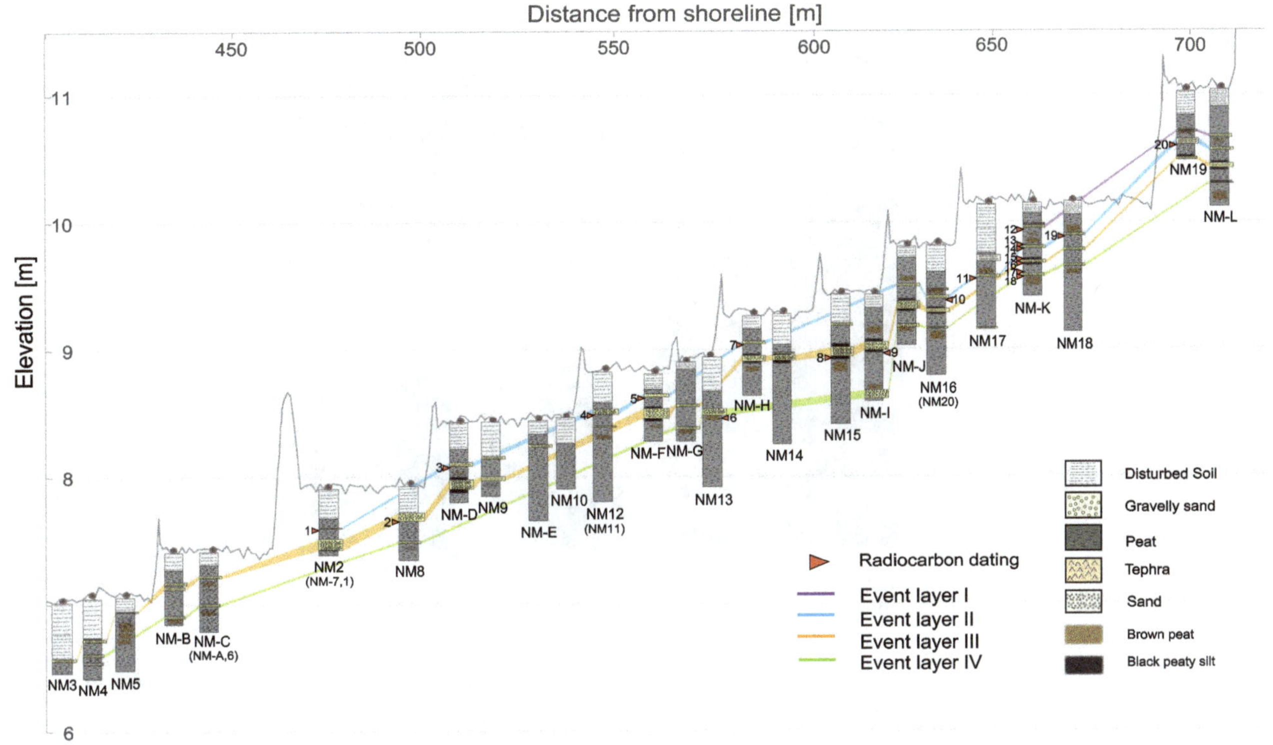

Fig. 4 Correlation of event layers (I, II, III, and IV) among the cores, based on lithology and chronological data. Red arrowheads indicate the locations of samples for radiocarbon dating and sample numbers (Table 2)

brown in color, and that at 34–47 cm depth is black, and the peat gradually becomes peaty silt with increasing depth below the gravelly sand layer. At 47–53 cm depth, the peat is mainly black in color, except for the lowest part, which is brown. The peat at 55–58 cm depth is brown. We used these characteristics to identify key beds for stratigraphic correlation.

The number and thickness of gravelly sand layers vary among the cores. At some sites, no gravelly sand layers were found (see Additional file 1, Table A1). The stratigraphic correlation of the gravelly sand layers between cores was based on the following criteria: (1) layer I is deposited immediately below patchy tephra or a layer of tephra, (2) brown peat occurs between layers II and III, (3) layer III is thicker than the other layer I, II, and IV, and interbedded within black peaty silt with sharp boundaries, and (4) layer IV is finer in grain size than layer III and brown peat occurs below layer IV.

Radiocarbon ages, as described below, are used to improve the accuracy of the correlation, as described in the "Discussion" section.

Grain size and thickness of gravelly sand layers

A typical example of the vertical variation in grain size is shown for the NM2 core (Fig. 5). The gravelly sands are generally poorly sorted (Fig. 5), and some show a bimodal grain-size distribution. The coarsest grain size is − 3.41 ϕ (in core NM-B). The lower event layer in core NM2 shows macroscopically normal grading. However, grain size analysis further revealed that variations in the mode of the diameter in the lower event layer represent repeated normal and inverse grading. In other event layers, the vertical variation in the mode of the diameter indicates various trends such as normal and inverse grading, and massive structure (Fig. 7). The NM-L core is the landward limit at which we identified gravelly sands along the transect. The maximum grain size at this site was − 2.83 ϕ.

Radiocarbon dating

Radiocarbon ages were used to confirm the continuity of sand layers and to constrain depositional ages. The ages range from 1200 to 2700 cal BP (Table 2; Fig. 4). Detailed age measurements were undertaken for the gravelly sand layers in NM-K, yielding ages of (in descending order): ~ 1290, 2150–1930, 2310–2070, and 2750–2385 cal BP (Fig. 6).

Tephra analysis

Radiocarbon dating results from the NM-K core show that the tephra layer was deposited after 1290 cal BP. Around the study site, the AD 915 Towada-a (To-a) and AD 946 Baedusan-Tomakomai (B-Tm) tephra are candidates during this period (Machida and Arai 2003; Aoki

Fig. 5 **a** Photographs, stratigraphy, CT images, and grain size of core NM2. For the grain size diagram, the horizontal axis is grain size (Φ) and the vertical axis is volume frequency (%). The *gray scale* indicates the mode of grain diameter. The volume frequency (%) of each grain size is indicated by the color scale. **b** Photograph of particles of layer III in core NM2 (depth 38–46 cm). *Red arrowheads* indicate the location of a sample for radiocarbon dating and the sample number (Table 2)

and Machida 2006; Nakamura 2016). The major element composition (TiO_2–KiO_2) of the tephra layers at the study site is similar to that of reference samples from the To-a and B-Tm tephras investigated at Tagajo (38° 18′24′′N, 140°59′18′′E) and Tomakomai (42°37′53′′N, 141°36′34′′E) (Aoki and Machida 2006), respectively (Fig. 8). Therefore, we suggest that the tephras at the study site are the To-a and B-Tm tephras.

Although a single-tephra layer was observed, and two different kinds of tephra glass were identified in the layer, probably because the tephra layers were formed during a very short time interval. Moreover, the tephra is locally intermixed, showing patch-like features, suggesting that minor reworking occurred after sedimentation.

Numerical simulation of storm waves

Cases 1C and 2C in Fig. 9 are assumed to the most extremely large storm waves and sea-level rise due to low atmospheric pressure among the study cases of the simulated waves. The results show that the maximum significant wave height (5.46 m) and flow speed (1.79 m/s) near the shoreline were both recorded in case 2C. The inundation limit of storm waves is up to 450 m from the shoreline (7.33 m in elevation) if we assume an extreme case (i.e., case 2C; Fig. 9d). The inundation limit covers a small portion of the seaward section of the transect, in which the seaward end was set 410 m inland of the coastline and 7 m in elevation. The modeled inundation by storm waves is much shorter and lower than the distance and elevation of the landward end of the site with gravelly sand layers (705 m inland and 11 m in elevation). It is important to note that this inundation can be explained mainly by the assumed sea-level rise (5 m) rather than the storm wave.

Discussion

Correlation of event deposits

The gravelly sand layers have clear upper and lower boundaries. Some layers show macroscopic normal grading. These features suggested that the gravelly sands were allochthonous and deposited by high-energy

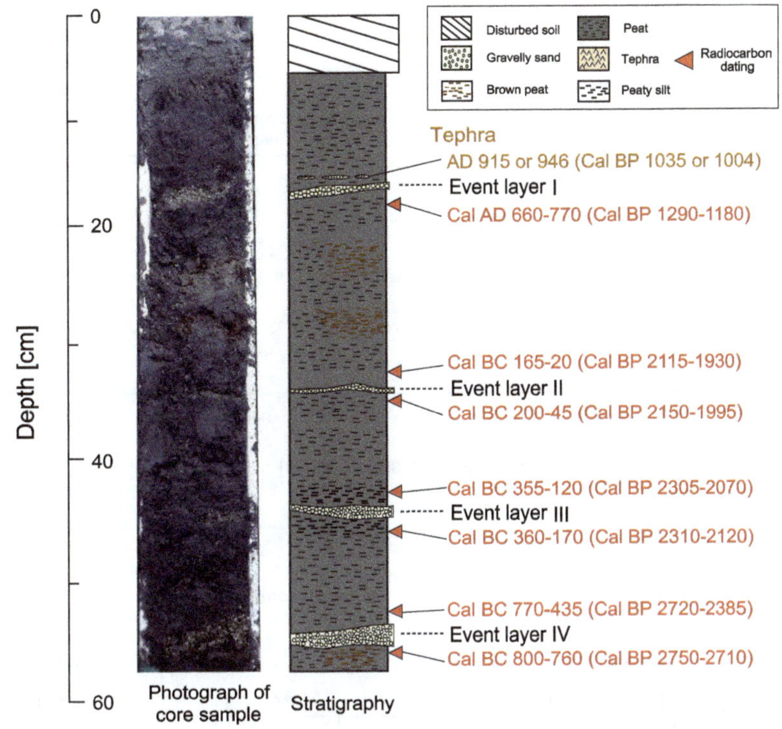

Fig. 6 Photograph, stratigraphy, and the radiocarbon dating results of core NM-K. From top to bottom, the deposits in cores NM-K are labeled as event layers I, II, III, and IV. *Red arrowheads* indicate the locations of samples for radiocarbon dating

events. Hereinafter, we designate these gravelly sand layers as event layers (Figs. 4 and 6).

Each event layer was correlated between cores based on lithology and radiocarbon ages (see Table A1 for a description of the event layers and criteria for stratigraphic correlation). In the seaward half of the transect (from NM-3 to NM-9), the upper and lower boundaries of the event layers are clear. The event layers along NM3 to NM-C display a remarkable difference in grain size (Table A1), which made it possible to correlate each event layer based on observations. In contrast, correlation of event layers in the middle part of the transect (from NM-E to NM12) was not easy because some layers were unclear or not visible in some cores. In the landward half of the transect (from NM-F to NM-L), the thickness of the event layers decreased landward. Consequently, correlation using only sedimentological features was not straightforward. However, radiocarbon dating results, tephra layers, brown peats, and the black peaty silt were useful as key beds to correlate event layers. Figure 4 shows the correlation of the event layers between different cores.

Based on the correlation, we clarified that event layer III is the most distinct layer in terms of lateral continuity and thickness (Fig. 4). Conversely, event layer I was only observed in the landward area, 650 m inland. Event layers II and IV were deposited widely but lacked continuity.

Sources and identification of tsunami deposits

Given that event layer III is the most distinct and continuous event layer, we discuss this layer first. In each coring site, some gravelly sand layers show overall normal grading (Figs. 5a and 7), indicating that event layer III was deposited under the influence of flow. Moreover, a coarser fraction (particles larger than -3 ϕ) was not observed in the landward part of the transect (> 588 m from the shoreline), as shown by the maximum grain sizes and their thicknesses of the layers. Grain size becomes finer inland, and layers become thinner (Fig. 7), which indicates the transportation of material from the sea-side such as beach and/or ocean (Dawson 1994).

Sediment supply from river flooding or the surrounding cliffs can be ruled out for the following reasons. Firstly, the event layer particles are well rounded (Fig. 5b) and similar to particles on the beach (Fig. 2b), which supports the possibility of sediment transportation from the sea. In contrast, sediments from terrestrial cliffs are usually composed of brecciated particles (e.g., Nishina et al. 2013). Thus, the sediments of nearby cliffs are unlikely to have been the source of the event layers. Secondly, although flooding from the Maita River might have occurred in the past, such flooding was probably insufficient to explain the distribution of event layers because of the small size of the river and the small volume of material carried by the river. Deposition from river

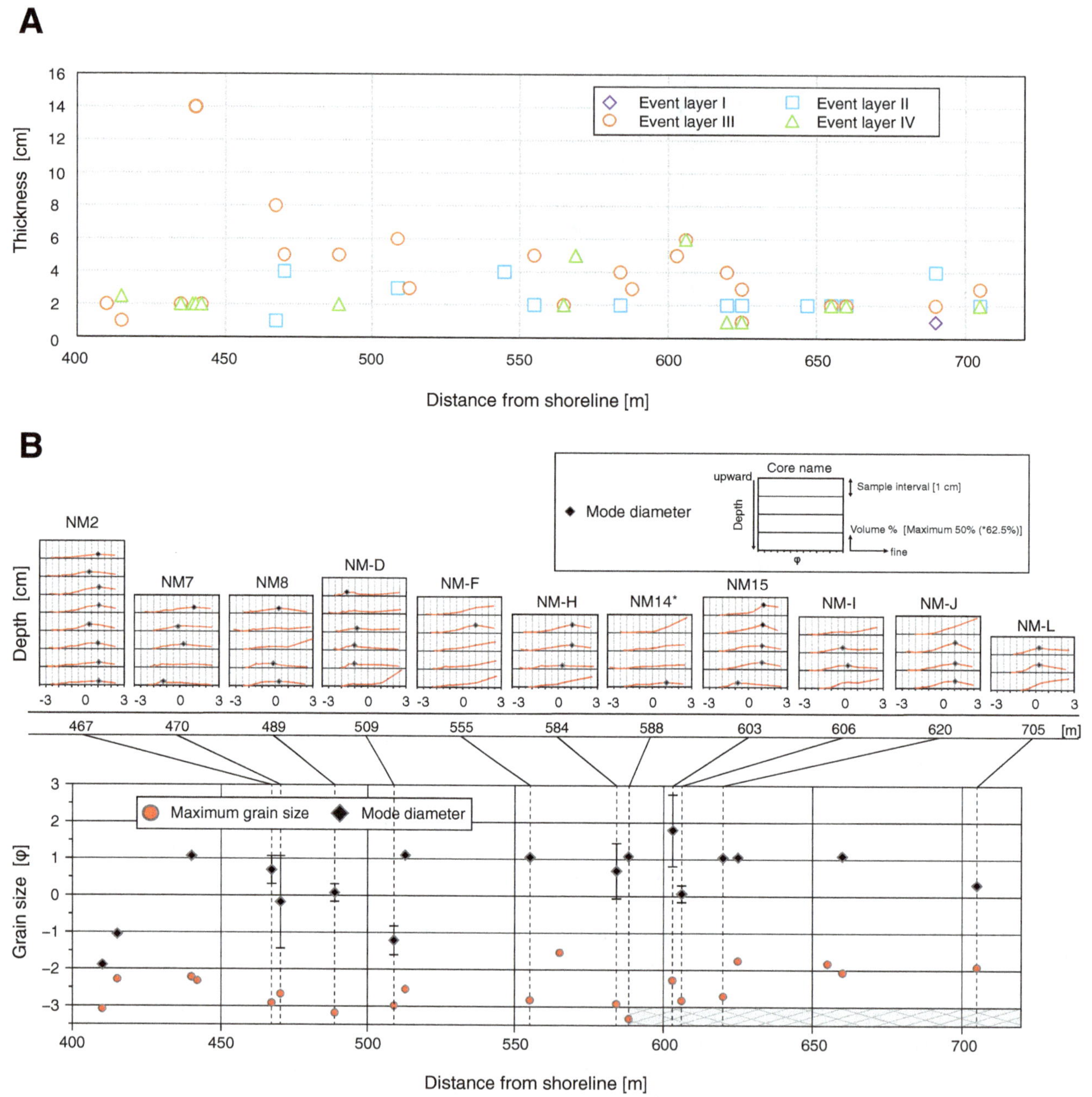

Fig. 7 a Thicknesses of event layers I, II, III, and IV with distance from shoreline. **b** Vertical (*upper panel*) and horizontal (*lower panel*) variations in grain size (φ) of event layer III with distance from the shoreline. Each error bar shows the range of the mode of the diameter. *Gray* hatching in the *bottom panel* indicates that the coarser fractions are not observed in the landward part of the transect

flooding is also contradictory to the landward fining and thinning of layer III. Therefore, we infer that it is highly likely that event layer III was formed by an extreme wave event.

Regarding the other event layers (I, II, and IV), event layers II and IV are consistently thin (1–2 cm) throughout the transect (Fig. 7). The interpretation of event layer I is uncertain because little evidence of the layer was found in the survey area. Hence, although evidence is limited for event layers I, II, and IV, we interpret that these layers were also formed by extreme wave events, considering the similarities in lithology and distribution to event layer III.

We further examined the type of wave that could have formed these event layers; i.e., tsunami or storm waves. In case 2C (Table 1), the size of the input waves was set considerably larger than the measured largest waves in Japan. The numerical simulation results for case 2C

Table 2 Radiocarbon dating results

Sample number	Core	Depth (cm)	Material	Conventional radiocarbon age (yrBP)	Calibrated age (2σ) cal BP	$\delta^{13}C$ (‰)
1	NM2	30–32	Organic sediment	1910 ± 30	1920–1815	− 27.2
2	NM8	25–27	Organic sediment	2100 ± 30	2150–1995	− 27.0
3	NM-D	35–37	Organic sediment	1740 ± 30	1715–1565	− 27.6
4	NM12	30–32	Organic sediment	2030 ± 30	2055–1900	− 24.4
5	NM-F	17–19	Organic sediment	1910 ± 30	1920–1815	− 26.9
6	NM13	48–50	Organic sediment	2610 ± 30	2765–2730	− 24.1
7	NM-H	21–23	Organic sediment	1940 ± 30	1945–1825	− 27.6
8	NM15	48–50	Organic sediment	2170 ± 30	2305–2070	− 27.6
9	NM-I	41–43	Organic sediment	2000 ± 30	2000–1885	− 27.3
10	NM16	42–44	Organic sediment	2070 ± 30	2120–1950	− 26.8
11	NM17	50–52	Organic sediment	2000 ± 30	2000–1885	− 26.8
12	NM-K	18–20	Organic sediment	1300 ± 30	1290–1180	− 27.4
13	NM-K	30–32	Organic sediment	2050 ± 30	2115–1930	− 26.6
14	NM-K	34–36	Organic sediment	2100 ± 30	2150–1995	− 26.5
15	NM-K	40–42	Organic sediment	2170 ± 30	2305–2070	− 26.2
16	NM-K	44–45	Organic sediment	2190 ± 30	2310–2120	− 24.3
17	NM-K	51–53	Organic sediment	2480 ± 30	2720–2385	− 25.2
18	NM-K	54–55	Organic sediment	2570 ± 30	2750–2710	− 26.6
19	NM18	25–27	Organic sediment	2150 ± 30	2300–2055	− 26.9
20	NM19	41–43	Organic sediment	2160 ± 30	2305–2065	− 24.7

All data were obtained by accelerator mass spectroscopy (AMS). Sample numbers correspond to those in Figs. 4 and 5

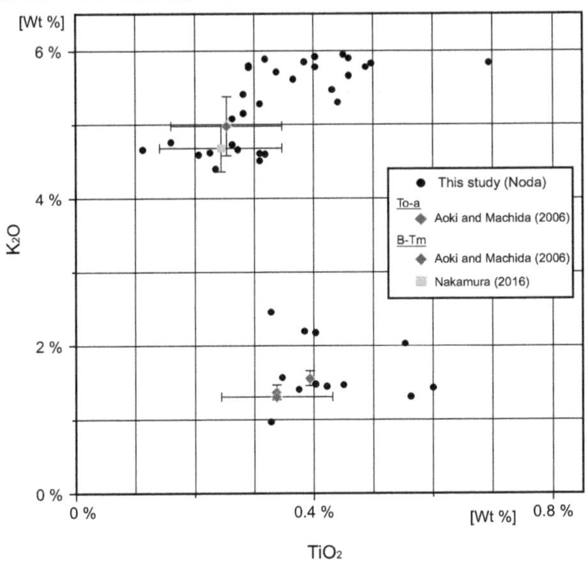

Fig. 8 K_2O–TiO_2 diagram for glass shards. The *red*, *blue*, and *green* symbols indicate marker tephras from Aoki and Machida (2006) and Nakamura (2016). Error bars show the range of values

show that storm waves reached only 450 m inland of the shoreline. The storm waves failed to inundate most of the transect (distance from the shoreline, 410–710 m). Therefore, it is likely that such storm conditions did not affect the study area. As for the strength of storms, Mishima et al. (2009) proposed that seawater temperature is the most important factor. According to Koizumi and Sakamoto (2010), the sea surface temperatures around Japan have been almost constant for the past 3000 years. Multi-decadal climate fluctuations in the NW Pacific Ocean also affect seawater temperature. However, Mishima et al. (2009) showed that seawater temperature, which has a cycle of 30–40 years, has had less of an influence on multi-decadal climate fluctuations, based on records from the last several hundred years. Therefore, it is likely that there has been no clear change in storm strength over the past 3000 years. Goto et al. (2013) also suggested, based on a study of coastal boulders deposited on fringing reefs by storm waves in the Ryukyu Islands, southern Japan, that storm wave impacts during at least the past 2300 years have remained stable. Therefore, it is reasonable to infer that the extremely large storm waves assumed in this study are not likely to have occurred over the past several thousand years.

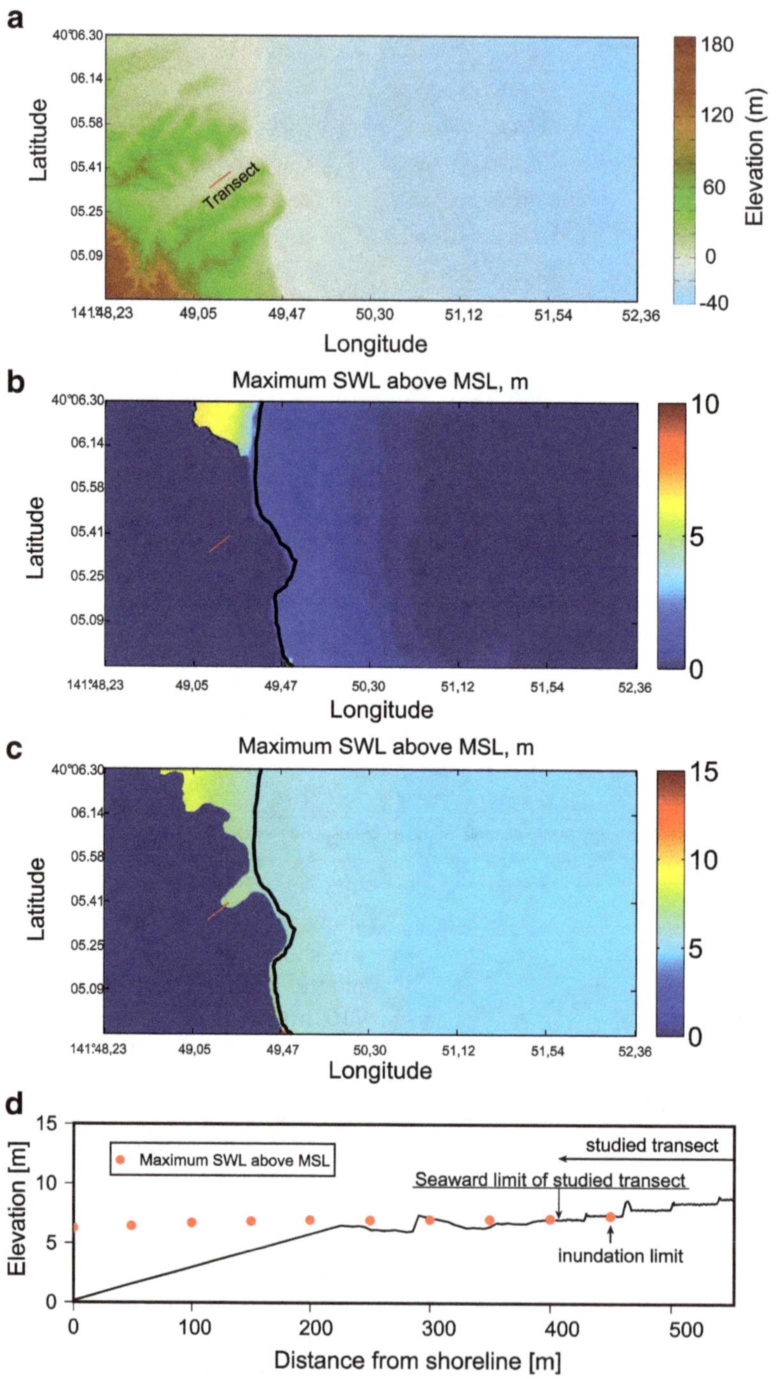

Fig. 9 **a** Domain of the numerical experiment. The *red line* denotes the transect. Numerical results are shown for maximum seawater level of **b** case 1C (20 m wave height, 20 s period, and water level 0 m) and **c** case 2C (20 m wave height, 20 s period, and water level 5 m). *Black line* denotes the shoreline. **d** Cross-sectional profile of the maximum seawater level (red points) above the mean sea level in case 2C. A minor discrepancy between the red points and altitude data occurs because of the difference between our own measured altitude data along the transect and the topographic data used for the numerical simulation

The inundation distance of former large storm waves can be related to the paleo-shoreline position and paleo-elevation. The shoreline position of the study area has changed as the Sanriku coast has been eroded by disproportionate littoral drift, resulting in a reduction in sediment supply from rivers, and by the settlement of artificial embankments (Tanaka 1980; Kinokuni and Udo 2015). This coastal erosion has occurred in the past 20–

100 years, so it may not be applicable directly to millennium-scale environmental change. However, our survey site is a backland located behind a pocket beach surrounded by rocky cliffs (Fig. 2a). Therefore, the shoreline might have been stable or shifted landwards in response to past erosion of the rocky cliffs. This shift suggests that the paleo-shoreline would have been located farther offshore than its present position if we consider coastal erosion. If we assume that the shoreline position was farther offshore than the present position, then the calculated inundation limits of the storm waves in this study should be shifted seaward.

Pleistocene marine terraces along the Sanriku coast, especially along the northern Sanriku coast, have been interpreted as evidence of long-term crustal uplift over several ~ 200,000 years (Koike and Machida 2001). The crustal uplift may change our numerical results of inundation distance by a storm wave. The marine terrace (MIS 5e, 125 ka) stretches continuously from Hachinohe to Kuji, which is located just north of Noda (Niwa and Sugai 2016). The uplift rate at Kuji is estimated to be ~ 0.2 mm/year based on the elevation of a marine terrace (Koike and Machida 2001). On the southern Sanriku coast (south of Miyako), where there is little information on the marine terraces, Niwa et al. (2014, 2015) reported that the Rikuzentakata and Kesennuma locations have undergone subsidence during the Holocene, yielding subsidence rates of 0.6–1.3 and 0.9–1.8 mm/year, respectively. Niwa and Sugai (2016) also pointed out that the Sanriku coast tilts from north (uplift) to south (subsidence). It is unclear whether Noda Village has been affected by this tilting. As Noda Village is located between Kuji and the southern Sanriku coast, it is reasonable to assume that no great crustal uplift has occurred near the study area during the past 2700 years, although a future study of crustal movement in this area is required.

Based on this evidence, it is unlikely that the event layers were formed by storm waves. The 2011 Tohoku-oki and 1896 Meiji Sanriku tsunamis inundated far beyond the possible maximum limit of storm waves and fully covered the distribution area of the event layers (Fig. 3b). Therefore, while the event layers do not show the laminations or shell fragments that are apparent in the 2011 tsunami deposits, we infer that the event layers were probably formed by tsunami waves.

Paleo-tsunami history and size

In core NM-K, the To-a (AD 915) and B-Tm (AD 946) tephras lie immediately above layer I. The absence of tsunami deposits above the tephra layers might be the result of recent agricultural disturbance. In fact, historical documents describe damage resulting from the AD 1611 Keicho tsunami near the study area (Imamura 1934; Tsuji and Ueda 1995). Hirakawa (2012), Ishizawa et al. (2015), and Takada et al. (2016) described a possible AD 1611 tsunami deposit near the present study site. Moreover, the AD 1896 tsunami certainly inundated beyond the inland limit of the 2011 tsunami, but no geological evidence remains because of anthropogenic disturbance. Therefore, we discuss the tsunami history for the period before deposition of the To-a and B-Tm tephras in the tenth century.

The radiocarbon dating and tephra results show that the ages of layers I–IV in core NM-K are 1290–1004, 2150–1930, 2310–2070, and 2750–2385 cal BP, respectively, (Fig. 6). The age of layer I can probably be correlated with the AD 869 Jogan tsunami event. No historical documentation has been found from this area that describes the AD 869 Jogan earthquake and tsunami. In recent studies, possible AD 869 Jogan tsunami deposits at Koyadori have been regarded as the northern limit of its distribution (Ishimura and Miyauchi 2015; Fig. 1). The identification of possible AD 869 Jogan tsunami deposits at the Noda Village site suggests that the tsunami affected areas farther north of the Sanriku region. If this is the case, then our study site currently defines one of the northern extents of the AD 869 Jogan tsunami. This is an important step towards improving the tsunami source model in future work. However, it should be mentioned that Takada et al. (2016) suggested another possible AD 869 Jogan tsunami deposit at Harashinai, ~ 25 km north of Noda, but the age of the deposit is not well constrained by tephra chronology.

Our results are also useful for refining our estimation of the tsunami recurrence interval along the Sanriku coast. At the study site, it is likely that four large tsunamis inundated the area between approximately 1100 and 2700 cal BP. It is noteworthy that four tsunami deposits are identified within the cores at approximately even intervals, although it might be necessary to consider the influence of sedimentation rates and erosion. The recurrence interval is estimated at 400–600 years.

Ishimura (2017) reported that five tsunamis (E4 (the 869 Jogan tsunami) to E8) occurred during the 1779 years between 1081 and 2860 cal BP at Koyadori (Fig. 1), compared with the four events at the present study site. Although the correlation of tsunami deposits between the two sites is not easy because of the wide age range, it is likely that the same tsunamis affected both sites. One possible reason for the one-fewer event bed at our study site is that it is located at a higher elevation than Koyadori. Therefore, it is possible that smaller tsunamis did not leave deposits in the study area. The location of the paleo-tsunami source and the difference in the direction of bay entrances at Noda and Koyadori may have also influenced the number of event layers.

It is noteworthy that all four tsunami deposits of the present study occur close to the inundation limit of the 2011 Tohoku–oki tsunami. The 2011 tsunami inundation was affected by artificial embankments. For that reason, it is difficult to compare the extent of inundation of the 2011 tsunami and older tsunamis directly. However, it is reasonable to speculate that the local sizes of paleo-tsunamis were equivalent to those of the 2011 Tohoku-oki or 1896 Meiji Sanriku tsunamis when we consider the landward extent of each tsunami deposit. Future numerical modeling is needed to elucidate the sources of the pre-historical tsunami events.

Conclusions

We identified four tsunami deposits in Noda Village, Iwate Prefecture, along the northern Sanriku coast, based on the distribution and sedimentary features of deposits and numerical modeling results. The results show that even unusually large storm waves with high sea-level rise caused by low atmospheric pressure fail to explain the distribution of event deposits. Numerical methods can be a useful tool in identifying tsunami deposits if the maximum possible storm waves are assumed. The ages of the tsunami deposits suggest that four large tsunamis occurred in this area during the 1600 years between 1100 and 2700 cal BP, the youngest of which might have been the AD 869 Jogan event. Considering the landward extents of the respective tsunami deposits, the local inundation extent of the tsunamis might have been similar to that of the 2011 or 1896 tsunamis, although the tsunami sources are uncertain.

Abbreviations

B-Tm: Baedusan-Tomakomai; MIS: Marine isotope stage; To-a: Towada-a

Acknowledgements

This research was supported by the JSPS research funds (no. 26242033) and funds from IRIDeS, Tohoku University. CT image analysis was conducted at the Center for Advanced Marine Core Research in Kochi University (accept no. 14B041). We especially thank a landowner for giving us permission to conduct surveys and for providing useful information. We thank T. Ishizawa for his assistance during the field survey and M. Yamada and S. Fujino for their assistance during grain size measurements. We thank J. D. Bricker for help with the numerical simulation and K. Edwards for helpful comments on editing the English version of the manuscript. J. Matsumoto, K. Ikehara, W. Szczuciński, and an anonymous reviewer provided valuable comments that improved the manuscript. We thank all of these people for their support.

Funding

Not applicable

Authors' contributions

All authors conducted field surveys, collected samples, and participated in discussions during the research. Additionally, TI mainly performed the experiments and wrote the paper. KG proposed the study topic, and conceived and designed the study. YN conducted the survey related to the 2011 tsunami deposits and tephra analysis. MW and TI performed the numerical simulation related to storm waves using Delft 3D. All authors read and approved the final manuscript.

Competing interests

The authors declare that they have no competing interests.

Author details

[1]Department of Earth Sciences, Graduate School of Science, Tohoku University, Sendai 980-8578, Japan. [2]International Research Institute of Disaster Science, Tohoku University, 468-1 Aza-Aoba, Aramaki, Aoba-ku, Sendai 980-0845, Japan. [3]Institute of Seismology and Volcanology, Faculty of Science, Hokkaido University, Sapporo, Hokkaido 060-0810, Japan. [4]Department of Civil and Environmental Engineering, Graduate School of Engineering, Tohoku University, 468-1-305 Aza-Aoba, Aramaki, Aoba-Ku, Sendai 980-0845, Japan. [5]Department of Earth Sciences, Graduate School of Science, Tohoku University, Sendai 980-8578, Japan. Present address: INPEX Corporation, Akasaka Biz Tower 5-3-1, Akasaka, Minato-ku, Tokyo 107-6332, Japan. [6]Department of Earth Sciences, Graduate School of Science, Tohoku University, Sendai 980-8578, Japan. Present address: Museum of Natural and Environmental History, 5762 Oya, Suruga-ku, Shizuoka, Shizuoka Prefecture 422-8017, Japan.

References

Abe H, Sugeno Y, Chigama A (1990) Estimation of the height of the Sanriku Jogan 11 earthquake-tsunami (AD 869) in the Sendai Plain. Zisin J Seis Soc Japan Second Ser 43(4):513–525 (in Japanese, with English abstract)

Aoki K, Machida H (2006) Major element composition of volcanic glass shards in the late Quaternary widespread tephras in Japan—distinction of tephras using K_2O-TiO_2 diagrams. Bull Geol Surv Japan 57:239–258

Booij N, Ris RC, Holthuijsen LH (1999) A third-generation wave model for coastal regions: 1. Model description and validation. J Geophys Res Oceans 104(C4):7649–7666

Bricker JD, Takagi H, Mas E, Kure S, Adriano B, Yi C, Roeber V (2014) Spatial variation of damage due to storm surge and waves during typhoon Haiyan in the Philippines. J Japan Soc Civ Eng Ser B2 (Coastal Engineering) 70(2):I_231–I_235

Dawson AG (1994) Geomorphological effects of tsunami run-up and backwash. Geomorphology 10(1):83–94

Dawson AG, Shi S (2000) Tsunami deposits. Pure Appl Geophys 157(6–8):875–897

Deltares (2011) Delft3D-FLOW user manual. Version 3.15

Goff J, Chagué-Goff C, Nichol S, Jaffe B, Dominey-Howes D (2012) Progress in palaeotsunami research. Sedi Geol 243:70–88

Goto K, Chagué-Goff C, Fujino S, Goff J, Jaffe B, Nishimura Y, Richmond B, Suguwara D, Szczucinski W, Tappin DR, Witter R, Yulianto E (2011) New insights of tsunami hazard from the 2011 Tohoku-oki event. Mar Geol 290:46–50

Goto K, Fujino S, Sugawara D, Nishimura Y (2014) The current situation of tsunami geology under new policies for disaster countermeasures in Japan. Episodes 37:258–264

Goto K, Miyagi K, Imamura F (2013) Localized tsunamigenic earthquakes inferred from preferential distribution of coastal boulders on Ryukyu Islands, Japan. Geology 41:1139–1142

Goto T, Satake K, Sugai T, Ishibe T, Harada T, Murotani S (2015) Historical tsunami and storm deposits during the last five centuries on the Sanriku coast, Japan. Mar Geol 367:105–117

Hatori T (2009) Regional deviation of inundation heights due to the large Sanriku tsunamis. Histo Earthq 24:41–48

Henphill-Haley E (1996) Diatoms as an aid in identifying late-Holocene tsunami deposits. The Holocene 6(4):439–448

Hirakawa K (2012) Large tsunami history along the Chishima and Japan trenches. Iwanami-Kagaku 82:172–181 (in Japanese)

Imamura A (1934) Past tsunamis of the Sanriku District. Bull Earthq Res Inst Tokyo Univ 1:1–16 (in Japanese, with English abstract)

Ishimura D (2017) Re-examination of the age of historical and paleo-tsunami deposits at Koyadori on the Sanriku Coast, Northeast Japan. Geosci Lett 4(1):11

Ishimura D, Miyauchi T (2015) Historical and paleo-tsunami deposits during the last 4000 years and their correlations with historical tsunami events in Koyadori on the Sanriku Coast, northeastern Japan. Prog Earth Planet Sci 2(1):16

Ishizawa T, Goto K, Yokoyama Y, Miyairi Y, Sawada C, Nishimura Y, Sugawara D (2015) High-precision ^{14}C dating of tsunami deposits at northern Sanriku, Japan. In: Abstracts of the Arthur Holmes meeting. The Geological Society, London 25 September 2015

Japan tsunami trace database. 2018. http://tsunami-db.irides.tohoku.ac.jp. Accessed 28 Sept 2016.

Kanamori H (1971) Seismological evidence for a lithospheric normal faulting--the Sanriku earthquake of 1933. Phys Earth Planet Interiors 4(4):289–300

Kawaguchi K, Inomata T, Seki K, Fujiki T (2015) Annual report on nationwide ocean wave information network for ports and harbors (NOWPHAS 2013). Tec note Port Airport Res Inst 1305:2–119. http://www.pari.go.jp/unit/kaisy/nowphas/annuareport/nowp2013.html. Accessed 25 Nov 2016

Kinokuni R, Udo K (2015) Characteristics of long term shoreline change in Japan. J Japan Soc Civ Eng Ser B2 (Coastal Engineering) 71(2):I_613–I_618

Koike K, Machida H (2001) Atlas of quaternary marine terraces in the Japanese islands. University of Tokyo Press, Tokyo

Koizumi I, Sakamoto T (2010) Synchronous Td'-derived SSTs (°C) off Japan with climatic in the northern hemisphere. Jour Tokyo Geogra Soc 119(3):489–509

Machida H, Arai F (1992) Atlas of Tephra in and around Japan. University of Tokyo press, Tokyo

Machida H, Arai F (2003) Atlas of Tephra in and around Japan [revised edition]. University of Tokyo Press, Tokyo

Matsumoto H, Yoshida M, Kumagai M (2012) Paleo-tsunami and the 2011 Tsunami Deposits on the Sendai Coastal Lowland, Northeast Japan. J Geomor 33(4):385–389

Minoura K, Hirano S, Yamada T (2013) Identification and possible recurrence of an oversized tsunami on the Pacific coast of northern Japan. Nat Haz 68(2):631–643

Minoura K, Imamura F, Sugawara D, Kono Y, Iwashita T (2001) The 869 Jogan tsunami deposit and recurrence interval of large-scale tsunami on the Pacific coast of northeastern Japan. J Nat Dis Sci 23(2):83–88

Minoura K, Nakaya S (1991) Traces of tsunami preserved in inter-tidal lacustrine and marsh deposits: some examples from northeastern Japan. J Geol 99(2):265–287

Minoura K, Nakaya S, Uchida M (1994) Tsunami deposits in a lacustrine sequence of the Sanriku coast, northeastern Japan. Sedi Geol 89(1–2):25–31

Mishima T, Yamashita T, Matsuoka K, Komaguchi T (2009) Study on the multi-decadal fluctuation of climate change and typhoon. J Japan Soc Civ Eng Ser B2 (Coastal Engineering) 56:1276–1280

Morton RA, Gelfenbaum G, Jaffe BE (2007) Physical criteria for distinguishing sandy tsunami and storm deposits using modern examples. Sedi Geol 200(3):184–207

Nakamura Y (2016) Stratigraphy, distribution, and petrographic properties of Holocene tephras in Hokkaido, northern Japan. Quat Int 397:52–62

Namegaya Y, Satake K (2014) Reexamination of the AD 869 Jogan earthquake size from tsunami deposit distribution, simulated flow depth, and velocity. Geophys Res Lett 41:2297–2303

Namegaya Y, Satake K, Yamaki S (2010) Numerical simulation of the AD 869 Jogan tsunami in Ishinomaki and Sendai plains and Ukedo river-mouth lowland. Annual Report of Active Fault and Paleoearthquake Researches 10: 1–21 (in Japanese, with English abstract)

Nishina K, Kawakami G, Tajika J (2013) Gravel deposits formed by the 2011 Tohoku earthquake tsunami along the Sanriku coast. Rep Geol Sur Hokkaido 85:27–44

Niwa Y, Sugai T (2016) Geomorphology and tectonic setting of the Sanriku coast, northeastern Japan, and introduction of recent studies on the formation of alluvial plains and Holocene crustal movements along the coast. In: Haruyama S, Sugai T (eds) Natural disaster and coastal geomorphology. Springer, Cham, pp 15–36

Niwa Y, Sugai T, Matsushima Y (2015) Holocene crustal movement in the Kesennuma Okawa Plain, southern Sanriku Coast, northeastern Japan, estimated from coastal geology. J Geogr 124:545–560

Niwa Y, Toda S, Sugai T, Matsushima Y (2014) Holocene sedimentary facies, sediment accumulation rate and coastal subsidence estimated from a sediment core in the Rikuzentakata Plain, northeast Japan. Quat Res (Daiyonki-Kenkyu) 53:311–322

Ozawa S, Nishimura T, Suito H, Kobayashi T, Tobita M, Imakiire T (2011) Coseismic and postseismic slip of the 2011 magnitude-9 Tohoku-Oki earthquake. Nature 475(7356):373–376

Peterson CD, Carver GA, Cruikshank KM, Abramson HF, Garrison Laney CE, Dengler LA (2011) Evaluation of the use of paleotsunami deposits to reconstruct inundation distance and runup heights associated with prehistoric inundation events, Crescent City, southern Cascadia margin. Earth Surf Proc Land 36(7):967–980

Pinegina TK, Bourgeois J (2001) Historical and paleo-tsunami deposits on Kamchatka, Russia: long-term chronologies and long-distance correlations. Nat Haz Earth Sys Science 1(4):177–185

Reimer PJ, Bard E, Bayliss A, Beck JW, Blackwell PG, Bronk RC, Buck CE, Cheng H, Edwards RL, Friedrich M, Grootes PM, Guilderson TP, Haflidason H, Hajadas I, Hatte C, Heaton TJ, Hoffmann DL, Hogg AG, Reimer RW, Richards DA, Scott EM, Southon JR, Staff RA, Turney CSM, van der Plicht J (2013) IntCal13 and Marine13 radiocarbon age calibration curves 0–50,000 years cal BP. Radiocarbon 55(4):1869–1887

Satake K, Namegaya Y, Yamaki S (2008) Numerical simulation of the AD 869 Jogan tsunami in Ishinomaki and Sendai plains. Annual Report of Active Fault and Paleoearthquake Researches 8:71–89 (in Japanese, with English abstract)

Sawai Y, Namegaya Y, Okamura Y, Satake K (2012) Challenges of anticipating the 2011 Tohoku earthquake and tsunami using coastal geology. Geophys Res Lett 39:L21309

Sawai Y, Namegaya Y, Tamura T, Nakashima R, Tanigawa K (2015) Shorter intervals between great earthquakes near Sendai: scour ponds and a sand layer attributable to AD 1454 overwash. Geophys Res Lett 42(12):4795–4800

Sawai Y, Shishikura M, Komatsubara J (2008) Study of paleotsunami using hand corer in Sendai plain (Sendai City, Natori City, Iwanuma City, Watari Town, Yamamoto Town), Miyagi, Japan. Annual Report on Active Fault and Paleoearthquake Researches 8:17–70 (in Japanese, with English abstract)

Sawai Y, Shishikura M, Okamura Y, Takada K, Matsuura T, Aung TT, Komatsubara J, Fujii Y, Fujiwara O, Satake K, Kamataki T, Sato N (2007) A study of paleotsunami using handy geoslicer in Sendai Plain (Sendai, Natori, Iwanuma, Watari, and Yamamoto), Miyagi, Japan. Annual Report on Active Fault and Paleoearthquake Research 7:47–80 (in Japanese, with English abstract)

Shishikura M, Sawai Y, Okamura Y, Komatsubara J, Aung TT, Ishiyama T, Fujiwara O, Fujino S (2007) Age and distribution of tsunami deposit in the Ishinomaki Plain, Northeastern Japan. Annual Report on Active Fault and Paleoearthquake Researches 7:31–46 (in Japanese, with English abstract)

Sugawara D, Goto K, Imamura F, Matsumoto H, Minoura K (2012) Assessing the magnitude of the 869 Jogan tsunami using sedimentary deposits: prediction and consequence of the 2011 Tohoku-oki tsunami. Sedi Geol 282:14–26

Sugawara D, Imamura F, Goto K, Matsumoto H, Minoura K (2013) The 2011 Tohoku-oki earthquake tsunami: similarities and differences to the 869 Jogan tsunami on the Sendai plain. Pure Appl Geophys 170(5):831–843

Sugawara D, Minoura K, Imamura F (2001) Sedimentation and numerical reconstruction of AD869 Jogan tsunami. DCRC Tsunami Eng Tech Rep 18:1–10 (in Japanese)

Takada K, Nakata T, Miyagi T, Haraguchi T, Nishitani Y (2002) Handy Geoslicer--new soil sampler for Quaternary geologist. Chisitsu News 579:12–18 (in Japanese with English abstract)

Takada K, Shishikura M, Imai K, Ebina Y, Goto K, Koshiya S, Yamamoto H, Igarashi A, Ichihara T, Kinoshita H, Ikeda T, River Division Department of prefectural Land Development, Iwate Prefecture Government (2016) Distribution and ages of tsunami deposits along the Pacific Coast of the Iwate Prefecture. Annual Report on Active Fault and Paleoearthquake Research 16:1–52 (in Japanese with English abstract)

Tanaka N (1980) Regional division of seaside in Japan from a coastal engineering perspective. Japan Soc Civ Eng 27:211–215 (in Japanese)

Tanigawa K, Sawai Y, Shishikura M, Namegaya Y, Matsumoto D (2014) Geological evidence for an unusually large tsunami on the pacific coast of Aomori, northern Japan. J Quat Sci 29(2):200–208

Tanioka Y, Satake K (1996) Tsunami generation by horizontal displacement of ocean bottom. Geophys Res Lett 23(8):861–864

Tanioka Y, Seno T (2001) Sediment effect on tsunami generation of the 1896 Sanriku tsunami earthquake. Geophys Res Lett 28(17):3389–3392

Terui K, Nagahama H (1995) Depositional facies and sequences of the Upper Cretaceous Kuji Group, northeastern Japan. Mem Geol Soc Japan 45:238–249

The 2011 Tohoku Earthquake Tsunami Joint Survey (TTJS) Group. http://www.coastal.jp/ttjt/. Accessed 10 May 2016

Tsuji Y, Ueda K (1995) Evaluation of AD 1611 Keicho, AD 1677 Enpou, AD 1763 Houreki, AD 1793 Kansei, and AD 1856 Ansei Sanriku tsunamis. Histo Earthq 11:75–106

Utsu T (2004) Catalog of damaging earthquakes in the world (through 2002), Tokyo. http://iisee.kenken.go.jp/utsu/index_eng.html. Accessed 1 Feb 2016

Vogel JC, Fulf A, Visser E, Becker B (1993) A simplified approach to calibrating ^{14}C dates. Radiocarbon 33:73–86

Watanabe M, Bricker JD, Goto K, Imamura F (2017) Factors responsible for the limited inland extent of sand deposits on Leyte Island during 2013 typhoon Haiyan. J Geophys Res Oceans 122:2795–2812

Yamada M, Fujino S (2013) Sedimentary characteristics of the onshore tsunami deposits formed by the 2011 Tohoku-oki tsunami in coastal lowlands, Ibaraki and Chiba prefectures. J Sedi Soc Japan 72:13–25

14

Interplay of irreversible reactions and deformation: a case of hydrofracturing in the rodingite–serpentinite system

Tadao Nishiyama[1*], Chisato Yoshida- Shiosaki[1], Yasushi Mori[2] and Miki Shigeno[2]

Abstract

The interplay of irreversible reactions and deformation during regional metamorphism was analyzed in the reaction zones between rodingite and serpentinite. Rodingites are leucocratic rocks found commonly in serpentinites and are considered metasomatic products of Ca-rich fluid. Rodingites occur ubiquitously in serpentinite from the Nomo metamorphic rocks, western Kyushu, a Cretaceous accretionary complex of greenschist to epidote–amphibolite facies condition. We used the singular value decomposition (SVD) method to analyze rodingitization reactions based on mineral compositions of a rodingite sample and a possible protolith (clinopyroxene gabbro). The resultant reaction implied that the rodingite was formed due to the addition of considerable amounts of CaO and H_2O to the protolith, whereas other components such as SiO_2, AF ($Al_2O_3 + Fe_2O_3$), and FM (FeO + MgO) were conserved. The Nomo rodingites are associated with reaction zones between serpentinite. A careful study on the compositional variations of diopside and chlorite in the rodingite, serpentinite, and the reaction zones between them showed a disequilibrium crystallization of diopside in the reaction zone. Diopside shows a serrated variation in composition across the reaction zone together with disequilibrium signature within grains (composite grain consisting of Fe-rich and Fe-poor parts). These features clearly indicate that diopside crystallized in a relatively short period compared to the metamorphism by irreversible reactions. Another conspicuous feature is that perovskite occurs in some reaction zones, whereas titanite is common in rodingites; this indicates that the reaction zone is poorer in SiO_2. The tremolite veins are typically derived from the reaction zone into serpentinite; showing this is a good example of hydrofracturing associated with the formation of the reaction zone. All the reactions obtained by the SVD method that contribute to the formation of the reaction zones consumed CaO and evolved H_2O, strongly suggesting that the hydrofracturing is caused by a rapid increment of fluid pressure during the progression of the reactions. The reactions consumed considerable amounts of SiO_2 in the reaction zone, which is consistent with the occurrence of perovskite, and the conservation of AF and FM components.

Keywords: Rodingite, Singular value decomposition method, Hydrofracturing, Perovskite, Reaction zone, Serpentinite, Irreversible reactions

Introduction

Metamorphism is a significant crustal process that results in the reconstruction of crustal rocks through various chemical reactions and deformation. Although chemical reactions and deformation in metamorphism have been studied separately in most existing research; however, their interplay has an important role in metamorphism. This study examines hydrofracturing caused by reactions between rodingite and serpentinite during regional metamorphism. The same type of interaction between irreversible reactions and hydrofracturing has been analyzed in a metabasite–marble system from a contact metamorphic aureole in the Hirao Limestone in Kitakyushu, Japan (Nishiyama 1989a). Hydrofracturing is driven by rapid increments of fluid pressure and is therefore likely to occur in contact metamorphic aureoles owing to rapid and irreversible reactions. This paper aims to show the same type of hydrofracturing can occur even in a regional metamorphic terrane, provided the reactions occur irreversibly

* Correspondence: tadao@sci.kumamoto-u.ac.jp
[1]Department of Earth and Environmental Science, Graduate School of Science and Technology, Kumamoto University, 2-39-1, Kurokami, Chuo-ku, Kumamoto 860-8555, Japan
Full list of author information is available at the end of the article

under the constant solid volume conditions. For this purpose, the reaction zones between rodingite and serpentinite in the Nomo metamorphic rocks, western Kyushu, Japan, were used as a case study.

Rodingites are common leucocratic rocks found in serpentinites and are considered metasomatic products of Ca-rich fluid. Rodingites are not petrologically significant because their typical mineral assemblage of diopside + grossular provides little information about their formation conditions, such as pressure and temperature. However, rodingites have attracted attention of petrologists because of their role in fluid–rock interaction in oceanic lithospheres (Austrheim and Prestvik 2008; Frost et al. 2008; Bach and Klein 2009) and subduction zone settings (Koutsovitis et al. 2013; Li et al. 2014). Rodingitization commonly occurs in Ca-rich and Si-deficient environments, although each rodingite has a unique formation setting. The chemical interactions between rodingites and serpentinites and the timing of rodingitization with serpentinization are also important aspects to be considered in the interpretation of the tectonic history of rodingite-bearing ophiolites, especially in subduction zone settings. Studies of rodingites have the potential to provide new information on these issues. In an example of mass transfer analysis associated with rodingitization, Barriga and Fyfe (1983) compared the bulk compositions of rodingites and protolith gabbros from East Liguria, Italy and reported significant enrichment in CaO and H_2O in rodingites with remarkable depletion in SiO_2, Na_2O, K_2O, and TiO_2. They suggested that these results were caused by interactions of the rocks with highly alkaline Ca–$Mg(OH)_2$-rich solutions derived from serpentinites. The isocon method (Grant 2005) is a powerful tool for clarifying the extent of mass transfer due to metasomatism because it compares the bulk compositions. This isocon method utilizes a plot of protolith composition versus altered rock composition and defines the isocon, which is a line connecting the origin of the plot and coordinates of inert components. Because the isocon is occasionally difficult to define in the isocon plot, this method is not universally applicable. The analysis of reaction relations among minerals by the singular value decomposition (SVD) method is an alternative way by means of determining the mass transfer due to metasomatism (Fisher 1989; Lang et al. 2004; Shigeno et al. 2012; Yuguchi et al. 2015). The SVD method is a type of least square method that systematically surveys reaction relations among minerals. Further, the SVD method can show which components are inert components because linear dependence among minerals for a set of components implies the conservation of those components through the reaction produced by the linear dependence. Shigeno et al. (2012) applied both the isocon and SVD methods to jadeitite reaction zones in the Nagasaki Metamorphic Rocks and determined that both the methods produce consistent results.

The purposes of this paper are as follows: First, we discuss rodingitization reaction by comparing the mineral assemblages and chemistries of the protolith gabbro with those of the rodingite. Second, we describe the extent of equilibrium in reaction zones between rodingite and serpentinite to emphasize the significance of irreversible reactions in the formation of reaction zones. Finally, we attempt to clarify the interplay of irreversible reactions and resulting hydrofracturing by using the SVD analysis of the reaction relations.

Methods/Experimental

Geological setting

The Nomo metamorphic rocks (Miyazaki and Nishiyama 1989), located in western Kyushu, Japan (Fig. 1), are ideal for studying rodingite–serpentinite reactions as excellent exposures of serpentinites are present along the coast. The Nomo metamorphic rocks are a member of the Nagasaki Metamorphic Rocks, which are distributed in three discrete regions: the Nishisonogi Peninsula (Nishisonogi metamorphic rocks), Nomo Peninsula (Nomo metamorphic rocks), and Amakusa–Takahama region (Amakusa–Takahama metamorphic rocks). The Nishisonogi and Amakusa–Takahama metamorphic rocks represent ancient subduction complexes of the Cretaceous Period (e.g., Hattori and Isomi 1976; Nishiyama 1989b; Miyazaki et al. 2013). The Nomo metamorphic rocks consist of three geological units of different ages that are in fault contact with one another: 590-460 Ma metagabbro (the older metagabbro; Igi et al. 1979), 250-180 Ma crystalline schists of the Suo unit (Nishimura 1998), and 90-70 Ma crystalline schists and serpentinites of the Sanwa Formation (Hattori and Shibata 1982; Miyazaki and Nishiyama 1989). The studied serpentinite belongs to the Sanwa Formation, in which the crystalline schists overlie the serpentinite body by low-angle thrusts. Some parts of serpentinite have been strongly sheared to form a serpentinite mélange consisting of a matrix (composed of sheared serpentinite or actinolite schist) with tectonic blocks of various types of rocks such as albitites, metagabbros, amphibolites, and meta-volcanics. The crystalline schists show greenschist to epidote–amphibolite facies without high-pressure metamorphism signatures, whereas metabasites (metagabbros and metadorelites) in the serpentinites contain omphacite and alkali amphiboles (magnesio-riebeckite) indicating high P/T metamorphism (Nishiyama and Miyazaki 1987; Miyazaki and Nishiyama 1989). The metabasites occur as metamorphosed dykes and tectonic blocks, both of which are accompanied by reaction zones consisting of chlorite, tremolite, and diopside or omphacite (Nishiyama and Miyazaki 1987). The rodingites occur ubiquitously in the serpentinite. Fukuyama et al. (2014) conducted geochemical and geochronological studies on these rodingites and determined that the U-Pb age of the zircons from the rodingites was 108-105 Ma, indicating that the formation

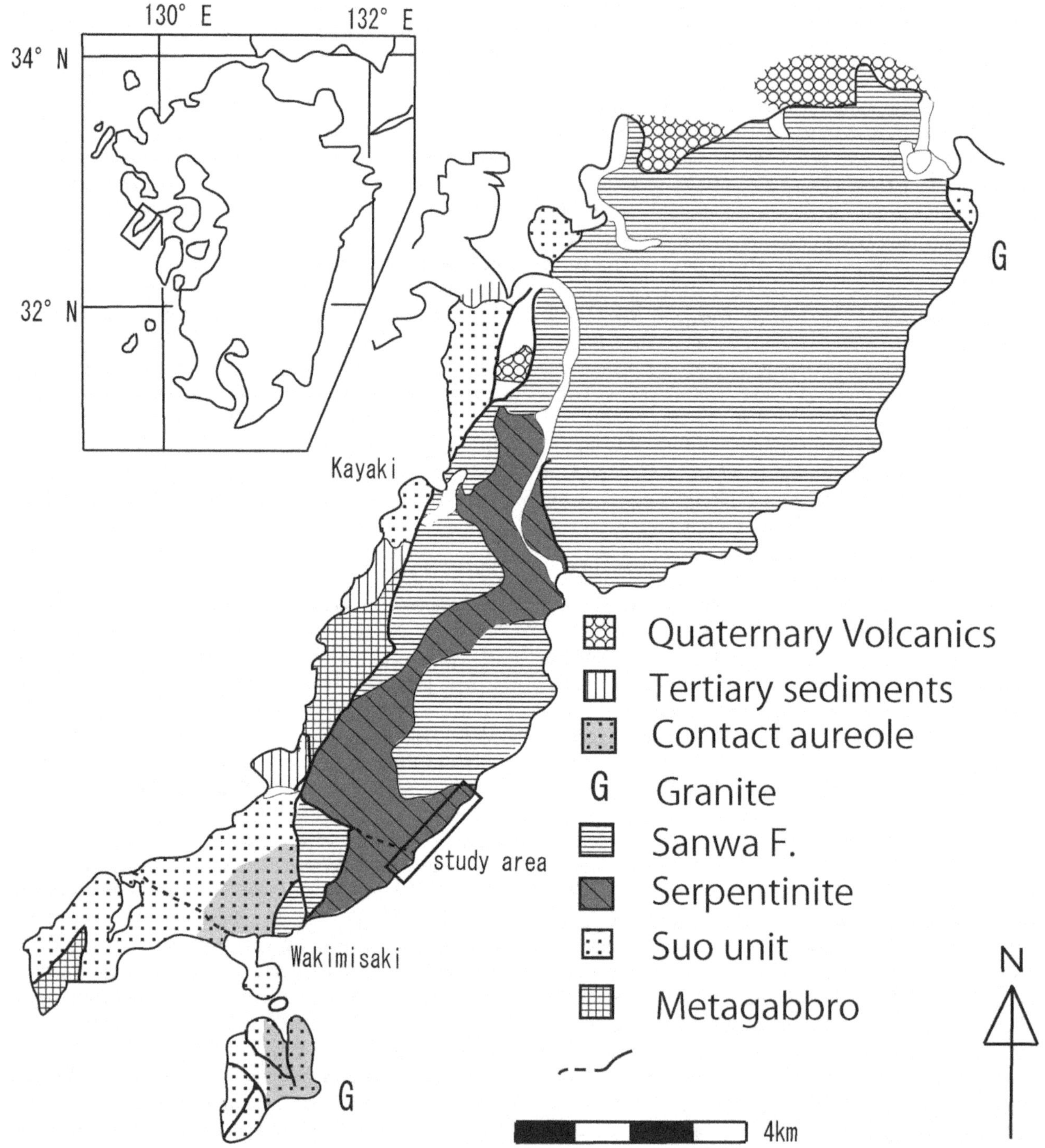

Fig. 1 Geological map of the Nomo Peninsula showing the study area of serpentinite. Compiled after Miyazaki and Nishiyama (1989), Nishimura (1998), and Nishimura et al. (2004)

of the rodingites predates the peak metamorphism of the Sanwa Formation. The rodingites mostly occur as metamorphosed dykes, typically 20 to 50 cm thick in the serpentinite, and occasionally as tectonic blocks with chlorite-rich reaction zones in a serpentinite mélange. Some rodingites occur in the serpentinite as thin veins of several centimeters in width, with a meandering structure. The rodingites show a gabbroic texture in some cases; however, a transition from metagabbro (diopside + clinozoisite) to rodingite (grossular + diopside) was not observed in outcrops from the study area. Clinopyroxene metagabbros rarely occur as dykes, of several meters in thickness, in the serpentinite and are accompanied by reaction zones consisting mainly of chlorite, tremolite, and diopside.

Petrography

This section describes the petrography and mineral chemistry of representative samples of rodingites, serpentinites, and the reaction zones between them. A possible equivalent to the rodingite protolith (gabbro), a clinopyroxene metagabbro, is also described for comparison. Rodingites are categorized into two types: dyke-type and vein-type. Dyke-type rodingites (Fig. 2a) occur as platy dykes of constant width ranging from 20 to 100 cm that commonly show coarse-grained gabbroic textures (Fig. 2b). Vein-type rodingites (Fig. 2c) occur as winding or meandering veins of variable width ranging from 1 to 10 cm, sometimes showing branching and no distinct internal textures. Both types of rodingites are associated with the reaction zone between the host serpentinite. Table 1 shows mineral assemblages of rodingites, serpentinites, associated reaction zones, and a clinopyroxene metagabbro. Serpentinites are massive rocks with metamorphosed clinopyroxenite layers of several tens of centimeters in thickness. They consist mainly of serpentine (antigorite) with small amounts of diopside and chlorite. Chromite surrounded by magnetite

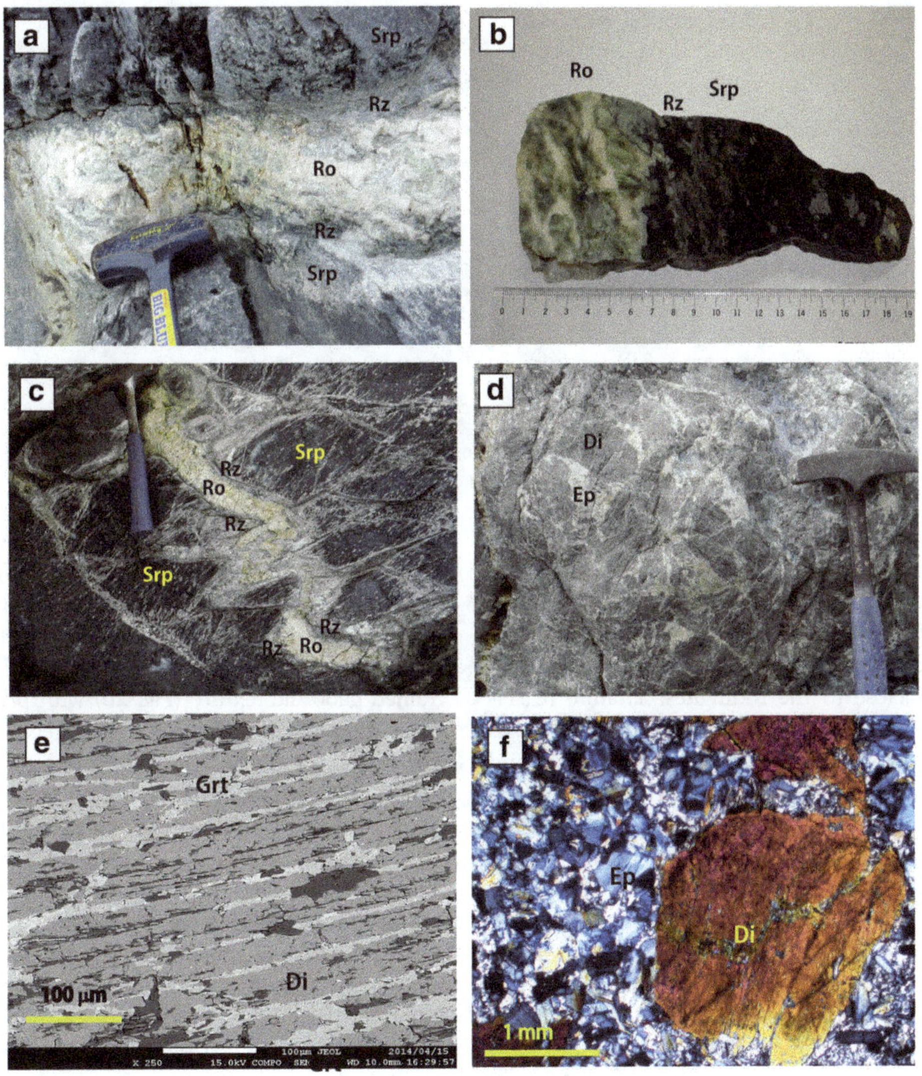

Fig. 2 Photographs of rodingite and metagabbro, back-scattered electron (BSE) image of rodingite, and photomicrograph of metagabbro. **a** Dyke-type rodingite (Ro, 131018-B) in serpentinite (Sp), consisting of green diopside and white aggregates of garnet and chlorite. The reaction zones (Rz) are developed between them. **b** Gabbroic texture of the dyke-type rodingite (131018-B), showing large crystals of green diopside embedded in white garnet. **c** Vein-type rodingite (Ro) in serpentinite (Sp), showing meandering structure, and the reaction zones (RZ) between them. **d** Clinopyroxene metagabbro consisting of large crystals of brown diopside and fine-grained aggregates of epidote, chlorite, and titanite. **e** BSE image of diopside from the rodingite, showing metasomatic development of lamellae-like garnet along the cleavage of diopside. **f** Photomicrograph of the clinopyroxene metagabbro under crossed polars

Table 1 Mineral assemblages of rodingites, serpentinites, associated reaction zones, and a metagabbro

Sample number	Rock type	Minerals															
		Di	Grt	Chl	Srp	Chr	Ep	Ves	Pmp	Amp	Prv	Ttn	Ilm	Mag	Ap	Cal	Zrn
131018-2A	Rodingite(v)[a]	○	○	○				○				○					
131018-2B	Rodingite(v)	○	○	○								○					
131018-4A	Rodingite(v)	○	○	○								○	○		○		
	Reaction zone	○		○							○				○		
131018-4B	Rodingite(v)	○	○	○								○			○		
	Reaction zone	○	○	○				○			○	○	○				
131019-2	Rodingite(d)	○	○	○				○									
	Reaction zone	○		○				○			○						
	Serpentinite	○		○	○									○			
TNKoba	Rodingite(v)	○	○					○									
	Reaction zone	○	○	○							○						
	Serpentinite	○			○	○								○			
08315-0	Rodingite(v)	○	○	○								○				○	○
	Reaction zone			○								○		○	○		○
131018-1B	Rodingite(d)[b]	○	○	○												○	
131019-1core	Rodingite(d)	○	○	○			○		○	○		○			○		○
131019-1rim	Rodingite(d)	○	○	○				○				○	○		○		
	Reaction zone	○		○								○	○	○			
	Serpentinite	○			○									○			
131019MG	Metagabbro	○		○			○					○				○	

[a]Rodingite(v) : vein-type rodingite
[b]Rodingite (d) : dyke-type rodingite. Mineral abbreviations after Whitney and Evans (2010)

occasionally occurs as an accessory mineral. Tremolite and diopside commonly occur as monomineralic veins in serpentinites. Chlorite and diopside are the main constituents in the reaction zones between the rodingites and serpentinites. Small amounts of garnet and vesuvianite occasionally occur close to the rodingites. Titanite ($CaTiSiO_5$), ilmenite, magnetite, apatite, and zircon occur as accessory minerals. Notably, perovskite ($CaTiO_3$) was found to occur in several samples of reaction zones, but not in the rodingites or serpentinites. The significance of perovskite occurrence will be discussed subsequently in detail. The minerals were analyzed with using a JEOL JSM7001F field-emission electron microscope equipped with an Oxford INCA X-Max energy dispersive X-ray analyzer at an accelerating voltage of 15 kV and a beam current of 1.02 nA. Some minerals were identified with using a Horiba Jobin Yvon LabRAM HR800 Raman spectrometer equipped with a 633 He–Ne laser at Kumamoto University. Mineral abbreviations in tables, figures, and equations follow those given in Whitney and Evans (2010).

Dyke-type rodingite and clinopyroxene metagabbro

A dyke-type rodingite (131018-1B) with a gabbroic texture and a clinopyroxene metagabbro (131019MG) are described in this subsection, and are subsequently discussed in conjunction with the rodingitization of gabbro. Figure 2 shows photographs and photomicrographs of the samples. The rodingite (Fig. 2a) occurs as a 20 cm-thick dyke, with a 1 cm wide reaction zone, in serpentinite. It shows a coarse-grained gabbroic texture (Fig. 2b) with prismatic crystals of green diopside and fine-grained aggregate of white garnet. The relative abundance of diopside and garnet is roughly 50:50, although the exact determination of modal abundance is hampered by the fine-grained and turbid texture of the rodingite. Microscopic analysis reveals that the large crystals of diopside are partially replaced by aggregates of fine-grained garnet along the cleavage (Fig. 2e). Table 2 summarizes the average compositions of the minerals. An analysis of 10 grains of diopside showed that they contain small amounts of Al_2O_3 (1.18 wt%) and have an average X_{Mg} ($Mg/(Mg + Fe^{2+})$) of 0.92. The garnets are all fine-grained, showing a local heterogeneity with Fe^{3+}-rich and Fe^{3+}-poor aggregates, $Sps_1Grs_{82}Adr_{17}$ and $Grs_{96}Adr_4$ in compositions, respectively. The total iron in garnets is assumed to be ferric, making the total cation close to 8.0. The chlorite shows $X_{Mg} = 0.93$, which is almost identical to that of diopside. The clinopyroxene metagabbro (Fig. 2d) occurs as a dyke in serpentinite, with ~20 cm-thick reaction

Table 2 Average compositions of minerals from dyke-type rodingite and metagabbro

	Dyke-type rodingite131018-1B				Metagabbro 131019-MG			
	Di	Grt1[a]	Grt 2[b]	Chl	Di	Chl	Ttn	Ep
	$n^c = 10$	$n = 7$	$n = 10$	$n = 6$	$n = 5$	$n = 3$	$n = 3$	$n = 13$
SiO_2	54.66	39.26	39.95	29.69	53.32	29.81	30.75	39.26
TiO_2	–	0.15	–	–	–	–	38.47	–
Al_2O_3	1.18	18.04	21.49	21.18	2.01	20.25	1.19	29.10
Fe_2O_3	–	5.85	1.34	–	–	–	–	5.40
FeO	2.59	–	–	3.77	4.93	4.55	0.41	–
MnO	–	0.50	–	–	–	0.19	–	–
MgO	16.81	–	–	29.70	15.61	29.48	–	–
CaO	24.80	36.17	36.96	0.09	23.72	–	29.10	24.68
Na_2O	–	–	–	–	0.44	–	–	–
Total	100.04	99.97	99.74	84.43	100.03	84.28	99.92	98.44
		apfu				apfu		
O	6	12	12	28	6	28	5	12.5
Si	1.98	3.01	3.01	5.76	1.95	5.82	1.00	3.02
Ti	–	0.01	–	–	–	–	0.95	–
Al	0.05	1.63	1.91	4.84	0.09	4.66	0.05	2.64
Fe^{3+}	–	0.34	0.08	–	–	–	–	0.31
Fe^{2+}	0.08	–	–	0.61	0.15	0.74	0.01	–
Mn	–	0.03	–	–	–	0.03	–	–
Mg	0.91	–	–	8.59	0.86	8.58	–	–
Ca	0.96	2.97	2.98	0.02	0.93	–	1.02	2.03
Na	–	–	–	–	0.03	–	–	–
Σcation	3.98	7.99	7.98	19.82	4.01	19.83	3.03	8.00
X_{Mg}	0.92	–	–	0.93	0.85	0.92	–	–
X_{Sps}	–	0.01	–	–	–	–	–	–
X_{Grs}	–	0.82	0.96	–	–	–	–	–
X_{Adr}	–	0.17	0.04	–	–	–	–	–

[a]Grt1: Fe-rich garnet
[b]Grt2: Fe-poor garnet
[c]n: number of analyzed grains

zones between them consisting of chlorite and diopside. The rock is massive and very coarse-grained, consisting of dark-brown clinopyroxene crystals, up to 10 cm in length, and white finer-grained aggregates of chlorite, epidote, and titanite (Fig. 2f). The mode of the rock was determined as Di = 65 vol.%, Ep = 28 vol.%, and Chl = 7 vol.% by the point-counting method. The diopside contains small amounts of Al_2O_3 and Na_2O (Table 2), with $X_{Mg} = 0.85$, which is slightly lower than that of the diopside from the rodingite. The chlorite, apart from its low content of MnO, has almost the same composition as that of the chlorite from the rodingite. Ferric iron content in epidote is 0.31 atoms per formula unit (apfu).

Dyke-type rodingite with reaction zones

A set of samples of dyke-type rodingite, serpentinite, and rocks from the reaction zone between them were collected from an outcrop block. This particular example is described here for subsequent discussion on the degree of attainment of equilibrium during the formation of the reaction zones. The rodingite (131019-2) occurs as a massive leucocratic rock surrounded by thin (~4 cm thick) reaction zones and further by serpentinite (Fig. 3a). The rodingite consists mainly of diopside, vesuvianite, and garnet with a small amount of chlorite. Microscopic analysis revealed that the presence of very fine grains gives it with a turbid appearance, preventing the identification of all minerals (Fig. 3b). This feature is prevalent in most of the rodingite samples. Figure 3c is a back-scattered electron (BSE) image of the rodingite under the FE-SEM, showing coarser-grained diopside and vesuvianite with finer-grained garnet. Table 3 shows the average compositions of minerals in these rocks. The garnet in the rodingite

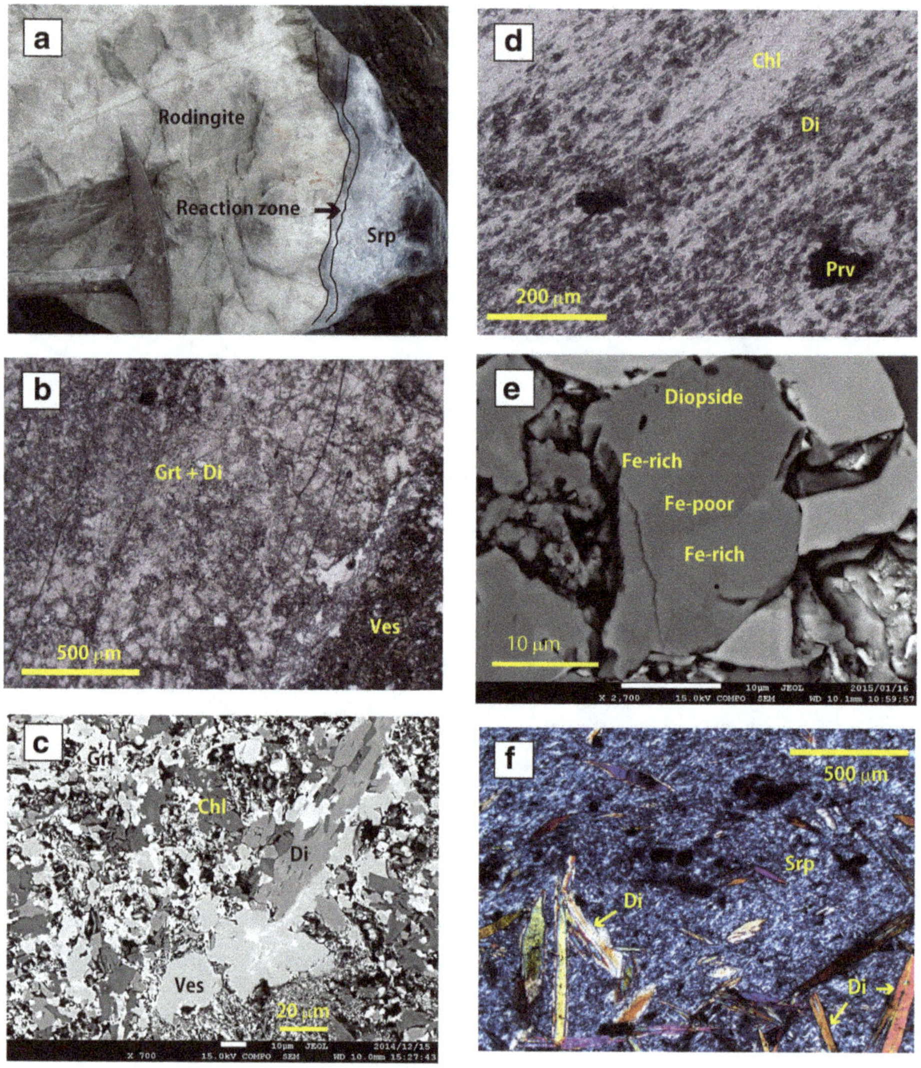

Fig. 3 Photographs, photomicrographs and BSE images of dyke-type rodingite, the reaction zone, and serpentinite. **a** Photograph of a block of dyke-type rodingite (131019-2) with thin reaction zones between serpentinite (Srp). **b** Photomicrograph of rodingite (131019-2) consisting of fine-grained aggregates of garnet, diopside, and vesuvianite in plane-polarized light. **c** BSE image of rodingite (131019-2) consisting of garnet, chlorite, diopside, and vesuvianite. **d** Photomicrograph of the reaction zone consisting of fine-grained chlorite and diopside with minor perovskite in plane-polarized light. **e** BSE image of heterogeneous diopside consisting of Fe-rich and Fe-poor parts. **f** Photomicrograph of serpentinite consisting of diopside and serpentinite under crossed polars

can be classified into two groups: one rich in Fe^{3+} (0.68 apfu) and the other poor in Fe^{3+}(0.27 apfu), considering total iron as Fe_2O_3. A small amount of TiO_2 (2–3 wt%) is present. The Raman spectrum of the garnets poor in Fe^{3+} represents a broad band at around 3590 cm^{-1}, possibly indicating the presence of H_2O in the crystal structure. The slightly low total (~99 wt%) of the garnet analysis is consistent with this result; thus, the garnet is hydrogrossular. The diopside is magnesian with $X_{Mg} = 0.96$. The chlorite is richer in Al_2O_3 and FeO than the chlorites from the reaction zones and serpentinite. The systematic compositional change in chlorite composition from rodingite to serpentinite through reaction zones is subsequently discussed. The reaction zones consist of two layers: a dark green, 13 mm-thick layer adjacent to the rodingite (reaction zone 1) and a light green, 6 mm-thick layer adjacent to the serpentinite (reaction zone 2). Reaction zone 1 consists mainly of chlorite with small amounts of diopside, vesuvianite, and perovskite (Fig. 3d), whereas reaction zone 2 is composed mostly of chlorite with a very small amount of diopside. The diopside crystals in the reaction zones are very fine grained (at tens of micrometers in size) and heterogeneous, consisting of Fe-rich and Fe-poor parts (Fig. 3e). The relation between the Fe-rich and Fe-poor parts is

Table 3 Average compositions of minerals from dyke-type rodingite(131019–2), serpentinite and reaction zones

	Dyke-type rodingite (131019–2)					Reaction zone 1				Reaction zone 2		Serpentinite		
	Di	Chl	Grt[a]	Grt[b]	Ves	Di	Chl	Ves	Prv	Di	Chl	Srp	Di	Chl
	$n=12$	$n=8$	$n=6$	$n=4$	$n=22$	$n=4$	$n=4$	$n=13$	$n=12$	$n=2$	$n=9$	$n=4$	$n=8$	$n=9$
SiO_2	55.07	28.17	38.14	37.05	36.53	54.10	31.70	34.34	–	54.92	32.04	42.75	55.41	34.85
TiO_2	–	–	1.91	2.75	2.00	–	–	3.89	57.05	–	–	–	–	–
Al_2O_3	–	19.05	18.61	12.29	14.82	0.38	15.95	6.06	–	–	14.56	1.39	–	11.50
Cr_2O_3	–	–	–	–	–	–	–	–	–	–	–	–	–	0.42
Fe_2O_3	–	–	4.67	11.45	–	–	–	–	–	–	–	–	–	–
FeO	1.38	12.33	–	–	4.18	4.80	5.56	15.95	0.51	1.26	4.51	3.11	0.64	3.81
MgO	17.40	23.62	–	–	2.45	15.31	30.86	0.85	–	17.5	31.06	37.12	17.98	34.43
CaO	26.19	0.27	35.62	35.89	36.11	25.30	–	34.00	40.55	25.89	–	–	26.29	–
Total	100.04	83.44	98.95	99.43	96.09	99.89	84.07	95.09	98.11	99.57	82.17	84.37	100.32	85.01
			apfu				apfu			apfu			apfu	
O	6	28	12	12	78	6	28	78	3	6	28	116	6	28
Si	2.00	5.81	2.94	2.94	19.24	2.00	6.24	19.66	–	2.00	6.41	34.01	2.00	6.73
Ti	–		0.11	0.17	0.79	–	–	1.69	0.99	–	–	–	–	–
Al	–	4.63	1.69	1.15	9.20	0.02	3.70	3.96	–	–	3.44	1.33	–	2.62
Cr	–	–	–	–	–	–	–	–	–	–	–	–	–	0.07
Fe^{3+}	–	–	0.27	0.68	–	–	–	–	–	–	–	–	–	–
Fe^{2+}	0.04	2.13	–	–	1.84	0.15	0.92	7.75	0.01	0.04	0.76	2.07	0.02	0.61
Mg	0.94	7.26	–	–	1.92	0.85	9.05	0.71	–	0.95	9.26	44.00	0.97	9.91
Ca	1.02	0.06	2.94	3.05	20.37	1.00	–	20.87	1.00	1.01	–	–	1.02	–
Σcation	4.00	19.83	7.95	7.99	53.36	4.02	19.91	54.64	2.00	4.00	19.87	81.41	4.01	19.94
X_{Mg}	0.96	0.77				0.86	0.91			0.96	0.92	0.96	0.98	0.94
X_{Grs}			0.86	0.63										
X_{Adr}			0.14	0.37										

[a]Fe3 + –poor garnet
[b]Fe3 + –rich garnet

neither a core–rim relationship nor a sector zoning but heterogeneity in individual grains, implying disequilibrium crystallization; this issue is addressed in later discussion subsequently. The serpentinite consists of serpentine (antigorite), diopside, chlorite, and magnetite (Fig. 3f). The diopside shows idiomorphic prismatic crystals with maximum lengths of 500 μm, which is significantly larger than those in the reaction zones.

Vein-type rodingite with reaction zones

A vein-type rodingite (131048-4A) associated with reaction zones is described in this subsection in detail and is subsequently discussed in conjunction with the interplay of irreversible reactions and hydrofracturing, which is the main topic of this paper. A conspicuous feature of this particular example is that many tremolite veins are derived from the reaction zones into the serpentinite (Fig. 4a). Similar tremolite veins are commonly observed in wall rock serpentinites that are close to rodingite dykes and veins. Average mineral analyses are given in Table 4. The rodingite consists mainly of diopside, garnet, and chlorite with small amounts of ilmenite, titanite, and apatite. The garnet is a fine-grained aggregate with a composition of $Grs_{87}Adr_{13}$. The diopside occurs as prismatic crystals several hundred micrometers in length within the garnet aggregates. The chlorite occurs in the interstices of the garnet aggregates (Fig. 4b) and has an X_{Mg} of 0.70. The titanite occasionally contains ilmenite in its core (Fig. 4b). The reaction zone is about 1 cm-thick and consists mainly of chlorite and diopside with minor amounts of vesuvianite, perovskite, and apatite. The perovskite shows a symplectic texture with diopside (Fig. 4c), which implies a pseudomorph after titanite. The tremolite veins in the serpentinite are characterized by acicular crystals of tremolite aligned normal to or at high angles to the vein wall (Fig. 4d). The X_{Mg} values of ferromagnesian minerals (diopside, chlorite, serpentine, and tremolite) are higher in the serpentinite and the reaction zones than they are in the rodingite (Table 4).

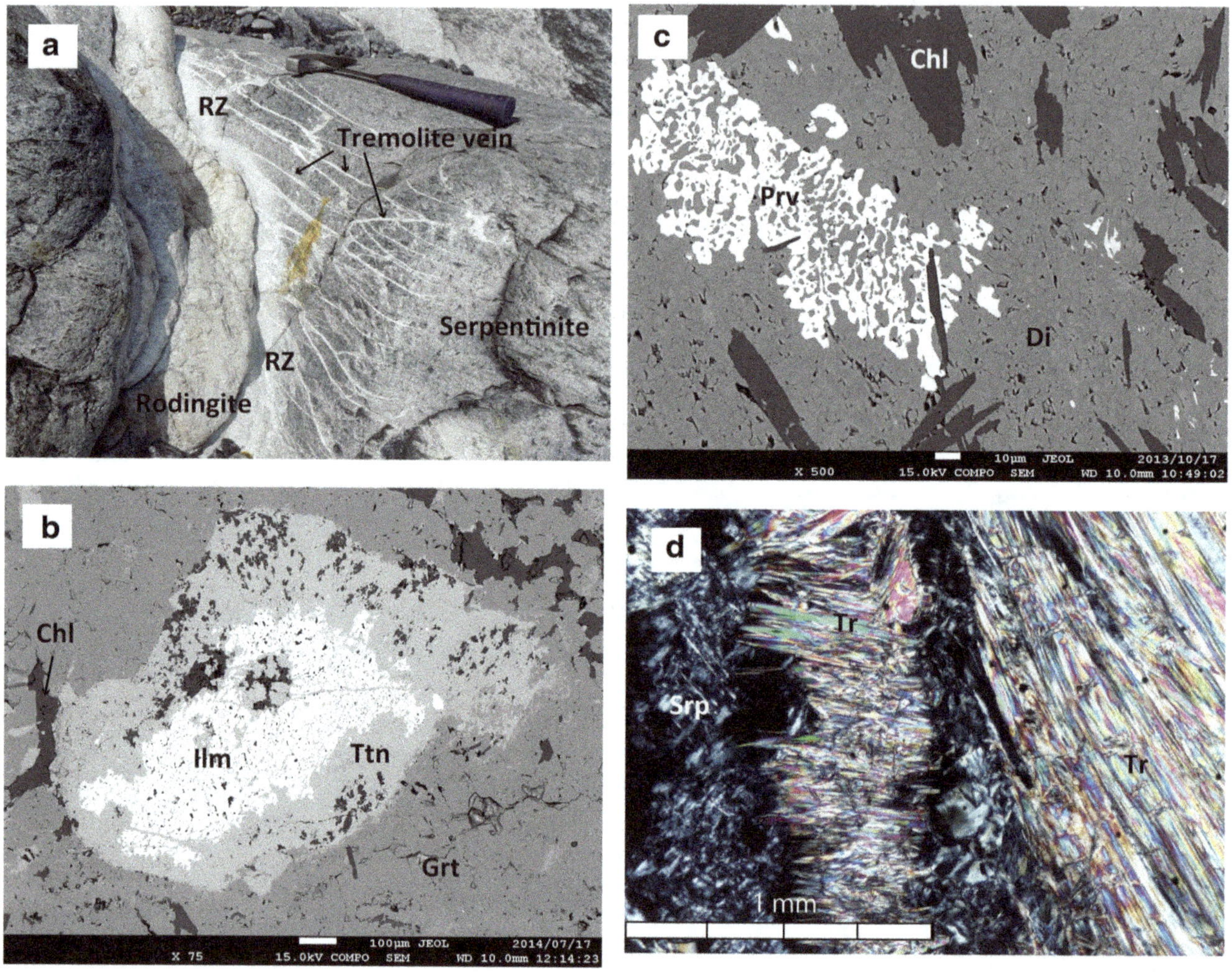

Fig. 4 Photographs, BSE images and photomicrograph of vein-type rodingite, and serpentinite. **a** Outcrop showing a vein-type rodingite (131018-4A) and tremolite veins in serpentinite derived from the reaction zone (RZ). **b** Titanite crystal containing ilmenite in its core, which was taken from the rodingite (131018-4A). **c** Perovskite crystal showing a symplectic texture with diopside, likely implying a pseudomorph after titanite. **d** Photomicrograph of tremolite veins in serpentinite under crossed polars

Results and Discussion

Rodingite protolith and rodingite-forming reactions

Rodingite can be formed from various protoliths such as gabbroic rocks (Coleman 1967; O'Brien and Rodgers 1973; Honnorez and Kirst 1975; Leach and Rodgers 1978; Muraoka 1985; Kobayashi and Kaneda 2010), basaltic rocks (Coleman 1967; Evans et al. 1979; Barriga and Fyfe 1983), amphibolite (Karakida 1980), diorite (Hatzipanagioutou and Tsikouras 2001), quartzite (El-Shazly and Al-Belushi 2004), and granite (Wares and Martin 1980). In the study area, gabbro (now metagabbro) is the only possible protolith of rodingite as no other rock types are present as dykes in the serpentinite. The gabbroic texture (Fig. 2b) in the dyke-type rodingite (131018-1B) described above also supports this conclusion. This subsection derives rodingite-forming reactions by the SVD method (Fisher 1989) applied to mineral assemblages of the metagabbro (131019-MG) and of the dyke-type rodingite (131018-1B) and discusses the mass transfer associated with rodingite formation in the study area. The metagabbro may have more or less undergone metasomatism but not as much as the rodingites. Here, we assume the metagabbro retained its original composition of the protolith gabbro. The SVD method is a simple but powerful type of total least square method (Van Huffel and Vanderwalle 1991), and that can systematically survey the reaction relations among minerals. First, we construct a matrix of mineral compositions given in atomic proportions per formula unit in the system CaO–SiO_2–AF ($Al_2O_3 + Fe_2O_3$)–FM (FeO + MgO) as shown in Table 5. Combinations of two oxides into a single component such as AF and FM are simplification in the analysis

Table 4 Average compositions of minerals from vein-type rodingite (131018-4A), serpentinite, and reaction zones

	Rodingite (131018-4A)					Reaction zone				Serpentinite	
	Di	Grt	Chl	Ilm	Ttn	Chl	Di	Prv	Ves	Srp	Tr
	n = 4	*n* = 11	*n* = 1	*n* = 4	*n* = 5	*n* = 21	*n* = 3	*n* = 10	*n* = 5	*n* = 9	*n* = 7
SiO_2	54.56	39.04	27.76	–	30.48	34.15	53.30	0.06	35.97	41.19	58.20
TiO_2	–	1.13	–	53.28	39.12	–	1.61	57.91	4.97	–	–
Al_2O_3	–	20.19	19.83	–	1.10	16.18	–	–	12.00	4.79	–
Fe_2O_3	–	4.96	–	–	–	–	–	–	–	–	–
FeO	5.78	–	15.87	44.68	–	3.55	1.59	0.68	5.83	2.86	2.07
MnO	–	–	–	1.62	–	–	–	–	–	–	–
MgO	14.50	–	21.18	0.42	–	33.86	16.89	–	2.12	36.19	23.34
CaO	25.47	34.70	–	0.05	28.89	–	26.38	40.95	35.99	–	13.17
Total	100.31	100.02	84.64	100.05	99.59	87.74	99.77	99.60	96.88	85.03	96.78
O	6	12	28	3	5	28	6	3	78	116	23
Si	2.01	2.96	5.74	–	1.00	6.36	1.95	–	19.05	32.40	8.02
Ti	–	0.06	–	0.76	0.96	–	0.05	0.98	1.99	–	–
Al	–	1.81	4.83	–	0.04	3.55	–	–	7.44	4.43	–
Fe^{3+}	–	0.28	–	–	–	–	–	–	–	–	–
Fe^{2+}	0.18	–	2.74	0.94	–	0.55	0.05	0.01	2.65	1.91	0.24
Mn	–	–	–	0.03	–		–	–	–	–	–
Mg	0.80	–	6.53	0.02	–	9.40	0.92	–	1.65	42.46	4.79
Ca	1.01	2.82	–	0.25	1.01	–	1.04	0.99	20.45	–	1.94
Σcation	4.00	7.93	19.84	2.00	3.01	19.86	4.01	1.98	53.23	81.20	14.99
X_{Mg}	0.82		0.70			0.94	0.95			0.96	0.95
X_{Grs}		0.87									
X_{Adr}		0.13									

because combined oxides take part in exchange reactions if treated separately, rather than in net-transfer reactions. The reactants are diopside (Di), chlorite (Chl), and epidote (Ep) in the metagabbro, and diopside (Di^R), garnet (Grt^R), and chlorite (Chl^R) in the rodingite. The superscript R in abbreviated names denotes minerals in rodingite. The SVD analysis gives linear dependence among these minerals, corresponding to the stoichiometry in reaction relations. Among them, only those satisfying proper sign convention (negative for reactants and positive for products) are judged as proper reactions. We first tried to determine the reaction relations from the four-component system but were unable to get proper reaction relations. We then tried the three-component subsystems, in which the arbitrary component is subtracted. Table 5 shows the three reactions, reactions 1 to 3, in the SiO_2–AF–FM subsystem. All these reactions, however, cannot be considered as proper reactions because reactions 1 and 2 do not satisfy the sign convention, and reaction 3 produces low quantities of garnet. We then examined linear combinations of these reactions, which can also give reaction relations among these minerals. As a result, four reactions in Table 5, [Di^R]2&3, [Grt^R]2&3, [Di]3&1, and [Grt]3&1, were obtained as proper reactions, where the mineral name in parentheses stands for mineral phase absent from the reaction, and 2&3 following the parenthesis denotes linear combination of the reactions, e.g., reactions 2 and 3 and so on. Among the four reactions, only [Di^R]2&3 can be the reaction describing rodingite formation because [Grt^R]2&3 and [Grt^R]3&1 do not produce garnet and [Di]3&1 describes the compositional change of chlorite with participation of minor amounts of other minerals. Thus we get the following reaction:

$$1.00\ \mathrm{Di} + 0.74\ \mathrm{Chl} + 1.56\ \mathrm{Ep} + 1.88\ \mathrm{CaO} + 0.18\mathrm{H_2O} = 2.00\ \mathrm{Grt^R} + 0.86\ \mathrm{Chl^R}\ \text{(R4, Table 4)},$$

where moles of open components, CaO and H_2O, are calculated using the data in Table 5. This reaction implies that rodingite formation can be explained by the addition of considerable amounts of CaO and H_2O to the protolith (gabbro), whereas other components such as SiO_2, AF, and FM are preserved. The subsystem CaO–SiO_2–FM does not yield a proper reaction. The subsystems CaO–AF–FM and CaO–SiO_2–AF each yield only one proper reaction, which are given as follows:

Table 5 Matrix for SVD analysis of rodingite-forming reactions and the results

	Reactants (131019-MG)			Products (131018-1B)		
	Di	Chl	Ep	Di^{Ra}	Grt^{Ra}	Chl^{Ra}
Ca	0.93	0	2.03	0.96	2.98	0.02
Si	1.95	5.82	3.02	1.98	3.01	5.76
$Al + Fe^{3+}$	0.09	4.66	2.95	0.05	1.99	4.84
$Mg + Fe^{2+}$	1.01	9.32	0	0.99	0	9.20
SiO2–AF–FM subsystem						
	Di	Chl	Ep	Di^R	Grt^R	Chl^R
Reaction 1	−0.669	0.158	0.107	0.689	−0.125	−0.161
Reaction 2	−0.316	0.062	−0.566	−0.101	0.752	−0.017
Reaction 3	−0.127	−0.685	−0.049	0.204	0.010	0.686
Linear combination of reactions 2&3						
[Di^R]2&3(R4)	−1.000	−0.736	−1.557	0.000	1.996	0.855
[Grt^R]2&3	−1.000	−5.590	−0.344	1.670	0.000	5.592
Linear combination of reactions 3&1						
[Di]3&1	0.000	−1.000	−0.097	0.102	0.046	1.002
[Grt^R]3&1	−1.000	−3.789	−0.232	1.441	0.000	3.793
CaO–AF–FM subsystem						
	Di	Chl	Ep	Di^R	Grt^R	Chl^R
Reaction 5	−1.000	−2.677	−1.051	0.000	1.009	2.822
$CaO–SiO_2–AF$ subsystem						
	Di	Chl	Ep	Di^R	Grt^R	Chl^R
Reaction 6	−1.000	0.000	−1.530	0.131	1.309	0.411

[a]R denotes minerals in rodingite

$$1.00\ Di + 2.68\ Chl + 1.05\ Ep + 0.64\ H_2O = 1.01\ Grt^R + 2.82\ Chl^R + 1.41\ SiO_2 (R5),$$

and

$$1.00\ Di + 1.53\ Ep + 2.91\ FM + 2.53\ H_2O = 0.13\ Di^R + 1.31\ Grt^R + 0.41\ Chl\ (R6),$$

respectively. Large amounts of chlorites (Chl and Chl^R) appear on both sides of the reaction (R5), whereas the mode of chlorite (7 vol.%) in the metagabbro is small compared to that of diopside (65 vol.%). The reaction (R6) produces both diopside and garnet, whereas in the rodingite, diopside is replaced by garnet (Fig. 2e). For these reasons, reactions (R5) and (R6) cannot be considered representative reactions in rodingite formation. Therefore, the reaction (R4) is the most appropriate reaction describing rodingite formation. This result supports the theory that the infiltration of Ca-rich fluids leads to the formation of rodingite (e.g., O'Brien and Rodgers 1973; Leach and Rodgers 1978; Wares and Martin 1980; Schandl et al. 1989; Python et al. 2011) rather than loss of SiO_2 from the protolith and the resultant enrichment of CaO in rodingite (e.g., Coleman, 1967). Serpentinization can produce Ca-rich fluid (e.g., Honnorez and Kirst 1975; Evans et al. 1979), and present-day serpentinization accompanies Ca^{2+}–OH^--type waters (Barnes and O'Neill 1978), thus validating our conclusion.

Extent of equilibrium in reaction zones

This section describes the compositional variations of diopside and chlorite in the rodingite to serpentinite through reaction zones, showing disequilibrium crystallization of diopside in the reaction zones. Figure 5 shows the compositional variation of chlorite across the reaction zone from serpentinite to rodingite. The chlorite shows a monotonous compositional change, and Mg and Si decrease from serpentinite to rodingite with reverse trends of Al and Fe. This change is due to the reaction between the rodingite (or its protolith) that is richer in Al_2O_3 and poorer in MgO, and the serpentinite with a higher Mg# (MgO/(MgO + FeO) in weight percent). A decrease of Si in the chlorite from serpentinite to rodingite indicates Tschermak's substitution, rather than SiO_2 transfer from serpentinite to rodingite. Figure 6 shows another compositional cross section indicating variations of Ca, Al, Mg, and Fe in the diopside across the reaction zone. The diopside is homogeneous in serpentinite but becomes quite heterogeneous in the reaction zone, showing serrated variations in Ca, Al, and Fe. In close proximity to the rodingite, diopside occurs as composite grains of Fe-rich and Fe-poor compositions as described in the "Method/Experimental" section. Because this feature is also observed in the rodingite, the region of composite diopside grains is named "two-diopside zone" as shown in Fig. 6. The two-diopside zone does not coincide with reaction zone 1. Reaction zones 1 and 2 in this rock are defined by their modal compositions, whereas the two-diopside zone is defined by the occurrence of composite diopside grains. Figures 6a, b show compositional variations of Fe-rich diopside and Fe-poor diopside, respectively. In the region outside the two-diopside zone, the data in Fig. 6b are the same as those in Fig. 6a. Thus, diopside shows a disequilibrium relation not only across the reaction zone (serrated variation) but also within grains, where the composite gains consist of Fe-rich and -poor parts. These compositional heterogeneities, together with the occurrence of fine-grained diopside, clearly indicate that disequilibrium crystallization occurred in a relatively short-time period compared to that of the metamorphism. That is, the two diopsides crystallized simultaneously, reflecting the local bulk compositions. The reason why the chlorite is more homogeneous than diopside may be possible because diffusion in chlorite was significantly faster than that in diopside, due to its crystal structure.

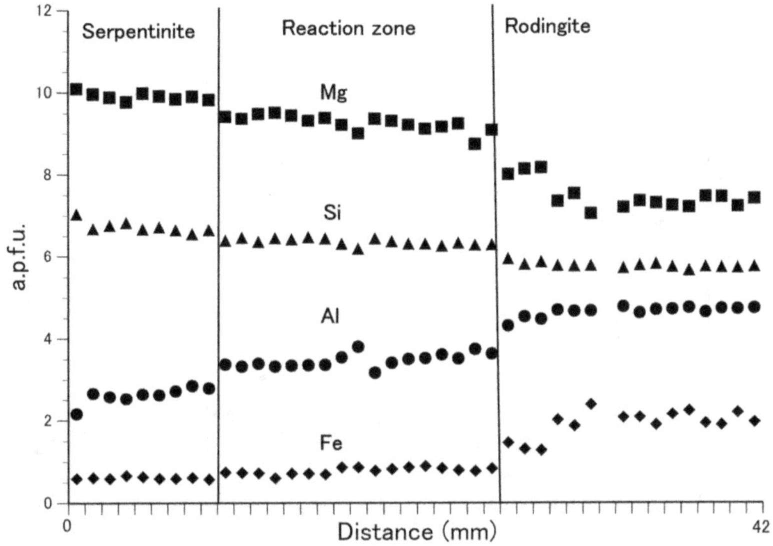

Fig. 5 Compositional cross section across the reaction zone between serpentinite and rodingite (131019-2). In this plot showing variations of Si, Al, Mg, and Fe (apfu) of chlorite, Mg and Si decrease monotonously from serpentinite to rodingite, whereas Al and Fe show reverse trends

Significance of perovskite

Perovskite ($CaTiO_3$) occurs in several reaction zones between serpentinite and rodingite as described in "Method/ Experimental" section (Table 1). Perovskite associated with rodingitization has also been reported in Slovakia by Li et al. (2014). They recognized two types of perovskite: one in rodingite and the other in the chlorite-rich blackwall zone (reaction zone). They used the U/Pb secondary iron mass spectrometer (SIMS) to determine the concordia ages of 137-135 Ma for the former and 134 Ma for the latter type of perovskite. They concluded that such a negligible age difference implies that a relatively short-lived rodingitization event is responsible for the crystallization of both the perovskite generations. Perovskite is a relatively rare mineral because it is stable only in silica-deficient environments. If excess silica is present, perovskite will form titanite ($CaTiSiO_5$). As shown in Table 1, titanite occurs in some of the rodingites in the study area, whereas perovskite occurs only in the reaction zones. Perovskite occasionally shows a symplectic structure with diopside (Fig. 4c), indicating a pseudomorph after titanite. This relation clearly indicates that perovskite is formed by the breakdown of titanite in the silica-deficient environment of the reaction zone. This constraint is critical in subsequent discussions on reaction relations forming the reaction zone as derived by the SVD method.

Interplay of irreversible reactions and hydrofracturing

This subsection derives reaction relations that formed the reaction zone between the serpentinite and rodingite, to explain why and how the tremolite veins develop in serpentinites located close to the reaction zone (Fig. 4a). Taking the vein-type rodingite 131048-4A as an example, the CaO–SiO_2–AF ($Al_2O_3 + Fe_2O_3$)–FM (FeO + MgO)–H_2O five-component system is considered. Antigorite from serpentinite, diopside, garnet, and chlorite from rodingite, and diopside^{RZ} and chlorite^{RZ} from the reaction zone are considered. Diopside is quite heterogeneous in composition, and the average composition of Fe-poor diopside is used because it is much more dominant than Fe-rich diopside. The heterogeneity in the diopside is mostly due to Fe^{2+}–Mg exchange, and it does not affect our analysis because Fe^{2+} and Mg are treated as a single component (FM). The superscript RZ denotes minerals from the reaction zone because they have compositions different from those in the rodingite. Titanite, perovskite, and vesuvianite were not included in the analysis because they are present only in small amounts. We determined whether the reactions between serpentinite and rodingite that lead to the formation of the reaction zone evolve water or not under a constraint of constant solid volume before and after the reactions. The constant solid volume condition was incorporated because no deformation was observed in the reaction zone. If the reactions evolved water, it inevitably caused hydrofracturing (Nishiyama, 1989a) because serpentinite has low permeability. First, we attempted to find reaction relations that hold the conservation of pseudoquarternary components CaO–SiO_2–AF ($Al_2O_3 + Fe_2O_3$)–FM (FeO + MgO) among the mineral assemblage containing all minerals occurring in the serpentinite, rodingite, and the reaction zone. Because no reaction relation was found, we considered CaO to be a mobile component (i.e., a component that was not preserved before and after the reaction) and omitted it from the system. We then attempted to find reaction relations

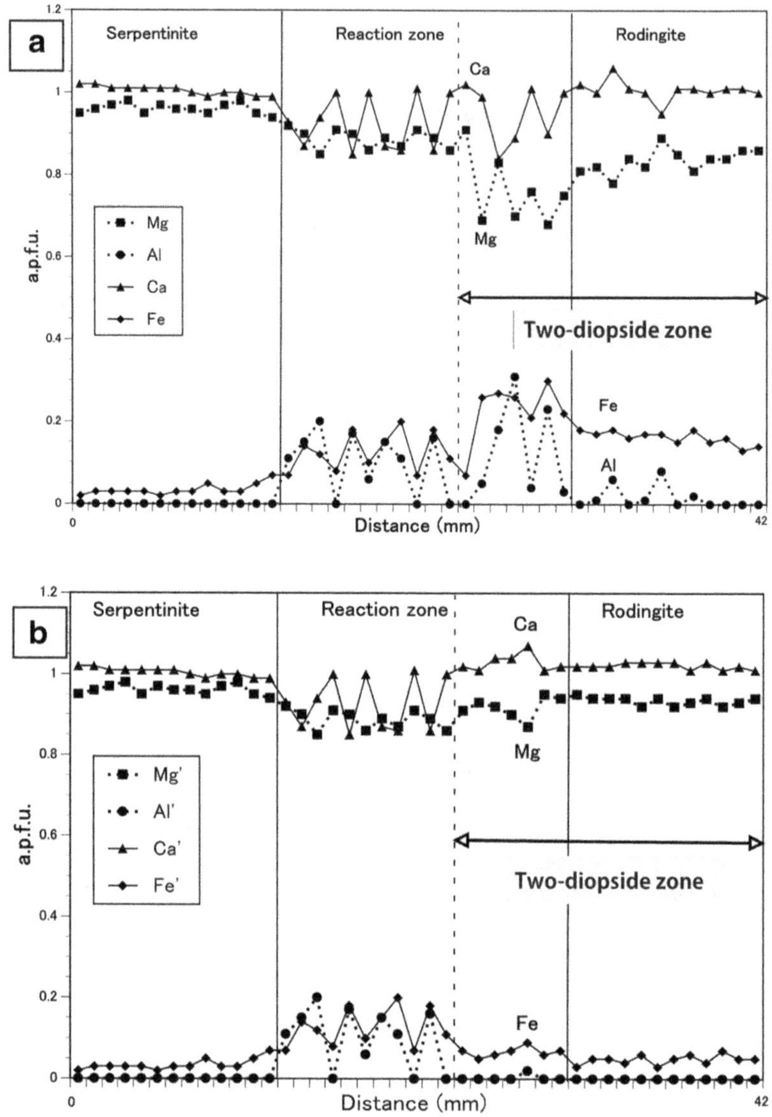

Fig. 6 Compositional cross section across the reaction zone between serpentinite and rodingite (131019-2). **a** Compositional variation of Fe-rich diopside in the "two-diopside zone". **b** Compositional variation of Fe-poor diopside in the "two-diopside zone"; data outside this zone are the same as those in **a**. The diopside is homogeneous in the serpentinite, whereas it becomes quite heterogeneous both in the reaction zone and in the rodingite. Serrated variations in Ca and Al are notable in the reaction zone, and Fe-rich and -poor diopsides coexist in the rodingite and also in the reaction zone close to the rodingite (in the region entitled "two-diopside zone" in the figure)

that hold the conservation of pseudoternary components SiO_2–AF–FM for the same mineral assemblage. Table 6 shows the results for the SiO_2–AF–FM system. It should be noted that a row consisting of molar volumes of corresponding minerals was added to the composition matrix for SVD analysis to ensure the constant solid volume condition, as per the method in Yuguchi et al. (2015). The molar volume data were taken from Helgeson et al. (1978), considering the chlorite as clinochlore and the garnet as grossular. Two reactions, reaction 7 and reaction 8 in Table 6, were found in the system SiO_2–AF–FM. Only reaction 7 satisfied the sign convention relation, which is negative for reactants and positive for products. Reaction 7 can be rewritten so as to normalize the stoichiometric coefficient of antigorite as unity as follows:

$$1\text{Atg} + 10.23\text{Di} + 3.14\text{Grt} + 0.66\text{Chl} + 4.39\text{CaO} = 51.80\text{Di}^{RZ} + 4.01\text{Chl}^{RZ} + 4.26\text{H}_2\text{O}. \quad (R7)$$

Two more reactions were obtained from the linear combinations of reaction 7 and reaction 8:

Table 6 Matrix for SVD analysis of 131018-4A and the results

	Reactants				Products	
	Atg	Di	Grt	Chl	Di^{RZ}	Chl^{RZ}
Si	32.397	2.01	2.96	5.74	1.95	6.36
AF	4.433	0	2.09	4.83	0	3.55
FM	44.37	0.98	0	9.27	0.97	9.95
V[a]	41.805	1.5796	2.9947	9.9	1.5796	9.9
Results in Si–AF–FM system						
Reaction 7	−0.015	−0.612	−0.048	−0.010	0.787	0.061
Reaction 8	−0.032	−0.197	0.191	−0.590	−0.206	0.730
Linear combination of reactions 7&8						
[Grt](R9)	−1.000	−28.590	0.000	−6.830	31.840	10.550
[Chl](R10)	−1.000	−41.480	−3.480	0.000	53.950	3.300
Results in AF–FM subsystem						
Reaction 11	−0.022	0.792	−0.276	0.405	0.037	−0.361
Reaction 12	−0.054	−0.144	−0.231	0.097	0.953	0.071
Reaction 13	−0.089	0.515	−0.087	−0.421	0.040	0.735
Linear combination of reactions 11&12						
[Grt](R14)	−1.000	−22.714	0.000	−6.790	23.013	10.487
[Chl](R15)	−1.000	−6.882	−3.386	0.000	19.460	3.243
Liner combination of reactions 12&13						
[Di](R16)	−1.000	0.000	−3.235	−0.264	12.245	3.514
[Chl](R17)	−1.000	−0.342	−3.369	0.000	12.945	3.233

[a]V denotes molar volume (cal/bar) from Helgeson et al.(1978)

$1Atg + 28.59Di + 6.83Chl + 4.24CaO = 31.84\,Di^{RZ} + 10.55Chl^{RZ} + 1.27H_2O$ [Grt] (R9) and
$1Atg + 41.49Di + 3.48Grt + 4.40CaO = 53.95Di^{RZ} + 3.30Chl^{RZ} + 4.59H_2O$ [Chl] (R10).

All three reactions consume CaO and evolve H_2O, strongly suggesting an association between the increment of fluid pressure with the progress of the reactions. We have further examined the AF–FM system (Table 6) to get the following reactions:

$1Atg + 22.71Di + 6.79Chl + 4.11CaO + 0.40SiO_2 = 26.01Di^{RZ} + 10.49Chl^{RZ} + 1.43H_2O$ [Grt] (R14);
$1Atg + 6.88Di + 3.39Grt + 3.74CaO + 3.32SiO_2 = 19.46Di^{RZ} + 3.24Chl^{RZ} + 5.06H_2O$ [Chl] (R15);
$1Atg + 3.24Grt + 0.26Chl + 3.61CaO + 2.74SiO_2 = 12.25Di^{RZ} + 3.51Chl^{RZ} + 5.01H_2O$ [Di] (R16); and
$1Atg + 0.34Di + 3.37Grt + 3.62CaO + 2.75SiO_2 = 12.95Di^{RZ} + 3.23Chl^{RZ} + 5.14H_2O$ [Chl] (R17).

All of these reactions consume SiO_2 and CaO and evolve H_2O. This relation holds if we take any linear combination of the above four reactions. When we consider the perovskite formation from titanite in the reaction zone, reactions in the AF–FM system are preferable because SiO_2 liberated by the titanite breakdown will be consumed by the above reactions in the AF–FM system. To conclude, we have successfully derived reaction relations in the rodingite–serpentinite system under constant solid volume conditions. All reactions evolve H_2O, which strongly suggests that hydrofracturing is associated with these reactions. The results are consistent with the field observation that the tremolite veins are derived from the reaction zone into the serpentinite (Fig. 4a).

The above discussions are based on the assumption of conservation of solid volume during the formation of the reaction zone, which will be a strong constraint for deriving the reaction relations. Apart from vein formation, there is no evidence of deformation due to strain accumulation around the reaction zones in the outcrops. This observation may not be strong enough to support the assumption; therefore, we examined the reactions without the volume constraint for comparison. Table 7 summarizes the results, with values of ΔV representing the relative change in solid volume due to the reaction. The four-component $CaO–SiO_2–AF–FM$ system does not yield proper reaction relations. We examined the reactions in the four subsystems: $CaO–SiO_2–AF$, $CaO–SiO_2–FM$, CaO–AF–FM, and $SiO_2–AF–FM$. We found two possible reactions in each subsystem, all of which are dehydration reactions. The two reactions in the $CaO–SiO_2–AF$ subsystem consume large amounts of the FM component, whereas the two reactions in the $CaO–SiO_2–FM$ subsystem evolve large amounts of the AF component. If this actually occurred, some chlorite veins would have been present around the reaction zones; however, these were none. Thus, there is little evidence to support the theory

Table 7 Reaction relations without volume constraints

$CaO–SiO_2–AF$ subsystem	ΔV
$1.00Atg + 12.98Di + 4.28Grt + 3.92FM = 24.21Di^{RZ} + 3.77Chl^{RZ} + 0.83H_2O$	0.60%
$1.00Atg + 3.84Di + 4.11Grt + 2.77FM = 14.86Di^{RZ} + 3.67Chl^{RZ} + 1.67H_2O$	−0.60%
$CaO–SiO_2–FM$ subsystem	
$1.00Atg + 12.67Di + 8.06Grt = 34.16Di^{RZ} + 2.39Chl^{RZ} + 61.38AF + 11.90H_2O$	−70.80%
$1.00Atg + 7.14Di + 7.27Grt = 26.65Di^{RZ} + 2.57Chl^{RZ} + 45.34AF + 10.41H_2O$	−61.69%
CaO–AF–FM subsystem	
$1.00Atg + 9.47Di + 3.81Grt = 19.53Di^{RZ} + 3.49Chl^{RZ} + 2.43SiO_2 + 3.07H_2O$	−4.10%
$1.00Atg + 1.80Di + 3.77Grt = 11.98Di^{RZ} + 3.47Chl^{RZ} + 1.75SiO_2 + 3.24H_2O$	−4.80%
$SiO_2–AF–FM$ subsystem	
$1.00Atg + 7.11Di + 3.59Grt + 1.85CaO = 18.43Di^{RZ} + 3.36Chl^{RZ} + 4.08H_2O$	−2.20%
$1.00Atg + 0.34Di + 3.62Grt + 1.35CaO = 11.44Di^{RZ} + 3.38Chl^{RZ} + 3.97H_2O$	−3.10%

of the mass transfer of these components. In the CaO–AF–FM subsystem, SiO_2 is liberated, which is contrary to the occurrence of perovskite in the reaction zone. As a result, no reaction without volume constraint is consistent with our observation of the reaction zone. These considerations strongly support the validity of our treatment of the solid volume conservation.

Condition for hydrofracturing

Hydrofracturing requires the accumulated fluid pressure to be higher than the lithostatic pressure. The condition for hydrofracturing can be written as

$$P_{\text{fluid}} > \sigma_3 + T_0, \tag{1}$$

where σ_3 denotes the minimum principal stress, and T_0 is the tensile strength of the rock. Such an accumulation of fluid pressure can be possible if the following two conditions are fulfilled (Nishiyama, 1989a): (1) fluid production rate is significantly faster than the fluid advection rate and (2) the rock deformation rate is small enough to prevent the relaxation of fluid pressure. These conditions are more likely to be satisfied by contact metamorphism rather than regional metamorphism because the reaction rate is significantly faster in contact metamorphism because heat transfer is more rapid than that it is in regional metamorphism. The rock deformation is also limited in contact metamorphism. However, our case study showed highly irreversible conditions for the formation of the reaction zone, implying rapid reaction rates. In such a case, if the country rocks (serpentinite in our case) are impermeable, the fluid pressure can be accumulated until the conditions required for hydrofracturing are met. The serpentinite in the study area is massive and not porous, implying very low permeability. These conditions made it possible for hydrofracturing to occur in the serpentinite located close to the reaction zone.

Tremolite vein formation

Tremolite vein formation can be explained by considering real chemical species in metamorphic fluid rather than oxide components used above for mass balance calculations. The hydrolysis reaction of a mineral can be expressed in a general form (Helgeson, 1968) as

$$MO_{z/2} + zH^+ = M^{z+} + (z/2)H_2O, \tag{2}$$

where M represents a cation with positive valency z. For example, the hydrolysis reaction of diopside can be written as

$$CaMgSi_2O_6 + 12H^+ = Ca^{2+} + Mg^{2+} + 2Si^{4+} + 6H_2O. \tag{3}$$

This reaction implies that diopside is in equilibrium with cations in the metamorphic fluid such as H^+, Ca^{2+}, Mg^{2+}, and Si^{4+}. The equilibrium condition can be written as follows by assuming the activities of pure solid (diopside) and water are unity:

$$\mathrm{dln}\left(\frac{a_{ca^{2+}}}{a^2_{H^+}}\right) + \mathrm{dln}\left(\frac{a_{Mg^{2+}}}{a^2_{H^+}}\right) + 2\mathrm{dln}\left(\frac{a_{Si^{4+}}}{a^4_{H^+}}\right) = 0. \tag{4}$$

Similarly, we get the following equilibrium conditions for antigorite and tremolite:

$$48\mathrm{dln}\left(\frac{a_{Mg^{2+}}}{a^2_{H^+}}\right) + 34\mathrm{dln}\left(\frac{a_{Si^{4+}}}{a^4_{H^+}}\right) = 0. \tag{5}$$

for antigorite and

$$2\mathrm{dln}\left(\frac{a_{Ca^{2+}}}{a^2_{H^+}}\right) + 5\mathrm{dln}\left(\frac{a_{Mg^{2+}}}{a^2_{H^+}}\right) + 8\mathrm{dln}\left(\frac{a_{Si^{4+}}}{a^4_{H^+}}\right) = 0 \tag{6}$$

for tremolite, respectively. Figure 7 shows a schematic relation of these equilibrium conditions in a $\mathrm{In}\left(\frac{a_{Ca^{2+}}}{a^2_{H^+}}\right)$ vs. $\mathrm{In}\left(\frac{a_{Mg^{2+}}}{a^2_{H^+}}\right)$ diagram under a fixed value of $\mathrm{In}\left(\frac{a_{Si^{4+}}}{a^4_{H^+}}\right)$.

Because activities of Mg^{2+} and Si^{4+} are buffered by a large amount of antigorite, they can be considered to be constant during the formation of the reaction zone. Consider the initial condition as A in Fig. 7, in which both antigorite and

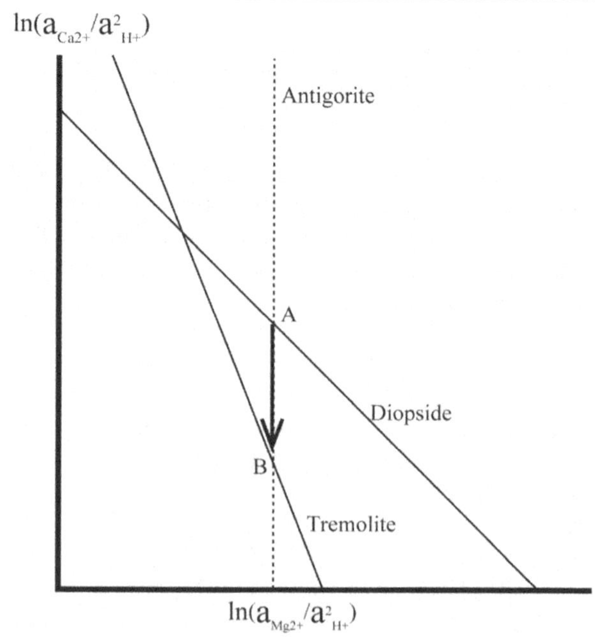

Fig. 7 Activity diagram showing the saturation lines of diopside and tremolite (*solid lines*) and that of antigorite (*broken line*), showing the equilibrium conditions of the minerals under constant activity of Si^{4+}. The *arrow* indicates a change from the equilibrium point A of diopside and antigorite to that B of tremolite and antigorite within a hydrofractured vein

diopside are in equilibrium with the fluid in the serpentinite. Once a fracture is formed in the serpentinite by hydrofracturing, the activity of Ca^{2+} in the fluid filling the fracture will decrease owing to the addition of water evolved from the reaction forming the reaction zone. Then, the fluid composition will shift down from the point A to the new equilibrium point B shown in Fig. 7, resulting in the formation of tremolite. This is possible because the activity of Ca^{2+} will not be buffered by diopside owing to its small amount in serpentinite.

Tectonic setting of rodingite formation

Rodingites have been reported in ophiolites from various tectonic settings, mostly from the seafloor and rarely from the subduction zones (Koutsovitis et al. 2013). Honnorez and Kirst (1975) reported rodingites from the mid-Atlantic fracture zones together with serpentinized ultramafics and fresh and metamorphosed basalts and gabbros. They discussed that serpentinization took place in the deeper part of the oceanic crust and that rodingitization occurred simultaneously with the host serpentinization. Such rodingites from the ocean floor or from the oceanic lithosphere have also been reported by Frost et al. (2008), Austrheim and Prestvick (2008), and Bach and Klein (2009). Subduction-related rodingites have been reported in East Othris, Greece, by Koutsovitis et al. (2013) and in the Western Carpathians, Slovakia by Li et al. (2014). Koutsovitis et al. (2013)) discussed that the rodingitization took place in an intra-oceanic subduction system in three successive stages during the exhumation of the mafic–ultramafic mantle wedge rocks in a fore-arc setting within a serpentinite subduction channel. Li et al. (2014) found perovskite-bearing rodingites in a mélange type blueschist-bearing accretionary wedge complex. In our study area, Fukuyama et al. (2014) carried out geochemical studies on the rodingite. They found two types of zircons, prismatic, and porous types, from dyke-type rodingites. This indicates that Hafnium isotopic compositions of prismatic zircons are close to or identical to the Hf isotope ratio of the mid-ocean ridge basalt (MORB). They also showed that the high field strength element (HFSE) concentrations in the rodingites are similar to those of MORBs. They further determined that the low ε_{Hf} values (11.8–18.9) of porous zircons indicate the incorporation of a small amount of Hf from fluid contaminated by subducted sediments. These lines of evidence clearly show that the protoliths of the rodingites in the Nomo metamorphic rocks are MORB-like basic rocks, suggesting the tectonic setting of the oceanic lithosphere. The rodingitization may have occurred in the oceanic lithosphere simultaneously with serpentinization and may have continued to the stage of the subduction as indicated by the occurrence of high-pressure minerals such as omphacite and alkali amphiboles from the serpentinite mélanges. Because antigorite is rare in oceanic serpentinites (Bach and Klein 2009), the abundant occurrence of antigorite in the Nomo serpentinite supports the continuing rodingitization in the subduction setting. The synchronous rodingitization with serpentinization is attributed to the hydrolysis reaction of diopside in serpentinite:

$$CaMgSi_2O_6 + 12H^+ = Ca^{2+} + Mg^{2+} + 2Si^{4+} + 6H_2O. \quad (7)$$

This reaction will evolve the Ca^{2+} necessary for rodingitization of the basic rocks in the serpentinite. It should be noted that the above reaction also evolves Si^{4+}, but the activity of Si^{4+} will be buffered by the hydrolysis reaction of antigorite. Therefore, the silica-deficient environments for rodingitization will not contradict with the above reaction.

Conclusions

The singular value decomposition (SVD) method is a powerful tool by which reaction relations among minerals with complex compositions can be systematically derived and the amount of mobile (open) components evaluated. Incorporation of molar volume data into the SVD matrix can assure the constant solid volume condition. The resulting reactions evolve H_2O, which strongly suggests that it is the cause for hydrofracturing leading to the formation of tremolite veins in the serpentinite. Heterogeneities in mineral compositions in the reaction zone imply that rapid irreversible reactions occurred during the formation of the reaction zone. This rapid progress of reactions is a necessary condition for hydrofracturing to occur. If the reaction proceeds slowly, the liberated water will escape from the reaction zone into serpentinite by porous flow, resulting in the relaxation of excess fluid pressure. Thus, tremolite vein associated with the reaction zone between the serpentinite and rodingite is a good example of the interplay between irreversible reactions and hydrofracturing. Our analysis clearly indicates that hydrofracturing can occur even in regional metamorphism if the reactions liberating H_2O occur irreversibly.

Abbreviations

apfu: Atoms per formula unit; SVD: Singular value decomposition

Acknowledgements

The authors appreciate two anonymous reviewers for their constructive reviews and the editor T. Hirajima for his valuable suggestions and editorial handling. We thank D. Shiosaki for his help in our fieldwork and also in the EDS analysis. Discussions with the members of "Henseigan-nado (Metamorphic rocks etc.) Symposium" are acknowledged. Co-works in the SVD analysis with T. Yuguchi are appreciated.

Funding

This work was financially supported by JSPS KAKENHI Grant Number JP16H02238 to TN.

Authors' contributions

TN designed the study. All authors carried out fieldworks together. Petrography and mineral chemistry were chiefly done by CY-S and TN with supports by YM and MS. The SVD analysis was conducted by TN. All authors read and approved the final manuscript.

Competing interests

The authors declare that they have no competing interests.

Author details

[1]Department of Earth and Environmental Science, Graduate School of Science and Technology, Kumamoto University, 2-39-1, Kurokami, Chuo-ku, Kumamoto 860-8555, Japan. [2]Kitakyushu Museum of Natural History and Human History, 2-4-1, Higashida, Yahatahigashi-ku, Kitakyushu 805-0071, Japan.

References

Austrheim H, Prestvik T (2008) Rodingitization and hydration of the oceanic lithosphere as developed in the Leka ophiolite, north-central Norway. Lithos 104:177–198

Bach W, Klein F (2009) The petrology of seafloor rodingites: insights from geochemical reaction path modeling. Lithos 112:103–117

Barnes I, O'Neill JR (1978) Present day serpentinization in New Caledonia, Oman and Yugoslavia. Geochim Cosmochim Acta 42:144–145

Barriga F, Fyfe WS (1983) Development of rodingite in basaltic rocks in serpentinites, East Liguria, Italy. Contrib Mineral Petrol 84:146–151

Coleman RG (1967) Low-temperature reaction zones and alpine ultramafic rocks of California, Oregon, and Washington. US Geol Surv Bull 1247:1–49

El-Shazly AK, Al-Belushi M (2004) Petrology and chemistry of metasomatic blocks from Bawshir, northeastern Oman. Int Geol Rev 46:904–938

Evans BW, Trommsdorff V, Richter W (1979) Petrology of an eclogite-metarodingite suite at Cima di Ganone, Ticino, Switzerland. Amer Mineral 64:15–31

Fisher GW (1989) Matrix analysis of metamorphic mineral assemblages and reactions. Contrib Mineral Petrol 102:69–77

Frost BR, Beard JS, McCraig A, Condliffe E (2008) The formation of micro-rodingites from IODP hole U1309D: key to understanding the process of serpentinization. J Petrol 49:1579–1588

Fukuyama M, Ogasawara M, Dunkley D, Wang K-L, Lee D-C, Hokada T, Maki K, Hirata T, Kon Y (2014) The formation of rodingite in the Nagasaki metamorphic rocks at Nomo Peninsula, Kyushu, Japan–zircon U–Pb and Hf isotopes and trace element evidence. Island Arc 23:281–293

Grant JA (2005) Isocon anlysis : a brief review of the method and applications. Phys Chem Earth 30:997–1004

Hattori J, Isomi H (1976) Geology of western part of Amakusa-Shimoshima and metamorphic rocks in northwest Kyushu. Monthly Rept Geol Surv Japan 27: 665–682 (in Japanese with English abstract)

Hattori J, Shibata K (1982) Radiometric dating of pre-Neogene granitic and metamorphic rocks in northwest Kyushu, Japan—with emphasis on geotectonics of the Nishisonogi zone. Bull Geol Surv Japan 33:57–84

Hatzipanagiotou K, Tsikouras B (2001) Rodingite formation from diorite in the Samothraki ophiolite, NE Aegean, Greece. Geol J 36:93–109

Helgeson HC (1968) Evaluation of irreversible reactions in geochemical processes involving minerals and aqueous solutions. Geochim Cosmochim Acta 32: 853–877

Helgeson HC, Delany JM, Nesbitt HW, Bird DK (1978) Summary and critique of the thermodynamic properties of rock-forming minerals. Amer J Sci 278-A:1–229

Honnorez J, Kirst P (1975) Petrology of rodingites from the equatorial mid-Atlantic fracture zones and their geotectonic significance. Contrib Mineral Petrol 49:233–257

Igi S, Hattori J, Shibata K (1979) Nomo metagbbro complex and their 450 m.y. ages as a clue to the basement geology. In: Editorial committee of 'The Professor Hiroshi Kano Memorial Volume' (ed). The Basement of the Japanese Islands, Akita University, Akita, pp 261–280, in Japanese with English abstract

Karakida Y (1980) An amphibolite - rodingite xenolith from Tsubokinohana, Kumamoto Prefecture. Sci Rep Dept Geol Kyushu Univ 13:85–97 (in Japanese with English abstract)

Kobayashi S, Kaneda H (2010) Rodingite with Ti- and Cr- rich vesuvianite from the Sartuohai chromium deposit, Xinjiang, China. J Mineral Petrol Sci 105:112–122

Koutsovitis P, Magganas A, Pomonis P, Ntaflos T (2013) Subduction-related rodingites from East Othris, Greece: mineral reactions and physicochemical conditions of formation. Lithos 172–173:139–157

Lang HM, Wachter AJ, Peterson VL, Ryan JQ (2004) Coexisting clinopyroxene/spinel and amphibole/spinel symplectites in metatroctolites from the Buck Creek ultramafic body, North Carolina Blue Ridge. Amer Mineral 89:20–30

Leach TM, Rodgers KA (1978) Metasomatism in the Wairere serpentinite, King Country, New Zealand. Mineral Mag 42:45–62

Li X-H, Putiš M, Yang Y-H, Koppa M, Dyda M (2014) Accretionary wedge hartzburgite serpentinization and rodingitization constrained by perovskite U/Pb SIMS age, trace elements and Sm/Nd isotopes: case study from the Western Carpathians, Slovakia. Lithos 205:1–14

Miyazaki K, Nishiyama T (1989) Petrological Study of the Nagasaki metamorphic rocks in the Nomo Peninsula—with special reference to biotite zone. Mem Geol Soc Japan 33:217–236 (in Japanese with English abstract)

Miyazaki K, Ikeda T, Arima K, Fukuyama M, Maki K, Yui T-F, Grove M (2013) Pressure–temperature structure of a mylonitized metamorphic pile, and the role of advection of the lower crust, Nagasaki Metamorphic Complex, Kyushu, Japan. Lithos 162–163:14–26

Muraoka H (1985) Serpentinization reaction responsible for rodingite formation of the Ashidachi ultramafic complex, Southwest Japan. J Japan Assoc Min Petr Econ Geol 80:413–428

Nishimura Y (1998) Geotectonic subdivision and areal extent of the Sangun belt, Inner Zone of Southwest Japan. J Metamorphic Geol 16:129–140

Nishimura Y, Hirota Y, Shiosaki D, Nakahara N (2004) The Nagasaki metamorphic rocks and their geotectonics in Mogi area, Nagasaki prefecture, Southwest Japan—juxtaposition of the Suo belt with the Sanbagawa belt. J Geol Soc Japan 110:372–383 (in Japanese with English abstract)

Nishiyama T (1989a) Kinetics of hydrofracturing and metamorphic veining. Geology 17:1068–1071

Nishiyama T (1989b) Petrological study of the Nagasaki metamorphic rocks in the Nishisonogi Peninsula—with special reference to the greenrock complex and the reaction enhanced ductility. Mem Geol Soc Japan 33:237–257 (in Japanese with English abstract)

Nishiyama T, Miyazaki K (1987) Omphacite-bearing reaction zone in the Nagasaki metamorphic rocks, the Nomo Peninsula. Sci Rept Dept Geol Kyushu Univ 15:89–101 (in Japanese with English abstract)

O'Brien JP, Rodgers KA (1973) Xonotlite and rodingites from Wairere, New Zealand. Mineral Mag 39:233–240

Python M, Yoshikawa M, Shibata T, Arai S (2011) Diopsidites and rodingtes: serpentinization and Ca-metasomatism in the Oman ophiolite mantle. In: Srivastava RK (ed) Dyke swarms: keys for geodynamic interpretation. Springer, Heidelberg, pp 401–435

Schandl ES, O'Hanley DS, Wicks FJ (1989) Rodingites in serpentinized ultramafic rocks of the Abitibi Greenstone belt, Ontario. Canad Mineral 27:579–591

Shigeno M, Mori Y, Shimada K, Nishiyama T (2012) Jadeitites with metasomatic zoning from the Nishisonogi metamorphic rocks, western Japan: fluid-tectonic block interaction during exhumation. Eur J Mineral 24:289–311

Van Huffel S, Vanderwalle J (1991) The total least squares problem: computational aspects and analysis, Frontiers in Appliede Mathematics Series 9. SIAM, Philadelphia

Wares RP, Martin RF (1980) Rodingitization of granite and serpentinite in the Jeffre Mine, Asbestos, Quebec. Can Mineral 18:231–240

Whitney DL, Evans BW (2010) Abbreviations for names of rock-forming minerals. Amer Mineral 65:185–187

Yuguchi T, Sasao E, Ishibashi M, Nishiyama T (2015) Hydorthermal chloritization processes from biotite in the Toki granite, Central Japan: temporal variations of the compositions of hydrothermal fluids associated with chloritization. Amer Mineral 100:1134–1152

15

Temporal changes in the internal stresses and pore pressures in a large-scale submarine mass transport deposit

Makoto Otsubo[1*], Hajime Naruse[2] and Ayumu Miyakawa[1]

Abstract

We examined the temporal changes in the internal stresses and pore fluid pressures of a submarine mass transport deposit (MTD) in the Akkeshi Formation of the Upper Cretaceous–Paleocene Nemuro Group, eastern Hokkaido Island, Japan. We first analyzed previous paleostress field results from meso-scale faults in the MTD blocks, which indicated two phases during the evolution of the debris flow: phase I, radial spreading of the flow body during downslope movement; phase II, the flow body underwent compression during deposition on the basin plain. We also estimated the pore fluid pressure ratio from the fault orientation distribution. There was a large increase in the pore fluid pressure ratio during the transition from phase I to phase II that continued to rise during the initial stage of phase II and then decreased in its latter stages, whereas the maximum horizontal compressive stress increased throughout phase II. This variation in pore fluid pressure relates to the dynamics and evolution of the debris flow, where the clasts in the central part of the flow were supported by the excess pore pressure due to the compression of the debris flow as the flow head decelerated. Although pore fluid pressure plays a critical role in the dynamics of debris flows, there was no previous methodology to quantify both the stress fields and pore fluid pressures in large debris flows and their resultant MTDs. Our results implemented for outcrop studies imply that meso-scale faults in MTDs can provide clues to better understand these paleoflow mechanisms.

Keywords: Stress, Meso-scale fault, Mass transport, Debris flow, Landslide, Subduction zone

Introduction

Mass transport deposits (MTDs) are the major components of depositional systems in deep-sea environments. For instance, when a large trench-type earthquake occurs in a subduction zone, it has the potential to initiate a large submarine mass transport process, which can in turn generate a hazardous tsunami (e.g., Gennesseaux et al. 1980; Ward 2001; Harbitz et al. 2006; Ogata et al. 2012; Festa et al. 2016). Understanding the development of MTDs is therefore an important goal for landslide hazard mitigation (e.g., Morgan et al. 2009). The concept of mass transport deposits contains a broad spectrum of depositional facies, including undeformed slide blocks facies, slump facies composed of folded and boudinaged successions, and debris flow facies characterized by block-in-matrix structures. Among these MTD facies, this study aims to understand the formative processes of the debris-flow type facies, focusing on a role of excess pore pressure in the flow. Various field observations of modern examples and outcrops have suggested that fluid overpressure represents the most significant mechanism to account for the long run-out and the high velocity of mass transport events (e.g., Canals et al. 2004). Here, we try to measure the extent of this excess pore pressure from outcrops by quantitative textural analysis.

As described above, the behavior of debris flow sediments can be largely controlled by their excess pore pressures (e.g., Iverson 1997; Ilstad et al. 2004a, 2004b; D'Agostino et al. 2013). For instance, the experiments of Iverson et al. (2011) indicated that the acceleration of a debris flow only occurs when there is a positive feedback between overpressure in the debris flow and the entrainment of basal sediment. Excess pore pressure helps to maintain the suspension of sediment in the debris flow,

* Correspondence: otsubo-m@aist.go.jp
[1]Geological Survey of Japan, AIST, AIST Tsukuba Central 7, Higashi-1-1-1, Tsukuba, Ibaraki 305-8567, Japan
Full list of author information is available at the end of the article

which can then scour the bed and reduce basal friction (Major and Iverson 1999). However, many approaches that model the behavior of debris flow sediments only adopt the simple viscoplastic (Bingham) fluid approximation with a constant pore pressure (e.g., Rickenmann et al. 2006). It is therefore necessary to examine the validity of applying this approximation to observed submarine debris flow processes, particularly since the relationships between pore pressure changes and fluid-flow behavior in debris flows are unclear.

Here, we focus on the fluid-driven processes of a submarine MTD, which are related to both its flow dynamics and deformation processes. Faulting is generally attributed to overpressurization of the fault plane (e.g., Nur and Booker 1972; Sibson 2007). Previous studies have indicated that a zone of high pore fluid pressure is an important factor in the activation of faults, owing to the resultant reduction in frictional strength, which in turn leads to an instability on the fault plane (Terzaghi and Peck 1967; Rice 1992; Ogata et al. 2014a). Although basal shear stress and pore fluid pressure measurements have been made in MTDs (e.g., Iverson 1997; Ilstad et al. 2004a, 2004b; D'Agostino et al. 2013), it is difficult to directly measure pore fluid pressures. Previous studies have therefore employed stress tensor inversions to reconstruct the paleostresses within gravitational slope failures (e.g., Husein et al. 2010; Naruse and Otsubo 2011a; Baroň et al. 2013, 2017).

Naruse and Otsubo (2011a) documented the internal structure of a submarine MTD in the Akkeshi Formation of the central section of the Upper Cretaceous–Paleocene Nemuro Group, eastern Hokkaido Island, northern Japan (Figs. 1 and 2). Paleostress analysis of meso-scale faults (Yamaji 2000; Otsubo and Yamaji 2006) within this large-scale MTD revealed that the evolution of the debris flow was defined by two phases: (1) an extensional stress regime due to radial spreading of the debris flow during downslope movement (phase I) and (2) a compressional stress regime during deposition of the debris on the basin plain (phase II) (Fig. 2c; Naruse and Otsubo 2011a).

Here, we examine the temporal changes in the internal stresses and pore fluid pressures in the above-

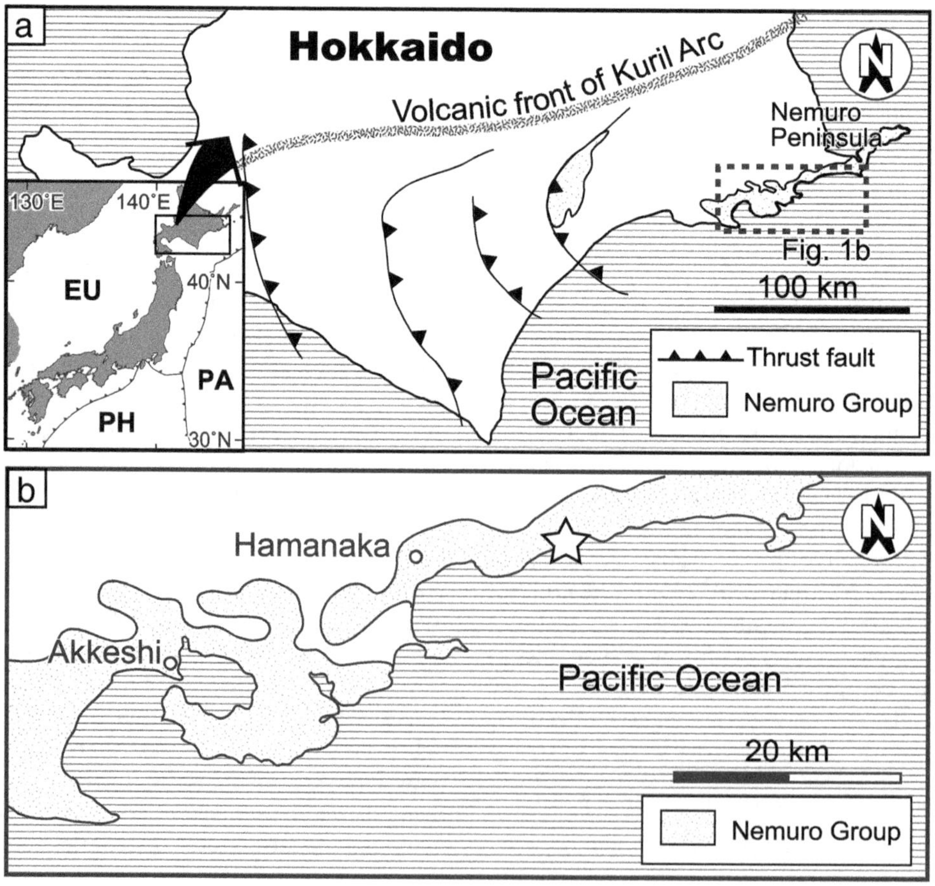

Fig. 1 Geological setting of the study area (modified from Naruse 2003). **a** Distribution of the Nemuro Group. **b** Location of the study area. Star indicates the coastal outcrop. Inset map: EU, Eurasian Plate. PA, Pacific Plate. PH, Philippine Sea Plate

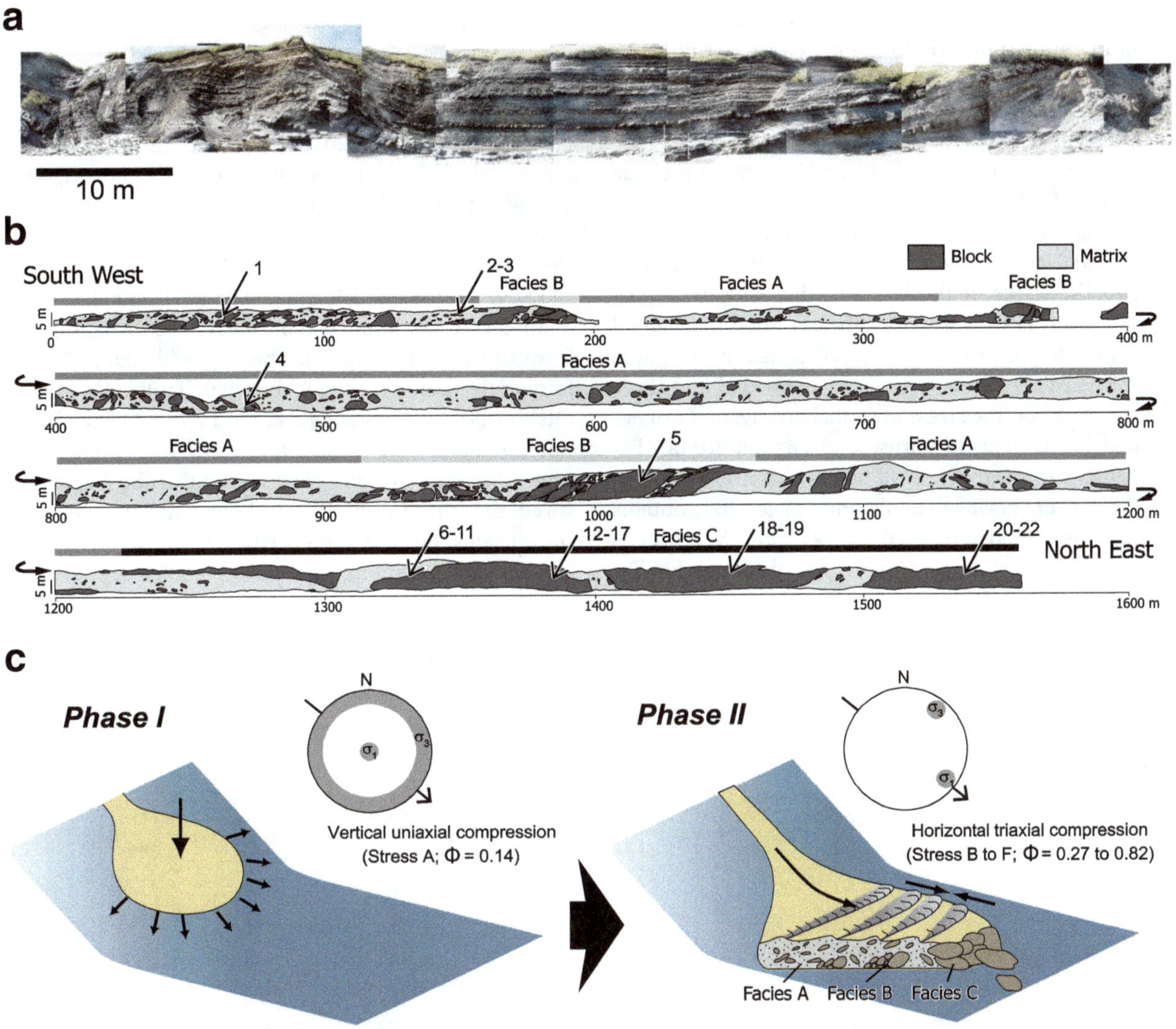

Fig. 2 MTD in the Akkeshi Formation in the Nemuro Group (modified from Naruse and Otsubo 2011a). **a** Mosaic photograph of the MTD observed in the coastal outcrop (modified from Naruse and Otsubo 2011a). The location of the outcrop is shown in Fig. 1. **b** Sketch of the mass transport deposit from the mosaic photograph of the ~ 1.6 km-long continuous outcrop in the coastal outcrop (modified from Naruse and Otsubo 2011a). The sketch shows the typical zone that we can recognize the facies A, B, and C. Arrows indicate outcrops from which fault data were obtained. **c** Schematic diagram of the transport, depositional processes, and internal stress changes of a mass flow, interpreted from the coastal outcrop. The flow was experienced two different stress fields: (1) extensional stress created by the radial spreading of the flow during its downcurrent movement (phase I), and (2) compressional stress during deposition on the basin plain (phase II) (Naruse and Otsubo 2011a). Stereograms (lower-hemisphere, equal-area projection) show stresses from a MTD reported by Naruse and Otsubo (2011a). Arrows show the direction of paleo-transport

mentioned submarine MTD using the relationships between its principal stress axes and fault plane attitudes. We first introduce the internal structures and stresses in the submarine MTD. Secondary, we explain how to estimate the pore fluid pressure ratio. Thirdly, we then estimate the pore fluid pressure ratio in this submarine MTD. Finally, we conclude by discussing its dynamic processes. On the internal structures and stresses in the submarine MTD in the Akkeshi Formation, we introduce the results by Naruse and Otsubo (2011a).

Geological background

The submarine MTD in the Akkeshi Formation is from the central section of the Upper Cretaceous–Paleocene Nemuro Group in the southern Nemuro Peninsula, eastern Hokkaido Island, Japan (Fig. 1), and is interpreted to have been deposited in the fore-arc basin of the Kuril Arc (Kiminami 1983). The Nemuro Group is composed of the Nokkamappu, Otadai, Monshizu, Hamanaka, Akkeshi, and Kiritappu formations, in ascending order, and is unconformably overlain by the Eocene Urahoro

Group. The Nemuro Group is composed of submarine fan deposits that contain hemipelagic mudstone, as well as turbidites and submarine slump deposits (Kiminami 1978; Naruse 2003). The Akkeshi Formation has been interpreted as a series of submarine channel–levee complexes with large-scale MTDs (Naruse 2003).

Here, we analyze a coastal outcrop that shows a single continuous MTD bed over a distance of > 1600 m (Naruse and Otsubo 2011a). Although the upper and lower boundaries are not exposed, the outcrop extends parallel to the strike of the bed and the paleo-transport directions (from northwest to southeast), such that lateral variations in the bed can be observed at nearly the same horizon. The MTD in this outcrop is subdivided into three lithofacies based on the size and spatial arrangement of accumulated blocks (Fig. 2; Naruse and Otsubo 2011a): Facies A consists of relatively small blocks that are supported by a matrix of gravelly mudstone, facies B contains clast-supported, moderate-sized blocks, and facies C is composed mainly of large (up to 100 m long) blocks. Facies A and B occur primarily toward the proximal end of the outcrop, while facies C occurs only at the distal end.

Syn-sedimentary stress analysis of the mass transport deposits

Naruse and Otsubo (2011a) analyzed 22 syn-sedimentary meso-scale faults in this submarine MTD and assessed the associated stress fields during the evolution of the debris flow (Table 1). Here, we briefly summarize the results of their analysis. The observed sedimentary blocks in the MTD are generally deformed. The small blocks (< 5 m in diameter) commonly exhibit heavy deformation in the form of slump folding, and the larger blocks (~ 5–100 m) frequently contain small internal faults (Naruse and Otsubo 2011a). The small faults in the MTD blocks do not extend into the flow matrix, suggesting a syn-sedimentary origin. Naruse and Otsubo (2011a) measured the strike and dip of these small faults with the slickensides on the fault planes (Fig. 3). NW–SE and NNW–SSW trending meso-scale faults are commonly

Table 1 Fault-slip dataset of the meso-scale faults obtained by Naruse and Otsubo (2011a) and misfit angles on the meso-scale faults. When a stress with a misfit of < 30° is judged to be compatible (e.g., Nemcok and Lisle 1995), then stresses A, C, D, and F can explain slip vectors on 20 faults (excluding the data for faults 4 and 7). This result indicates that only stresses A, C, D, and F can be adopted as the optimum solution in this study. In other words, stresses B and E are not needed to explain the data

No.	Faults				Misfit angle (°)					
	Strike (°)	Dip (°)	Rake (°)	Sense	Stress A	Stress B	Stress C	Stress D	Stress E	Stress F
1	N23W	74S	R85E	N	2	3.3	93.5	146.2	149.4	3.6
2	N24W	62S	R80E	N	2.3	2.3	90.2	140.7	117.2	34.6
3	N34W	48S	R90W	N	8.2	25.2	95.8	172.7	122.3	144.2
4	N14W	88S	R13W	N	92.4	91.3	164.7	142.2	31.9	101.5
5	N40E	68N	R74E	R	156.8	146.5	13.3	9.4	136.9	34.7
6	N18W	64S	R87E	N	5.8	12.3	97.4	141.7	110.8	6.8
7	N30W	80S	R85E	N	2.3	4.9	97.2	155.4	164.9	0.3
8	N45W	72N	R73E	N	1.3	120.7	8	19.3	65.5	161.8
9	N50W	66S	R86E	N	3.4	14.5	131.1	155.1	151.4	32.3
10	N16W	40S	R25W	R	102.9	41.9	10	3.2	2.2	92.6
11	N4E	40N	R16E	R	93.4	8.3	15.6	17.7	7	104.4
12	N23W	68S	R76E	N	6.6	3.8	85.6	136.8	118.6	12.3
13	N22W	50S	R52E	R	151.5	167.9	116.9	77.7	100.8	4.3
14	N16W	27S	R60E	R	167.6	55.2	105.6	143.4	99.5	1.2
15	N12E	40N	R30E	R	108.1	31.5	9.4	7.5	5.8	92.1
16	N8W	60S	R45W	R	124.6	111.9	30.6	2.8	42.7	65.6
17	N33W	62S	R62E	R	157.7	157.9	107	40.1	68.4	122.6
18	N10W	20S	R20W	R	85.6	60.8	0.8	57.9	2.6	104
19	N68W	58S	R54E	R	144.9	135.2	3.7	19	68.3	53.6
20	N30W	82N	R34W	N	76	131	23.6	29.7	148.5	112.2
21	N64W	84N	R50E	N	3.6	91.9	8.1	5	27.7	121.9
22	N42W	82S	R20W	R	117.9	123.4	17	73.1	78.7	115.5

Fig. 3 Photograph of the fault observed in the coastal outcrop of the Akkeshi Formation (modified from Naruse and Otsubo 2011a). A minor deformed structure of beddings at the right end of the fault and the fact that the fault does not continue to the external matrix indicate that this fault is syn-sedimentary origin. The white circle shows a scale (1 m)

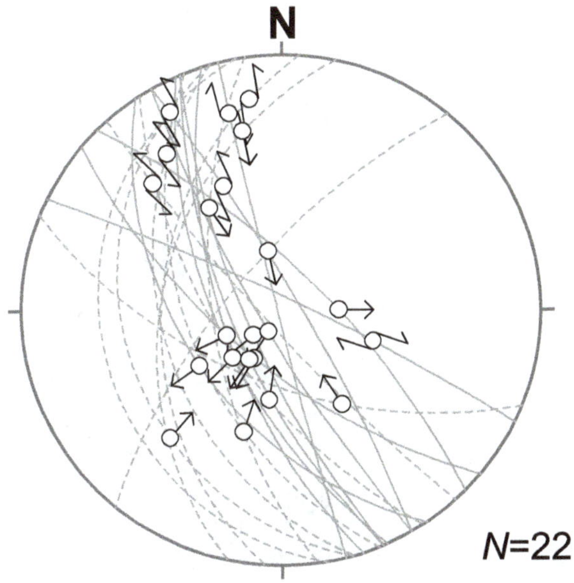

Fig. 4 Fault-slip data obtained from the coastal outcrop of the Akkeshi Formation (modified from Naruse and Otsubo 2011a). Great circles shown by lower-hemisphere, equal-angle projection represent the orientation of normal or oblique normal faults. Solid and dotted great circles represent normal or oblique normal and reverse or oblique reverse faults, respectively. Open circles on the great circles show the direction of slickenside striations on the fault planes. Arrows attached to the open circles show the sense of shear. Arrows pointing to the center of stereonet and those pointing to the opposite direction represent reverse and normal senses, respectively. Paired arrows attached to the open circles indicate a sinistral or dextral sense of shear

observed in the blocks (Fig. 4). Most of the faults are reverse faults, and there are few normal faults (Fig. 4). Furthermore, six paleo-stresses (stresses A–F) were detected from the meso-scale fault data in the MTD blocks using the multiple inverse method (Yamaji 2000; Otsubo and Yamaji 2006). These stresses were classified into two groups: (1) a vertical uniaxial compression stress field with the axis of maximum compressive stress (σ_1) oriented normally to bedding (stress A) and (2) horizontal triaxial compression stresses with the σ_1-axis oriented parallel to the paleocurrent direction (stresses B–F) (Figs. 2c and 5; Table 2; Naruse and Otsubo 2011a).

Methods/Experimental

Here, we attribute the different fault orientations to the variable pore fluid pressure regimes along the individual faults, which then allow us to estimate the pore fluid pressure ratio (Fig. 6) defined as an indicator of the pore fluid pressure regime (Hubbert and Rubey 1959). This approach can be explained using Mohr circles, where the normal and shear stresses acting on a fault of any attitude (i.e., strike and dip) within a given stress regime are described by a point on the Mohr diagram (Jaeger and Cook 1979). We assume that the shear stress (τ) acting on a fault is described by Byerlee's law, such that

$$\tau = \mu(\sigma_n - P_f), \tag{1}$$

where μ is the friction coefficient (usually 0.6–0.85; Byerlee 1978), σ_n is the normal stress (with compression being positive), and P_f is the pore fluid pressure. We can therefore determine the orientation of a fault plane, along with its associated normal and shear stresses, by identifying where it is located on the Mohr diagram. Any variation in the attitudes of the faults is attributed to heterogeneities in the fault strength that are due to changes in the effective friction coefficient, which is represented by the ratio of the normal stress to shear stress on each fault (Angelier 1989). Equation 1 is therefore rewritten as

$$\tau = \mu(1-\lambda)\sigma_n, \tag{2}$$

where λ is the pore fluid pressure ratio, which is defined by $P_f = \lambda\sigma_n$. This implies that λ ($0.0 \le \lambda \le 1.0$) for a given fault plays an important role in determining τ. An increase in P_f weakens the fault, such that the Coulomb failure criterion (Eq. 1) can still be satisfied if the faults are misoriented or the pore fluid is overpressurized (Terakawa et al. 2010) (Fig. 6). The normal and shear stresses acting on a given plane depend on the orientation of the plane when the principal stresses are known (Jaeger 1969), such that

$$\sigma_n = \sigma_1 l^2 + \sigma_2 m^2 + \sigma_3 n^2 \tag{3}$$

and

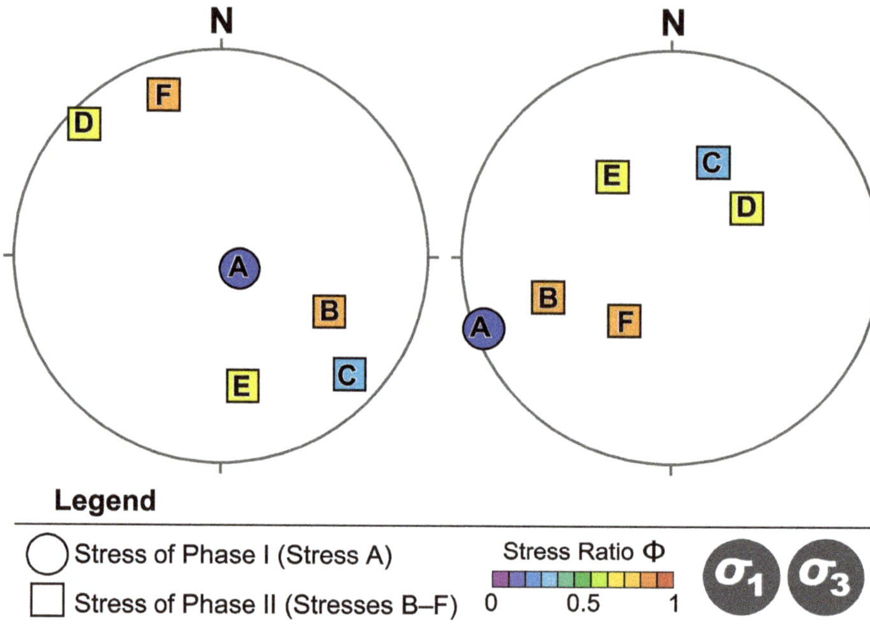

Fig. 5 Paleostresses inferred from fault-slip data obtained from the coastal outcrop of the Akkeshi Formation (modified from Naruse and Otsubo 2011a). Lower-hemisphere, equal-area projections are shown. The symbols indicate the optimal stress solutions acquired by the multiple inversion method (Yamaji 2000; Otsubo and Yamaji 2006) and *k*-means clustering (Otsubo et al. 2006) with $k = 6$. Left and right columns correspond to σ_1 and σ_3 axes, respectively. The color of the symbol represents the stress ratio, $\Phi = (\sigma_2 - \sigma_3)/(\sigma_1 - \sigma_3)$

$$\tau = \sqrt{(\sigma_1 - \sigma_2)^2 l^2 m^2 + (\sigma_2 - \sigma_3)^2 m^2 n^2 + (\sigma_3 - \sigma_1)^2 l^2 n^2}, \quad (4)$$

where l, m, and n are the direction cosines of the normal to the plane with respect to the principal stress axes, and σ_1, σ_2, and σ_3 are the maximum, intermediate, and minimum compressive principal stresses, respectively. Here, the shear and normal stresses are normalized by the maximum shear stress (τ_{max}). We prepared two synthetic fault planes (A and B) and a reverse faulting stress regime with the σ_1-axis oriented NW–SE and the sub-vertical σ_3-axis to demonstrate the model (Fig. 6a). We obtained λ values of 0.00 and 0.42 for the fault planes A and B, respectively, when we set $\mu = 0.6$ (Fig. 6b). The result means that P_f for fault plane B is larger than that for fault plane A under same stress regime.

Table 2 List of paleostresses inferred from the meso-scale faults obtained by Naruse and Otsubo (2011a)

Stress	σ_1-axes		σ_3-axes		Stress ratio Φ
	Azimuth	Plunge	Azimuth	Plunge	
A	130	81	248	4	0.14
B	119	40	250	38	0.79
C	133	16	23	49	0.27
D	314	9	45	55	0.58
E	171	36	323	51	0.62
F	340	19	219	56	0.82

Results

We examined whether all the inferred stresses are required to parameterize the 22 MTD faults, with the six sets of stresses (stresses A–F) used as candidates for determining the optimal solution to the data (Yamaji et al. 2006). We calculated the angular misfits between the slip directions that were predicted from the inferred stresses and those observed from the fault data based on the Wallace–Bott hypothesis (Wallace 1951; Bott 1959), which states that the slip vector of a fault is parallel to the resolved shear stress on the fault. Stresses A, C, D, and F can explain the slip vectors on 20 faults (not faults 4 and 7; Table 1) when we assume that a stress with a misfit of < 30° is sufficient to characterize a fault (e.g., Nemcok and Lisle 1995). These results indicate that only stresses A, C, D, and F are needed to explain the data.

The faults characterized by stresses A, C, D, and F were plotted on Mohr diagrams, where we applied the Mohr circles to these 20 faults to estimate λ (Fig. 7). All representative points (faults) for each stress are shown in the gray shaded regions in Fig. 7. The points to the left of each black dashed line in the Mohr diagrams are within the 95% confidence limits of the fault data for every stress (Fig. 7). In this study, we define the 95% confidence limits that represent the optimal λ for the stresses. The λ values for stresses A, C, D, and F are 0.14, 0.39, 0.41, and 0.71, respectively, for $\mu = 0.6$ (Fig. 7). In the case of stress C, we excluded fault 5 from the λ discussions, because it lies far outside of the 95%

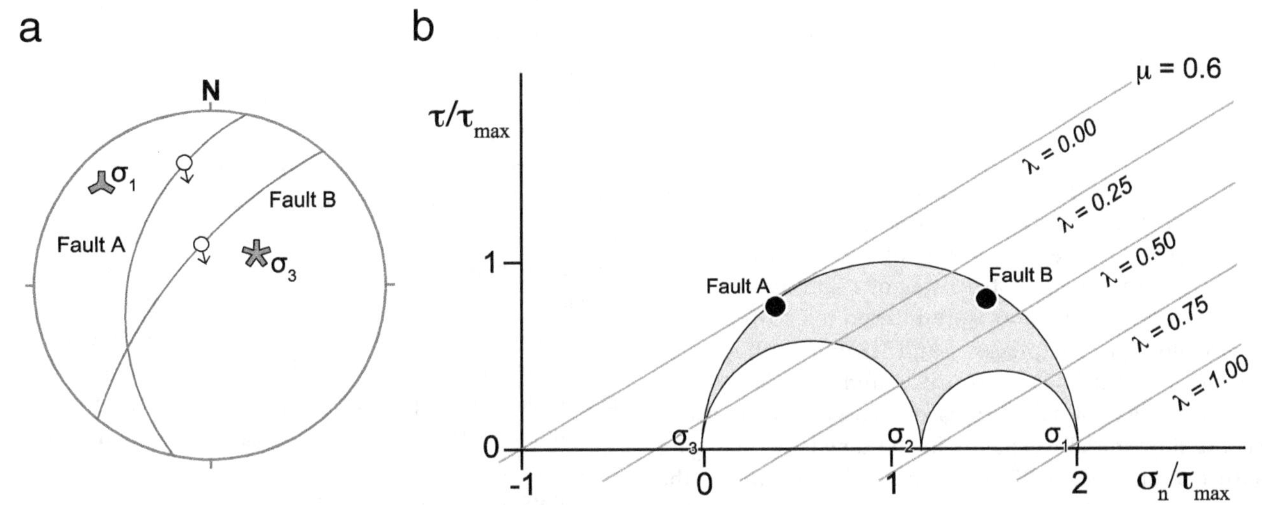

Fig. 6 Schematic illustrations to estimate pore fluid pressure ratio by using the Mohr circles. **a** Equal-area projection showing stress axes and fault plane with the slip direction. The stress is the reverse faulting type of stress regime. **b** Normal stress magnitudes on the fault planes calculated according to the optimal stresses. σ_1, σ_2, and σ_3 are the maximum, intermediate, and minimum compressive principal stresses, respectively. The stress ratio is 0.6. This figure shows the Mohr's diagrams that depicts the fault data. Horizontal and vertical axes show the normal and shear stresses acting on fault planes. Gray lines show the fault strength when $\lambda = 0.00$, 0.25, 0.50, 0.75, and 1.00. When we used a value of 0.6 for the friction coefficient μ, the values of λ for the fault planes A and B are 0.00 and 0.42, respectively

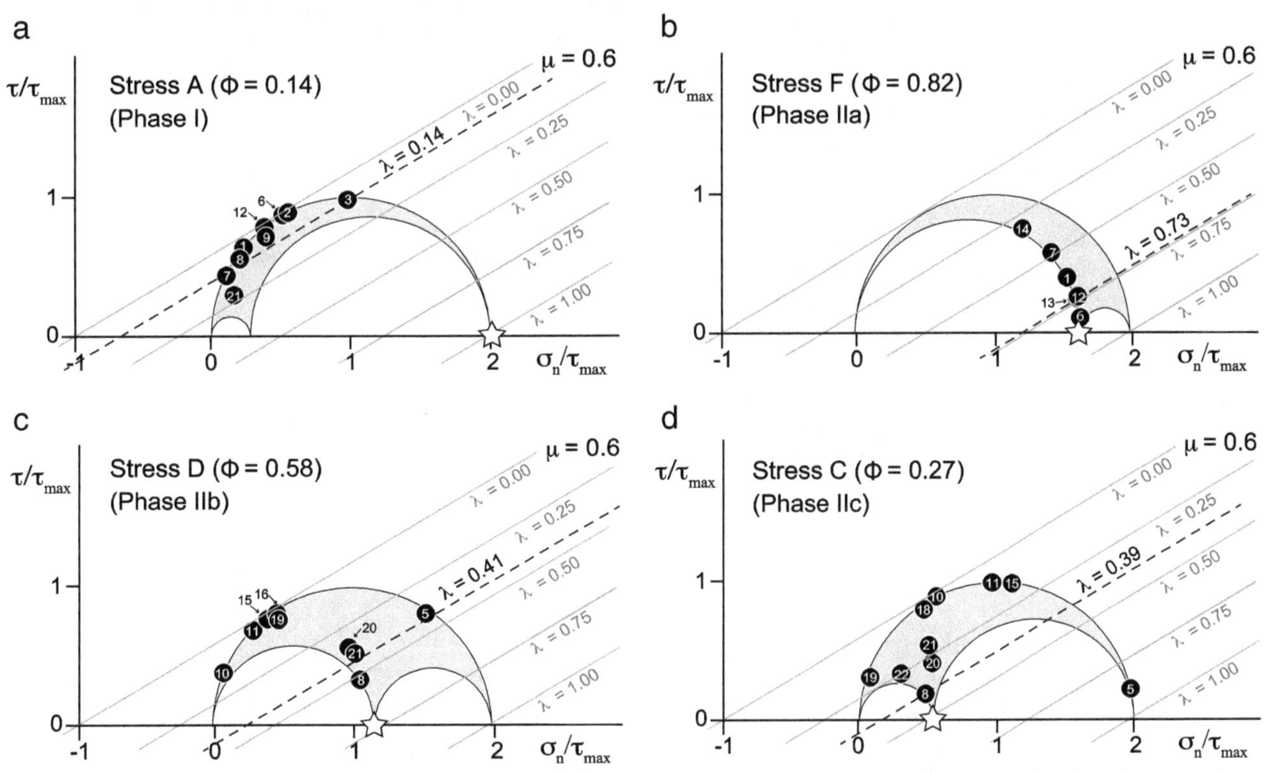

Fig. 7 Pore fluid pressure ratio in a mass transport deposit estimated by using Mohr circles for stresses (**a**) A, (**b**) F, (**c**) E, and (**d**) C. σ_1, σ_2, and σ_3 are the maximum, intermediate, and minimum compressive principal stresses, respectively. Gray lines show the fault strength when $\lambda = 0.00$, 0.25, 0.50, 0.75, and 1.00. Stars indicate vertical stress axes. Number indicates the faults reported by Naruse and Otsubo (2011a). Numbers in outlined black-filled circles in Mohr circles indicate those faults for which stresses A, C, D, and F can explain the slip vectors. The circles from the black dotted line to the left region are contained within 95% confidence limits of the fault data for every stress

confidence limit for stress C and plots to the far right of the Mohr diagram (Fig. 7d). We can then calculate the pore fluid pressure coefficient for the MTDs from the Mohr diagrams.

Discussion

Here, we discuss temporal changes in the stress fields detected in the submarine MTD based on the outcrop-based approach. The two groups of the stress fields detected from the internal structures of a submarine MTD in the Akkeshi Formation were interpreted to represent the two flow phases (phases I and II) during the evolution of the debris flow (Naruse and Otsubo 2011a, 2011b). Numerical simulations of debris flows revealed that the flow initially expands in the downstream region (Naruse and Otsubo 2011b), where the blocks in the flow undergo vertical compression and lateral extension during this flow phase (phase I). This phase I stress field, which indicates uniaxial vertical compression, is best represented by stress A. The head of the debris flow reaches the low-gradient downstream region and then decelerates rapidly after lateral expansion of the flow, whereas the main body of the flow continues to move. This results in the shortening of the slide body due to the velocity difference between the head and the body, which produces compressional ridges on the surface of the deposit (Naruse and Masuda 2006; Naruse and Otsubo 2011b). This phase II stress field, which indicates triaxial compression with the σ_1-axis oriented parallel to the horizontal flow direction (from northwest to southeast; Naruse and Otsubo 2011a), is best represented by stresses B–F (Naruse and Otsubo 2011a, 2011b). In this study, we ignore the rotation of block during the phase II because the orientation of σ_1-axis coincides with the paleo-current direction (Naruse and Otsubo 2011a).

The numerical experiments suggest that the maximum horizontal compressive stress (σ_{hmax}) increases during phase II, when the debris flow is deposited on the basin plain (Naruse and Otsubo 2011b). This may cause a variation in the stress ratio (Φ), which is defined as:

$$\Phi = \frac{\sigma_2 - \sigma_3}{\sigma_1 - \sigma_3}. \tag{5}$$

We therefore subdivided phase II into three subphases on the basis of the stress ratio: phases IIa, IIb, and IIc, which have high ($\Phi = \sim 0.8$), intermediate ($\Phi = \sim 0.6$), and low ($\Phi = \sim 0.3$) stress ratios, respectively (Fig. 7). Phase IIa underwent an axial tension with $\sigma_1 \approx \sigma_2 (=\sigma_v) > \sigma_3$, where σ_v is the overburden pressure, whereas phase IIc underwent an axial compression with $\sigma_1 > \sigma_2 (=\sigma_v) \approx \sigma_3$. Therefore, the increase in Φ represents the increase in σ_{hmax} during phase II. We interpreted that stresses F, D, and C must be the primary stresses of phases IIa, IIb, and IIc, respectively, to ensure a consistent relationship between σ_v and σ_{hmax}.

We estimated σ_{hmax} in phase II normalized by the differential stress of phase I (stress A; $\sigma_v^A - \sigma_{hmin}^A$) (Fig. 8), where σ_{hmin} is the minimum horizontal compressive stress. We observe that σ_v is highest in phase I and then decreases by a factor of two during phase II, as inferred from the temporal changes in the basal normal stress that are revealed in Fig. 5 of Iverson (1997). Here, the direction of σ_{hmin} is approximately orthogonal to the MTD paleocurrent direction (Naruse and Otsubo 2011b). σ_{hmax} during phase IIc is estimated to be approximately four times larger than σ_{hmax} during phase IIa (Fig. 8) when σ_{hmin} is held constant relative to σ_v and σ_{hmax}.

The estimated λ in phase II is larger than that in phase I (Fig. 8), which indicates the important relationship between pore fluid pressure and faulting during phase II. λ reaches a maximum value during phase IIa and then decreases to a stable value by phase IIb, whereas σ_{hmax} increases throughout phase II (Fig. 8). We interpret that the increased pore fluid pressure during phase IIa is reduced during the faulting in phases IIb and IIc. These temporal changes in the MTD pore fluid pressures are consistent with the temporal changes in the basal pore pressure revealed by Iverson (1997).

Figure 9 shows a schematic illustration of the dynamic processes during phase II of the submarine MTD

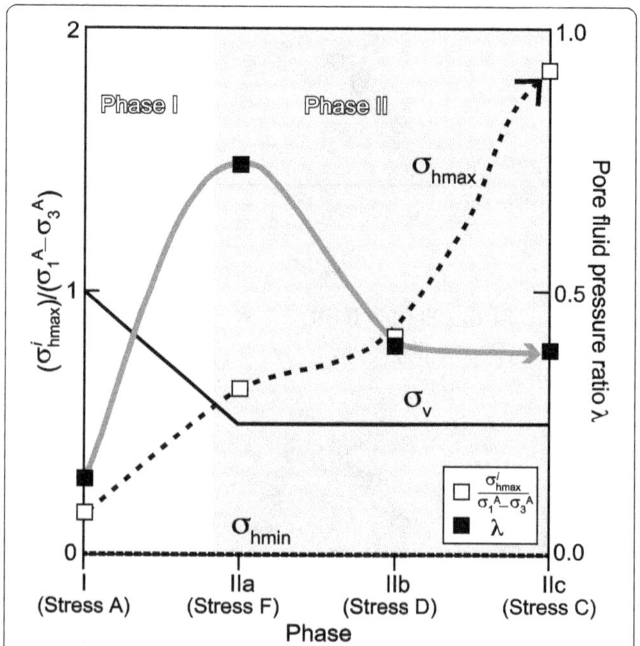

Fig. 8 Temporal changes in horizontal maximum compressional stresses and pore fluid pressure ratio in a MTD in the Akkeshi Formation. The horizontal axis shows the phases in the MTD. Left and right vertical axes show normalized σ_{hmax} and λ, respectively

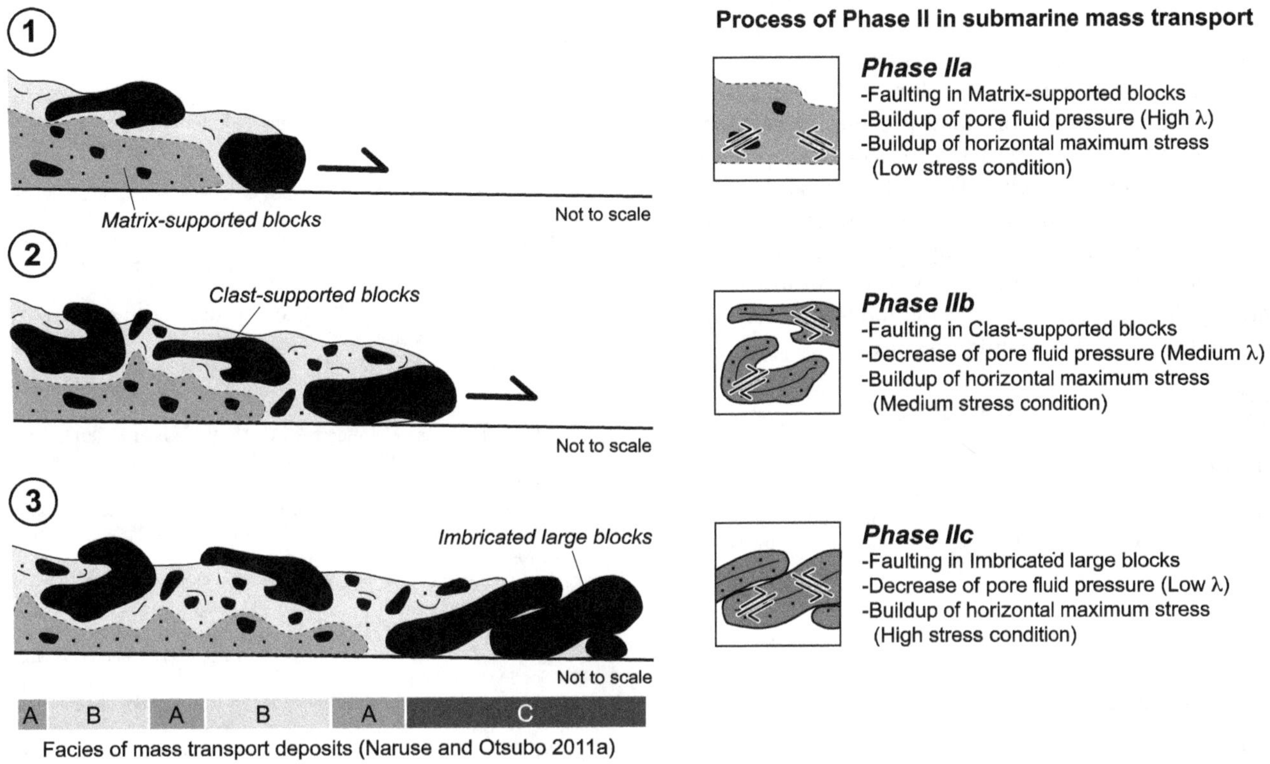

Fig. 9 Schematic cartoon showing the dynamic processes of phase II in a MTD in the Akkeshi Formation. The dynamic processes in a MTD is modified from Ogata et al. (2012)

depositional sequence. We infer that the buildup of pore fluid pressure during phase IIa was caused by the rapid and undrained shortening deformation of the debris flow under matrix-supported conditions (Fig. 9a; Ogata et al. 2012, 2014a, 2014b). However, the decrease in pore fluid pressure caused an increase in the effective stress during the drained deformation of the debris flow following phase IIb, which resulted in the variable faulting patterns from phase IIa to phase IIc (Fig. 9b, c). These phenomena may have promoted shortening deformation of the debris flow during phase IIc (Fig. 9c), because faults 18–22 (observed at the distal end of the outcrop) were activated by the phase IIc stress field (Fig. 9d). The above results therefore indicate that it is possible to estimate the temporal changes in the internal stresses and pore fluid pressures in MTDs from the relationships between the principal stress axes and the orientations of fault planes. Although the entire path of the flow may exceed tens of kilometers, this study implies that the 1.6 km-long outcrop was large enough for examining sufficient number of blocks that can record the whole history of stress in the flow.

Our results suggest that a drop in pore fluid pressure plays a significant role in the deceleration and subsequent deposition of large-scale submarine debris flows (Ogata et al. 2012, 2014a, 2014b). This is the first known documented attempt to infer pore fluid pressures from outcropping submarine MTDs. Although previous studies have reported that the behavior of subaerial debris flows is controlled primarily by pore pressure (Iverson 1997, 2005), numerical modeling of submarine debris flows (e.g., Imran et al. 2001) generally assumes that the pore fluid pressure in such flows is constant, because they generally contain large amounts of clay particles that impede a drop in excess pore fluid pressure. However, the large-scale MTDs investigated here indicate the necessity to develop models that incorporate the process of pore fluid pressure. Future work will focus on comprehensive studies of the process by integration of the outcrop investigation and numerical study.

Conclusions

We examined temporal changes in the internal stresses and pore fluid pressures of a submarine mass transport deposit in the Akkeshi Formation from the central section of the Upper Cretaceous–Paleocene Nemuro Group, eastern Hokkaido Island, Japan. The evolution of this debris flow comprised of two phases: phase I consisted of radial spreading of the debris flow during downslope movement, and phase II consisted of compression during deposition on the basin plain. The distribution of fault orientations was sensitive to the internal stress phases, which can be attributed to the variable pore fluid pressure along the individual faults. There was a large increase in the pore fluid pressure ratio during the transition from phase I to phase II from the fault orientation distribution. Our analysis suggests that this drop in pore fluid pressure plays a significant role in the deceleration and subsequent deposition of large-scale submarine debris flows. The dynamics of submarine debris flow processes are poorly known, due to the extreme difficulties in directly observing them. The large-scale submarine mass transport deposit investigation here highlights the need to develop models that incorporate the dynamics of pore fluid pressure.

Abbreviation

MTD: Mass transport deposit

Acknowledgements

We thank Dr. Ogata and an anonymous reviewer for the improvement of the manuscript. We are also grateful to M. Utsunomiya and N. Yamaguchi for the helpful discussions. Some of the figures were drawn using the GMT software package (Wessel and Smith 1995).

Funding

This study was supported by MEXT KAKENHI (number 26109003).

Authors' contributions

MO proposed the topic and conceived and designed the study. MO carried out the estimation of the internal pore pressure ratio, drafted the figures, and contributed to the assessment and discussion of the results. HN and AM contributed to the discussion of the results. All authors read and approved the final manuscript.

Competing interests

The authors declare that they have no competing interests.

Author details

[1]Geological Survey of Japan, AIST, AIST Tsukuba Central 7, Higashi-1-1-1, Tsukuba, Ibaraki 305-8567, Japan. [2]Division of Earth and Planetary Sciences, Kyoto University, Kitashirakawa Oiwakecho, Kyoto 606-8502, Japan.

References

Angelier J (1989) From orientation to magnitudes in paleostress determinations using fault slip data. J Struct Geol 11:37–50

Baroň I, Kernstocková M, Faridi M, Bubík M, Milovský R, Melichar R, Sabouri J, Babůrek J (2013) Paleostress analysis of a gigantic gravitational mass movement in active tectonic setting: the Qoshadagh slope failure, Ahar, NW Iran. Tectonophysics 605:70–87. https://doi.org/10.1016/S0040-1951(13)00569-6

Baroň I, Kernstockovább M, Melicharb R (2017) Stress field reconstruction in an active mudslide. Geomorphology. https://doi.org/10.1016/j.geomorph.2017.04.020

Bott MHP (1959) The mechanics of oblique slip faulting. Geol Mag 96:109–117

Byerlee J (1978) Friction of rocks. Pure Appl Geophys 116:615–626

Canals M, Lastras G, Urgeles R, Casamor JL, Mienert J, Cattaneo A, De Batist M, Haflidason H, Imbo Y, Laberg JS, Locat J, Long D, Longva O, Masson DG, Sultan N, Trincardi F, Bryn P (2004) Slope failure dynamics and impacts from seafloor and shallow sub-seafloor geophysical data: case studies from the COSTA project. Mar Geol 213:9–72

D'Agostino V, Bettella F, Cesca M (2013) Basal shear stress of debris flow in the runout phase. Geomorphology 201:272–280

Festa A, Ogata K, Pini GA, Dilek Y, Alonso JL (2016) Origin and significance of olistostromes in the evolution of orogenic belts: a global synthesis. Gondwana Res 39:180–203

Gennesseaux M, Mauffret A, Pautot G (1980) Les glissements sous-marins de la pente continentale niçoise et la rupture de câbles en mer Ligure (Méditerranée occidentale). C R Acad Sci Paris 290:959–962

Harbitz CB, Løvholt F, Pedersen G, Masson DG (2006) Mechanisms of tsunami generation by submarine landslides: a short review. Norwegian J Geology Nor Geologisk Foren 86:255–264

Hubbert MK, Rubey WW (1959) Role of fluid pressure in mechanics of overthrust faulting. Bull Geol Soc Am 70:115–166

Husein S, Sudarno I, Pramumijoyo S, Karnawati D (2010) Paleostress analysis to interpret the landslide mechanism: a case study in Parangtritis, Yogyakarta. J Southeast Asian Appl Geology 2:104–109

Ilstad T, De Blasio FV, Elverhøi A, Harbitz CB, Engvik L, Longva O, Marr JG (2004a) On the frontal dynamics and morphology of submarine debris flow. Mar Geol 213:481–497

Ilstad T, Marr JG, Elverhoi A, Harbitz CB (2004b) Laboratory studies of subaqueous debris flows by measurements of pore-fluid pressure and total stress. Mar Geol 213:403–414

Imran J, Parker G, Locat J, Lee H (2001) 1D numerical model of muddy subaqueous and subaerial debris flow. J Hydraul Eng 127:959–968

Iverson RM (1997) The physics of debris flows. Rev Geophys 35:245–296

Iverson RM (2005) Debris-flow mechanics. In: Jakob M, Hungr O (eds) Debris-flow hazards and related phenomena. Springer, New York

Iverson RM, Reid ME, Logan M, LaHusen RG, Godt JW, Griswold JP (2011) Positive feedback and momentum growth during debris-flow entrainment of wet bed sediment. Nat Geosci 4:116

Jaeger J (1969) Elasticity fracture and flow, 3rd edn. Methuen & Co. Ltd., London

Jaeger JC, Cook NG (1979) Fundamentals of rock mechanics, 3rd edn. Chapman & Hall, New York

Kiminami K (1978) Stratigraphic re-examination of the Nemuro Group. Chikyu Kagaku 32:120–132

Kiminami K (1983) Sedimentary history of the Late Cretaceous–Paleocene Nemuro Group, Hokkaido, Japan: a forearc basin of the Paleo-Kuril arc-trench system. J Geol Soc Jpn 89:607–624

Major JJ, Iverson RM (1999) Debris-flow deposition: effects of pore-fluid pressure and friction concentrated at flow margins. Geol Soc Am Bull 111:1424–1434

Morgan JK, Silver E, Camerlenghi A, Dugan B, Kirby S, Shipp C, Suyehiro K (2009) Addressing geohazards through ocean drilling. Sci Drilling 7:15–30. https://doi.org/10.2204/iodp.sd.7.01.2009

Naruse H (2003) Cretaceous to Paleocene depositional history of North-Pacific subduction zone, reconstruction from the Nemuro Group, eastern Hokkaido, northern Japan. Cretac Res 24:55–71

Naruse H, Masuda F (2006) Visualization of the internal structure of the massive division in experimental sediment-gravity-flow deposits by mapping of grain fabric. J Sediment Res 76:854–865

Naruse H, Otsubo M (2011a) Heterogeneity of internal structures in a mass-transport deposit, Upper Cretaceous to Paleocene Akkeshi Formation, Hokkaido Island, northern Japan. SEPM Spec Publ 96:279–290

Naruse H, Otsubo M (2011b) Internal stress fields of a large-scale submarine debris flow. In: Lykousis V, Sakellarious D, Locat J (eds) Submarine mass movements and their consequences. Springer, Heidelberg

Nemcok M, Lisle RJ (1995) A stress inversion procedure for polyphase fault/slip data sets. J Struct Geol 17:1445–1453

Nur A, Booker JR (1972) Aftershocks caused by pore fluid flow. Science 175:885–887

Ogata K, Mountjoy JJ, Pini GA, Festa A, Tinterri R (2014a) Shear zone liquefaction in mass transport deposit emplacement: a multi-scale integration of seismic reflection and outcrop data. Mar Geol 356:50–64

Ogata K, Mutt E, Pini GA, Tinterri R (2012) Mass transport-related stratal disruption within sedimentary mélanges: examples from the northern Apennines (Italy) and south-Central Pyrenees (Spain). Tectonophysics 568–569:185–199

Ogata K, Pogačnik Ž, Pini GA, Tunis G, Festa A, Camerlenghi A, Rebesco M (2014b) The carbonate mass transport deposits of the Paleogene Friuli Basin (Italy/Slovenia): internal anatomy and inferred genetic processes. Mar Geol 356:88–110

Otsubo M, Sato K, Yamaji A (2006) Computerized identification of stress tensors determined from heterogeneous fault-slip data by combining the multiple inverse method and K-means clustering. J Struct Geol 28:991–997

Otsubo M, Yamaji A (2006) Improved resolution of the multiple inverse method by eliminating erroneous solutions. Comput Geosci 32:1221–1227

Rice JR (1992) Fault stress states, pore pressure distributions, and the weakness of the San Andreas Fault. In: Evans B, Wong TF (eds) Fault mechanics and transport properties of rock: a Festchrift in honor of W.F. Brace. Academic Press, London

Rickenmann D, Laigle DM, McArdell BW, Hübl J (2006) Comparison of 2D debris-flow simulation models with field events. Comput Geosci 10:241–264

Sibson RH (2007) An episode of fault-valve behavior during compressional inversion? The 2004 M(J) 6.8 mid-Niigata Prefecture, Japan, earthquake sequence. Earth Planetary Sci Lett 257:188–199

Terakawa T, Zoporowski A, Galvan B, Miller SA (2010) High-pressure fluid at hypocentral depths in the L'Aquila region inferred from earthquake focal mechanisms. Geology 38:995–998

Terzaghi K, Peck RB (1967) Soil mechanics in engineering practice. Wiley, New York

Wallace RE (1951) Geometry of shearing stress and relationship to faulting. J Geol 59:111–130

Ward SN (2001) Landslide tsunami. J Geophys Res 106:11201–11215

Wessel P, Smith WHF (1995) Free software helps map and display data. Eos Trans Am Geophysical Union 72:445–446

Yamaji A (2000) The multiple inverse method: a new technique to separate stresses from heterogeneous fault-slip data. J Struct Geol 22:441–452

Yamaji A, Otsubo M, Sato K (2006) Paleostress analysis using the Hough transform for separating stresses from heterogeneous fault-slip data. J Struct Geol 28:980–990

ESR signal intensity of quartz in the fine-silt fraction of riverbed sediments from the Yangtze River: a provenance tracer for suspended particulate matter

Keita Saito[1*], Ryuji Tada[1], Hongbo Zheng[2], Tomohisa Irino[3], Chao Luo[4], Mengying He[4], Ke Wang[1] and Yoshiaki Suzuki[1]

Abstract

The Asian summer monsoon exerts a strong influence on the hydrologic cycle in East Asia. Moreover, the distribution of heavy precipitation in the Yangtze River basin, which covers a large area of South China, is sensitive to changes in monsoon intensity. Thus, the paleo-distribution of heavy precipitation in the Yangtze basin is key in reconstructing the paleo-monsoon intensity. In this study, we established proxies for distinguishing sediments from the northwestern and southeastern parts of the Yangtze River basin, with the goal of using them to reconstruct the paleo-distribution of heavy precipitation in the basin based on changes in the provenance of detrital sediments, as determined using sediment records at the river mouth. The proxies included the electron spin resonance (ESR) signal intensity and the crystallinity index (CI) of quartz in modern riverbed sediments along the mainstream and tributaries of the Yangtze River. The data revealed that the fine-silt fraction of the sediments (which represents suspended particulate matter, SPM) from each major tributary can be distinguished on the basis of ESR signal intensity and CI values. The values for each tributary reflect the age and type of bedrock in the tributary basin and show distinct regional variations according to the distribution of geological blocks within the Yangtze basin. In addition, the analyzed ESR signal intensity and CI values of the fine-silt fraction of riverbed sediments below mainstream–tributary junctions agree well with values calculated according to a simple mixing model of SPM discharged from the tributaries and the mainstream.

Keywords: Yangtze River, Riverbed sediment, Suspended particulate matter, Provenance, Electron spin resonance, Crystallinity index

Introduction

The Yangtze River is the longest river in Asia (length >6300 km), draining an area of 1.81×10^6 km^2 (Hu et al. 2009) and with an annual water discharge of 910 km^3/year, which is the fifth largest in the world (Yang et al. 2002). More than 400 million people live in the Yangtze River drainage basin, and many large cities are situated along its mainstream (Zong and Chen 2000). According to historical records, more than 200 large floods have occurred on the Yangtze River over the last 2000 years, severely affecting the economies and lives of people in the region (Dai and Lu 2010). Because the drainage basin occupies a substantial portion of South China, which is under a temperate climate, more than 60% of the total precipitation within the drainage basin is discharged through the Yangtze River (Xu et al. 2010). Also, a positive correlation exists between annual and monthly precipitation in the basin and discharge of the Yangtze River (Xu et al. 2010). Water discharge of the Yangtze River during summer accounts for 71% of the annual discharge (Chen et al. 2001). During summer, precipitation in the middle to lower parts of the Yangtze drainage basin is affected by the East Asian summer monsoon (EASM), whereas precipitation in the uppermost part of the Yangtze River basin is sourced from the Indian summer monsoon (ISM), whose frontal position is connected

* Correspondence: keitas@eps.s.u-tokyo.ac.jp
[1]Department of Earth and Planetary Science, University of Tokyo, 7-3-1 Hongo, Bunkyo, Tokyo 113-0033, Japan
Full list of author information is available at the end of the article

to the EASM front (Wang and Lin 2002); thus, water discharge of the Yangtze River during summer reflects the amount of precipitation sourced from these monsoons. The EASM is characterized by a seasonal northwestward migration of its precipitation front, and the timing and extent of the migration are associated with migration of the westerly jet axis (Tada et al. 2016). Zhang et al. (2008) reported that inter-annual changes in the intensity of the EASM, which are defined by the strength of the northward wind around China, are associated with spatial changes in the location of the high-precipitation area, with its northwestward penetration being greatest during times of strong EASM convection. Thus, the reconstruction of past spatiotemporal changes in areas of heavy precipitation and flood within the Yangtze drainage basin is critical for understanding the long-term behavior of the EASM and ISM.

Suspended particulate matter (SPM) constitutes more than 95% of the sediments discharged from the Yangtze River (Chen et al. 2001). Because discharge of the Yangtze River reflects precipitation amounts in the drainage basin (Xu et al. 2010), and because a strong positive relationship exists between discharge and sediment load (Chen et al. 2001), we hypothesize that changes in the SPM provenance reflect changes in the distribution of high-precipitation areas. Based on this reasoning, changes in the provenance of sedimentary materials recorded at the Yangtze rivermouth could enable us to reconstruct the spatiotemporal variability of heavy-precipitation areas within the Yangtze drainage basin. Application of this technique to sediment archives collected from an area immediately outside the rivermouth is the ultimate goal of this study.

To identify the provenance of SPM, we analyzed the electron spin resonance (ESR) signal intensity and the crystallinity index (CI) of detrital quartz in three size fractions of riverbed sediments: 4–16, 16–63, and >63 μm. The ESR signal intensity and CI values have been used to distinguish sources of sediments, such as eolian dusts from various deserts in western China (e.g., Nagashima et al. 2007; Sun et al. 2007). We focused on the quartz fraction because it is the most common mineral in detrital materials and is resistant to physical and chemical weathering. Moreover, its specific gravity is close to the average of detrital materials, meaning that size-sorting effects are minimal. The objectives of this study were to (i) characterize quartz in the sediments discharged from each tributary of the Yangtze River based on ESR signal intensity and CI values and (ii) test whether the relative contribution of detrital materials derived from different tributaries is accurately reflected in ESR signal intensity and CI values of quartz in sediments deposited along the mainstream and discharged from the rivermouth.

Geographical setting

The headwaters of the Yangtze River are on the Tibetan Plateau at an altitude of >6000 m. From the plateau, the river flows through the Sichuan Basin, the Three Gorges, and across the Jianghan Plain before draining into the East China Sea. In this study, we subdivided the mainstream of the Yangtze River into four parts using a modified version of the definition provided by Chen et al. (2001): uppermost, upper, middle, and lower (Fig. 1). The uppermost part is defined as the mainstream between the headwaters on the Tibetan Plateau and Yibin, the latter being located at the river's entrance to the Sichuan Basin. The uppermost part is a major contributor of SPM to the Yangtze River and is characterized by steep slopes of the riverbed and deep and steep valleys (Yang et al. 2006). The length of the uppermost part is ~3100 km (Liu et al. 2011; Wu et al. 2009) and its drainage area is 4.6×10^5 km^2 (Hu et al. 2009), constituting 25% of the entire Yangtze drainage basin. The difference in altitude from the headwaters to the outlet of the uppermost part is >4000 m (USGS 2004), and the average riverbed slope is >1‰. The depth of the river valley reaches 2000 m. A major tributary, the Yalongjiang flows into the mainstream from the north in the lower section of the uppermost part (Fig. 1). The "Big Bend" in Yunnan Province is situated where the mainstream turns from a southeastward to northeastward flow direction (Fig. 1).

The upper part, or the Chuanjiang, is defined as the mainstream between Yibin and Yichang, the latter being located just below the outlet of the Three Gorges. In the upper part, the mainstream runs through the southern part of the Sichuan Basin, then flows into the Three Gorges (Fig. 1). The length of the upper part is ~800 km and its drainage area is 5.2×10^5 km^2 (Hu et al. 2009), constituting 29% of the entire Yangtze River drainage basin. In the Sichuan Basin, the riverbed slope is 0.1–0.2‰ and the valley depth is 200–300 m; both the riverbed slopes and the depths of valleys are smaller in the upper part than in the uppermost part. In the Three Gorges, the maximum riverbed slope and valley depth increase to 0.4‰ and 1000 m, respectively (Chen et al. 2001; USGS 2004). The channel width in the upper part is 0.5–1.5 km and water depth is 5–20 m (Chen et al. 2001). In the Sichuan Basin portion of the upper part, two major tributaries, the Minjiang and Jialingjiang, flow into the mainstream from the north and one major tributary, the Wujiang, flows into the mainstream from the south. Construction of the Three Gorges Dam (TGD) was completed in 2003 on the mainstream at the outlet of the Three Gorges, upstream of Yichang.

The middle part, or the middle Yangtze, is defined as the mainstream between Yichang and Hukou, the latter being located just downstream of the Poyang Lake outlet. The length of the middle part is 950 km and its

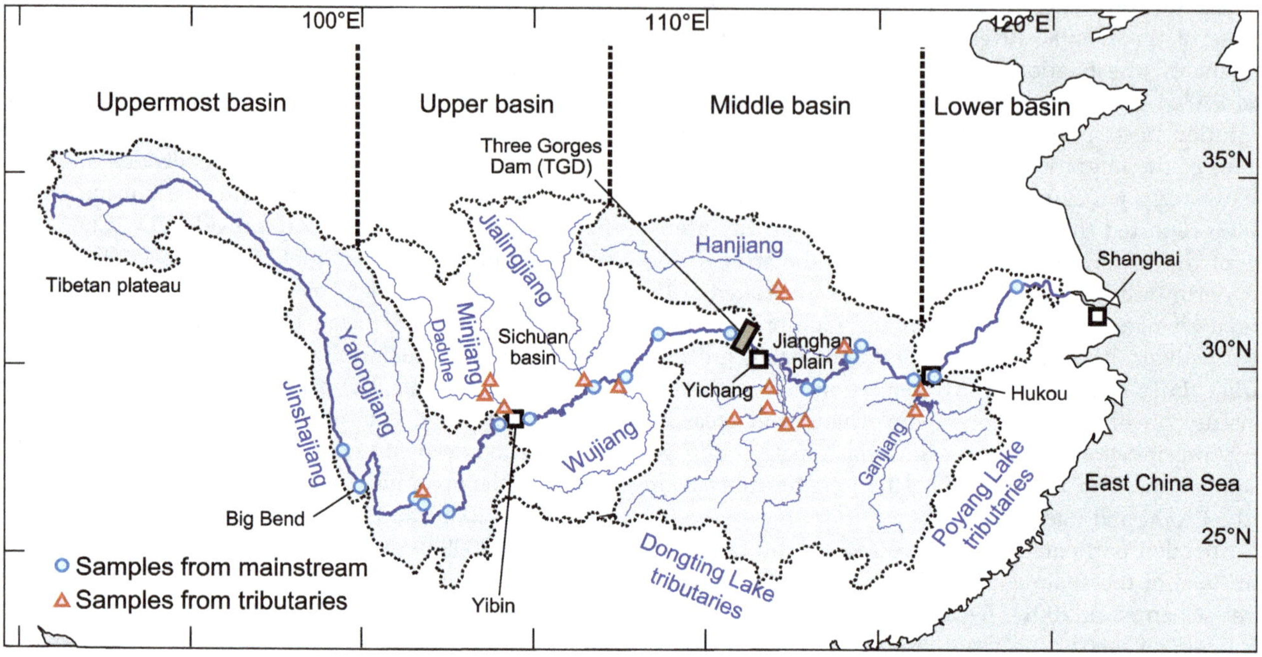

Fig. 1 Map of the Yangtze River drainage basin showing the locations of sampling sites in this study, and delineation of the uppermost, upper, middle, and lower parts of the mainstream and their drainage basins

drainage area is 6.8×10^5 km^2, constituting 38% of the entire Yangtze River drainage basin (Chen et al. 2001). As it flows across the Jianghan Plain, the mainstream is characterized by a 10 km scale meandering channel, and the plain includes many lakes that originated as cutoff channels. The riverbed slope of the mainstream deceases dramatically in the middle part, to 0.02–0.03‰, and the channel widens to 1–2 km, with a depth of 6–15 m (Chen et al. 2001). Three major tributaries flow into the mainstream in the middle part, the Hanjiang from the north and tributaries through the Dongting Lake and Poyang Lake drainages from the south (Fig. 1). The Dongting Lake drainage, with an area of 2.7×10^5 km^2 (Hayashi et al. 2008), gathers waters from four major tributaries. In addition, three natural channels flow from the mainstream into Dongting Lake, known as the "Three Channels" or "Three Outlets" (Dai et al. 2005; Yin et al. 2007). Thus, some water and SPM in the upper part of the mainstream flow into Dongting Lake through these three channels (Dai et al. 2005). The area of the Poyang Lake drainage basin is 1.6×10^5 km^2, and its largest tributary is the Ganjiang (Hu et al. 2007).

The lower part, or the lower Yangtze, is defined as the mainstream between Hukou and the rivermouth. Its length is 930 km and its drainage area is 1.2×10^5 km^2 (Chen et al. 2001), constituting 7% of the entire Yangtze River drainage basin. The mainstream in this part flows across flat plains and past low hills and is characterized by numerous braided channels (Chen et al. 2001). The riverbed slope in the lower part decreases to 0.01‰, and the channel width widens to an average of 2–4 km, reaching a width of 15 km in the estuary region; the channel depth increases to 10–20 m (Chen et al. 2001). Although many lakes are situated near the mainstream, no large tributary exists in the lower part.

Geologic setting

A geological map of the Yangtze drainage basin is presented in Fig. 2. In the headwater region of the uppermost basin, the mainstream flows through the northern Qamdo Block, which is composed mainly of Triassic–Jurassic shallow-marine carbonate, basaltic, and fluvial sedimentary rocks (Yin and Harrison 2000; Xia et al. 2011). The mainstream then flows into the Songpan–Garze terrane, which is composed of Triassic turbiditic sedimentary rocks that were deposited in the Paleo-Tethys Ocean and folded during the Late Triassic (Enkelmann et al. 2007; Roger et al. 2008). The major tributary in the uppermost drainage basin, the Yalongjiang, also flows mainly through the Songpan–Garze terrane. The southwestern part of the uppermost Yangtze drainage basin, where the Jinshajiang flows southward parallel to the Lancanjiang (Mekong River) and Nujiang (Salween River), includes Paleozoic carbonates, mafic volcanics, and Permian–Paleogene shales and sandstones (Reid et al. 2005a). In the lower part of the uppermost basin below Big Bend, Permian–Triassic mafic igneous rocks known as the Emeishan large igneous province are extensively distributed in association with Jurassic sandstones and mudstones (Ali et al. 2005; Liu et al. 2011).

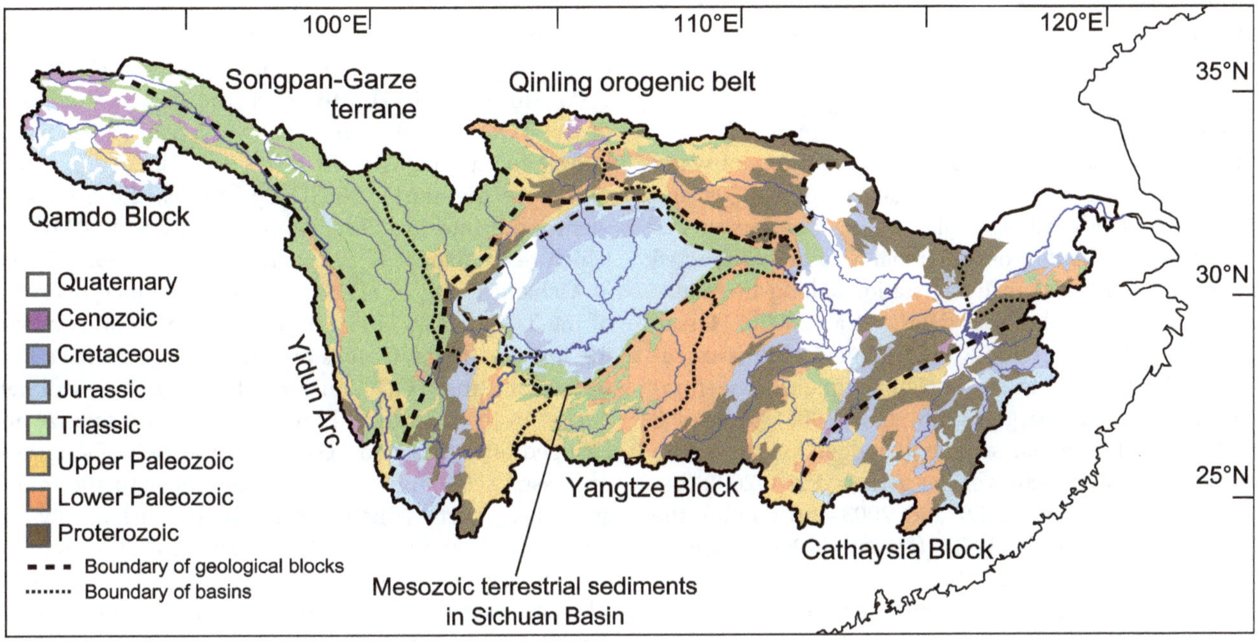

Fig. 2 Map showing ages of bedrock in the Yangtze drainage basin (after Teraoka and Okamura 2003, 2007)

The northwestern part of the upper Yangtze drainage basin, which is drained by the Daduhe (the largest tributary of the Minjiang; Fig. 1), is also underlain by the Songpan–Garze terrane, where it is composed mainly of Triassic turbiditic rocks that are partly intruded by Late Triassic or Early Jurassic granites (Reid et al. 2005a; Roger et al. 2008). The upper part of the Jialingjiang drainage basin belongs to the South Qinling orogenic belt, consisting of Meso–Neoproterozoic basement that was weakly metamorphosed during the Neoproterozoic, shallow-marine deposits of Neoproterozoic–Devonian age, and Triassic granitoids (Meng and Zhang 2000; Qin et al. 2010; Dong et al. 2011). The central to southeastern Sichuan Basin, which is drained by the middle to lower part of the Jialingjiang and the lower part of the Wujiang, is widely covered by thick terrestrial mudstone and sandstone deposited during the Jurassic–Cretaceous (Hao et al. 2008; Ma et al. 2008). The upper part of the Wujiang drainage basin is underlain by Permian–Triassic limestones and dolomitic limestones with minor basalts. Cambrian and Ordovician carbonates and Silurian sandstones are exposed mainly in the lower part of the Wujiang drainage basin (Han and Liu 2004; Wang et al. 2010).

In the middle Yangtze drainage basin, Quaternary sediments transported by the Yangtze are widely deposited along the mainstream, particularly across the Jianghan Plain between the mainstream and the Hanjiang. The upper part of the Hanjiang drainage basin belongs to the Qinling orogenic belt and is composed of Neoproterozoic basement and Neoproterozoic–Devonian sediments (Qin et al. 2010). Quaternary loess deposits also occur in some parts of the upper Hanjiang drainage basin (Zhang et al. 2012). The Jianghan Plain forms the middle to lower part of the Hanjiang drainage basin. Most of the Dongting Lake drainage basin belongs to the eastern part of the Yangtze Block. The Proterozoic basement of the Yangtze Block has been subjected to low-grade metamorphism. Cambrian and Ordovician carbonate rocks, Devonian sandstone and shale, and Cretaceous red-colored sandstone are the main rocks exposed in this region (Chen and Jahn 1998; Wang et al. 2010, 2013). The drainage basin of Poyang Lake and the easternmost part of the Dongting Lake drainage basin belong to the Cathaysia Block, which is composed of Neoproterozoic conglomerate and sandstone, lower Paleozoic siliciclastics, Jurassic granites, and Cretaceous red-colored sandstone (Wang et al. 2007a, 2013).

In the lower Yangtze drainage basin, Quaternary sediments derived from the upper and middle Yangtze were deposited along the mainstream. The southern part of the lower Yangtze drainage basin belongs to the Cathaysia Block (Wang et al. 2007a, 2013).

In summary, the bedrock of the uppermost and upper Yangtze drainage basins is relatively young and is composed mainly of Mesozoic strata, while the northernmost and southern parts of the upper Yangtze drainage basin are composed mainly of Paleozoic strata, and the middle and lower Yangtze drainage basins are composed mainly of Paleozoic–Proterozoic strata (Fig. 2). Thus, the strata in the Yangtze drainage basin tend to increase in age downstream. Regarding rock types, sedimentary and low-grade metasedimentary rocks are exposed mainly in the uppermost and upper parts of the basin, whereas crystalline plutonic rocks are more common in the middle and lower parts of the basin.

Hydrologic setting

In the Yangtze basin, hydrological observations of water discharge and sediment load began in the 1870s and 1950s, respectively (Chen et al. 2001). Measurements of water flow speed and cross-sectional area are conducted daily by ships at major gauging stations along the mainstream and major tributaries (Wang et al. 2007b; Hassan et al. 2010). At each station, the sediment load is calculated as the product of suspended sediment concentration and water discharge (Yang et al. 2006). Figures 3 and 4 show the annual sediment budget for the entire drainage basin averaged over the periods 1950–2002 (before operation of the TGD) and 2003–2012 (after the start of operation of the TGD) (Dai et al. 2005; Yang et al. 2006, 2014; Hu et al. 2009).

Dongting Lake, which drains into the middle part of the mainstream, has played an important role in the sediment budget of the Yangtze (Dai et al. 2005). In particular, below the Three Gorges, turbid water from the upper part flows not only into the mainstream of the Yangtze but also into channels flowing into Dongting Lake, known as the Three Outlets (Dai et al. 2005; Yin et al. 2007). Therefore, Dongting Lake receives sediments from the mainstream above the Three Gorges through the Three Outlets, in addition to sediment originating from its own tributaries. Thus, the sediments discharged from Dongting Lake to the mainstream are a mixture of sediments from the upper part of the mainstream and from tributaries flowing into Dongting Lake itself. According to observational data, the volume of sediments accumulating in Dongting Lake is proportional to the inflow volume from the Three Outlets: it is estimated that 87% of sediments flowing into the Three Outlets were deposited in the Dongting Lake area during 1950–2002 (Dai et al. 2005).

Since the 1950s, when observations of sediment load began, sediment discharge in the mainstream at Datong has shown a decreasing trend, owing to reservoir construction and the National Soil Conservation Project, particularly in the Jialingjiang drainage basin (Yang et al. 2006; Hu et al. 2011). Moreover, the start of operations of the TGD in 2003 significantly changed the sediment budget in the mainstream below the TGD (Figs. 3 and 4). Before the start of TGD operations, the total sediment input into the Three Gorges area was 489 Mt/year, which nearly balanced the sediment load of 492 Mt/year at Yichang (Yang et al. 2007). However, since operation of the TGD, 80.3% (188 Mt/year) of the sediments flowing into the Three Gorges area (234 Mt/year) have been trapped by the TGD (Yang et al. 2014). The sediment budget in the middle and lower parts of the Yangtze also changed after 2003 on account of the TGD, because of decreased sediment loads from the upper to the middle and lower parts (Chang et al. 2010). Specifically, the middle and lower parts have experienced erosion of 63 Mt/year since the start of TGD operations in 2003, whereas prior to 2003, these regions experienced a net sediment accumulation of 36.1 Mt/year (Dai et al. 2005; Yang et al. 2014). In the 2010s, more than 10 hydropower reservoirs were proposed in the uppermost part of the basin. Five dams had been constructed on the mainstream by 2012, including the Xiangjiaba and Xiluodu Dams between the Yibin and Yalongjiang tributaries, and the Longkaikou, Jinangqiao, and Ahai Dams between the Yalongjiang and Big Bend (Hennig et al. 2013). Operation of the Jinangqiao Dam began prior to 2010, while construction of the other dams started in the early 2010s. These dams are expected to influence the sediment budget in the uppermost Yangtze in the near future.

Methods/Experimental

Samples

Riverbed sediments were sampled mainly from the riverside of the mainstream and major tributaries of the Yangtze in July and September 2011 and in February and October 2012. Sediments from the mainstream were

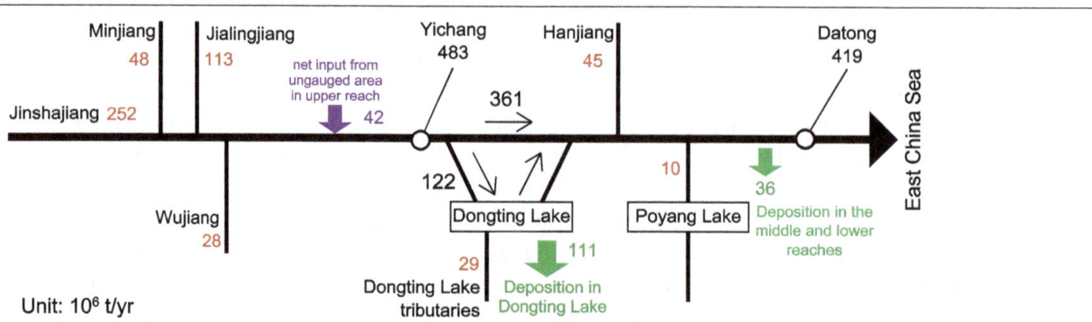

Fig. 3 Average annual sediment budget in the Yangtze drainage basin during 1950–2002, before construction of the Three Gorges Dam (TGD) (based on Dai et al. 2005; Yang et al. 2006; Hu et al. 2009). *Green arrows* show deposition in the mainstream and lakes. *Purple arrow* shows sediment inputs from ungauged areas

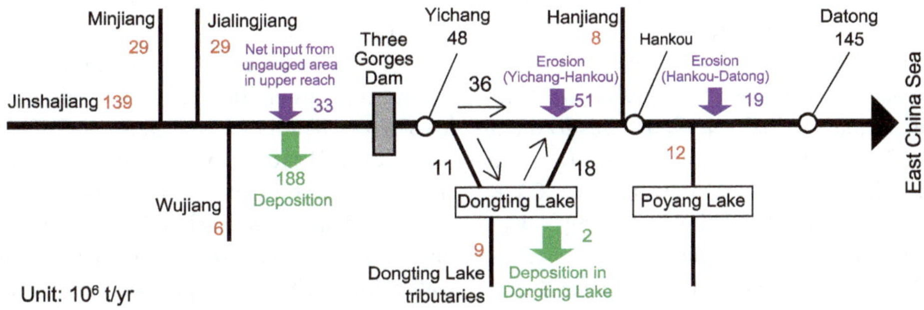

Fig. 4 Average annual sediment budget in the Yangtze drainage basin during 2003–2012, after construction of the Three Gorges Dam (TGD) (based on Yang et al. 2014). *Green arrows* show deposition in the mainstream and lakes. *Purple arrows* show sediment inputs from ungauged areas

sampled at locations a few hundred meters to a few kilometers above and below the junctions with major tributaries to evaluate the effects of sediment mixing between the mainstream and the tributaries. Sediments from the tributaries were sampled at locations several kilometers upstream of the junction with the mainstream so as to prevent contamination of sediments from the mainstream. Our data show that in the lower part, the SPM consists of fine silt particles with a mean diameter of ~10 μm throughout the year (Hassan et al. 2010; Mao et al. 2011). Furthermore, SPM accounts for 98% of the sediment discharge from the Yangtze rivermouth (Yang et al. 2002). For these reasons, we sampled silty sediments wherever possible and otherwise sampled sandy sediments.

We also analyzed three SPM samples (JSJ-GLS, JSJ110701, and YZ110703-03; Additional file 1) obtained at the same locations and on the same days as the riverbed samples were taken. Then, we examined if the composition of the fine-silt fraction of the riverbed samples could be used as a proxy of the SPM.

Sample preparation

Samples were pretreated prior to grain size separation based on methods presented by Rea and Janecek (1981) and Tada et al. (2000). Approximately 3 g of dry sample was placed into each of two plastic centrifuge tubes. In the first step, carbonates were removed by immersion in 50 ml of 20% acetic acid at room temperature for 8 h. After removing carbonates, the supernatants were drained. Approximately 50 ml of deionized water was added, and the tubes were centrifuged twice at 3000 rpm to remove any remaining acetic acid. In the second step, organic matter was removed by immersion in ~50 ml of 10% hydrogen peroxide at 60 °C for more than 20 h until all bubbling ceased. After the supernatant was drained, ~50 ml of deionized water was added and the tubes were centrifuged twice at 3000 rpm. In the third step, Fe and Mn oxides were removed using a mixture of 0.3 mol/l sodium citric acid and 0.1 mol/l sodium hydrogen carbonate at a ratio of 8:1, followed by the addition of 3 g of sodium hydrosulfite. The tubes were heated at 80 °C for 8 h. Then, the supernatant was removed, 50 ml of deionized water was added, and the tubes were centrifuged twice.

These pretreated samples were kept wet until the grain-size separation procedure. Grain size separation was conducted by the settling method based on Stokes' law at 4 and 16 μm. Separation at 63 μm for the >16 μm fraction was also conducted using a sieve. Although we did not sieve the upper limit of the >63 μm fraction, the grain size in the sand fraction was not larger than 2–3 mm, based on macroscopic observation. The riverbed sediments were separated into four size fractions: <4 μm (clay), 4–16 μm (fine-silt), 16–63 μm (coarse-silt), and >63 μm (sand); the latter three fractions were analyzed. The SPM samples were separated into two size fractions: <4 and >4 μm, and the latter was analyzed. The <4 μm (clay) fraction was not analyzed, either in the riverbed sediments or in the SPM. After completion of the separation procedures, each fraction was dried and weighed.

ESR analysis

The ESR signal intensity of quartz was used to constrain the provenance of quartz in the sediments. The ESR signal intensity of the E_1' center of quartz reflects the number of unpaired electrons accompanied by an oxygen vacancy (Feigl et al. 1974; Toyoda and Naruse 2002). As these oxygen vacancies in quartz are formed mainly by natural radiation generated by radioactive elements in the host rock, the ESR signal intensity increases with the age of the host rock (Toyoda and Naruse 2002).

Prior to the ESR signal intensity measurement, electron holes were created in dried sediment grains by irradiating with 2.5 kGy ^{60}Co γ-rays at room temperature, at the Takasaki Advanced Radiation Research Institute, National Institutes for Quantum and Radiological Science and Technology, Takasaki, Japan, according to the method of Toyoda and Hattori (2000). Then, ~0.1 g of sample was heated at 300 °C for 15 min to convert the oxygen vacancy of quartz to the E_1' center. The ESR measurements were conducted using an ESR spectrometer (FA-100, JEOL) at the Department of Earth and Planetary Science, University of Tokyo, Tokyo, Japan. The measurement conditions were a microwave strength of 0.01 mW, a measurement range of 5 × 1 mT, a measurement time at room temperature of 2 min, and a sample weight of 0.1 ± 0.001 g. The ESR signal intensity of the bulk sample was normalized by the weight percent of quartz in the sample to obtain the ESR signal intensity of the quartz (unit, 1.3×10^{15} spins/g) (Toyoda and Hattori 2000).

To evaluate the reproducibility of the ESR signal intensity data, we selected two γ-ray-irradiated samples (consisting of the fine-silt fractions of samples YZ1210-4 and YZ110703-04m2; Additional file 1), split each sample into three subsamples, packed each subsample into a discrete holder, and then measured the ESR signal intensity of quartz in each subsample. The standard error (95%) for each sample was calculated based on three measurements on each of the three subsamples (nine measurements in total). The reproducibility of the ESR signal intensity results was represented by a standard error of 1.3 for sample YZ110703-04m2 and 3.0 for sample YZ1210-4. We adopted the average of these standard errors (2.2) as the reproducibility of the ESR signal intensity data obtained in this study.

X-ray powder diffraction analysis

The quartz content (QC) was determined using X-ray power diffraction (XRD) to calculate the ESR signal intensity of quartz. The CI was also determined by XRD, to correct for crystallinity effects on the quartz peak height. The CI is an index of the degree of crystallization of quartz, as it reflects conditions of crystallization, such as cooling speed and crystallization temperature (Murata and Norman 1976). The CI values were calculated as the degree of the quartz peak split at 67.74° 2θ, based on the definition of Murata and Norman (1976). The CI was also used to constrain the provenance of quartz. Prior to the XRD measurements, two size fractions (16–63 and >63 μm) were ground to powder for ~15 min using an agate mortar; the fine-silt fraction (4–16 μm) was not ground, as it was considered sufficiently small for XRD analysis. To measure the QC, 10.0 wt.% of silicon powder was added to the samples as an internal standard. The XRD measurement (X'pert Plus, PANalytical) was conducted at the Department of Earth and Planetary Science, University of Tokyo. The XRD measurements were repeated three times with repacking, each time using CuKα X-rays, at a voltage of 45 kV and a current of 40 mA; results for each sample were averaged. The divergence and anti-scatter slits were 1.52 and 3 mm in width, respectively. The measurement ranges were 20–30° 2θ for QC and 66–71° 2θ for CI. The QC was calculated as the ratio of the silicon and quartz peak heights at 27.9° 2θ and 20.9° 2θ. The QC was also corrected for the effect of crystallinity based on the method of Isozaki (2009).

The reproducibility of the CI values was evaluated for two γ-ray-irradiated samples (the fine-silt fractions of samples YZ1210-4 and YZ110703-04m2; Additional file 1). We collected three subsamples from each of the two irradiated samples. Each subsample was split and packed into an XRD holder and measured by XRD. The CI value for each subsample was calculated as the average of three measurements. Therefore, we collected nine CI measurements for each irradiated sample, and the standard error for each irradiated sample was based on the three CI measurements of the subsamples. The standard error was 0.23 for sample YZ1210-4 and 0.34 for sample YZ110703-04m2. We adopted the average of these standard errors (0.29) as the reproducibility of the CI values.

Results and Discussion

ESR signal intensity of quartz in riverbed sediments

Fine-silt (4–16 μm) fraction

Figure 5a shows the geographical distribution of the ESR signal intensity of quartz in the fine-silt fraction in the Yangtze drainage basin. Samples collected from the mainstream show ESR signal intensities of 0.0–9.4, with an average of 3.9. The ESR signal intensities of samples collected from the mainstream increase from 0.0–2.3 in the uppermost part to 3.1–5.7 in the lower part; however, values in the uppermost part above Big Bend are ~8. Samples collected from the Yalongjiang and Minjiang show ESR signal intensities of 0.0–5.5, with an average of 2.2, which is similar to values from the upper part of the mainstream, while samples collected from the Jialingjiang and Wujiang show ESR signal intensities of 1.5–15.5, with an average of 10.3, which is higher than values from the upper part of the mainstream. Samples collected from the tributaries in the middle part show a wide range of ESR signal intensities, from 0.0 to 23.4 with an average of 9.2, which is higher than values from the middle part of the mainstream.

Coarse-silt (16–63 μm) fraction

Figure 5b shows the geographical distribution of ESR signal intensities of quartz in the coarse-silt fraction in the Yangtze drainage basin. Samples collected from the mainstream show ESR signal intensities of 0.0–5.8, with an average of 2.6. Samples collected from the Yalongjiang and Minjiang show ESR signal intensities of 0.0–2.4, with an average of 1.2, which is similar to values from the upper part of the mainstream, while samples collected from the Jialingjiang and Wujiang show ESR signal intensities of 3.4–9.5, with an average of 5.2, which is higher than values from the upper part of the mainstream.

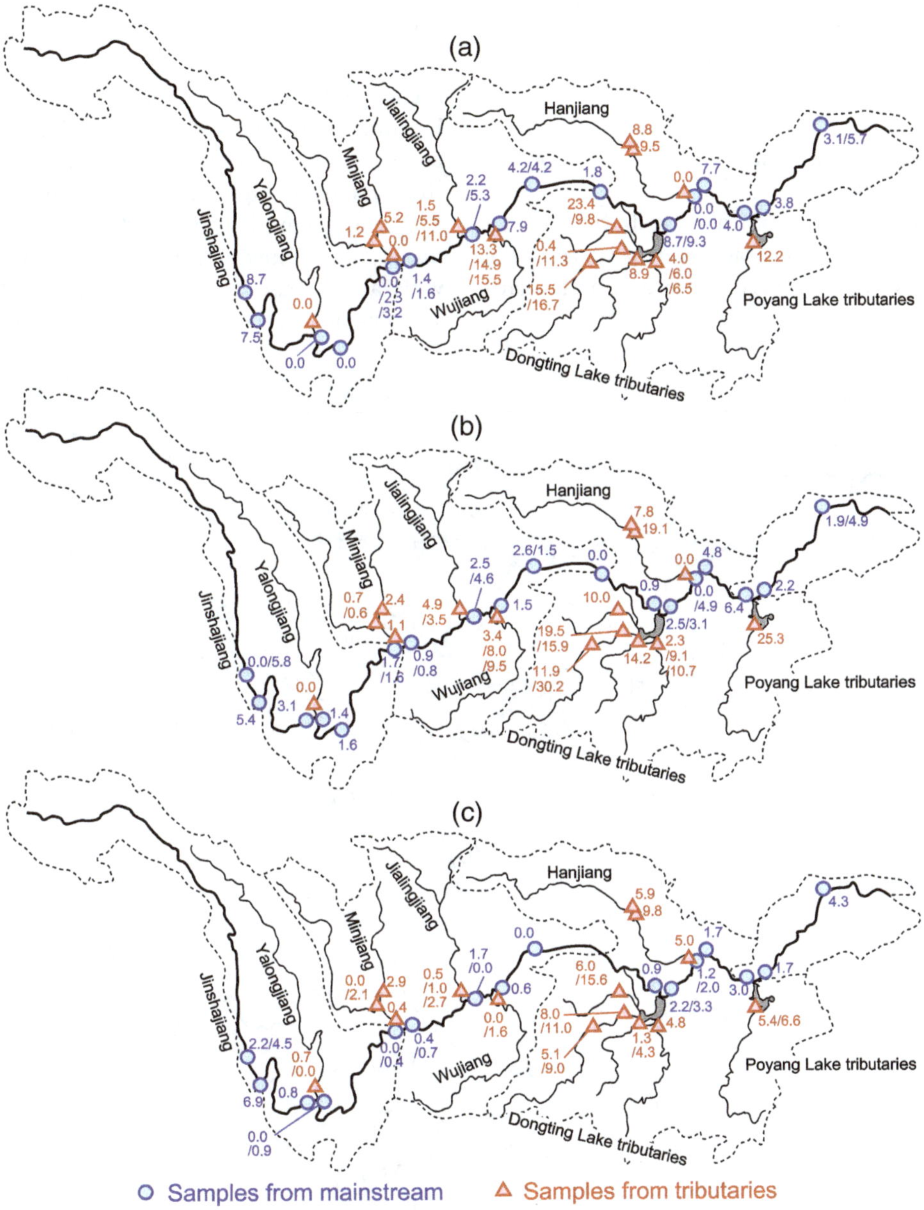

Fig. 5 Geographical distribution of electron spin resonance (ESR) signal intensity of quartz in riverbed samples. **a** Sand (>63 μm) fraction. **b** Coarse-silt (16–63 μm) fraction. **c** Fine-silt (4–16 μm) fraction. The "/" separates the values of different samples collected from sites in close proximity to one another

Samples collected from tributaries in the middle part show a wide range of ESR signal intensities, from 0.0 to 30.2, with an average of 11.2, which is higher than values from the middle part of the mainstream.

Sand (>63 μm) fraction

Figure 5c shows the geographical distribution of ESR signal intensities of quartz in the sand fraction of riverbed samples from the Yangtze drainage basin. Samples collected from the mainstream show ESR signal intensities of 0.0–6.9, with an average of 1.6. The ESR signal intensities of samples collected from the mainstream increase from 0.0–0.9 in the uppermost part to 1.8–4.8 in the lower part, except in the uppermost part above Big Bend, where values are in the range of 2.2–6.9. Samples collected from tributaries in the uppermost and upper parts show lower ESR signal intensities (0.0–2.9, with an average of 1.1) than those observed in the uppermost and upper parts of the mainstream. Samples collected from tributaries in the middle part show a wide range of ESR signal intensities, from 0.0 to 15.6, with an average of 5.4, which is higher than values from the middle and lower parts of the mainstream.

CI of quartz in riverbed sediments

Fine-silt (4–16 μm) fraction

Figure 6a shows the geographical distribution of CI values of quartz in the fine-silt fraction, which are in the range of 6.8–9.0, with an average of 8.5 and a standard deviation of 0.4.

Samples collected from the mainstream show CI values of 6.8–9.0, with an average of 8.4, and display no obvious downstream trends. In the uppermost and upper parts, samples collected from the Yalongjiang and Daduhe show low CI values of 8.2–8.3, while samples collected from the Minjiang, Jialingjiang, and Wujiang show relatively high CI values of 8.6 to 8.9, with an average of 8.7. Samples collected from the tributaries in the middle part show a wide range of CI values of 8.0–8.9, with an average of 8.5. Among the tributaries, some samples from the Minjiang, Wujiang, Dongting Lake

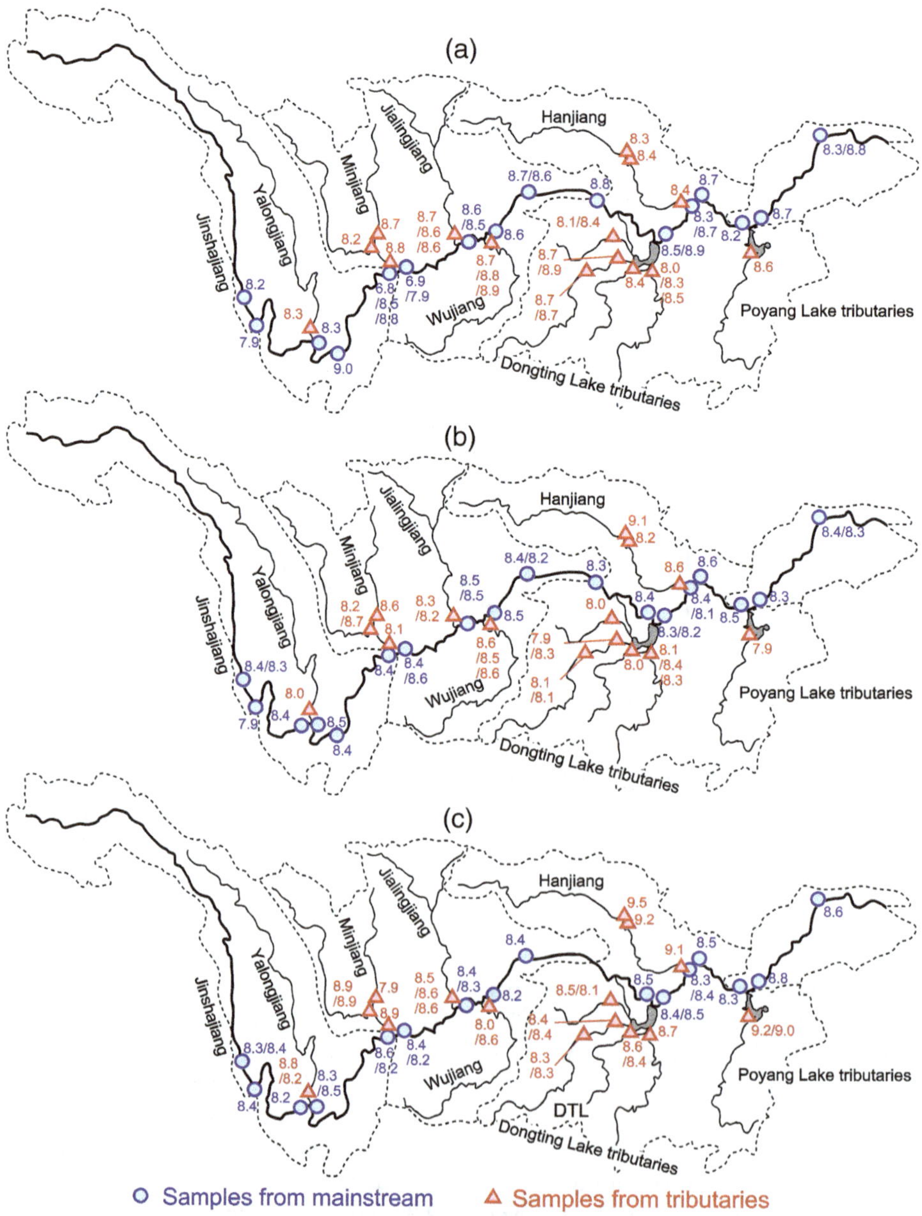

Fig. 6 Geographical distribution of crystallinity index (*CI*) values of quartz in riverbed samples. **a** Sand (>63 μm) fraction. **b** Coarse-silt (16–63 μm) fraction. **c** Fine-silt (4–16 μm) fraction. The "/" separates values of different samples collected from sites in close proximity to one another

tributaries, and Poyang Lake tributaries show relatively high CI values of ~9.

Coarse-silt (16–63 μm) fraction

Figure 6b shows the geographical distribution of CI values of quartz in the coarse-silt fraction in the Yangtze drainage basin. Samples collected from the mainstream show CI values of 7.9–8.6, with an average of 8.4 and show no clear downstream trends. Samples collected from the tributaries in the uppermost and upper parts show a wide range of CI values, from 7.9 to 8.7, with an average of 8.4. Samples from tributaries in the middle part also show a wide range of CI values, from 7.9 to 9.1, with an average of 8.3. Among the tributaries, a sample from the Hanjiang is characterized by relatively high CI values of ~9, while samples from the Dongting Lake and Poyang Lake tributaries are characterized by low CI values of ~8.

Sand (>63 μm) fraction

Figure 6c shows the geographical distribution of CI values of quartz in the sand fraction in the Yangtze drainage basin. Samples collected from the mainstream show CI values of 8.2–8.8 with an average of 8.4 and a slightly increasing downstream trend from 8.3 to 8.7, although with some scatter. Samples collected from the tributaries in the uppermost and upper parts show a wide range of CI values, from 7.8 to 8.9, with an average of 8.5. Samples from tributaries in the middle part also show a wide range of CI values, from 8.1 to 9.5, with an average of 8.7. Among the tributaries, some samples from the Minjiang, Dongting Lake, Hanjiang, and Poyang Lake tributaries are characterized by relatively high CI values of ~9.

Comparisons of ESR signal intensities among the three size fractions

Figure 7 shows differences between ESR signal intensities for the various size fractions in the same sample collected from the major tributaries. A comparison of the fine-silt and sand fractions reveals that ESR signal intensities for samples from the Jinshajiang, Minjiang, and Hanjiang are similar to one another, whereas intensities for samples from the Jialingjiang, Wujiang, Dongting Lake, and Poyang Lake tributaries are higher in the fine-silt fraction than in the sand fraction (Fig. 7a). On the other hand, ESR signal intensities in the fine-silt and coarse-silt fractions in samples from the Jinshajiang and Minjiang are similar. Samples from the Jialingjiang and Wujiang show higher ESR signal intensities in the fine-silt fraction than in the coarse-silt fraction. Many of the samples from the Hanjiang, Dongting Lake, and Poyang Lake tributaries show similar values for the fine-silt and coarse-silt fractions. However, some samples from the Dongting Lake drainage basin show higher ESR signal intensities in the fine-silt fraction, while some samples collected from the Hanjiang, Dongting Lake, and Poyang Lake drainage basins show higher ESR signal intensities in the coarse-silt fraction (Fig. 7b). The different ESR signal intensities among the three size fractions suggest origins from different source rocks. Thus, grain size should be considered when discussing the provenance of sediments.

Comparison of ESR signal intensities between the fine-silt fraction of riverbed samples and that of SPM samples

Although the modal diameter of SPM is within the range of the fine-silt (4–16 μm) fraction, the coarse-silt (16–63 μm) and very fine sand (63–125 μm) fractions are also transported as suspension in addition to the clay (<4 μm) and fine-silt (4–16 μm) fractions (Mao et al. 2010; Gao et al. 2015). Therefore, we examined whether the fine-silt fraction of riverbed samples could be used as a proxy for SPM, by comparing their ESR signal intensities.

Three SPM samples were analyzed for comparison of their ESR signal intensities with those of riverbed sediments; the SPM and riverbed samples were collected at the same locality and on the same day. The ESR signal intensities of the SPM samples of 7.9, 5.6, and 3.4 (JSJ-GLS, JSJ110701, and YZ110703-03, respectively; Additional file 1) agree within errors with the intensities of the corresponding riverbed samples of 8.7, 2.3, and 3.9 (samples UYZ110911-01m, YZ110701-01c, and YZ110703-04m, respectively; Additional file 1). Thus, the ESR signal intensities of quartz in the fine-silt (4–16 μm) riverbed sediments basically could be used as a proxy for the SPM samples.

Relationship between ESR signal intensity and CI values of riverbed sediments, and the ages and types of bedrock in tributary basins

We summarize ESR signal intensity and CI values of sediments collected from each tributary (presented in Fig. 8) and examine their relationship with ages and types of bedrocks exposed in the respective drainage basins. As the basement rocks in each geological block are generally considered as the source of younger detrital sediments in the block, each sample is categorized with respect to the geological block(s) exposed in its drainage basin (Fig. 8). The geological blocks that make up the basement rocks in the Yangtze River drainage basin are classified into the Qamdo Block and Yidun Arc (QY), Songpan–Garze terrane (SG), Qinling orogenic belt (QL), Yangtze Block (YZ), and Cathaysia Block (CX). In addition, Quaternary sediments in the middle to lower parts are classified as a sixth type (Q). Mesozoic terrestrial sediments in the Sichuan Basin, which unconformably overlie basement rocks, are considered to be derived from

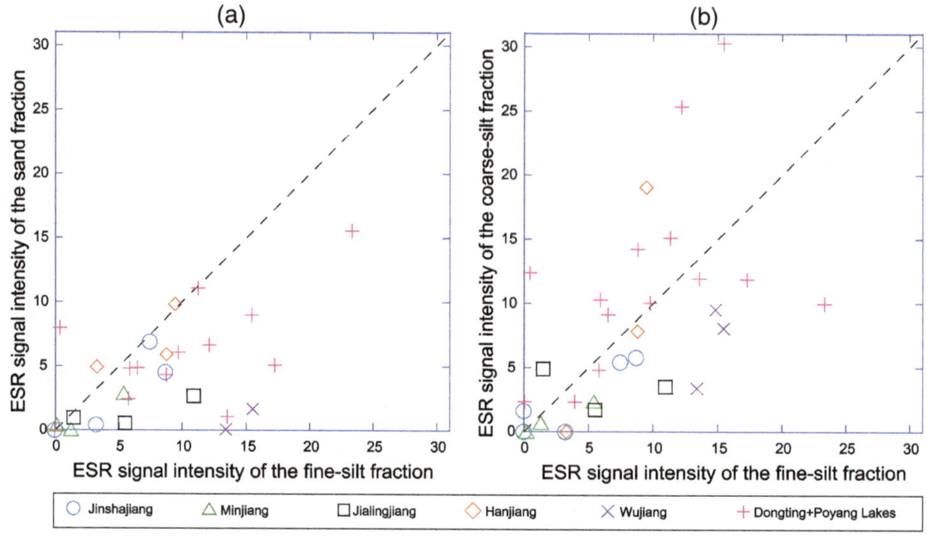

Fig. 7 Comparison of electron spin resonance (ESR) signal intensities among the three size fractions of riverbed samples collected in this study. **a** Fine-silt versus sand fractions. **b** Fine-silt versus coarse-silt fractions

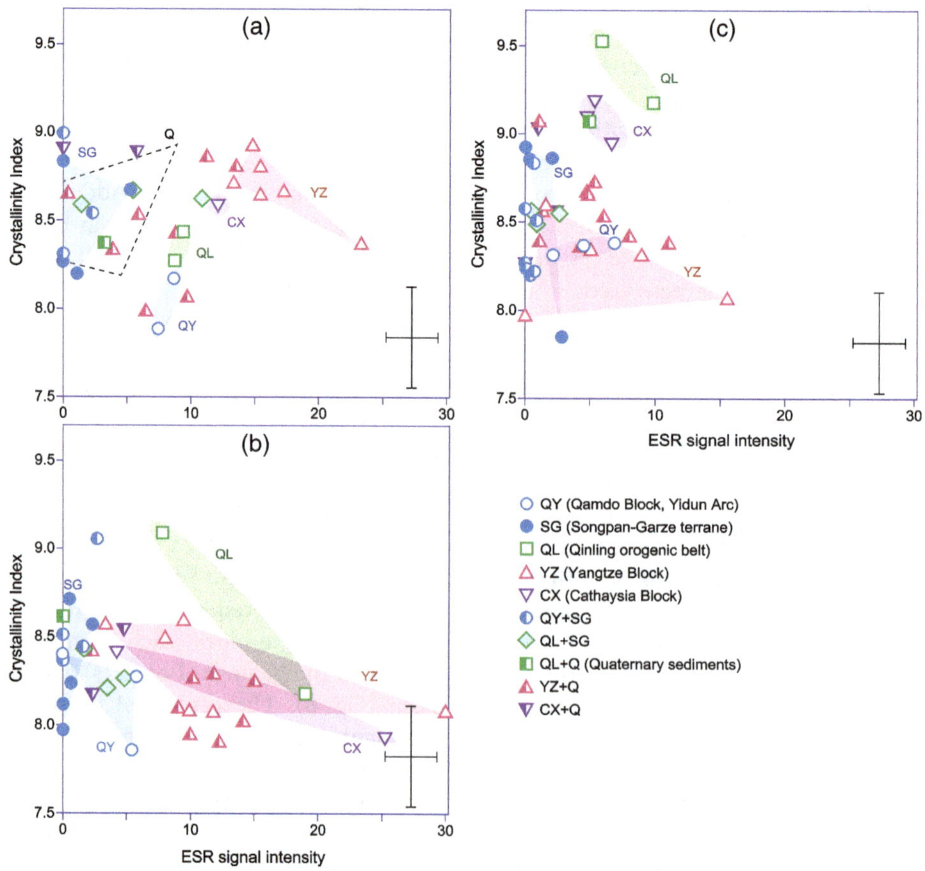

Fig. 8 An ESR–CI plot for riverbed sediments from each of the geological blocks in the Yangtze drainage basin. **a** Fine-silt (4–16 μm) fraction. **b** Coarse-silt (16–63 μm) fraction. **c** Sand (>63 μm) fraction. *QY* Qamdo Block and Yidun Arc, *SG* Songpan–Garze terrane, *QL* Qinling orogenic belt, *YZ* Yangtze Block, *CX* Cathaysia Block, *Q* Quaternary sediments in the middle–lower parts of the mainstream

both the SG and the QL, although the basement of the Sichuan Basin belongs to the YZ (Wang et al. 2013; Luo et al. 2014).

The ESR signal intensity increases with increasing age of the host rock (Toyoda and Naruse 2002). According to Toyoda (1992), the Cenozoic granites show E_1' center densities of $<1.3 \times 10^{15}$ spins/g while the Proterozoic granites show E_1' center densities of $>1.3 \times 10^{16}$ spins/g. Because the units of ESR signal intensity are defined as 1.3×10^{15} spins/g, the ESR signal intensities of the Cenozoic rocks are less than ~1, those of Mesozoic rocks are from ~1 to ~3, those of upper Paleozoic rocks are from ~3 to ~10, and those of Proterozoic rocks are greater than ~10 (Toyoda and Hattori 2000). Based on these general patterns, we categorized ESR signal intensities into three classes: low (<3; roughly corresponding to Cenozoic and Mesozoic rocks), medium (3–10; roughly corresponding to Paleozoic rocks), and high (>10; roughly corresponding to Proterozoic rocks). However, we recognize the need to take into account the rock type dependency of the ESR signal intensity caused by differences in radionuclide activity in different rocks.

We also categorized CI values into three classes based on the 1st quartile (8.2) and 3rd quartile (8.7) of all fractions: low CI values are <8.2, medium CI values are from 8.2 to 8.7, and high CI values are >8.7.

Relationship between ESR signal intensity and CI values of the fine-silt (4–16 μm) fraction of sediments in tributaries and the geological blocks in their drainage basins

Quartz grains in riverbed sediments can be a mixture of different source rocks within a given drainage basin, and thus, ESR signal intensity and CI values of the riverbed sediments do not necessarily represent those of a single geological block or source rock. However, in the discussion that follows, we assume that the range of ESR signal intensity and CI values in riverbed samples collected from a given geological block represent the minimum range of values in the various source rocks within that block.

Figure 8a shows ESR signal intensity and CI values of quartz in the fine-silt fraction of river sediments, with symbols indicating the geological block(s) exposed in the drainage basin of the sample. As shown in the "Results and Discussion" section, the fine-silt fraction could well represent the composition of SPM in the Yangtze River. Here, we discuss the features of ESR signal intensity and CI values in the fine-silt fraction as proxies for the SPM provenance.

Songpan–Garze terrane Samples derived from SG exhibit low to medium ESR signal intensities (0.0–5.3) and medium CI values (8.2–8.8). The SG is composed mainly of Mesozoic shallow-marine sediments, including sandstones and carbonates (Enkelmann et al. 2007; Liu et al. 2011), which is consistent with low ESR signal intensities.

Qamdo Block and Yidun Arc Samples derived from QY exhibit medium ESR signal intensities (~8) and low CI values (7.8–8.2). The Yidun Arc, which occupies a large part of QY, is composed of Paleozoic melange and Permian–Tertiary shales and sandstones that yield upper Paleozoic to Proterozoic zircon ages (Wang et al. 2000; Reid et al. 2005b; Deng et al. 2014); these ages are consistent with medium ESR signal intensities of ~8.

Qinling orogenic belt Samples derived from QL exhibit medium ESR signal intensities (~9) and medium CI values (~8.4). The QL is composed mainly of Neoproterozoic turbidites, phyllite, and sandstone; Silurian slate and sandstone; Devonian schist; and Triassic granite (Meng and Zhang 2000; Dong et al. 2011; Dong and Santosh 2016). Because the QL block in the Yangtze drainage basin was not metamorphosed after the Neoproterozoic (Dong et al. 2011; Dong and Santosh 2016), the rocks in QL are assumed to yield quartz showing an upper Paleozoic–Proterozoic age when quartz grains were originally formed. This age is consistent with ESR signal intensity of ~9.

Yangtze Block Samples derived from YZ exhibit high ESR signal intensities (12–23) and medium to high CI values (8.3–9.0). The YZ is composed mainly of Proterozoic basement that experienced low-grade metamorphism before the Neoproterozoic and lower Paleozoic sandstones, mudstones, and carbonates (Chen and Jahn 1998; Wang et al. 2010, 2013). As the lower Paleozoic sandstones in YZ are assumed to be derived from the Proterozoic basement and/or granites in the Cathaysia Block (Wang et al. 2010), both Proterozoic basement and Paleozoic clastic rocks seem to have a Proterozoic source rock age. This inference is consistent with the high ESR signal intensity values observed in the block (>10).

Cathaysia Block A sample from CX shows a high ESR signal intensity of 12.2 and a medium CI value of 8.6. The CX is composed mainly of Proterozoic basement, lower Paleozoic siliciclastics, and Cretaceous red-colored sandstones (Wang et al. 2007a, 2013). Cretaceous sandstones were probably derived from the YZ basement, which are Proterozoic in age (as mentioned above), which is consistent with the observed ESR signal intensity of >10.

Quaternary sediments in the middle to lower parts These types of sediment are thought to have been transported by the Yangtze mainstream (Yin et al. 2007; Dai and Lu 2010). Therefore, the Quaternary (Q) sediments in the middle and lower parts could be represented by riverbed sediments deposited in the middle part of the mainstream, where ESR signal intensities are low (~3) and CI values are moderate (~8.4).

Mixing of various source rocks Samples derived from QY + SG (from the Jinshajiang) show low to medium ESR signal intensities (0.0–3.2) and medium to high CI values (8.3–9.0), which are close to the values observed in SG. Because 80% of the sediments discharged from the Jinshajiang are derived from the SG region in the lower part of the Jinshajiang (Liu et al. 2011), it is reasonable that ESR signal intensity and CI values of samples derived from QY + SG are strongly affected by SG values. Samples derived from QL + SG (from the Jialingjiang) show a wide range of ESR signal intensities (1.4–10.9) and medium CI values (~8.6). These values lie between those of SG and QL, supporting the idea that the sediments of QL + SG are mixtures of sediments derived from QL and SG. A sample derived from QL + Q (from the Hanjiang) shows a medium ESR signal intensity of 3.3 and a medium CI value of 8.4. Samples derived from YZ + Q obtained from the Dongting Lake tributaries show low to high ESR signal intensities of 0–14. Samples derived from CX + Q obtained in the Poyang Lake tributaries show low to medium ESR signal intensities of 0–6. Because sediments of Q have similar ESR signal intensity and CI values to samples from the middle part of the mainstream, the ESR signal intensity and CI values of QL + Q, YZ + Q, and CX + Q lie along the mixing lines between QL and Q, YZ and Q, and CX and Q, respectively. Therefore, the ESR and CI values of the samples can be explained by a mixing of the values of individual geological blocks.

In summary, our data demonstrate that the ESR signal intensity of the fine-silt fraction (4–16 μm) extracted from riverbed samples collected on each geological block are consistent with the ages of the dominant rock types in that block.

Differences in ESR signal intensity and CI values among the three size fractions representing each geological block

Figure 8 shows clear differences in ESR signal intensity and CI values among the three sediment size fractions, with the nature of the differences depending on the composition of the geological blocks. Thus, in this subsection, we discuss the differences in ESR signal intensity and CI values among the three size fractions representing each geological block.

Qamdo Block and Yidun Arc Compared with the fine-silt fraction, the sand fraction of samples from QY shows lower or similar ESR signal intensities (2.2–6.9) and slightly higher CI values (~8.3), while the coarse-silt fraction shows values intermediate between those of the sand and fine-silt fractions (Fig. 8a–c). Clastic sediments from QY yield Proterozoic to upper Paleozoic zircon ages (Wang et al. 2000; Deng et al. 2014) that are consistent with the medium ESR signal intensities observed in the fine-silt fraction. On the other hand, coarse clastic sediments or Mesozoic granite–granodiorite may have contributed to low ESR signal intensities in the sand fraction, although it is difficult to specify the source rock of each fraction due to the structural complexity of QY (Reid et al. 2005a).

Songpan–Garze terrane The coarse-silt fraction of samples derived from SG shows low ESR signal intensities (0.0–2.4) and low to medium CI values (8.0–8.7) as compared with the sand fraction, which also shows low ESR signal intensities (0.0–2.9) but a wide range of CI values (7.8–8.9) (Fig. 8b, c). As the three size fractions derived from SG show similar ESR signal intensity and CI values, they are inferred to be derived from the same rocks.

Qinling orogenic belt Compared with the fine-silt fraction, the coarse-silt fraction of samples derived from QL shows similar to higher ESR signal intensities (7.8–19.3) and lower to similar CI values (7.8–8.4), whereas the sand fraction shows similar to higher ESR signal intensities (5.9–9.8) and higher CI values (9.2–9.5) (Fig. 8a–c). The high CI values of the sand and coarse-silt fractions suggest that Neoproterozoic and Triassic granite in the South Qinling area or Quaternary loess in the Hanjiang drainage basin contributed to the coarse-silt and sand fractions (Meng and Zhang 2000; Dong et al. 2011; Zhang et al. 2012; Dong and Santosh 2016).

Yangtze Block Compared with the fine-silt fraction, the sand and coarse-silt fractions of samples derived from YZ show lower to similar ESR signal intensities (3.4–30.0) and lower to similar CI values (8.1–8.6) as (Fig. 8a–c). Silicate rocks of YZ are dominantly composed of Proterozoic basement that was metamorphosed during the Neoproterozoic, lower Paleozoic slate and sandstone, Mesozoic sandstone, and Mesozoic granite (Wang et al. 2013). Based on the high ESR signal intensity and medium to high CI values, the fine-silt fraction seems to have been derived from the Proterozoic basement. In the YZ area, lower Paleozoic sandstones seem to be derived from the Proterozoic basement of the Cathaysia Block (CX) and they show high ESR signal intensity and low CI values, while Mesozoic granites show low ESR signal intensity and high CI values (Wang et al. 2010, 2013). Therefore, the lower ESR signal intensities and CI values of the coarse-silt and sand fractions than the fine-silt fraction in YZ could be explained by contributions from Mesozoic granites and Paleozoic sandstones.

Cathaysia Block The sand fraction of samples derived from CX shows lower ESR signal intensities (4.8–6.6) and higher CI values (9.0–9.2) than those of the fine-silt fraction. The presence of granites that intermittently intruded into CX during the Neoproterozoic–Jurassic (Wang et al. 2013) might explain the higher CI values in the coarser sediment fractions.

We suggested that detrital quartz grains of different sizes are derived from different sources. To test this possible explanation for differences in ESR signal intensity and CI values, it is necessary to perform an ESR analysis of quartz, combined with petrological observations of quartz grains in all varieties of potential source rocks in each geological block.

Classification of tributaries based on ESR signal intensity and CI values of the fine-silt fraction of sediments discharged from each tributary

We examine the relationship between ESR signal intensity and CI values of the fine-silt fraction of sediments discharged from each tributary, in terms of the geological block(s) exposed in each drainage basin. The major tributaries can be categorized into four groups based on the distribution of the geological block(s) in each drainage (Fig. 9): (1) the upper part of the Jinshajiang, including the Jinshajiang above Big Bend; (2) western tributaries, including the lower parts of the Jinshajiang, Yalongjiang, and Minjiang; (3) northeastern tributaries, including the Jialingjiang and Hanjiang; and (4) southeastern tributaries, including the Wujiang, Dongting Lake, and Poyang Lake tributaries.

The upper part of the Jinshajiang is characterized by medium ESR signal intensities (7.5–8.7) and low CI values (7.9–8.2), reflecting the presence of clastic sediments with upper Paleozoic to Proterozoic ages. The characteristics of this group are nearly equivalent to those of the QY. The western tributaries are characterized by low ESR values (0.0–3.2) and a wide range of CI values (7.8–9.0), reflecting the presence of Mesozoic shallow-marine sediments in the basins, including sandstone and carbonates. This group is characterized by a dominant contribution from SG. The northeastern tributaries are characterized by a wide range of ESR values (1.5–11.0) and medium CI values (8.3–8.7), reflecting

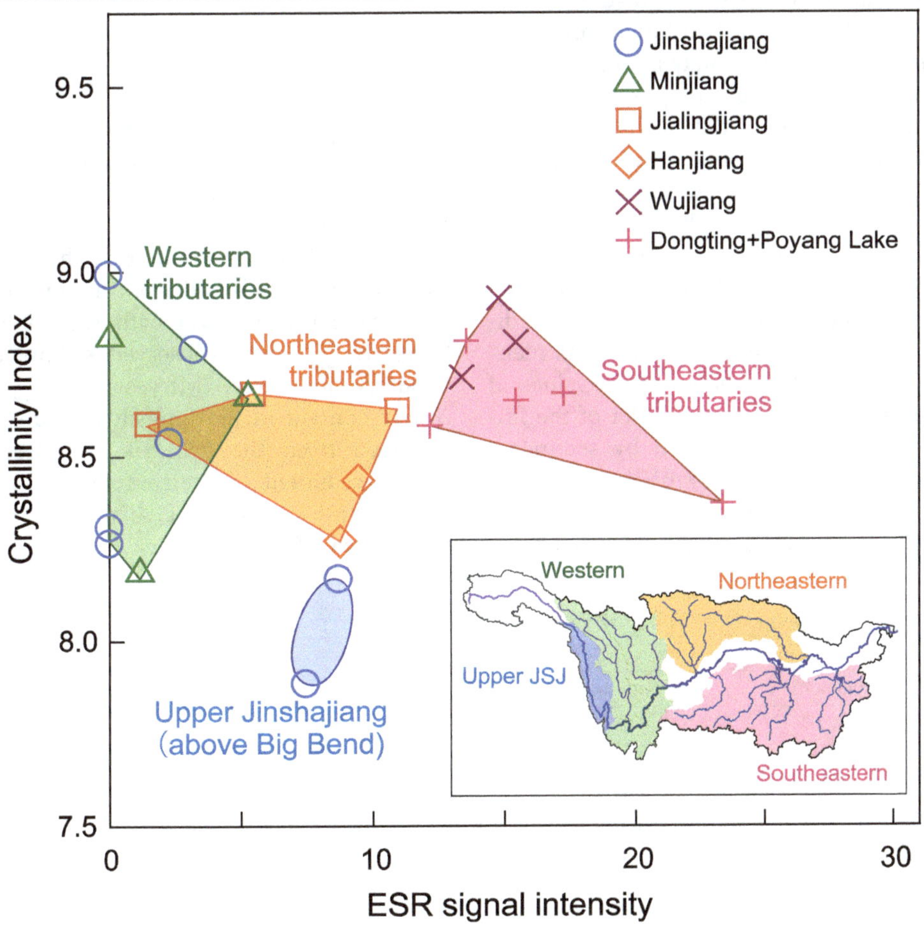

Fig. 9 An ESR–CI plot of quartz in the fine-silt fraction of riverbed samples from the major tributaries

the presence of both Mesozoic shallow-marine sediments derived from SG and upper Paleozoic clastic rocks and Proterozoic basement from QL. The southeastern tributaries are characterized by high ESR values (11.3–23.4) and low–medium CI values (8.4–8.9), representing the presence of Proterozoic basement in YZ and CX. Because contributions of sediment from the lower part of the Yangtze are negligible, they are not discussed here.

Predicted changes in the ESR signal intensity in sediments of the Yangtze mainstream below its junction with major tributaries

We calculated the theoretical (predicted) ESR signal intensity of quartz in the fine-silt fraction of the sediments in the mainstream immediately below junctions with major tributaries by applying a simple mixing ratio approach, to examine whether ESR signal intensity values of sediments reflect mixing ratios of sediments discharged from the mainstream and tributary immediately above the junctions. We used the sediment (SPM) budget at the junctions as determined from modern (post-TGD) observational data (Fig. 4), in addition to the analyzed ESR signal intensities of quartz in the fine-silt fraction of the sediments from each tributary and from sites along the mainstream immediately above the junctions, as end-members for determining mixing ratios and for calculating the predicted ESR signal intensities of quartz downstream from the junctions. The results are compared with the analyzed ESR signal intensities for the fine-silt fraction of samples collected from mainstream sites immediately below the junctions.

In our calculations, we used ESR signal intensity and CI values of quartz in the fine-silt fractions of the sediments from each tributary, as listed in Table 1. As mentioned above, the ESR signal intensity and CI values of the sediments collected at Yibin, at the outlet of the Jinshajiang, appear to be strongly affected by sediments supplied from the lower part of the Jinshajiang. Therefore, in the calculations of the average ESR signal intensity and CI values of the sediments from the Jinshajiang, we used only samples from the lower part of the Jinshajiang. For the Hanjiang, Dongting Lake, and Poyang Lake drainages, we used samples collected from the middle parts of the drainages, as samples from the lower parts and along the shorelines of lakes are likely contaminated by sediments from the mainstream during periods of flood.

The ESR signal intensity and CI values for the fine-silt fraction of quartz in sediments from below the junctions of the mainstream and tributaries were calculated as the weighted averages of the values in sediments from immediately above the junction in the mainstream and the tributary, respectively. In the calculations, we assume that sediments from the mainstream and the tributary are sufficiently mixed below the junction to apply the simple mixing ratio approach. Underlined variables in the following equations indicate observed or analyzed data.

First, we calculated the fine-silt fraction of the quartz load from the mainstream (Q_{main}; Mt/year) and a tributary (Q_{tri}; Mt/year) as

$$Q_{main(aj)} = \underline{S_{main(aj)}} \times \underline{f_{fs}} \times \underline{C_{main(aj)}}, \text{ and} \quad (1)$$

$$Q_{tri} = \underline{S_{tri}} \times \underline{f_{fs}} \times \underline{C_{tri}}, \quad (2)$$

where $\underline{S_{main(aj)}}$ is the observed sediment load in the mainstream just above the junction with the tributary of interest (Mt/year), $\underline{f_{fs}}$ is the proportion of the fine-silt fraction with respect to the total SPM, and $\underline{C_{main(aj)}}$ is the analyzed quartz content in the fine-silt fraction of the sediments collected from a site on the mainstream just above the junction with the tributary of interest. Similarly, $\underline{S_{tri}}$ is the observed sediment load from the tributary of interest (Mt/year) and $\underline{C_{tri}}$ is the analyzed quartz content in the fine-silt fraction of the sediments collected from the tributary of interest. The average annual sediment loads from the mainstream and each tributary are shown in Fig. 4; all data were collected after the start of TGD operations.

Table 1 Ranges of ESR and CI values for each tributary in the fine-silt fraction (4–16 μm)

Tributary	ESR signal intensity			CI values		
	Minimum	Maximum	Average	Minimum	Maximum	Average
Jinshajiang	0.00	3.23	1.39	8.31	8.79	8.66
Minjiang	0.00	5.26	2.14	8.83	8.67	8.57
Jialingjiang	1.46	10.9	5.96	8.59	8.67	8.63
Wujiang	13.4	15.5	14.6	8.71	8.81	8.81
Dongting Lake	11.3	23.4	15.9	8.87	8.68	8.37
Hanjiang	8.76	9.45	9.10	8.27	8.44	8.35
Poyang Lake	12.2	12.2	12.2	8.58	8.58	8.58

As the grain size distribution of SPM changes seasonally and in different parts of the Yangtze (Mao et al. 2010), the value of $\underline{f_{fs}}$ could also change depending on the season and the part of interest. We estimated the value of $\underline{f_{fs}}$ as follows.

Regarding seasonal variations, we assumed that summertime values of $\underline{f_{fs}}$ provide a reasonable representation of the annual average, as roughly 90% of the annual sediment yield from the Yangtze River occurs during the summer (Hu et al. 2011). To evaluate the value of $\underline{f_{fs}}$, we analyzed the grain size distribution of SPM in samples taken at Yichang and Nanjing during the same season and year as samples collected for our riverbed sampling. The values of $\underline{f_{fs}}$ at Yichang and Nanjing were 32 and 34%, respectively. These ratios of the fine-silt fraction in SPM are consistent with those at Yichang and Datong, respectively, as determined by Gao et al. (2015). Because the values of $\underline{f_{fs}}$ at Yichang and Nanjing are similar, we adopted the average value of $\underline{f_{fs}}$ obtained at Yichang and Nanjing, of 33%, as the value of $\underline{f_{fs}}$ in Eqs. (1) and (2).

Next, we calculated the quartz load of the fine-silt fraction of the mainstream below the junction ($Q_{\mathrm{main(bj)}}$; Mt/year) as

$$Q_{\mathrm{main(bj)}} = Q_{\mathrm{main(aj)}} + Q_{\mathrm{tri}}, \qquad (3)$$

and then calculated the contribution ratio of the fine-silt-sized quartz load of the mainstream above the junction to that of the mainstream below the junction ($r_{\mathrm{Q,main}}$), as well as the contribution ratio of the fine-silt-sized quartz load from the tributary of interest to that of the mainstream below the junction (r_{Qtri}), using Eqs. (4) and (5), respectively:

$$r_{\mathrm{Q,main}} = Q_{\mathrm{main(aj)}}/Q_{\mathrm{main(bj)}}, \text{ and} \qquad (4)$$

$$r_{\mathrm{Q,tri}} = Q_{\mathrm{tri}}/Q_{\mathrm{main(bj)}}. \qquad (5)$$

Finally, we calculated the ESR signal intensity of quartz in the fine-silt fraction of the sediment from the mainstream below the junction ($\mathrm{ESR}_{\mathrm{main(bj)}}$) as

$$\mathrm{ESR}_{\mathrm{main(bj)}} = \underline{\mathrm{ESR}_{\mathrm{main(aj)}}} \times r_{\mathrm{Q,main}} + \underline{\mathrm{ESR}_{\mathrm{tri}}} \times r_{\mathrm{Q,tri}}, \qquad (6)$$

where $\underline{\mathrm{ESR}_{\mathrm{main(aj)}}}$ is the analyzed ESR signal intensity of quartz in the fine-silt fraction of the sediments from the mainstream above the junction and $\underline{\mathrm{ESR}_{\mathrm{tri}}}$ is that in the fine-silt fraction of the sediments from the tributary. The value of $\underline{\mathrm{ESR}_{\mathrm{tri}}}$ is assumed to vary between the minimum and maximum ESR signal intensities listed in Table 1.

Using Eqs. (1) to (6), the ESR signal intensities below each junction were predicted at the mainstream junctions with the Minjiang, Jialingjiang, Wujiang, and Poyang Lake. The junctions with the Yalongjiang and Dongting Lake were excluded from the analysis, as riverbed sediments in the mainstream are lacking above the junctions with these tributaries. The junction with the Hanjiang was also excluded, as the sample collected from the mainstream above the junction showed an unusual ESR signal intensity of 0.0. Because this sample was collected from meter-scale bedded sediments exposed above the river surface, this sample could represent past sediments deposited during a flood and could thus show a low ESR signal intensity reflecting a specific source area, such as the SG block.

Results of the calculations are listed in Table 2. Figure 10 shows the correlation between the predicted and analyzed ESR signal intensities in the mainstream below the junctions with each major tributary. The analyzed ESR signal intensities agree well with the predicted values (within the reproducibility of the analyzed ESR values; ±2.2) below the junctions with the Minjiang, Jialingjiang, and Poyang Lake tributaries, indicating that the analyzed ESR signal intensity properly reflects the relative sediment contributions from the mainstream and the tributary from above the junctions.

On the other hand, the analyzed ESR signal intensity below the junction with the Wujiang is larger than predicted (and beyond the analytical error). At this junction, the sampling site on the mainstream was located 40 km below the Wujiang (LYZ120212-01a; Additional file 1), and it is possible that a local river, the Longhe, which drains into the mainstream from the same side as the Wujiang at a location less than 1 km above the sampling site, influenced the results. As the geology of the Longhe drainage basin is similar to that of the Wujiang, it is likely that detrital material from the Longhe is also characterized by high ESR values (as are the Wujiang sediments). This material may have affected the ESR signal intensity values of sample LYZ120212-01a, on account of the proximity of the Longhe to the sampling site. Thus, the analyzed ESR signal intensity of sample LYZ120212-01a is likely to be higher than predicted because of the effects of the high ESR signal intensities of sediments from the Longhe.

Estimated ESR signal intensity and CI values of quartz in the fine-silt fraction of modern sediments along the mainstream of the Yangtze

As discussed above, the analyzed ESR signal intensities of samples collected from sites below junctions with

Table 2 Predicted and analyzed ESR signal intensity of the mainstream below the junction with tributaries

Tributary	ESR signal intensity	
	Predicted	Analyzed
Minjiang	1.34	1.49
Jialingjiang	2.23	3.88
Wujiang	4.27	7.92
Poyang Lake	5.19	3.74

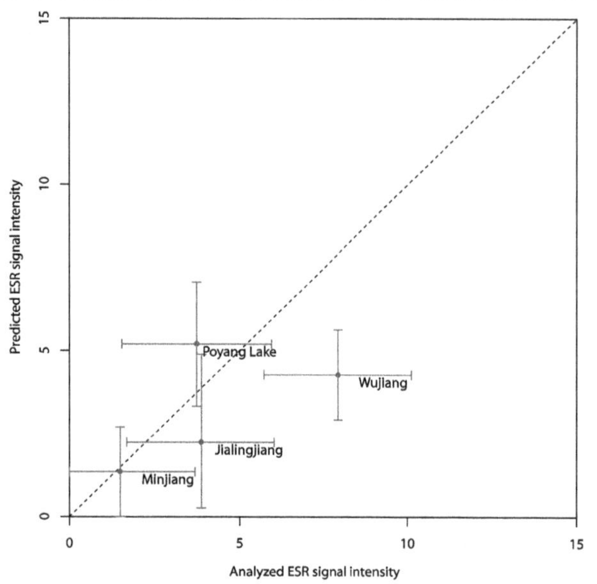

Fig. 10 Predicted and analyzed ESR signal intensities of quartz below junctions with major tributaries

major tributaries reflect the relative contributions of sediments from the mainstream and the tributary above the junction. Using such an approach, we can also estimate changes in ESR signal intensities of the fine-silt fraction of sediments along the mainstream from the uppermost part to the rivermouth. This approach could further validate whether ESR signal intensities in the fine-silt fraction of the sediments at the rivermouth can be used to predict sediment budgets in the Yangtze drainage basin in the past.

The following assumptions were made in performing the calculations which predict the ESR signal intensity along the mainstream: (a) sediments from the mainstream and tributaries are sufficiently mixed below junctions, (b) sediment input from small tributaries is negligible, and (c) quartz contents do not change during transport as SPM along the mainstream.

We first performed calculations for the junction of the Jinshajiang and Minjiang and predicted the ESR signal intensity below the junction using Eqs. (1) to (6) (see above). Next, we predicted the ESR signal intensity at the junction of the Yangtze mainstream and the Jialingjiang using the same equations; however, the ESR signal intensity of the mainstream above the junction was assumed to be the same as the predicted value below the previous junction (i.e., the junction of the Jinshajiang and Minjiang). Then, similar calculations were repeated at every junction with a major tributary (i.e., at the junctions with the Wujiang, Dongting Lake, Hanjiang, and Poyang Lake tributaries).

The amount of eroded sediments on the mainstream between Yichang and Hankou and between Hankou and Datong were assumed to be 47 and 17 Mt/year, respectively (Fig. 4). In our calculations, erosion between Yichang and Hankou was split equally into two parts: from Yichang to Yueyang (outlet of Dongting Lake) and from Yueyang to Hankou. The ESR signal intensity and CI values for the sediments eroded between Yichang and Yueyang, Yueyang and Hankou, and Hankou and Datong were assumed to be the same as those of sediments passing through Yichang, Yueyang, and Hankou before TGD construction, respectively, as the sediments were deposited along the middle to lower part of the mainstream before TGD construction (Fig. 3). The ESR signal intensity and CI values at Yichang, Yueyang, and Hankou prior to the TGD were estimated using the same calculations (Eqs. (1) to (6)) and the sediment budget shown in Fig. 3.

It is assumed that 80.3% of the sediment flowing into the TGD area between the junction of the mainstream and the Wujiang and the dam body is trapped in the TGD. Thus, we have

$$Q_{\mathrm{TGD,output}} = (1-0.803) \times Q_{\mathrm{TGD,input}}, \tag{7}$$

where $Q_{\mathrm{TGD,input}}$ is the quartz load of the mainstream below the junction with the Wujiang and $Q_{\mathrm{TGD,output}}$ is the quartz load of the mainstream below the TGD.

Similar to Eq. (7), we assumed that 10.0% of the sediment flowing into Dongting Lake is deposited within Dongting Lake, based on the sediment budget presented in Fig. 4:

$$Q_{\mathrm{DTL,output}} = 0.9 \times Q_{\mathrm{DTL,input}}. \tag{8}$$

The calculation in Eqs. (1)–(6) was repeated at every junction from the upper part to the lower part along the mainstream (Table 3; Fig. 11). The results of these calculations using averaged ESR signal intensity and CI values are listed in Table 3 as "predicted averages" of the ESR signal intensity and CI values for each tributary (data in Table 1). Similarly, results of the calculations using maximum and minimum values of the ESR signal intensity and CI are listed in Table 3 as "predicted maximum" and "predicted minimum" values for each tributary (see Table 1). As described above, the sediments collected from the site below the junction with the Wujiang and the site above the junction with the Hanjiang are assumed to be exceptional. Thus, the characteristics of these two sediments are omitted from Table 3 and Fig. 11.

If the sediment budget shown in Fig. 4 properly reflects the volume of sediments transported along the mainstream, then we should expect that predicted average ESR signal intensity and CI values after mixing will be within the analytical error of the analyzed ESR signal intensities of the riverbed sediments. Figure 11 shows that the predicted average ESR signal intensities (solid circles)

Table 3 Predicted and analyzed ESR and CI values along the mainstream in the fine-silt fraction (4–16 μm)

Location	Predicted						Analyzed	
	ESR signal intensity			CI values			ESR	CI
	Minimum	Maximum	Average	Minimum	Maximum	Average		
Above MJ (Jinshajiang)	0.0	3.2	1.8	8.3	8.8	8.7	1.9	8.7
Below MJ/above JLJ	0.0	3.5	1.9	8.3	8.8	8.6	1.5	7.9
Below JLJ/above WJ	0.3	4.8	2.6	8.4	8.7	8.6	3.9	8.6
Below WJ/TGD input	0.9	5.3	3.1	8.4	8.8	8.7	–	–
Upper area of TGD	0.9	5.3	3.1	8.4	8.8	8.7	4.2	8.7
TGD output	0.9	5.3	3.1	8.4	8.8	8.7	1.8	8.8
Below DTL	2.8	9.1	5.6	8.5	8.7	8.7	9.1	8.7
Above HJ	2.6	8.9	5.3	8.5	8.7	8.7	–	–
Below HJ	3.1	8.9	5.6	8.5	8.7	8.7	7.7	8.7
Above PYL	3.0	9.0	5.6	8.5	8.7	8.6	4.0	8.2
Below PYL	4.0	9.7	6.5	8.5	8.7	8.6	3.8	8.7
Rivermouth	4.0	9.7	6.5	8.5	8.7	8.6	4.4	8.5

Note: *MJ* Minjiang, *JLJ* Jialingjiang, *WJ* Wujiang, *TGD* Three Gorges Dam, *DTL* Dongting Lake, *HJ* Hanjiang, *PYL* Poyang Lake

agree well with the analyzed ESR signal intensities (within the reproducibility of the ESR analysis of ±2.2; open circles with errors) at 8 out of 10 locations, indicating that the predicted average ESR signal intensities can be used to predict the ESR signal intensities of the riverbed sediments.

Two predicted ESR signal intensity values, below the junctions with the Dongting Lake and Poyang Lake tributaries, are beyond the acceptable error range of the analyzed values. Lane et al. (2008) reported that several to tens of kilometers of distance are necessary for river waters to completely mix after two rivers merge. Sampling sites on the mainstream below the junctions with Dongting Lake, which show higher than predicted ESR values, are located on the tributary-merging side of the mainstream 8 km below the junction. As previously described, the ESR signal intensities for sediments from Dongting Lake are characterized by values that are higher than those of the mainstream. Thus, it is likely that samples below the junction with Dongting Lake are influenced by sediments derived from the lake, which show ESR values higher than those of the

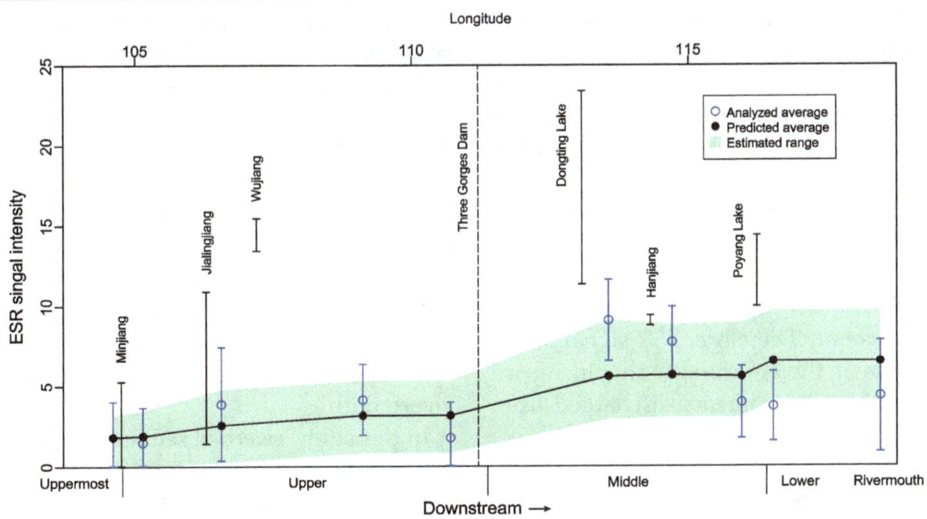

Fig. 11 Comparison of predicted and analyzed ESR signal intensities of quartz in the fine-silt fraction along the mainstream. *Green area* indicates the range of predicted ESR signal intensities. *Solid circles* indicate the averages of predicted ESR signal intensities along the mainstream. The *green-shaded area* indicates the ranges of predicted ESR signal intensities. The *blue open circles* indicate analyzed ESR signal intensities of the fine-silt fractions at sites along the mainstream. Samples that are identified as exceptions in Fig. 10 are omitted from this figure, namely a sample collected below the junction with the Wujiang and a sample collected above the junction with the Hanjiang (see text for details)

mainstream above the junction. This problem could be resolved by taking samples from sites at sufficient distances from the junctions with tributaries.

On the other hand, the analyzed ESR signal intensities in the middle to lower parts tend to be lower than predicted values. These lower values could be the result of erosion along the middle to lower parts of the mainstream. For example, the meter-scale bedded sediments above the junction with the Hanjiang are readily eroded and show ESR signal intensity values of 0.0; such sediments could strongly modify the signal in the SPM fraction in the mainstream. Our present predictions of ESR signal intensity along the mainstream assume that the ESR signal intensity of eroded sediments in the middle to lower parts corresponds to the average ESR signal intensity of ~4 that was observed at Yichang or Yueyang. However, if we assume that all eroded sediments have ESR signal intensities of 0, the predicted average ESR signal intensity would be 4.6, which is very close to the analyzed values. Thus, the effects of eroded sediments on the ESR signal intensity values observed in the middle to lower parts could explain the departures between predicted and analyzed ESR values in these sections of the mainstream.

He et al. (2015), in a study of the geochemistry of riverbed sediments in the Yangtze basin, reported that Nd and Sr isotope ratios varied in different parts of the Yangtze depending on the nature of the geological block(s) in each part of the basin and that the riverbed sediments in the middle to lower parts of the mainstream were derived mainly from the Songpan–Garze terrane and Yangtze Block; this result is consistent with our results shown in Fig. 8a. While they reported an absence of coherent downstream variations in isotope ratios in the mainstream, they also did not take into account the effects of eroding sediments along the middle to lower parts of the river, as discussed above. Thus, our results based on mixing calculations do not necessarily contradict their results.

In summary, the average predicted ESR signal intensities of the fine-silt fraction along the modern mainstream are consistent with analyzed values, when incomplete mixing of tributary waters and eroded materials are taken into account. Therefore, ESR signal intensities of the sediments at the rivermouth should record the relative contributions of sediments discharged from each part of the Yangtze River basin.

Distinguishing the provenance of the fine-silt fraction of sediments in the Yangtze drainage basin using ESR signal intensity data from sediments at the rivermouth

In this section, we examine whether it is possible to distinguish the provenance of the fine-silt fraction of sediments at the rivermouth using ESR signal intensity and CI values, especially the relative contributions of sediments from two tributary groups within the Yangtze drainage basin: the western plus northeastern tributaries and the southeastern tributaries, as shown in Fig. 9.

Using Eqs. (1) to (6), the ESR signal intensity of sediments discharged from each tributary group can be calculated as

$$R_{\mathrm{tri}} = Q_{\mathrm{tri}}/Q_{\mathrm{trig}} \tag{9}$$

where Q_{trig} is the total quartz load in the tributary group, R_{tri} is the contribution ratio of quartz from each tributary to the total quartz load within the tributary group, and Q_{tri} is the fine-silt fraction of the quartz load from the tributary, as defined in Eq. (1), such that

$$\mathrm{ESR}_{\mathrm{trig}} = \mathrm{ESR}_{\mathrm{tri1}} \times R_{\mathrm{tri1}} + \mathrm{ESR}_{\mathrm{tri2}} \times R_{\mathrm{tri2}} + \ \ldots, \tag{10}$$

where $\mathrm{ESR}_{\mathrm{trig}}$ is the predicted ESR signal intensity of sediments from each tributary group and $\mathrm{ESR}_{\mathrm{tri1}}$, $\mathrm{ESR}_{\mathrm{tri2}}\ldots$ are the ESR signal intensities of sediments from each tributary within the tributary group.

These calculations were performed for each of the minimum, average, and maximum ESR signal intensities listed in Table 1. The ESR signal intensity of sediments discharged from the western plus northeastern tributary group shows an average of 4.3, a minimum of 2.5, and a maximum of 6.3, while those from the southeastern tributary group show an average of 14.3, a minimum of 12.8, and a maximum of 16.3. Thus, the ESR signal intensities of the two groups do not overlap one another. Assuming that the sediments at the rivermouth represent a mixture of these two end-members, we can therefore use the ESR signal intensity to estimate the relative contributions of sediments from the western plus northeastern tributaries and the southeastern tributaries to the total sediment load at the rivermouth.

Conclusions

Meteorological data suggest that changes in the strength of the EASM correspond to changes in the spatial distribution of monsoon precipitation in the Yangtze drainage basin. Thus, reconstruction of temporal changes in the provenance of sediments in the basin could be a useful approach for reconstructing changes in EASM intensity in the past, based on sediment records preserved at the rivermouth.

In this study, riverbed sediments taken from the Yangtze mainstream and its major tributaries were analyzed in terms of their ESR signal intensity and CI values of quartz in three different size fractions of sediment, to distinguish the sources of sediments derived from the different regions of the Yangtze drainage basin. The results suggest that sediment from the different tributaries can be categorized into four groups based on the geological block(s) exposed within the drainage basins of each tributary and that each group exhibits a characteristic distribution of ESR signal intensity and CI values. The first group, corresponding to the Yidun Arc and the southwestern margin of the Yangtze Block, is characterized by intermediate ESR values (~7) and low CI values (7.9–8.2). The second group, corresponding to the Songpan–Garze terrain, is characterized by lower ESR values (<5) and a wide range of CI values (8–9). The third group, corresponding to the Qinling orogenic belt, is characterized by a wide range of ESR values (1–10) and a narrow range of CI values (8.5–8.9). The fourth group, corresponding to the southeastern parts of the Yangtze and Cathaysia Blocks, is characterized by high ESR values (>10) and CI values of 8–9.

Theoretical predictions of ESR signal intensities below the junctions of the mainstream and major tributaries were calculated as weighted averages based on quartz discharge levels of the mainstream and each tributary, measured just above the junction. These predicted ESR signal intensities were compared with the measured ESR signal intensities of sediments collected from the mainstream just below the junctions with the tributaries. The analyzed values are in good agreement with the predicted values, within the reproducibility of the ESR analysis, indicating that the analyzed ESR signal intensities of sediments accurately represent the relative contributions of quartz from the mainstream and the tributaries.

In addition, changes in ESR signal intensity and CI values along the mainstream were estimated based on values for quartz in the fine-silt fraction discharged from the major tributaries, assuming complete mixing of sediments from the mainstream and each tributary immediately below the junction. The predicted average ESR signal intensities of riverbed samples are in good agreement with analyzed ESR signal intensities of the same samples, within the reproducibility of the ESR analysis, except for some samples that are insufficiently mixed or affected by smaller tributaries. Thus, we conclude that ESR signal intensity and CI values can be used to estimate the mixing ratios of the fine-silt fraction of sediments from the different tributaries in the Yangtze drainage basin, thereby raising the possibility that changes in the spatial distribution of precipitation within the Yangtze drainage basin can be reconstructed from sediment records at the rivermouth.

Abbreviations

CI: Crystallinity index; CX: Cathaysia Block; EASM: East Asian summer monsoon; ESR: Electron spin resonance; ISM: Indian summer monsoon; QC: Quartz content; QL: Qinling orogenic belt; QY: Qamdo Block plus Yidun Arc; SG: Songpan–Garze terrane; SPM: Suspended particle matter; TGD: Three Gorges Dam; XRD: X-ray powder diffraction; YZ: Yangtze Block

Acknowledgements

The authors thank P. Wang, S. Li, and the car drivers for their assistance during the field trips in China and K. Nagashima and A. Karasuda for their assistance with the ESR and XRD analyses in Japan. The staff at the Takasaki Advanced Radiation Research Institute, National Institutes for Quantum and Radiological Science and Technology (QST), assisted with γ-ray irradiation. We also thank Y. Saito, an anonymous reviewer, and H. Kawahata (Editor of PEPS) for their critical reading of the manuscript and many constructive suggestions.

Funding

This study was supported by the Japan Society for the Promotion of Science (JSPS) KAKENHI Grant Number 23221002 to RT and a Japan–China bilateral joint research project awarded to RT from JSPS and also jointly supported by the National Basic Research Program of China (973 program 2015CB953804) to HZ.

Authors' contributions

RT designed and directed the projects on the Japan side, helped with the collection of samples, and supervised KS. KS collected the riverbed samples, helped collect the SPM samples, conducted ESR and XRD measurements (including pretreatments and irradiation), and wrote the draft manuscript. HZ designed and directed the projects on the China side and assisted in the collection of samples. TI helped collect samples in Jul 2011 and Feb 2012. CL and MH planned the field trips in 2011 and 2012, collected SPM samples, helped collect the riverbed samples, and arranged the transportation of samples. KW collected samples in Oct 2012 and helped arrange the transportation of samples. YS helped with the sampling in Sep 2011. All authors read and approved the manuscript.

Competing interests

The authors declare that they have no competing interests.

Author details

[1]Department of Earth and Planetary Science, University of Tokyo, 7-3-1 Hongo, Bunkyo, Tokyo 113-0033, Japan. [2]School of Resource, Environment and Earth Science, Yunnan University, Chenggong District, Kunming 650500, People's Republic of China. [3]Faculty of Environmental Earth Science, Hokkaido University, Kita 8, Nishi 5, Kita-ku, Sapporo, Hokkaido 060-0808, Japan. [4]College of Geographic Science, Nanjing Normal University, 1 Wen Yuan Lu, Qixia, Nanjing 210046, China.

References

Ali JR, Thompson GM, Zhou MF, Song X (2005) Emeishan large igneous province, SW China. LITHOS 79:475–489. doi:10.1016/j.lithos.2004.09.013

Chang J, Li J, Lu D, Zhu X, Lu C, Zhou Y, Deng C (2010) The hydrological effect between Jingjiang River and Dongting Lake during the initial period of Three Gorges Project operation. J Geog Sci 20(5):771–786. doi:10.1007/s11442-010-0810-9

Chen J, Jahn B (1998) Crustal evolution of southeastern China: Nd and Sr isotopic evidence. Tectonophysics 284:101–133

Chen Z, Li J, Shen H, Wang Z (2001) Yangtze River of China: historical analysis of discharge variability and sediment flux. Geomorphology 41:77–91. doi:10.1016/S0022-1694(02)00028-8

Dai S-B, Lu XX (2010) Sediment deposition and erosion during the extreme flood events in the middle and lower reaches of the Yangtze River. Quat Int 226:4–11. doi:10.1016/j.quaint.2010.01.026

Dai S-B, Yang S-L, Zhu J, Gao A, Li P (2005) The role of Lake Dongting in regulating the sediment budget of the Yangtze River. Hydrol Earth Syst Sci 9:692–698

Deng J, Wang Q, Li G, Li C, Wang C (2014) Tethys tectonic evolution and its bearing on the distribution of important mineral deposits in the Sanjiang region, SW China. Gondwana Res 26(2):419–437. doi:10.1016/j.gr.2013.08.002

Dong Y, Santosh M (2016) Tectonic architecture and multiple orogeny of the Qinling Orogenic Belt, Central China. Gondowana Res 29(1):1–40. doi:10.1016/j.gr.2015.06.009

Dong Y, Zhang G, Neubauer F, Liu X, Genser J, Hauzenberger C (2011) Tectonic evolution of the Qinling orogen, China: review and synthesis. J Asian Earth Sci 41:213–237. doi:10.1016/j.jseaes.2011.03.002

Enkelmann E, Weislogel A, Ratschbacher L, Eide E, Renno A, Wooden J (2007) How was the Triassic Songpan-Ganzi basin filled? A provenance study. Tectonics 26:TC4007. doi:10.1029/2006TC002078

Feigl FJ, Fowler WB, Yip KL (1974) Oxygen vacancy model for the E_1' center in SiO2. Solid State Commun 14:225–229

Gao JH, Jia J, Wang YP, Yang Y, Li J, Bai F, Zou X, Gao S (2015) Variations in quantity, composition and grain size of Changjiang sediment discharging into the sea in response to human activities. Hydrol Earth Syst Sci 19(2):645–655. doi:10.5194/hess-19-645-2015

Han G, Liu C-Q (2004) Water geochemistry controlled by carbonate dissolution: a study of the river waters draining karst-dominated terrain, Guizhou Province, China. Chem Geol 204:1–21. doi:10.1016/j.chemgeo.2003.09.009

Hao F, Guo T, Zhu Y, Cal X, Zou H, Li P (2008) Evidence for multiple stages of oil cracking and thermochemical sulfate reduction in the Puguang gas field, Sichuan Basin, China. Am Assoc Pet Geol Bull 92(5):611–637. doi:10.1306/01210807090

Hassan MA, Church M, Yan Y, Slaymaker O (2010) Spatial and temporal variation of in-reach suspended sediment dynamics along the mainstem of Changjiang (Yangtze River), China. Water Resour Res 46:W11551. doi:10.1029/2010WR009228

Hayashi S, Murakami S, Xu K-Q, Watanabe M (2008) Effect of the Three Gorges Dam Project on flood control in the Dongting Lake area, China, in a 1998-type flood. J Hydro-environ Res 2:148–163. doi:10.1016/j.jher.2008.10.002

He M, Zheng H, Clift PD, Tada R, Wu W, Luo C (2015) Geochemistry of fine-grained sediments in the Yangtze River and the implications for provenance and chemical weathering in East Asia. Progr Earth Planet Sci 2:32. doi:10.1186/s40645-015-0061-6

Hennig T, Wang W, Feng Y, Ou X, He D (2013) Review of Yunnan's hydropower development. Comparing small and large hydropower projects regarding their environmental implications and socio-economic consequences. Renewable Sustainable Energy Rev 27:595. doi:10.1016/j.rser.2013.07.023

Hu Q, Feng S, Guo H, Chen G, Jiang T (2007) Interactions of the Yangtze river flow and hydrologic processes of the Poyang Lake, China. J Hydrol 347:90–100. doi:10.1016/j.jhydrol.2007.09.005

Hu B, Yang Z, Wang H, Sun X, Bi N, Li G (2009) Sedimentation in the Three Gorges Dam and the future trend of Changjiang (Yangtze River) sediment flux to the sea. Hydrol Earth Syst Sci 13:2253–2264

Hu B, Wang H, Yang Z, Sun X (2011) Temporal and spatial variations of sediment rating curves in the Changjiang (Yangtze River) basin and their implications. Quat Int 230:34–43. doi:10.1016/j.quaint.2009.08.018

Isozaki Y (2009) Characterization of eolian dust and its sources in the Tarim Basin and their temporal changes during Plio-Pleistocene based on the ESR signal intensity and crystallinity index of quartz. Ph. D. thesis, The University of Tokyo, Tokyo, Japan.

Lane SN, Parsons DR, Best JL, Orfeo O, Kostaschuk RA, Hardy RJ (2008) Causes of rapid mixing at a junction of two large rivers: Rio Parana and Rio Paraguay, Argentina. J Geophys Res 113:F02019. doi:10.1029/2006JF000745

Liu H, Lan H, Liu Y, Zhou Y (2011) Characteristics of spatial distribution of debris flow and the effect of their sediment yield in main downstream of Jinsha River, China. Environ Earth Sci 64:1653–1666. doi:10.1007/s12665-009-0409-6

Luo L, Qi J-F, Zhang M-Z, Wang K, Han Y-Z (2014) Detrital zircon U-Pb ages of Late Triassic-Late Jurassic deposits in the western and northern Sichuan Basin margin: constraints on the foreland basin provenance and tectonic implications. Int J Earth Sci (Geol Rundsch) 103:1553–1568. doi:10.1007/s00531-014-1032-7

Ma Y, Zhang S, Guo T, Zhu G, Cai X, Li M (2008) Petroleum geology of the Puguang sour gas field in the Sichuan Basin, SW China. Mar Pet Geol 25:357–370. doi:10.1016/j.marpetgeo.2008.01.010

Mao CP, Chen J, Yuan XY, Yang ZF, Balsam W, Ji JF (2010) Seasonal variation in the mineralogy of the suspended particulate matter of the lower Changjiang River at Nanjing, China. Clay and Clay Minerals 58(5):691–706. doi:10.1346/CCMN.2010.0580508

Mao CP, Chen J, Yuan XY, Yang ZF, Ji JF (2011) Seasonal variations in the Sr-Nd isotopic compositions of suspended particulate matter in the lower Changjiang River: provenance and erosion constraints. Chin Sci Bull 56:2371–2378. doi:10.1007/s11434-011-4589-6

Meng Q-R, Zhang G-W (2000) Geologic framework and tectonic evolution of the Qinling orogen, central China. Tectonophysics 323:183–196. doi:10.1016/S0040-1951(00)00106-2

Murata KJ, Norman MB (1976) An index of crystallinity of quartz. Am J Sci 276:1120–1130

Nagashima K, Tada R, Tani A, Toyoda S, Sun Y, Isozaki Y (2007) Contribution of aeolian dust in Japan Sea sediments estimated from ESR signal intensity and crystallinity of quartz. Geochem Geophys Geosyst 8:Q02Q04. doi:10.1029/2006GC001364

Qin J-F, Lai S-C, Grapes R, Diwu C, Ju Y-J, Li Y-F (2010) Origin of Late Triassic high-Mg adakitic granitoid rocks from the Dongjiangkou area, Qinling orogen, central China: implications for subduction of continental crust. Lithos 120:347–367. doi:10.1016/j.lithos.2010.08.022

Rea DK, Janecek TR (1981) 25. Mass-accumulation rates of the non-authigenic inorganic crystalline (eolian) component of deep-sea sediments from the western Mid-Pacific Mountains, Deep Sea Drilling Project Site 463. Initial Reports of the Deep Sea Drilling Project 62:653–659. doi:10.2973/dsdp.proc.62.125.1981

Reid AJ, Wilson CJL, Liu S (2005a) Structural evidence for the Permo-Triassic tectonic evolution of the Yidun Arc, eastern Tibetan plateau. J Struct Geol 27(1):119–137. doi:10.1016/j.jsg.2004.06.011

Reid AJ, Folwer AP, Phillips D, Wilson CJL (2005b) Thermochronology of the Yidun Arc, central eastern Tibetan Plateau: constraints from $^{40}Ar/^{39}Ar$ K-feldspar and apatite fission track data. J Asian Earth Sci 25:915–935. doi:10.1016/j.jseaes.2004.09.002

Roger F, Jolivet M, Malavieille J (2008) Tectonic evolution of the Triassic fold belts of Tibet. Compt Rendus Geosci 340:180–189. doi:10.1016/j.crte.2007.10.014

Sun Y, Tada R, Chen J, Chen H, Toyoda S, Tani A, Isozak Y, Nagashima K, Hasegawa H, Ji J (2007) Distinguishing the sources of Asian dust based on electron spin resonance signal intensity and crystallinity of quartz. Atmos Environ 41:8537–8548. doi:10.1016/j.atmosenv.2007.07.014

Tada R, Sato S, Irino T, Matsui H, Kennett JP (2000) Millennial-scale compositional variations in late Quaternary sediments at site 1017, southern California. Proc Ocean Drill Program Sci Results 167:277–296. doi:10.2973/odp.proc.sr.167.222.2000

Tada R, Zheng H, Clift PD (2016) Evolution and variability of the Asian monsoon and its potential linkage with uplift of the Himalaya and Tibetan Plateau. Progr Earth Planet Sci 3:4. doi:10.1186/s40645-016-0080-y

Teraoka Y, Okamura K (2003) Geological map of East Asia. Geological Survey of Japan, AIST

Teraoka Y, Okamura K (2007) Geological map of Central Asia. Geological Survey of Japan, AIST

Toyoda S (1992) Production and decay characteristics of paramagnetic defects in quartz: applications to ESR dating. Ph.D. thesis, Osaka University, Toyonaka, Osaka, Japan.

Toyoda S, Hattori W (2000) Formation and decay of the E_1' center and of its precursor. Appl Radiat Isot 52:1351–1356

Toyoda S, Naruse T (2002) Eolian dust from the Asian deserts to the Japanese islands since the Last Glacial Maximum: the basis for ESR method. Transactions, Jpn Geomorphol Union 23:811–820

USGS (2004) Shuttle radar topography mission, 3 Arc Second, Version 2.1, Global Land Cover Facility, University of Maryland, College Park, Maryland, February 2000.

Wang B, Lin H (2002) Rainy season of the Asian-Pacific summer monsoon. J Clim 15(4):386–398. doi:10.1175/1520-0442(2002)015<0386:RSOTAP>2.0.CO;2

Wang X, Metcalfe I, Jian P, He L, Wang C (2000) The Jinshajiang-Ailaoshan Suture Zone, China: tectonostratigraphy, age and evolution. J Asian Earth Sci 18(6):675–690. doi:10.1016/S1367-9120(00)00039-0

Wang X-L, Zhou J-C, Griffin WL, Wang R-C, Qiu J-S, O'Reilly SY, Xu X, Liu X-M, Wang X-L (2007a) Detrital zircon geochronology of Precambrian basement sequences in the Jiangnan orogen: dating the assembly of the Yangtze and Cathaysia Blocks. Precambrian Res 159:117–131. doi:10.1016/j.precamres.2007.06.005

Wang Z-Y, Li Y, He Y (2007b) Sediment budget of the Yangtze River. Water Resour Res 43:W04401. doi:10.1029/2006WR005012

Wang Y, Zhanf F, Fan W, Zhang G, Chen S, Cawood PA, Zhang A (2010) Tectonic setting of the South China Block in the early Paleozoic: resolving intracontinental and ocean closure models from detrital zircon U-Pb geochronology. Tectonics 29:TC6020. doi:10.1029/2010TC002750

Wang Y, Fan W, Zhang G, Zhang Y (2013) Phanerozoic tectonics of the South China Block: key observations and controversies. Gondwana Res 23:1273–1305. doi:10.1016/j.gr.2012.02.019

Wu W, Xu S, Yang J, Yin H, Tao X (2009) Sr fluxes and isotopic compositions in the headwaters of the Yangtze River, Tongtian River and Jinsha River originating from the Qinghai-Tibet Plateau. Chem Geol 260:63–72. doi:10.1016/j.chemgeo.2008.12.007

Xia L, Li X, Ma Z, Xu X, Xia Z (2011) Cenozoic volcanism and tectonic evolution of the Tibetan plateau. Gondwana Res 19:850–866. doi:10.1016/j.gr.2010.09.005

Xu K, Milliman JD, Xu H (2010) Temporal trend of precipitation and runoff in major Chinese Rivers since 1951. Global Planet Change 73:219–232. doi:10.1016/j.gloplacha.2010.07.002
Yang S-L, Zhao Q-Y, Belkin IM (2002) Temporal variation in the sediment load of the Yangtze river and the influences of human activities. J Hydrol 263:56–71. doi:10.1016/S0022-1694(02)00028-8
Yang Z, Wang H, Saito Y, Milliman JD, Xu K, Qiao S, Shi G (2006) Dam impacts on the Changjiang (Yangtze) River sediment discharge to the sea: the past 55 years and after Three Gorges Dam. Water Resour Res 42:W004407. doi:10.1029/2005WR003970
Yang S-L, Zhang J, Xu XJ (2007) Influence of the Three Gorges Dam on downstream delivery of sediment and its environmental implications, Yangtze River. Geophys Res Lett 34:L10401. doi:10.1029/2007GL029472
Yang S-L, Milliman JD, Xu K-H, Deng B, Zhang XX, Luo XX (2014) Downstream sedimentary and geomorphic impacts of the Three Gorges Dam on the Yangtze River. Earth Sci Rev 138:469–486. doi:10.1016/j.earscirev.2014.07.006
Yin A, Harrison TM (2000) Geologic evolution of the Himalayan-Tibetan orogen. Annu Rev Earth Planet Sci 28:211–280. doi:10.1146/annurev.earth.28.1.211
Yin H, Liu G, Pi J, Chen G, Li C (2007) On the river-lake relationship of the middle Yangtze reaches. Geomorphology 85:197–207. doi:10.1016/j.geomorph.2006.03.017
Zhang Q, Xu C-Y, Zhang Z, Chen YD, Liu C-L, Lin H (2008) Spatial and temporal variability of precipitation maxima during 1960–2005 in the Yangtze River basin and possible association with large-scale circulation. J Hydrol 353:215–227. doi:10.1016/j.jhydrol.2007.11.023
Zhang Y, Huang CC, Pang J, Zha X, Zhou Y, Yin S, Wang J (2012) Comparative study of the modern flood slackwater deposits in the upper reaches of Hanjiang and Weihe River Valleys, China. Quat Int 282:184–191. doi:10.1016/j.quaint.2012.03.056
Zong Y, Chen X (2000) The 1998 Flood on the Yangtze, China. Nat Hazards 22:165–184. doi:10.1023/A:1008119805106

PERMISSIONS

All chapters in this book were first published in PEPS, by Springer; hereby published with permission under the Creative Commons Attribution License or equivalent. Every chapter published in this book has been scrutinized by our experts. Their significance has been extensively debated. The topics covered herein carry significant findings which will fuel the growth of the discipline. They may even be implemented as practical applications or may be referred to as a beginning point for another development.

The contributors of this book come from diverse backgrounds, making this book a truly international effort. This book will bring forth new frontiers with its revolutionizing research information and detailed analysis of the nascent developments around the world.

We would like to thank all the contributing authors for lending their expertise to make the book truly unique. They have played a crucial role in the development of this book. Without their invaluable contributions this book wouldn't have been possible. They have made vital efforts to compile up to date information on the varied aspects of this subject to make this book a valuable addition to the collection of many professionals and students.

This book was conceptualized with the vision of imparting up-to-date information and advanced data in this field. To ensure the same, a matchless editorial board was set up. Every individual on the board went through rigorous rounds of assessment to prove their worth. After which they invested a large part of their time researching and compiling the most relevant data for our readers.

The editorial board has been involved in producing this book since its inception. They have spent rigorous hours researching and exploring the diverse topics which have resulted in the successful publishing of this book. They have passed on their knowledge of decades through this book. To expedite this challenging task, the publisher supported the team at every step. A small team of assistant editors was also appointed to further simplify the editing procedure and attain best results for the readers.

Apart from the editorial board, the designing team has also invested a significant amount of their time in understanding the subject and creating the most relevant covers. They scrutinized every image to scout for the most suitable representation of the subject and create an appropriate cover for the book.

The publishing team has been an ardent support to the editorial, designing and production team. Their endless efforts to recruit the best for this project, has resulted in the accomplishment of this book. They are a veteran in the field of academics and their pool of knowledge is as vast as their experience in printing. Their expertise and guidance has proved useful at every step. Their uncompromising quality standards have made this book an exceptional effort. Their encouragement from time to time has been an inspiration for everyone.

The publisher and the editorial board hope that this book will prove to be a valuable piece of knowledge for researchers, students, practitioners and scholars across the globe.

LIST OF CONTRIBUTORS

Nao Yasuda and Ikuro Sumita
Earth and Environmental Sciences Course, Division of Natural System, Graduate School of Natural Science and Technology, Kanazawa University, Kakuma, Kanazawa 920-1192, Japan

Jason Lackey and Gregory Moore
Department of Earth Science, University of Hawai'i, 1680 East-West Rd, Honolulu, HI 96822, USA

Michael Strasser
Institute of Geology, University of Innsbruck, Innrain 52f, 6020 Innsbruck, Austria

Peter D. Clift
Department of Geology and Geophysics, Louisiana State University, Baton Rouge, LA 70803, USA
School of Geography Science, Nanjing Normal University, Nanjing 210023, China

Shota Yamashita
JX Nippon Oil & Gas Exploration Corporation, JX building, 1-1-2, Ote-Machi, Chiyoda-Ku, Tokyo, Japan

Hajime Naruse
Division of Earth and Planetary Sciences, Graduate School of Science, Kyoto University, Kitashirakawa Oiwakecho, Sakyo-ku, Kyoto 606-8502, Japan

Takeshi Nakajo
Osaka Museum of Natural History, Nagai Park 1-23, Higashi-Sumiyoshi-ku, Osaka 546-0034, Japan

Celine Thomachot-Schneider, Emilie Huby, Kevin Chalons, Xavier Drothière and Patricia Vazquez
GEGENAA EA 3795, University of Reims Champagne-Ardenne (URCA), Reims, France

Max W Schmidt, Anastasia Golubkova and James AD Connolly
Department of Earth Sciences, ETH, Zurich 8092, Switzerland

Changgui Gao
Department of Earth Sciences, ETH, Zurich 8092, Switzerland
State Key Laboratory of Geological Processes and Mineral Resources, School of Earth Science, China University of Geosciences, Wuhan 430074, China
School of Ocean and Earth Science, Tongji University, Shanghai 200092, China

Arno Rohrbach
Institut für Mineralogie, Westfälische Wilhelms-Universität Münster, Münster 48149, Germany

Masato Sato
Graduate School of Life and Environmental Sciences, University of Tsukuba, 1-1-1 Tennodai, Tsukuba, Ibaraki 305-8572, Japan

Tsuyoshi Hattanji
Faculty of Life and Environmental Sciences, University of Tsukuba, 1-1-1 Tennodai, Tsukuba, Ibaraki 305-8572, Japan

Akihisa Kitamura, Ryoya Ikuta and Masayuki Ikeda
Faculty of Science, Shizuoka University, 836 Ohya Suruga-ku, Shizuoka 422-8529, Japan
Center for Integrated Research and Education of Natural hazards, Shizuoka University, 836 Ohya, Suruga-ku, Shizuoka 422-8529, Japan

Mami Ito
Graduate School of Science, Shizuoka University, 836 Ohya Suruga-ku, Shizuoka 422-8529, Japan

Motoko Igisu
Department of Subsurface Geobiological Analysis and Research, Japan Agency for Marine-Earth Science and Technology (JAMSTEC), 2-15 Natsushima-cho, Yokosuka 237-0061, Japan

Yuichiro Ueno
Department of Subsurface Geobiological Analysis and Research, Japan Agency for Marine-Earth Science and Technology (JAMSTEC), 2-15 Natsushima-cho, Yokosuka 237-0061, Japan
Department of Earth and Planetary Sciences, Tokyo Institute of Technology, 2-12-1 O-okayama, Meguro-ku, Tokyo 152-8551, Japan
Earth-Life Science Institute, Tokyo Institute of Technology, 2-12-1 O-okayama, Meguro-ku, Tokyo 152-8551, Japan

Ken Takai
Earth-Life Science Institute, Tokyo Institute of Technology, 2-12-1 O-okayama, Meguro-ku, Tokyo 152-8551, Japan

Hagar Hecht
Department of Natural Environmental Studies, The University of Tokyo, 5-1-5 Kashiwanoha, Kashiwa City, Chiba 277-8563, Japan

Takashi Oguchi
Department of Natural Environmental Studies, The University of Tokyo, 5-1-5 Kashiwanoha, Kashiwa City, Chiba 277-8563, Japan
Center for Spatial Information Science, The University of Tokyo, 5-1-5 Kashiwanoha, Kashiwa City, Chiba 277-8568, Japan

Timothy D. Jones and Peter E. van Keken
Department of Terrestrial Magnetism, Carnegie Institution for Science, 5241 Broad Branch Road NW, Washington, DC, 20015 USA

Ross R. Maguire
Department of Geology, University of Maryland, 8000 Regents Drive, College Park, MD, 20742, USA
Department of Earth and Planetary Sciences, University of New Mexico, 221 Yale Blvd NE, Albuquerque, NM, 87131, USA

Jeroen Ritsema
Department of Earth and Environmental Sciences, University of Michigan, 1100 North University Avenue, Ann Arbor, MI, 40819, USA

Paula Koelemeijer
Department of Earth and Sciences, Royal Holloway University of London, Egham Hill, Egham, Surrey, TW20 0EX, UK

Valérie Beaumont, Eric Deville and Daniel Pillot
1IFP Energies Nouvelles, 1 et 4 avenue de Bois Préau, 92852 Rueil-Malmaison Cedex, France

Viacheslav Zgonnik
IFP Energies Nouvelles, 1 et 4 avenue de Bois Préau, 92852 Rueil-Malmaison Cedex, France
Natural Hydrogen Energy Ltd, 819-35th Ave., Greeley, CO 80634, USA

Nikolay Larin
Russian Academy of Science, Schmidt Institute of Physics of the Earth, B. Gruzinskaya St. 10, 123995 Moscow, Russia

Kathleen M. Farrell
North Carolina Geological Survey, Coastal Plain Office and Core Repository, MSC 1620, Raleigh, NC 27699-1620, USA

Taiga Inoue
Department of Earth Sciences, Graduate School of Science, Tohoku University, Sendai 980-8578, Japan

Yasutaka Iijima
Department of Earth Sciences, Graduate School of Science, Tohoku University, Sendai 980-8578, Japan
Department of Earth Sciences, Graduate School of Science, Tohoku University, Sendai 980-8578, Japan
INPEX Corporation, Akasaka Biz Tower 5-3-1, Akasaka, Minato-ku, Tokyo 107-6332, Japan

Kazuhisa Goto
International Research Institute of Disaster Science, Tohoku University, 468-1 Aza-Aoba, Aramaki, Aoba-ku, Sendai 980-0845, Japan

Daisuke Sugawara
International Research Institute of Disaster Science, Tohoku University, 468-1 Aza-Aoba, Aramaki, Aoba-ku, Sendai 980-0845, Japan
Department of Earth Sciences, Graduate School of Science, Tohoku University, Sendai 980-8578, Japan
Museum of Natural and Environmental History, 5762 Oya, Suruga-ku, Shizuoka, Shizuoka Prefecture 422-8017, Japan

Yuichi Nishimura
Institute of Seismology and Volcanology, Faculty of Science, Hokkaido University, Sapporo, Hokkaido 060-0810, Japan

Masashi Watanabe
Department of Civil and Environmental Engineering, Graduate School of Engineering, Tohoku University, 468-1-305 Aza-Aoba, Aramaki, Aoba-Ku, Sendai 980-0845, Japan

Tadao Nishiyama and Chisato Yoshida- Shiosaki
Department of Earth and Environmental Science, Graduate School of Science and Technology, Kumamoto University, 2-39-1, Kurokami, Chuo-ku, Kumamoto 860-8555, Japan

Yasushi Mori and Miki Shigeno
Kitakyushu Museum of Natural History and Human History, 2-4-1, Higashida, Yahatahigashi-ku, Kitakyushu 805-0071, Japan

Makoto Otsubo and Ayumu Miyakawa
Geological Survey of Japan, AIST, AIST Tsukuba Central 7, Higashi-1-1-1, Tsukuba, Ibaraki 305-8567, Japan

Hajime Naruse
Division of Earth and Planetary Sciences, Kyoto University, Kitashirakawa Oiwakecho, Kyoto 606-8502, Japan

Keita Saito, Ryuji Tada, Ke Wang and Yoshiaki Suzuki
Department of Earth and Planetary Science, University of Tokyo, 7-3-1 Hongo, Bunkyo, Tokyo 113-0033, Japan

Hongbo Zheng
School of Resource, Environment and Earth Science, Yunnan University, Chenggong District, Kunming 650500, People's Republic of China

Tomohisa Irino
Faculty of Environmental Earth Science, Hokkaido University, Kita 8, Nishi 5, Kita-ku, Sapporo, Hokkaido 060-0808, Japan

Chao Luo and Mengying He
College of Geographic Science, Nanjing Normal University, 1 Wen Yuan Lu, Qixia, Nanjing 210046, China

Index

R

S

T

www.ingramcontent.com/pod-product-compliance
Lightning Source LLC
LaVergne TN
LVHW080727210926
848485LV00003B/528